# Microorganisms for Sustainability

Volume 31

**Series Editor**

Microorganisms perform diverse roles on our planet most of which are important to make earth a habitable and sustainable ecosystem. Many properties of microorganisms are being utilized as low input biotechnology to solve various problems related to the environment, food security, nutrition, biodegradation, bioremediation, sustainable agriculture, bioenergy and biofuel, bio-based industries including microbial enzymes/ extremozymes, probiotics etc. The book series covers all the wider aspects and unravels the role of microbes towards achieving a sustainable world. It focuses on various microbial technologies related to sustenance of ecosystems and achieving targets of Sustainable Development Goals. Series brings together content on microbe based technologies for replacing harmful chemicals in agriculture, green alternatives to fossil fuels, use of microorganisms for reclamation of wastelands/ stress affected regions, bioremediation of contaminated habitats, biodegradation purposes. Volumes in the series also focus on the use of microbes for various industrial purposes including enzymes, extremophilic microbes and enzymes, effluent treatment, food products.

The book series is a peer reviewed compendium focused on bringing up contemporary themes related to microbial technology from all parts of the world, at one place for its readers, thereby ascertaining the crucial role of microbes in sustaining the ecosystems.

More information about this series at http://www.springer.com/series/14379

Pankaj Kumar Arora
Editor

# Microbial Products for Health, Environment and Agriculture

Springer

*Editor*
Pankaj Kumar Arora
Department of Environmental Microbiology
Babasaheb Bhimrao Ambedkar University
Lucknow, Uttar Pradesh, India

ISSN 2512-1901 ISSN 2512-1898 (electronic)
Microorganisms for Sustainability
ISBN 978-981-16-1946-5 ISBN 978-981-16-1947-2 (eBook)
https://doi.org/10.1007/978-981-16-1947-2



This Springer imprint is published by the registered company Springer Nature Singapore Pte Ltd.
The registered company address is: 152 Beach Road, #21-01/04 Gateway East, Singapore 189721, Singapore

# Contents

# About the Series Editor and About the Editor

## About the Series Editor

**Naveen Kumar Arora**, PhD in Microbiology Fellow of International Society of Environmental Botanists (FISEB), is Professor and Head, Department of Environmental Science at Babasaheb Bhimrao Ambedkar University (A Central University), Lucknow, Uttar Pradesh, India. He is a renowned researcher in the field of environmental microbiology and biotechnology. His specific area of research is plant–microbe interactions, particularly plant growth promoting rhizobacteria. He has more than 75 research articles published in premium international journals and several articles published in magazines and dailies. He is an editor of 25 books, published by Springer. He is a member of several national and international societies, Secretary General of Society for Environmental Sustainability, in editorial board of 4 journals, and reviewer of several international journals. He is also the editor in chief of the journal "Environmental Sustainability" published by Springer Nature. He has delivered lectures in conferences and seminars around the globe. He has a long-standing interest in teaching at the PG level and is involved in taking courses in bacteriology, microbial physiology, environmental microbiology, agriculture microbiology, and industrial microbiology. He has been advisor to 134 postgraduate and 11 doctoral students. He has been awarded for excellence in research by several societies and national and international bodies/organizations.

Although an academician and researcher by profession, he has a huge obsession for the wildlife and its conservation and has authored a book, *Splendid Wilds*. He is the President of Society for Conservation of Wildlife and has a dedicated website www.naveenarora.co.in for the cause of wildlife and environment conservation.

## About the Editor

**Pankaj Kumar Arora** is currently an Assistant Professor and DBT-Ramalingaswami Re-entry Fellow at the Department of Environmental Microbiology, Babasaheb Bhimrao Ambedkar University, Lucknow, India. Dr. Arora is also an Editorial Board Member for the journal *Scientific Reports*, an Associate Editor for the journal *Frontiers in Microbiology*, and an Academic Editor for *PLOS ONE*. He is the recipient of several national awards and fellowships including a Young Botanist Award and Dr. Y. S. Murty Medal from the Indian Botanical Society. His major focus area is environmental microbiology, and is currently investigating the biodegradation and bioremediation of various xenobiotic compounds including nitrophenols, chlorinated nitrophenols, and indole. He has published a total of 40 papers in reputed journals, 3 edited books, and has 9 years of teaching and research experience at national and international institutes.

# Chapter 1
# The Good Side of Evil: Harnessing the Power of Helminths as Therapeutics

**Naina Arora and Amit Prasad**

**Abstract** The evolution of mankind has always aimed at better living conditions with constantly evolving urbanization and civilization. The persistent efforts in improving the life style and health regime have decreased the rate of infectious diseases but have imposed us at risk of autoimmune disorders. Our environment comprising of flora and fauna not only shape up the surroundings geographically but also evolve and mould immune system of human beings by continuous exposure to different allergens and pathogens. Helminths are known strong immune manipulator, with very well devised strategies to evade human immune system they are able to thrive within the host for long period without evoking an immune insult to the host. This immune suppression arises as a result of Th2 bias induced by the residing helminth and the helminth secreted products. Auto-immune disorders are associated with incessant inflammation and organ damage in the long run. There have been reports suggesting that exposure to helminths confers protection against these auto-immune disorders which are Th1 associated. Hence, helminth derived products might be useful in ameliorating the pathology linked to auto-immune disorders as they might help in restoring the homeostasis. Here, we discuss the potential of helminths and helminth derived products in therapy.

**Keywords** Helminth · Auto-immune disorders · Hygiene hypothesis · Immune modulation · Therapeutics

## 1.1 Introduction

The healthy state of an individual is constantly challenged and shaped up by the environment they are exposed to. The surrounding environment is shared with a large number of pathogens and many potent immune agents/allergens and hence our

---

N. Arora · A. Prasad (✉)
School of Basic Sciences, Indian Institute of Technology, Mandi, Himachal Pradesh, India
e-mail: amitprasad@iitmandi.ac.in



P. K. Arora (ed.), *Microbial Products for Health, Environment and Agriculture*, Microorganisms for Sustainability 31, https://doi.org/10.1007/978-981-16-1947-2_1

immune system is constantly evolving. Not only that, there are thousands of bacteria, fungi and parasites reside in our body all the time. The immediate micro/parasite/fungal biome is highly influential to our evolutionary existence; for instance, the gut immunity plays a vital role in our nutrition, development and protection (Johnson and Foster 2018). The large number of microbes inhabiting human body outnumber the total cells in human body; so it is justified to say we are more microbes than humans!

Our environment is constantly changing as a result of human activities and so is the microbiome surrounding us, the exposure and insult with these microbes are not always harmful, the foreign insult in some case confers protection from reinfection with such pathogenic agents and as is with the helminths; a more surprising advantage that they offer; confers protection against auto-immune/inflammatory disorders. This is contrary to the present-day status of infectious disease burden which is taking a toll on human health and soil -transmitted helminths itself infect 1.5 billion people worldwide as per World Health Organization. But, a limiting amount of infection with helminths is proving to offer an edge to many autoimmune disorders, allergies and inflammatory bowel disease.

Helminths are very skilful and powerful manipulator of immune system (Maizels et al. 2018). By an estimate, approximately one third of world population get infected with some kind of helminth infection in their life time. These infectious agents are responsible for many chronic and debilitating diseases or syndromes (Arora et al. 2019). These helminths not only affect humans but also are a major problem for livestock diseases and cause major loss to agriculture too. The unexpected pervasiveness of helminth infections indisputably reflects their special capability to influence the host immune system, subduing host immune responses that could end in their ejection from host body. Host immunity has also developed over times in this condition of coevolution with parasites defence mechanisms to constraint pathology and to balance resistance, susceptibility and immunopathogenesis. Hence, host immune responses habitually allow ongoing infection in preference to complete elimination of parasite and the collateral damage that would result to surrounding tissues.

The immune axis weighs between pro-inflammatory or anti-inflammatory scale to establish immune homeostasis (Fig. 1.1). The host–parasite interaction results in downregulation of host immune system and promotes parasite survival by moving the immune axis to anti-inflammatory scale (Amit et al. 2011). To inhabit and create comfortable niche for themselves they have adopted various immune regulatory roles, immune masking and immune evasion strategies like altering the expression of host microRNA which eventually alters protein expression of signalling cascades (Arora et al. 2017). The protective immunity against helminths develops at slow pace over a time period only, and the effector mechanisms for eliminating them in human host are not well defined; however, by animal models studies, it is defined as of Th2-dependent pathways that mediate protection, as successful parasites seek to blunt the host immune attack for their own survival. The cross talk between helminth and humans has revealed a number of excretory–secretory products from helminths which participate in this differential immune regulation to more permissive humoral

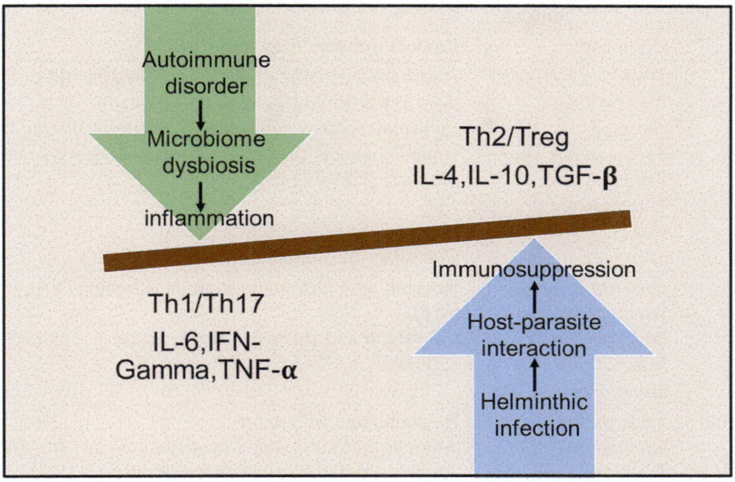


**Fig. 1.1** Immune axis strikes a balance between Th1 and Th2 response; autoimmune disorders weighs down the Th2 axis resulting in inflammation and systemic disorders, whereas the helminthic infections uplift the Th2 axis and restores the balance between Th1 and Th2. Thus, ameliorating the inflammation due to Th1 response

immune response (Harris and Gause 2011). This facilitates survival of parasite and does not cause immediate damage to host system. The more developed and aware part of the world has been successful in eradicating the incidences of the parasitic diseases but has become more prone to allergies, inflammatory bowel disease, metabolic disorders, etc. (Wiria et al. 2012). As evident from the rising status of inflammatory disorders in the west where the cleaner surroundings have restricted the prevalence of helminthic infections; suggestive of the protective status helminth confers from auto-immune disorders. These helminth derived products are proving to be beneficial to humans in protecting against auto-immune disorders where the immune axis weighs more on pro-inflammatory side (Table 1.1). Here, we discuss the helminth derived products in therapy and associated challenges.

## 1.2 Evolution of Biota Alteration Theory

Coined by David Barker; Hygiene hypothesis dates back to the 1970s where in urban communities were reported to have much higher allergic incidences as compared to rural communities which are exposed to many pathogenic viral, bacterial or helminthic agents (Bloomfield et al. 2006). Later on, this theory continued to develop; with increasing living standards, reduced family size, better personal hygiene and reduced exposure to common pathogens specially helminths which induce anti-inflammatory tolerance. Though the helminth infections are prevalent in major parts of the developing world but have been cleared out from high income and developed

**Table 1.1** Helminth derived products investigated for auto-immune disorders

| Disease | Organism | Product/immune response |
|---|---|---|
| IBD | *Trichinella spiralis*<br>*Ancylostoma caninum*<br>*Syphacia obvelata*<br>*Trichuris suis*<br>*S. japonicum* | Recombinant serine proteases and cystine protease inhibitors TsKaSPI and TsAdSPI; TsCystatin<br>Low-molecular-weight metabolites derived from both somatic extracts (LMWM-SE) and excretory-secretory products (LMWM-ESP)<br>Egg<br>Ova<br>Cercariae, recombinant Sj16 |
| Type 1 diabetes | *Dirofilaria immitis*<br>*Wuchereria bancrofti*<br>*Schistosoma japonicum* | Recombinant Dirofilaria immitis antigen (rDiAg)<br>rWbL2<br>rSjcystatin and fructose-1,6-bisphosphate aldolase<br>rSjFBPA |
| Rheumatoid arthritis | *Schistosoma japonicum*<br>*Trichinella spiralis*<br>*Fasciola spp*<br>*Acanthocheilonema viteae* | Recombinant SjCystatin<br>Autoclaved Schistosoma mansoni antigen (ASMA)<br>Autoclaved Trichinella spiralis antigen (ATSA)<br>C-terminal of Fasciola helminth defence molecule-1<br>(C-FhHDM-1 ES62) |

countries, thus western communities with low to no prevalence of helminthic infection are being reported to have higher risk and incidence of inflammatory disorder (Smallwood et al. 2017). But it still remains elusive as to what factors with urbanization and economic development in western lifestyle correlate to immune mediated inflammatory disorders. One factor that can be attributed to this shift in immune/protective paradigm is altered biome with environment as they have been co-existing and co-evolving together (Fig. 1.2). This altered biome affects the composition of symbionts than the chances of getting infections conferring to protection. It is a more complex relationship than it appears, not all infections or microbes are same and each has its own immune axis of induction which shapes up host immune status. Thus, the hygiene hypothesis translated/alter to "Biota alteration theory" which states that depletion of biome in industrialized world results in unstable immune status and over reactivity to self-antigens. The factors contributing to altered biome are not only clean surroundings, along with that lifestyle changes, sedentary life style over active life, diet more processed than fresh (inflammatory diet), stress, environmental stimuli, genetic factors are among many other factors contributing to inflammatory phenotype in west (Holick 2007; Brenner et al. 2015; Villeneuve et al. 2017).

The surge of more clean practices has seen an increase in number of autoimmune disorders, it has been in debate whether the cleaner environment; clearing the parasitic worms out of environment also affects the good bacteria status in our gut and whether more hygienic practices predispose human being to auto-immune disorders at some point of time. The present-day global disease burden of helminths shows an inverse relation to auto-immune disorders; is generating curiosity among researchers as to how the immune axis responds and protects the individuals against

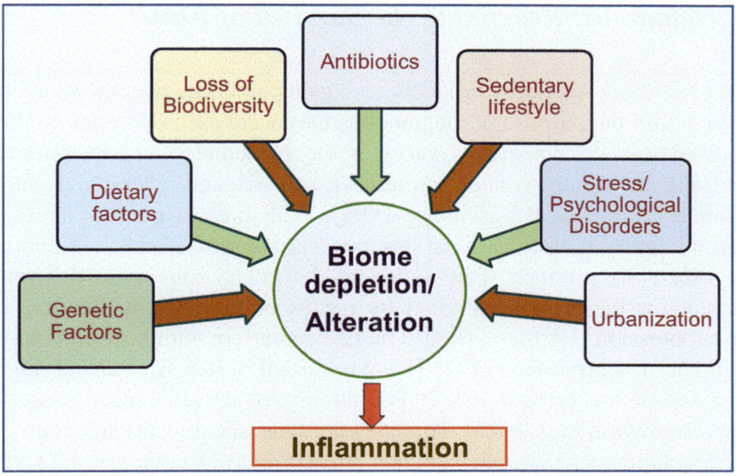


**Fig. 1.2** The multitude of factors that constantly alter the biome and comprise inflammatory insults to the human beings by disturbing the microbiome around us

auto-immune disorders in helminthic endemic regions and puts helminth derived products in a new spotlight for therapeutic interventions to address auto-immune therapies.

## 1.3 What Are Helminths?

Helminths are parasitic worms. They are classified into three types depending on the external and internal physiology. Most of the helminths cause gastrointestinal infection, some of them are tissue or blood vessel residents. They are both hermaphrodite and bisexual species. The life cycle comprises one or more intermediate host and a definitive host and transit from various life stages of egg to larvae to mature sexually active adult contributing to multiplicity of infection in the host. The three classes are Trematodes: flat leaf like digenetic worm, Cestode: flat tapeworm and Nematode: round soil transmitted worm. Among the helminths, soil transmitted helminths incur the maximum rate of infection to their share, affecting 24% of the world's population (WHO fact sheet 2020). The infections are transmitted through contaminated soil and water in areas of poor hygiene. Helminths colonizing the human host remain asymptomatic for a prolonged period; the messengers of helminths to human immune system which temper the host immune system to promote their survival are being exploited/evaluated to understand their therapeutic role in autoimmune disorders.

### *1.3.1 Helminths: Redefining the Survival of Fittest*

Helminths excretory–secretory products have gain popularity over the years, initially investigated with the purpose of diagnosis as they were easily detected in the blood of parasitized host, immunoassays, vaccines, etc. Nonetheless various active molecular functions like tissue invasion (proteases), anti-oxidants, chemotaxis, etc., were also reported (Harnett and Parkhouse 1995). With time, a number of interesting immunomodulatory molecules were described and it was established that parasite excretory–secretory products (ESPs) play an elementary role in creating/setting up environment for the parasite survival by means of immune modulation and not immune suppression (Hartnett 2014). The ESPs interfere with host immune regulation and induce the expression of (1) Th2 cytokines IL4, IL5, IL13 and (2) regulatory immune response involving IL-10, TGF-β, alternative activated macrophages, Tregs and Bregs (Hepworth et al. 2013). The ESPs are stage specific, the signature changes with the developing stage of parasite, it is interesting to observe that the ESPs from wide range of cestodes, nematodes and trematodes carry molecules capable of altering/alarming the antigen presenting cells to set up protective response. The immune modulation by ESPs also protects host from exacerbated inflammation in invading and residing tissue and help with wound healing. For example, well described and characterized ESPs come from nematode *Acanthocheilonema viteae* ES-62, 240 kDa tetrameric protein have shown to induce Th2 response with IL-4 secretion in naïve DCs (Whelan et al. 2000). Similarly, there are many more helminth derived products which pursue a protective immune response in host and hence confer protection against the auto-immune disorders which are primarily Th1. Since helminth infection strike a balance with immune regulation for clearance of pathology and survival, it is of increasing interest if this phenomenon can be used to ameliorate therapy for inflammatory and allergy disorders.

## 1.4 Helminths in Therapy

The most widely studied helminths for therapeutic purpose in auto-immune disorders are Nematodes: *Heligmosomoides polygyrus bakeri, Trichuris sp, Trichinella sp, Nippostrongylus brasiliensis, Acanthocheilonema viteae, Necator americanus;* Cestodes: *Hymenolepis diminuta, Taenia crassiceps;* Trematodes: *Schistosoma sp, Fasciola hepatica.*

In this review, we specifically discussed the patterns and pathways of helminth derived host immune modulation that has been used for treatment of other diseases specially autoimmune and allergic diseases, with the standpoint that this understanding of helminths use in immune modulation will not only provide new avenues for understanding parasite immunology but also offer routes to treat noncommunicable immune dysfunctions by utilizing the benefit of helminths or helminths derived products.

### 1.4.1 Inflammatory Bowel Disease (IBD)

Inflammatory bowel disease comprises two chronic gastrointestinal disorders, i.e. Ulcerative colitis (UC) and Crohn's disease (CD). The disease of the west affecting approximately one million people in the USA and more than 2.5 million people in Europe is becoming a global trend with "compounding prevalence" in every continent (Kaplan 2015). There is no single factor or microbe associated with incidence of disease; genetic susceptibility, environmental factors, immune status/ susceptibility, use of antibiotics and commensal dysbiosis in gut are multitude factors associated with risk of mucosal inflammation. The disease is population-varied in its association with genetic risk loci, for example: strongly associated gene loci with disease are; innate sensing of bacteria (NOD2), the inflammatory response to microbes (IL23R) and autophagy (ATG16L1). These genes result in elimination of some intracellular bacteria or bacterial peptide sensing, eventually resulting in intestinal inflammation. In Asian population ATG16L1 finds no correlation to Crohn's disease and NOD2 mutation found here is different from white European population at genetic predisposition. The environmental factors affecting the prevalence shows differential existence to regions; for example, smoking was considered to alter gut dysbiosis and was corelated to occurrence of CD or UC, but on the contrary, the newly industrialized countries do not show same trend, may be the trend associated with chance of occurrence of disease has not yet set in this population (Ng et al. 2015). The environmental factors affect the gut microbiota which affects the individual's immune status. The disease models have shown more abundance of pro-inflammatory bacteria than anti-inflammatory ones. This imbalance in microbial population disturbs the immune axis and induces Th1/Th17 immune response leading to inflammation in mucosal lining with IFN-gamma and IL-17 being signatory cytokines released in abundance. Eventually, this chronic disease though low in morbidity affects the quality of life. The advances in industrialized status of country come up with better health-care systems and as therapeutic interventions are needed to ameliorate the effect, helminth therapy comes out as a promising approach. The animal models for IBD; D/TNBS colitis, IL-10 deficient colitis and T-cell transfer colitis have been studied vastly for effect of helminths and it appears that the helminth therapy works at four levels; starting from suppressing Th1 cytokines by Th2 cytokines like IL-4, IL-5, IL-13 and reduces the severity of inflammation in colitis model. Further, helminth infections are associated with induction of dendritic cells, alternatively activated macrophage and regulatory T cells which release IL-10 and TGF-beta that restore homeostasis and avoid effector T cells response. Eventually helminth infection alters the intestinal flora, for example, *H. polygyrus* promotes lactobacilli family and thus increases the count of anti-inflammatory bacteria in the gut (Walk et al. 2010; Elliott and Weinstock 2012). Helminth therapy has proven to be safe, well tolerated and significantly effective in treating UC and CD since the first report which dates back to 20 years (Summers et al. 2005). Helminth products from *T. suis* and *N. americanus* have been probed for therapeutics perpetually, with *T. suis* being more safer than *N. americanus* as a long duration infection with later can have some deleterious effects (Helmby 2015; Sipahi and Baptista 2017). *T. suis* embryonated ova have been used in various studies, these ova when ingested orally hatch into larvae and this larvae impedes the TH1 axis and induces regulatory immune axis. Being a porcine whipworm *T. suis* does not infect humans and is morphologically different from *T. trichiura* which infects man. *T. suis* has been studied in UC and Cd with patient cohorts and compared to placebos has shown promising result and translatability to helminth therapy (Weinstock and Elliott 2013; Huang et al. 2018).

### *1.4.2 Rheumatoid Arthritis*

Rheumatoid arthritis is an auto-immune disorder marked by circulating auto-antibodies to IgG and citrullinated proteins, chronic in nature, its effects are systemic and arises as a result of inflammation due to immune dysfunction affecting joints, eventually impairing the life quality and limiting life expectancy of the affected. The global epidemiology is lacking and prevalence data available from western countries show 0.5–1.0% frequency in the white population (Myasoedova et al. 2010). Risk factors interlinked to RA are genetic predisposition (60% risk factor); which show approximately 100 loci to be associated to susceptibility or severity of RA, for example, gene loci HLA-DRB1*01 and HLA-DRB1*04 are significantly associated with RA, environmental factors; dust inhalation containing silica, asbestos, pulverized cement, textile dust; all these environmental hazards show significant correlation to RA (Webber et al. 2015), gender bias; women are 2–3 times at higher risk or 3.6% chances of women to develop RA during a lifetime compared to men at 1.7% (Ngo et al. 2014; Crowson et al. 2011) and microbiota are associated risk factors to develop RA. The gut microbiota plays an important role as observed there is reduction of predominant abundant taxa and expansion of a rarer taxa of Actinobacteria (Konig et al. 2016) and *Prevotella copri* appears to be early marker for onset of disease (Chen et al. 2016). Another example of microbiota in influencing disease outcome is observed in alteration of periodontal microbiota, *Porphyromonas gingivalis* and *A. actinomycetemcomitans*. It is known to cause common periodontal disease and can influence citrullination process as it expresses peptidyl arginase deaminase and hence increasing the anti-citrullination protein antibody (ACPA) (Dissick et al. 2010). These risk factors lay the foundation of disease or the susceptibility to disease with no sign and symptom of autoimmunity, it then enters into preclinical phase with onset of autoimmunity and there is transition from asymptomatic autoimmunity to symptomatic phase; this phase is associated with citrullination or modification of self-proteins generating anti-citrullinated protein antibodies or neo-epitopes (post-translational modification), loss of immune tolerance at mucosal surface/site and generation/formation of auto-antibody leading to cell activation as a result of antigen loading and migration activating secondary lymphoid tissue and production of B-cells and T-cells along with auto-antibodies by antigen presenting cells. The activation of APCs, macrophages, FLSs (fibroblast like synoviocytes) leads to release of various cytokines and chemokines which triggers synovial inflammation and formation of immune complexes, activation of complement leads to synovial vascular leakage, a second hit or perpetual cytokine storm due to activated adaptive immune response manifests into cartilage and bone destruction and damages joint and bone erosion by osteoclasts (release proteases). With the propagation/expansion of autoimmunity or auto-antibodies clinical RA transits from undifferentiated to classifiable RA. The associated symptoms are infiltration of immune cells at the joint, hyperplasia of lining layer, pannus formation. Synovium is a delicate structure which helps in maintaining the homeostasis by supplying nutrient to cartilage and secreting lubricant for cartilage to function smoothly. Destruction to synovium destructs the intimal lining which houses macrophages, FLS, adipocytes, fibroblasts, etc. The activated APC activates adaptive arm and cytokines pre-dominating are IL-1, IL-6, TNF-α, small molecules like MMPs (collagenases, stromelysins), leukotrienes, prostaglandins, miRNAs. This disturbs the matrix and pathways involved in cell migration, adhesion, cytokine signalling, etc., are implicated in pathogenesis of the disease. Thus, this pro-inflammatory set up forms a network loop with minimal apoptosis and due to paracrine and autocrine nature of synovium, inflammation spreads nearby, though the internal inhibitors/antagonist are produced like Il-10, IL-1R, Il-35 and Soluble-TNF, but they are not strong enough to mitigate the effect. With better understanding of the disease pathology, once damage that was considered irreversible now has remission hope with early diagnosis, better patient management and treatment strategies. The past two decades have seen a paradigm shift in understanding of RA, better instrumentation for clinical assessment and targeted therapy. Existing therapy is target based using synthetic or biological "Disease-modifying antirheumatic drugs for RA" (DMARDs). Conventionally used synthetic DMARD is methotrexate and glucocorticoids. There is ongoing debate on using biological DMARDs along with conventional ones, discussion of which is out of scope for this review/discussed elsewhere and there are various side effects associated with the present targeted therapy discussed in (Smolen et al. 2018). As observed gut microbiota dysbiosis plays major role in RA, number of animal studies have been undertaken to study the role of helminths in remission of RA by restoring microbiota in inflammation. Inflammatory murine model to study RA are collagen induced arthritis (CIA) (where DBA/I mice inoculated with bovine collagen), other murine models developed by inoculation of specific microbes develop T-cell mediated arthritis spontaneously, suggesting microbiota plays role in K/BxN (segmented filamentous bacteria) murine model, IL-1ra$^{-/-}$ (*Lactobacillus bifidus Helicobacter*) SKG (*Prevotella* spp) RA models (Maeda and Takeda 2019). RA therapeutic approach is target based, i.e. inhibiting or blocking the molecule directly associated in joint damage, papain-like cysteine proteases appear to play significant role in joint damage and bone remodelling (Trabandt et al. 1990). Recombinant S. *japonicum* cystatin protein is known to interfere with antigen presentation and downregulates immune activation by suppressing production of TNF-alpha and IL-6 (Zhu et al. 2014), it possesses enzymatic property of inhibiting proteolytic activity of papain. Murine model of CIA-RA showed prophylactic effect of rSjcystatin in reducing the structural joint pathology by augmenting Th2 cytokines IL-10,IL-4 and collagen specific IgG and T reg population of CD4+ CD25+ Foxo3+, but did not alter the RANKL level. RANKL is a receptor activator for NF-KB ligand and had been associated with osteoclast differentiation and hence pathology in RA (Liu et al. 2016). The rSjcystatin works by disarming TH1 response and not by its inhibition of proteases activity as it failed to block the cathepsin activity in synovium in an established osteoarthritis (Kyostio-Moore et al. 2015). Another helminth derived molecule that could inhibit/block RANKL was studied by Khan et al. (2020); C-terminal sequence of Fasciola helminth defence molecule-1 (C-FhHDM-1). This molecule impaired the macrophage differentiation to osteoclast, therefore, impacting the osteoclastogenesis and altering the RANKL/OPG ratio decreased bone resorption and prevention of bone loss due to CIA, without compromising on systemic immune surveillance. Another novel strategy of employing helminth in RA treatment is altering the gut microbiota to restore homeostasis. As discussed above it is evident that RA pathology arises of altered microbiota and Tuftsin–Phosphorylcholine a chimeric molecule with immunomodulatory property, when administered in CIA mouse model had gut microbiota composition similar to control with reduced score of RA, joint destruction and inflammation storm (Bashi et al. 2016). The gut microbiota of healthy group had more short chain fatty acid producing bacteria and PBS/CIA had loss of *enterobacteriaceae* and abundance of *Mucispirillum, Oscillospira* and unclassified Clostridiales genera (Ben-Amram et al. 2017). Thus, this study established and paved for another therapeutic intervention to RA therapy by restoring the gut microbiota.

Though helminth shows great potential in remission, it is difficult to convince patients in getting infection with helminth which causes altogether a different disease. The systemic effect of parasitic products is not well evaluated and hence the deleterious effect associated with it and this lack of knowledge offsets the therapeutic advantage they offer. Eissa et al. 2016 proposed and investigated autoclaved antigens from *S. mansoni* and *T. spiralis* in adjuvant induced arthritis model in rats and met with much success with upregulated Treg FOxo3+ population. Thus, counteracting the drawback of incurring helminth infection with these autoclaved extracts.

Langdon et al. (2019) performed a meta-analysis study to evaluate the effectiveness of helminth products in RA in CIA model and put forwards *Acanthocheilonema viteae's* ES-62 a phosphorylcholine and well characterized immunomodulatory molecule in therapeutic spotlight for RA. The *A. vitae* ES62 induces IL-10 expression via downregulating MYD88 expression which attenuates NF-κB pathway and hence pro-inflammatory signalling.

### *1.4.3 Type-1 Diabetes*

Another chronic auto-immune disorder is type 1 diabetes, where the immune system becomes destructive against self pancreatic-beta cells, which hampers insulin production and it results in high blood glucose level which leave the patients in debilitated state. It is highly prevalent in Asia-pacific and the available statistics suggest even higher prevalence of disease in Indian population in coming years. The mechanism behind beta-islet cell destruction remains intangible, factors associated with disease occurrence are genetic predisposition, lifestyle and diet. Epidemiological survey reports concomitant increase in diabetes prevalence with reducing helminth infection rates (Bashi et al. 2015). In support of this data, there have been many investigational studies with mouse models and parasites to establish this. The mouse model of streptozotocin induced diabetes (STZ) or non-obese diabetes have shown reduction/elimination of diabetes when treated with ESP of filarial parasites *Fasciola hepatica* or *Litomosoides sigmodontis* and recombinant *Dirofilaria immitis* antigen (rDiAg) (Lund et al. 2014; Imai et al. 2001; Chenthamarakshan et al. 1996). Amdare et al. investigated the effect of different antigenic preparation of *Brugia malayi* in multiple low-dose STZ-induced T1D mouse model; adult soluble antigen (Bm A S), microfilarial excretory–secretory antigens (Bm mf ES) and microfilarial soluble antigen (Bm mf S), respectively. They found low severity of disease with reduction of 37.5% incidence, with increase in number of healthy islets and reduced pancreatic inflammation. The destructive islet Beta cell phenomenon is a consequence of cell mediated Th1 immunity along with phagocyte-dependent protective responses producing IFN-gamma causing massive/heightened inflammation (Kikodze et al. 2013). There was a cytokine paradigm shift of inflammatory TNF-α/IFN-γ to IL-10 and IL-4 elevated levels in *B. malayi* treated mouse models, this was observed along with over expression of anti-insulin IgG1 and reduced levels of anti-insulin IgG2a antibodies and elevated levels of insulin specific IgE, which mediates protective effect. Other helminths; *T. spiralis, H. polygyrus and S. mansoni* have shown similar results in mouse diabetes models (Saunders et al. 2007; Mishra et al. 2013). *S. japonicum* recombinant proteins cystatin and fructose-1,6 bisphosphate which are excretory–secretory proteins and known for their immunomodulating properties ameliorated type-1 diabetes in NOD female mice model with elevated IL-10 and TGF-Beta and increase in number of Treg cells (Tang et al. 2019). Helminth can ameliorate the degradation of pancreatic islets and hyperglycaemia not only by induction of Tregs, but by expression of Arg1 and Ym1 as well (Osada et al. 2017). Soluble egg antigen (SEA) or soluble worm antigen (SWA) confers protection against development of Type-1 diabetes induced proliferation of NKT cells subtype, it induces phenotypical changes in dendritic cells which results in Tregs expansion and proliferation associated with IL-10 and TGF-β, SEA increases TGFB secretion in TLR2 dependent manner downstream leading to CXCL10 expression, this recruits T-cell macrophages, DCs to pancreatic islets. SEA also differentiates macrophages to M2 phenotype. Thus, depending on various conditions SEA can either prevent or ameliorate type1-diabetes (Burton et al. 2010; Parsa et al. 2012).

Thus, induction of regulatory immune response yet again suppressed disease associated inflammatory pathology (Yan et al. 2020) and there might be one more pathway affecting the pathology of the disease, it is not always induction of Th2 that can suppress disease pathology. The present-day therapy is insulin administration, immunosuppressive drugs, which are short term and cause toxicity on prolonged use. Thus, better therapeutics are demanded considering the increasing toll on disease burden and helminth therapy can prove to be beneficial.

### *1.4.4 Allergy/Asthma*

Unlike above mentioned/discussed auto-immune disorders which are Th1/Th17 driven, Allergy or asthma is Th2 dominant, i.e. arise because of excess of Th2, so implying helminth intervention may further worsen the scenario. But that is not the case as the helminth derives regulatory Th2 response and hence controls Th2-driven atopic inflammation. To study asthma/allergy mouse models with allergic airway inflammation with ovalbumin as allergen has been created and experimentally studied. In general, this allergic response is associated with exaggerated levels of IL-4 and IL-13, which results in infiltration of basophils and stimulation of goblet cells releasing more mucus and proliferation.

Helminth infection in mouse models has met with considerable success in suppressing chronic and acute inflammation. Impressively the gut associated parasites (*H. polygyrus)* induce regulatory immune phenotype in gut lymphoid tissue; Bregs (IL-10 + CD19+), Tregs (IL-10 + CD4+ and FoxP3 + CD4+) are able to ameliorate lung airway inflammation (Gao et al. 2019). Adoptive transfer of Treg FoxP3+ CD4 + CD25+ from *T. spiralis* infection model conferred protection by OVA-induced lung airway inflammation (Kang et al. 2014) and *T. spiralis* extracts of adult worm and larvae muscle reduced the infiltration of eosinophils in airway inflammation, thus, the worm based therapeutics helped in reducing Il-4 elevated levels and increased Il-10 and TGF-β and they also reduced IgE specific to OVA, proving their promising therapeutic efficacy.

## 1.5 Conclusions

Helminths including other microbes have co-evolved with humans and their immune system. Wiping these infectious agents out of the environment has resulted in increase in propensity of non-infectious autoimmune disorders. Although helminth infection is associated with protective immunity, not all helminth infection confers protection against autoimmune disorders, occurrence of concurrent infections and infection intensity influence the allergenicity status and hence the protective outcome associated with helminths as observed with hookworm infection (Helmby 2015). The Th2 bias induced by helminths in host system creates immune suppression to support parasite survival and parasite secreted products are of particular interest and their expansive repertoire from number of helminths signposts the scope of therapeutics (Fig. 1.3).

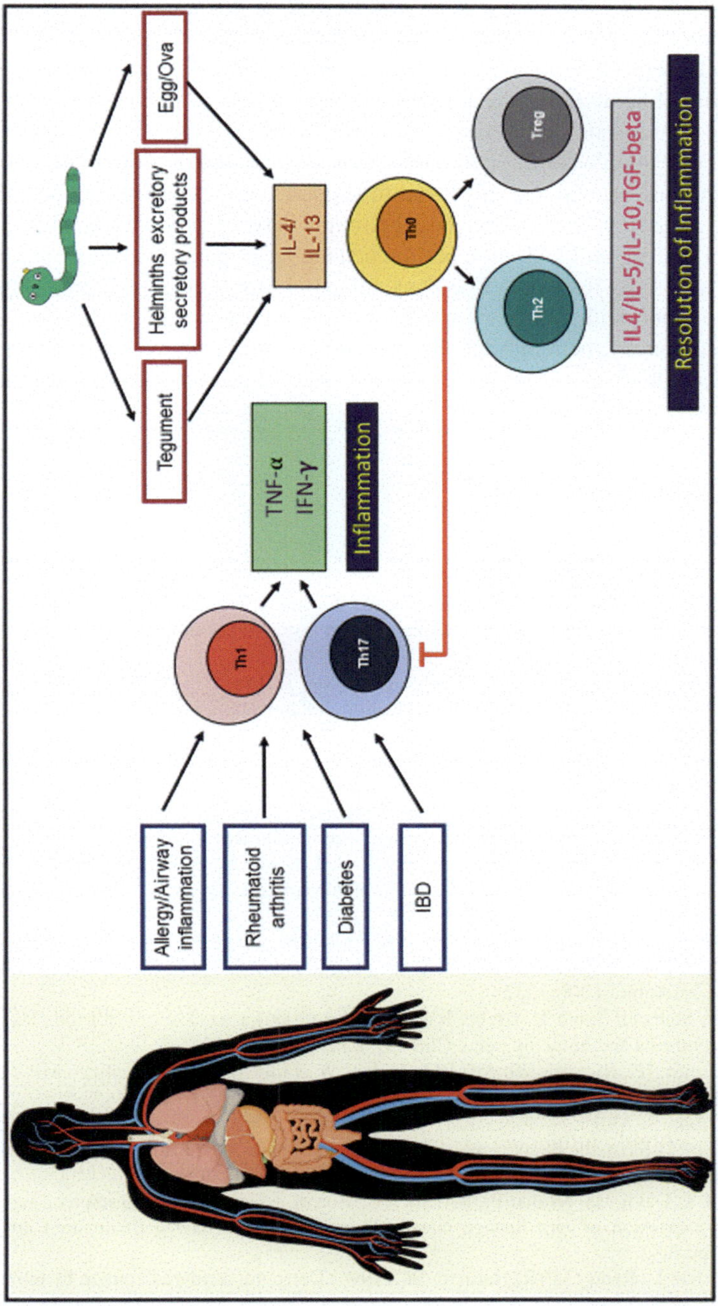


**Fig. 1.3** Helminth derived products (tegument, excretory–secretory or egg/ova) induce immune suppression and the helminth associated pathogenesis holds potential for therapeutics in auto-immune disorders

Though the animal models show promising results in suppressing or preventing inflammatory disorder; the use of helminth derived products may be used as supplement in modern diets in future. As of now there are certain challenges and limitations, the translation of mouse model results in clinical trials is slow, there is limitation with obtaining the pure form of helminth products and their downstream effect on the host physiology remains unknown, some of these identified helminth molecules (For example ES-62) could be systemic in abrogating inflammation than being tissue/organ specific and might fail to suppress or prevent disease pathology (Doonan et al. 2019). This is because helminth derived products are life stage development and we still lack in-depth understanding of parasite biology and their interaction with host response. The route of administration of helminth derived products and taboo with eating worms for therapy imposes challenges to effective therapy. Undoubtedly, helminth therapy has shown promising results in ameliorating disease pathology, nonetheless detrimental effects/risk associated with helminth infection remain a major concern.

## References


Amit P, Prasad KN, Kumar GR, Shweta T, Sanjeev J, Kumar PV, Mukesh T (2011) Immune response to different fractions of Taenia solium cyst fluid antigens in patients with neurocysticercosis. Exp Parasitol 127(3):687–692

Arora N, Kaur R, Anjum F, Tripathi S, Mishra A, Kumar R, Prasad A (2019) Neglected agent eminent disease: linking human helminthic infection, inflammation, and malignancy. Front Cell Infect Microbiol 9:402

Arora N, Tripathi S, Singh AK, Mondal P, Mishra A, Prasad A (2017) Micromanagement of immune system: role of miRNAs in helminthic infections. Front Microbio 8:586

Bashi T, Bizzaro G, Ben-Ami SD, Blank M, Shoenfeld Y (2015) The mechanisms behind helminth's immunomodulation in autoimmunity. Autoimmun Rev 14:98–104

Bashi T, Shovman O, Fridkin M, Volkov A, Barshack I, Blank M, Shoenfeld Y (2016) Novel therapeutic compound tuftsin-phosphorylcholine attenuates collagen-induced arthritis. Clin Exp Immunol 184:19–28

Ben-Amram H, Bashi T, Werbner N, Neuman H, Fridkin M, Blank M, Shoenfeld Y, Koren O (2017) Tuftsin-phosphorylcholine maintains normal gut microbiota in collagen induced arthritic mice. Front Microbiol 10(8):1222

Bloomfield SF, Stanwell-Smith R, Crevel RW, Pickup J (2006) Too clean, or not too clean: the hygiene hypothesis and home hygiene. Clin Exp Allergy 36(4):402–425

Brenner SL, Jones JP, Rutanen-Whaley RH, Parker W, Flinn MV, Muehlenbein MP (2015) Evolutionary mismatch and chronic psychological stress. J Evol Med 3:235885

Burton OT, Gibbs S, Miller N, Jones FM, Wen L, Dunne DW, Cooke A, Zaccone P (2010) Importance of TLR2 in the direct response of T lymphocytes to Schistosoma mansoni antigens. Eur J Immunol 40:2221–2229

Chen J, Wright K, Davis JM, Jeraldo P, Marietta EV, Murray J, Nelson H, Matteson EL, Taneja V (2016) An expansion of rare lineage intestinal microbes characterizes rheumatoid arthritis. Genome Med 8:43

Chenthamarakshan V, Reddy MVR, Harinath BC (1996) Detection of filarial antigen by inhibition enzyme linked immunosorbent assay using fractionated Brugia malayi microfilarial excretory secretory antigen. J Biosci 21:27–34

Crowson CS, Matteson EL, Myasoedova E, Michet CJ, Ernste FC, Warrington KJ, Davis JM III, Hunder GG, Therneau TM, Gabriel SE (2011) The lifetime risk of adult-onset rheumatoid arthritis and other inflammatory autoimmune rheumatic diseases. Arthritis Rheum 63:633–639

Dissick A, Redman RS, Jones M, Rangan BV, Reimold A, Griffiths GR, Mikuls TR, Amdur RL, Richards JS, Kerr GS (2010) Association of periodontitis with rheumatoid arthritis: a pilot study. J Periodontol 81(2):223–230

Doonan J, Tarafdar A, Pineda MA, Lumb FE, Crowe J, Khan AM, Hoskisson PA, Harnett MM, Harnett W (2019) The parasitic worm product ES-62 normalises the gut microbiota bone marrow axis in inflammatory arthritis. Nat Commun 10(1):1–4

Eissa MM, Mostafa DK, Ghazy AA, El Azzouni MZ, Boulos LM, Younis LK (2016) Anti-arthritic activity of Schistosoma mansoni and Trichinella spiralis derived-antigens in adjuvant arthritis in rats: role of FOXP3+ Treg cells. PLoS One 11(11):e0165916

Elliott DE, Weinstock JV (2012) Where are we on worms? Curr Opin Gastroenterol 28:551–556

Gao X, Ren X, Wang Q, Yang Z, Li Y, Su Z, Li J (2019) Critical roles of regulatory B and T cells in helminth parasite-induced protection against allergic airway inflammation. Clin Exp Immunol 2198(3):390–402

Harnett W (2014) Secretory products of helminth parasites as immunomodulators. Mol Biochem Parasitol 195(2):130–136

Harnett W, Parkhouse RME (1995) Structure and function of nematode surface and excretory-secretory products. In: Perspectives in nematode physiology and biochemistry. Narendra Publishing House, New Delhi, pp 207–242

Harris N, Gause WC (2011) To B or not to B: B cells and the Th2-type immune response to helminths. Trends Immunol 32:80–88

Helmby H (2015) Human helminth therapy to treat inflammatory disorders - where do we stand? BMC Immunol 16:12

Hepworth MR, Grencis RK, Artis D (2013) Regulation of immunity and inflammation following intestinal helminth infection in parasitic nematodes: molecular biology, biochemistry and immunology. CABI, Oxfordshire, UK, pp 106–129

Holick MF (2007) Vitamin D deficiency. N Engl J Med 357:266–281

Huang X, Zeng LR, Chen FS, Zhu JP, Zhu MH (2018) Trichuris suis ova therapy in inflammatory bowel disease: a meta-analysis. Medicine 97(34):e12087

Imai S, Tezuka H, Fujita K (2001) A factor of inducing IgE from a filarial parasite prevents insulin-dependent diabetes mellitus in nonobese diabetic mice. Biochem Biophys Res Commun 286:1051–1058

Johnson KV, Foster KR (2018) Why does the microbiome affect behaviour? Nat Rev Microbiol 6 (10):647–655

Kang SA, Park MK, Cho MK, Park SK, Jang MS, Yang BG, Jang MH, Kim DH, Yu HS (2014) Parasitic nematode-induced CD4+ Foxp3+ T cells can ameliorate allergic airway inflammation. PLoS Negl Trop Dis 8(12):e3410

Kaplan GG (2015) The global burden of IBD: from 2015 to 2025. Nat Rev Gastroenterol Hepatol 12(12):720–727

Khan YA, Maurya SK, Kulkarni C, Tiwari MC, Nagar GK, Chattopadhyay N (2020) Fasciola helminth defense molecule-1 protects against experimental arthritis by inhibiting osteoclast formation and function without modulating the systemic immune response. FASEB J 34 (1):1091–1106

Kikodze N, Pantsulaia I, Rekhviashvili K, Iobadze M, Dzhakhutashvili N, Pantsulaia N, Kukuladze N, Bikashvili N, Metreveli D, Chikovani T (2013) Cytokines and T regulatory cells in the pathogenesis of type 1 diabetes. Georgian Med News 222:29–35

Konig MF, Abusleme L, Reinholdt J, Palmer RJ, Teles RP, Sampson K, Rosen A, Nigrovic PA, Sokolove J, Giles JT, Moutsopoulos NM, Andrade F (2016) Aggregatibacter actinomycetemcomitans-induced hypercitrullination links periodontal infection to autoimmunity in rheumatoid arthritis. Sci Transl Med 8(369):ra176

Kyostio-Moore S, Piraino S, Berthelette P, Moran N, Serriello J, Bendele A, Sookdeo C, Nambiar B, Ewing P, Armentano D, Matthews GL (2015) Overexpression of cystatin C in synovium does not reduce synovitis or cartilage degradation in established osteoarthritis. Arthritis Res Ther 17:5

Langdon K, Phie J, Thapa CB, Biros E, Loukas A, Haleagrahara N (2019) Helminth-based therapies for rheumatoid arthritis: a systematic review and meta-analysis. Intl Immunopharmaco 66:366–372

Liu F, Cheng W, Pappoe F, Hu X, Wen H, Luo Q, Wang S, Deng F, Xie Y, Xu Y, Shen J (2016) Schistosoma japonicum cystatin attenuates murine collagen-induced arthritis. Parasitol Res 115 (10):3795–3806

Lund ME, O'Brien BA, Hutchinson AT, Robinson MW, Simpson AM, Dalton JP, Donnelly S (2014) Secreted proteins from the helminth Fasciola hepatica inhibit the initiation of autoreactive T cell responses and prevent diabetes in the nod mouse. PLoS One 9:e86289

Maeda Y, Takeda K (2019) Host–microbiota interactions in rheumatoid arthritis. Exp Mol Med 51 (12):1–6

Maizels RM, Smits HH, McSorley HJ (2018) Modulation of host immunity by helminths: the expanding repertoire of parasite effector molecules. Immunity 49(5):801–818

Mishra PK, Patel N, Wu W, Bleich D, Gause WC (2013) Prevention of type 1 diabetes through infection with an intestinal nematode parasite requires IL-10 in the absence of a Th2-type response. Mucosal Immunol 6:297–308

Myasoedova E, Crowson CS, Kremers HM, Therneau TM, Gabriel SE (2010) Is the incidence of rheumatoid arthritis rising?: results from Olmsted County, Minnesota, 1955–2007. Arthritis Rheum 62:1576–1582

Ng SC, Tang W, Leong RW, Chen M, Ko Y, Studd C, Niewiadomski O, Bell S, Kamm MA, De Silva HJ, Kasturiratne A (2015) Environmental risk factors in inflammatory bowel disease: a population-based case-control study in Asia-Pacific. Gut 64(7):1063–1071

Ngo ST, Steyn FJ, McCombe PA (2014) Gender differences in autoimmune disease. Front Neuroendocrinol 35:347–369

Osada Y, Fujiyama T, Kamimura N, Kaji T, Nakae S, Sudo K, Ishiwata K, Kanazawa T (2017) Dual genetic absence of STAT6 and IL-10 does not abrogate anti-hyperglycemic effects of Schistosoma mansoni in streptozotocin-treated diabetic mice. Exp Parasitol 177:1–12

Parsa R, Andresen P, Gillett A, Mia S, Zhang XM, Mayans S, Holmberg D, Harris RA (2012) Adoptive transfer of immunomodulatory M2 macrophages prevents type 1 diabetes in NOD mice. Diabetes 61(11):2881–2892

Saunders KA, Raine T, Cooke A, Lawrence CE (2007) Inhibition of autoimmune type 1 diabetes by gastrointestinal helminth infection. Infect Immun 75:397–407

Sipahi AM, Baptista DM (2017) Helminths as an alternative therapy for intestinal diseases. World J Gastroenterol 23(33):6009–6015

Smallwood TB, Giacomin PR, Loukas A, Mulvenna JP, Clark RJ, Miles JJ (2017) Helminth immunomodulation in autoimmune disease. Front Immunol 24(8):453

Smolen JS, Aletaha D, Barton A, Burmester GR, Emery P, Firestein GS, Kavanaugh A, McInnes IB, Solomon DH, Strand V, Yamamoto K (2018) Rheumatoid arthritis. Nat Rev Dis Primers 4:18001. https://doi.org/10.1038/nrdp.2018.1. PMID: 29417936

Summers RW, Elliott DE, Urban JF Jr, Thompson RA, Weinstock JV (2005) Trichuris suis therapy for active ulcerative colitis: a randomized controlled trial. Gastroenterology 128:825–832

Tang C, Zou J, Zhang R, Liu Z, Mao C (2019) Helminths protect against type 1 diabetes: effects and mechanisms. Parasitol Res 118(4):1087–1094

Trabandt A, Aicher WK, Gay RE, Sukhatme VP, Nilson-Hamilton M, Hamilton RT, Mcghee JR, Fassbender HG, Gay S (1990) Expression of the collagenolytic and Ras-induced cysteine proteinase cathepsin L and proliferation-associated oncogenes in synovial cells of MRL/I mice and patients with rheumatoid arthritis. Matrix 10(6):349–361

Villeneuve C, Kou HH, Eckermann H, Palkar A, Anderson LG, McKenney EA, Bollinger RR, Parker W (2017) Evolution of the hygiene hypothesis into biota alteration theory: what are the paradigms and where are the clinical applications? Microbes Infect 20(3):147–155
Walk ST, Blum AM, Ewing SA, Weinstock JV, Young VB (2010) Alteration of the murine gut microbiota during infection with the parasitic helminth Heligmosomoides polygyrus. Inflamm Bowel Dis 16:1841–1849
Webber MP, Moir W, Zeig-Owens R, Glaser MS, Jaber N, Hall C, Berman J, Qayyum B, Loupasakis K, Kelly K, Prezant DJ (2015) Nested case-control study of selected systemic autoimmune diseases in world trade center rescue/recovery workers. Arthritis Rheumatol 67:1369–1376
Weinstock JV, Elliott DE (2013) Translatability of helminth therapy in inflammatory bowel diseases. Int J Parasitol 43:245–251
Whelan M, Harnett MM, Houston KM, Patel V, Harnett W, Rigley KP (2000) A filarial nematode-secreted product signals dendritic cells to acquire a phenotype that drives development of Th2 cells. J Immunol 164:6453–6460
WHO fact sheet (2020). https://www.who.int/news-room/fact-sheets/detail/soil-transmitted-helminthinfections
Wiria AE, Djuardi Y, Supali T, Sartono E, Yazdanbakhsh M (2012) Helminth infection in populations undergoing epidemiological transition: a friend or foe? Semin Immunopathol 34 (6):889–901
Yan K, Wang B, Zhou H, Luo Q, Shen J, Xu Y, Zhong Z (2020) Amelioration of type 1 diabetes by recombinant fructose-1, 6-bisphosphate aldolase and cystatin derived from Schistosoma japonicum in a murine model. Parasitol Res 119(1):203–214
Zhu J, Zhou Y, Chen X, Li J (2014) A meta analysis of the increased risk of rheumatoid arthritis-related pulmonary disease as a result of serumanticitrullinated protein antibody positivity. J Rheumatol 41(7):1282–1289

# Chapter 2
# Microbes: An Integral Component of Flavor Production

**Himanish Dutta Choudhury, Pappu Deb, and Ravi Rajwanshi**

**Abstract** Natural flavoring industry has undergone an immense surge in production that has done more than enough to compete and even supersede the chemically synthesized flavoring industry, due to a rapid increase in demand from the consumer. The natural flavoring industry utilizes techniques like plant tissue culture, de novo synthesis, and biotransformation to increase their production. However, de novo synthesis and biotransformation have been found to be better suited for industrial production than plant tissue culture pertaining to the ease and efficiency in bioproduction. Especially, the propitious role played by microorganisms like bacteria, fungi, and algae in either production or enzymatic conversion into secondary metabolites has been an essential part for the production of flavoring compounds. The production of desired flavor from low cost precursor with a limited emission of harmful wastes is also a valued addition to industrialization of bioflavor production considering the detrimental impact of chemically synthesized flavoring compounds on environment. With the advancement in the field of biotechnology, the genetically engineered microorganisms or enzymes have ushered in a new era of the flavoring industry with economical production of flavorants. The present chapter reviews the different processes involving the diverse microbial entity in the efficient production of flavors and fragrances, and also enumerates a series of advantages that are conferred by it to the natural flavoring industry.

**Keywords** Bioflavor · Flavorant · PTC · de novo synthesis · Biotransformation · Elicitor

All authors contributed equally.

H. D. Choudhury · P. Deb · R. Rajwanshi (✉)
Department of Biotechnology, Assam University, Silchar, Assam, India

 
P. K. Arora (ed.), *Microbial Products for Health, Environment and Agriculture*, Microorganisms for Sustainability 31, https://doi.org/10.1007/978-981-16-1947-2_2

## 2.1 Introduction

The term flavor describes the impression of food or substance due to a chemoreception of olfactory and gustatory system. The effectiveness of a flavor depends upon the consumer's liking and acceptability where the presence of a flavorant alters the smell or taste and hence contributes to the value addition of food. More specifically, it is the flavor that determines the olfactory quality of the food which can be considered as the backbone of the economically important food flavoring industry involved in the production of flavorants. The flavorants used in the food, cosmetic, or pharmaceutical products can be natural or artificial as per the requirement (Scragg 2007). The chemical or biochemical reactions during food processing and storage along with the quality of the raw materials used may influence the flavor of the food. Flavor may also be changed by the addition of the organic or inorganic substances (Cabaroğlu and Yılmaztekin 2010; Hosoglu et al. 2018). The compounds mostly involved in generating flavor include aldehydes, alcohols, esters, ketones, lactones, free fatty acids, phenolic compounds, and sulfur compounds (Gupta et al. 2015). There are more than 6500 flavor compounds identified for food industries, but only a few (~400) are widely used for foods such as beverages, dairy products, and sauces (Güneşer et al. 2015; Scragg 2007). Flavors can be produced naturally from plants, by extraction or distillation process, considered as natural flavors or can be synthesized chemically, termed as artificial flavor compounds (Janssens et al. 1989). Flavor compounds extracted from plants are more consumer friendly. However, limited availability of the natural resources due to different environmental factors, high cost of extraction, and distillation process of precursors is a regular constrain in the overall productivity of natural flavoring agents (Hosoglu 2018). With the advancement of modern techniques, synthetic flavor development has become an integral part of flavor industry but flavor produced by biological processes is in demand due to health concerns. The chemically synthesized flavoring agents have an efficient and low cost approach to boost the production but their complex chemical processes and release of environmentally hazardous substances conclude to give more priority towards the efficient synthesis of natural flavoring compounds using biotechnological approaches. However, it has been observed that the biotechnological processes do not require extreme conditions and are environment friendly than the chemical processes (Willaert et al. 2005). Therefore, production of flavoring compounds by nature friendly processes can be considered as one of the most effective alternatives for production of flavor and fragrances. Microbial processes have been playing a vital role since long time in the production of high quality of foods and beverages with flavors liked by the mankind (Tamang 1998). Though the role of microorganism in the development of flavor was unclear in the ancient times but the advancement in the field of microbiology and biochemistry helped to understand the role of microbes in flavor generation. After the publication of the first article on microbial flavor in the year 1923 (Omelianski 1923; Schrader 2007), the analytical techniques to isolate and separate the volatile and nonvolatile compounds along with the identification of their structure and function become more systematic and integral part of flavor generating industry. The use of microorganisms became an integral part of different biotechnological approaches for the production of bioflavoring agents with enhanced shelf life of the product (Vandamme 2003). The present chapter highlights the role of microorganism in the production of flavoring compounds with special concern to different techniques involved in microbial bioflavoring.

## 2.2 Biotechnological Approach in Bioproduction of Flavorant

The production of commercially important compounds with the help of biological systems with an eco-friendly approach is considered as bioproduction (Willaert et al. 2005). The bioproduction of flavoring compounds and fragrances in a large scale is a necessary requirement to fulfill the demand of natural flavoring compounds for the associated industry. The intial step towards the bioproduction of flavoring compunds by microorganism is the formation of a complex flavor by mixing the microbial culture used for fermentation process. The perfect example for the same process is the flavor produced during the fermentation of gueuze beer, soy sauce, kefir (Hallé et al. 1994; Halm et al. 1993; Kumara 1989). Bioproduction sometimes leads to the poor yield, lower rate of biotransformation, self toxicity of microorganisms and causes trouble in recovery from the bioreaction mixture, however, the use of biotechnological approach has turned into a most appealing method for the production of natural flavors and fragrances as it requires lenient conditions with no release of harmful or toxic chemicals which has resulted in the production of more than 100 commercially available flavor compounds so far (Bicas et al. 2010; Krings and Berger 1998; Medeiros et al. 2000). Microorganisms are not only used for complex flavor production but they can also be used for the production of single flavoring compound from a well-chosen substrate (Willaert et al. 2005). The same can be achieved by various biotechnological approaches, viz. plant cell or tissue culture (PTC), microbial fermentation and microbial or enzymatic bioconversion and biotransformation (Bicas et al. 2010; Schrader et al. 2004).

### *2.2.1 Flavor Production Using Plant Cell or Tissue Culture*

Plant cells are totipotent that can rediffrentiate to develop into new plant, organ, and cell type at certain stages of development and perform various metabolic process under suitable environmental condition. Such metabolic processes involve the production of secondary metabolites which are considered as sources of flavor. Production of flavoring compounds by cell suspension culture is advantageous as it remains uneffected from harsh and differential environmental conditions

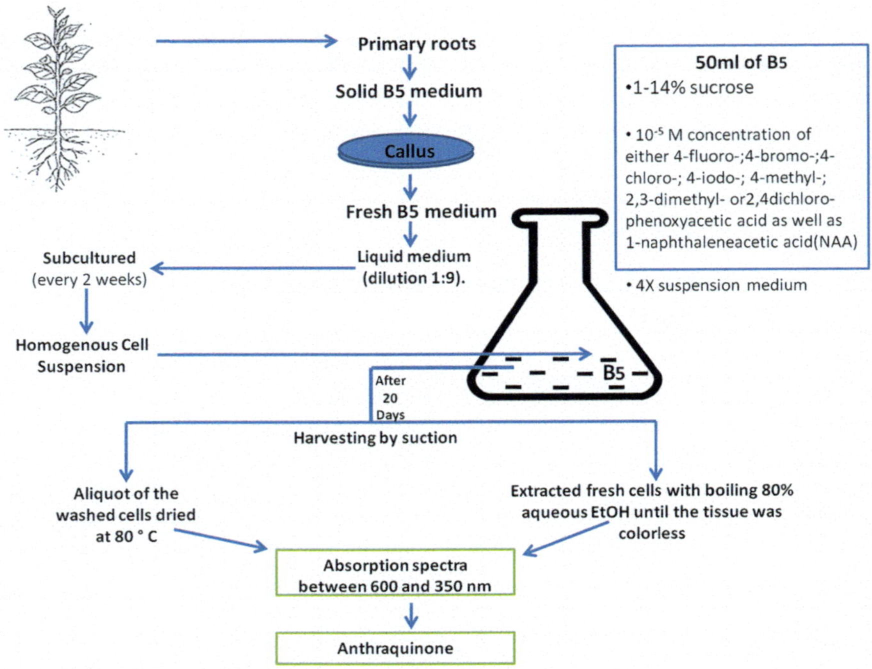


**Fig. 2.1** Production of anthraquinone using plant cell culture technique

compared to those isolation from plants grown in agricultural fields (Berger 1995; Harlander 1995; Medeiros et al. 2000). PCT helps in the direct production of flavoring compounds from specific plant cells or tissues in large bioreactors as it provides controlled condition for secondary metabolism. Secondary metabolite production requires standardization of controlled condition and manipulation of culture medium with respect to micronutrient, carbon source, and enzymes; addition of external elicitors along with the introduction of well differentiated cells which results into an efficient production of secondary metabolites (Zenk 1990).

Anthraquinone production from Rubiaceae species is a well-studied example of PTC based secondary metabolite production (Schulte et al. 1984). The study describes the effect of different concentration of sucrose and auxin quality (effector) on optimal production of anthraquinone by the development of the homogeneous cell suspension in B5 medium containing 1–14% sucrose concentration, 1-naphthaleneacetic acid and one of the effectors such as 4-fluoro-, 4-bromo-, 4-chloro-, 4-iodo-, 4-methyl-, 2,3-dimethyl-, and 2,4-dichloro-phenoxyacetic acid at $10^{-5}$ M concentration (Fig. 2.1). The same study also shows that on addition of elicitors or carbon component in the PTC medium, an increase in the level of secondary compounds is observed in comparison to the well differentiated plants. The industrial production of berberine from *Coptis japonica* and shikonin from *Lithospermum erythrorhizon* cell line by repeated cell aggregate selection and protoplast selection process are also well known examples of secondary metabolite production as reported by Fujita (1988). The environmental factors such as light, osmotic pressure, carbon dioxide also play an essential role in the secondary metabolite production. As reviewed by Harlander (1995) the exposure of light results in the formation of aliphatic ketones and esters in *Ruta graveolens* cell cultures whereas maintenance in the dark leads to the monoterpene formation in roots. The presence of mannitol increases the osmotic pressure that leads to the four to five fold increased production of indole alkaloid (Zhao et al. 2001). Increased production of ethyl butyrate, ethanol ethyl acetate, and isobutanol was observed in apple and grape cell culture due to increased level of carbon dioxide in the atmosphere (Harlander 1995).

The plant tissue culture technique is also utilized in the field of aroma production. Vanilla is produced from suspension culture of vanilla pod in MS media containing NAA (Funk and Brodelius 1990a). The study also reported the low concentration of extractable secondary compounds while testing the role of elicitors like 2,4-Dichlorophenoxyacetic acid (2,4D), cytokines, kinetin, and 2-benzyladenine. The production of vanillic acid was also observed by introduction of 4-hydroxycinnamate: CoA ligase inhibitor, 3,4-methylenedioxy cinnamic acid in cell cultures of *Vanilla planifolia* (Funk and Brodelius 1990b). The tissue culture technique also facilitates the production of p-Hydroxyphenylbutanone (raspberry ketone) from raspberry cell suspension in Anderson's medium in addition of 2,4-D, Indole-3-butyric acid (IBA) and 6-(γ,γ-Dimethylallylamino) purine (2iP) (Borejsza-Wysocki and Hrazdina 1994). An increase in raspberry ketone level was also monitored upon addition of methyl jasmonate under cell suspension; however, the yield is insufficient for large batch culture as reported by Pedapudi et al. (2000). The aroma is basically associated with different volatile compounds such as ketones, aldehydes esters, lactones, acetals, sulfur compounds, aromatic compounds (Salunkhe et al. 1976). Even the tissue culture technique cannot resolve the elevation of every compound; however, it can facilitate the increase in production of specific component of aroma. The same can be well understood by the experiments carried out on strawberry by different researchers across the world. For example, addition of 6-deoxy-D-fructose, a potential precursor of dimethyl-4-hydroxy-2H-furan-3-one (DHMF) triggers the biosynthesis of DMHF-glucoside or addition of small chain fatty acid responsible for flavor (Hong et al. 1990). Another example is α-keto-acid, a precursor to produce ethyl butyrate and butyl butyrate which facilitates the formation of fruity flavor compound under well balanced tissue culture system (Hong et al. 1990; Zabetakis and Holden 1995). Even though there are several applications of PTC in the field of natural flavor production but there is less scope for large-scale production of flavor compound using the same technique. The limitations include the higher doubling time of the plant cells resulting in higher batch cycle, difficulty in maintaining sterility, lower level of yield, higher chances of contamination, and loss of desirable metabolite during the growth of cell in the culture that results in the lower stability of metabolite making the PTC an unprofitable option for flavor industry.

### 2.2.2 Microorganisms Involved in Bioflavoring

Microorganisms involved in the production of flavors and fragrances are omnipresent in nature. Industrial production of flavors and fragrances by microorganisms for economical benefits is a new approach, but the concept behind this practice is very old such as production of wine by fermentation. Benzaldehyde responsible for the almond-like flavor was the first flavor compound to be recognized (Speelmans et al. 1998). In situ microbial culture of bacteria such as *Lactobacillus* sp. produces flavors in dairy products like butter, milk, cheese, yogurt, and curd. Some microorganisms are found to produce fruity and floral aroma in dairy products rather than buttery flavor. *Ceratocystis fimbriata,* a famous fungi is known to produce a very strong fruity fragrance in Solid State Fermentation (SSF) by utilizing some of the substrates like wheat bran, cassava bagasse, and sugar bagasse (Christen et al. 1997). Bioflavor production is classified into three major types according to the source of microorganisms (Roy and Kumar 2019). These include:

1. *Bacterial originated Bioflavors*: There are several species of bacteria that help in the bioproduction of flavor components either by de novo synthesis or transformation (Table 2.1). Vanillin is an active compound which is used for developing a unique flavor named as "Vanilla." *Nocardia iowensis* is a bacterium that produces vanillin by de novo synthesis and also by biotransformation (Bicas et al. 2010; Walton et al. 2000). Some other bacteria such as *Pseudomonas putida, Corynebacterium glutamicum, Arthrobacter globiformis* also produce vanillin by bioconversion of euginol and isoeuginol (Shimoni et al. 2003). *Pseudomonas gladioli* produces bioflavor present in essential oil such as α-terpinol (Cadwallader et al. 1989). *Bacillus lichiniformis* is another bacterium that produces isoamyl acetate from isoamyl alcohol (Torres et al. 2009). "Nookatone" is the grapefruit aroma produced by some soil bacteria of genera Enterobacteriaceae, having a very high market demand and used in beverages and perfumes (Gupta et al. 2015).
2. *Fungal originated bioflavors*: Number of fungal species including yeast and molds have been found to produce bioflavors naturally as well as by biotransformation. Fungal species such as *Geotrichum fragrans* produces various secondary metabolites such as alcohol, acids, and esters. Strong fruity fragrance of pineapple is produced by esters such as ethyl acetate and ethyl butyrate (Damasceno et al. 2003). Studies have been performed on the optimization of α-terpinol production by using some fungal molds such as *Fusarium oxysporum, Penicillium digitatum,* and *Cladosporium* sp. (Bicas et al. 2010). *Cladosporium* sp. produces the α-terpinol at a highest concentration of 1.0 g/L whereas *Fusarium oxysporum* produces at the lower concentration of 500 mg/L (Bicas et al. 2010; Maróstica Jr and Pastore 2007). Several aromatic compounds like isobutyl acetate, isoamyl acetate, geraniol, and citronella can also be synthesized by fungi *Ceratocystis moniliformis* (Longo and Sanromán 2006).
3. *Algal originated bioflavors*: Unicellular cyanobacterium such as *Synechococcus elongatus* can produce buttery bioflavor through acetone synthesis metabolic

**Table 2.1** List of microorganisms involved in the production of bioflavoring compounds

| Organisms | | Precursors | Products | | REFERENCES |
|---|---|---|---|---|---|
| Group | Species | | Compound Produced | Flavor | |
| Bacteria | *Pseudomonas oleovorans* | Fatty acids esterified in plant oils | Methylketones | Butter scotch | (Krings and Berger, 1998) |
| | *Pseudomonas gladioli* | α-pinene | α-terpineol | Anticonvulsant agent, perfume, monopolymer | (Cadwallader, et al. 1989) |
| | *Bacillus subtilis* | Glucose and xylose | Acetoin | Buttery | (Hua, et al. 2007) |
| | *Lactococcus lactis* | Glucose | Diacetyl | Buttery | (Nielsen, et al. 2010) |
| | *Pseudomonas putida* | Geraniol | Geranic acid | Leafy with citrus hint | (Mi, et al. 2014) |
| | *Nocardia iowensis* | Isoeugenol | Vanillin | Vanilla | (Carroll, et al. 2016) |
| | *Streptomyces griseus* | White potatoes and corn meal bread | Geosmin | Earthy | (Gupta, et al. 2015) |
| | *Streptomyces sp.* V-1 | Ferulic acid | Vanillin | Vanilla | (Xu, et al. 2009) |
| | *Arthobacter globiformis* | Ferulic acid | Vanillin | Vanilla | (Shimoni, et al. 2003) |
| | *Pediococcus pentosaceus* | Glucose | Acetoin | Buttery | (Escamilla-Hurtado, et al. 2005) |

(continued)

**Table 2.1** (continued)

| | | | | | |
|---|---|---|---|---|---|
| | *Saccharomyces cerevisiae* | Benzyl alcohol | Butyl butyrate | Pineapple | (Aggelopoulos, et al. 2014) |
| | *Pichia pastoris* | Agro-industrial wastes | Benzaldehyde | Cherry & Almond | (Craig and Daugulis, 2014) |
| | *Ceratocystis fimbriata* | Cassava wastewater | Esters | Fruity | (Soares, et al. 2000) |
| | *Geotrichum fragans* | Sucrose | Esters | Fruity | (Damasceno, et al. 2003) |
| | *Trichoderma viride* | Olive mill waste | 6-pentyl-α-pyrone | Coconut | (Fadel, et al. 2015) |
| **Fungi** | *Rhizopus oryzae & Candiada tropicalis* | Acetic acid, 2-methylpropyl ester | Lemonene | Citrus | (Guneser, et al. 2017) |
| | *Ceratocystis moniliformis* | Ferulic acid | Isobutyl acetate, Geraniol | Banana, rose | (Akacha and Gargouri 2015) |
| | *Pycnoporous cinnabarinus* | L-phenylalanine | Vanillin | Vanilla | (Falconnier, et al. 1994) |
| | *Ischnoderma benzoinum* | L-phenylalanines and glucose | Benzaldehyde | Nutty & Almond | (Longo and Sanromán, 2006) |
| | *Nidula niveo-tomentosa* | Acetone | 4-(4-Hydroxyphenyl)2-butanone | Raspberries | (Krings and Berger, 1998). |

pathway. Earthy aroma of Geosmin by de novo synthesis is also reported to be produced by some algae (Oliver et al. 2013).

#### 2.2.2.1 Flavor Production Using Microbial Fermentation/de novo Synthesis

Since the ancient times the microorganisms play a key role in the production of fermented products. The microorganisms like yeast, bacteria, and fungi are found to have inevitable role in the production of several flavoring compounds including aldehydes, ketones, esters, alcohol, lactones, and sulfur compounds as secondary metabolites (Hosoglu et al. 2018). With the advancement in the field of biotechnology, the fermentation technology was utilized for economical and industrial perspective. The production of complex mixture of flavoring compounds using the fermentation was the first step towards the progress of flavor industry. However, immobilization technique along with proper design leads to the development of single flavoring compound. The de novo fermentation is also advantageous as the process can be boosted up with limited supply of intermediate or precursor molecule (Vandamme and Soetaert 2002). The microbial fermentation involves Submerged Fermentation (SmF) and Solid State Fermentation (SSF) as a rapid approach towards the production of secondary metabolite which is used as a flavor along with bioactive compound (Subramaniyam and Vimala 2012). The microbial fermentation is also considered as an inexpensive technique that requires low cost supplies like carbon, nitrogen, along with vitamins, minerals, and micronutrients in the culture media (Harlander 1995). The use of agrowastes in bioproduction is preferred due to its contribution towards the low cost and rich content during the flavor and fragrance production. For example, the addition of tapioca bagasse, sugar beet, beet molasses, wheat bran, apple pomace, soy bean meal, rice bran, fruits and vegetables pomaces, and whey in the microbial fermentation has shown the efficient production of natural flavors and fragrances (Akacha and Gargouri 2015; Nigam and Pandey 2009). Due to high fermentation capability with low-growth requirements and high-enzyme catalyzed systems, it was noted that metabolism of yeasts and fungi are more efficient during fermentation of agrowastes (Häusler and Münich 1997).

Several studies have shown the usage of yeast or fungal strains in the production of flavor. One such study is about the production of aroma products like isobutyl acetate, ethyl acetate, propyl acetate, isoamyl acetate, citronellal, and geraniol from *Ceratocystis moniliformis* (Collins 1976). Several other fungal species like *Aspergillus niger*, *Aspergillus parasiticus*, *Aspergillus ochraceus*, *Aspergillus oryzae*, *Penicillium chrysogenum*, *Penicillium funiculosum*, *Penicillium citrinum*, *Penicillium raistrickii*, *Penicillium viridicatum*, *Cephalosporium*, *Alternaria*, and *Fusarium* sp. are found to produce 3-methylbutanol, 3-octanone, 3-octanol, 1-octen-3-ol, 1-octanol, and 2-octen-1-ol together with octane, isobutyl alcohol, butyl alcohol, butyl acetate, octyl acetate, pyridine, hexanol, nonanone, dimethylpyrazine, benzaldehyde, propylbenzene, and phenethyl alcohol in coarse wheat medium under culture condition (Kaminski et al. 1974). Degradation of $\beta,\beta$-carotene is an essential step for production of terpene components. Studies show that several fungi are responsible for degradation of $\beta,\beta$-carotene either directly or with extracellular enzyme activity in culture containing $\beta,\beta$-carotene (Zorn et al. 2003). Study performed by Zorn et al. (2003) showed that the production of dihydroactinidiolide which exhibits a pleasant hay like odor important for black tea flavor, from submerged culture of *Ganoderma applanatum*, *Hypomyces odoratus*, *Kuehneromyces mutabilis,* and *Trametes suaveolens*. The genera *Saccharomyces* are often associated with alcoholic fermentation and also found to produce esters, alcohols, and acetates. Some non-*Saccharomyces*, yeasts, such as *Hanseniaspora guilliermondii*, *Hanseniaspora uvarum,* and *Pichia anomala*, were also reported as a producers of ethyl acetate, geranyl acetate, isoamyl acetate, and 2-phenylethyl acetate which influence the sensory quality of wine (Rojas et al. 2001). Rojas et al. (2001) reported that *Pichia* is able to promote the esterification of ethanol, geraniol, isoamyl alcohol, and 2-phenylethanol which increases the concentration of esters having a fruity aroma. The strains of *Saccharomyces*, *Hansenula*, *Candida*, *Pichia*, and *Torulaspora* along with the strains of *Aspergillus*, *Mucor*, *Rhizopus*, *Monascus* and *Trichoderma* help in production, processing, and providing aroma to dacu, Maotai-flavor liquor (Wang et al. 2008). The fruity, floral flavor producing components such as citronellal, geraniol, and linalool were also reported to be synthesized de novo by *Kluyveromyces lactis* (Drawert and Barton 1978). The production of γ-lactone and γ-decalactone along with alcohol, esters, and carbonyl compounds was observed in the synthetic media containing glucose, L-asparagine, $MgSO_4$. $7H_2O$, $KH_2PO_4$, and deionized water of *Sporobolomyces odorus* culture (Tahara et al. 1972). The component produced was found to be responsible for the development of peach like flavor in the culture. The de novo synthesis and identification of the components produced by the culture of *S. odorus* are described in Fig. 2.2. Some bacterial strains like *Gluconobacter roseus* are found to produce methylbutyric acid, a common ester among the flavor compounds, on oxidation of methylbutanol (Gatfield and Sand 1995).

The production of tetramethylpyrazine which results into production of nutty, roasted, and toasty flavor was also reported to be produced by a mutant species of *Corynebacterium glutamicum* (Demain et al. 1967). Although the same compound can also be produced from the suspension culture of *Bacillus subtilis*, where Daqu or glucose is used as a precursor (Xiao et al. 2006; Zhu et al. 2010). SSF of *B. subtilis* was reported to produce aroma compounds such as 2,5-Dimethylpyrazine (2,5-DMP) and tetramethylpyrazine (TTMP) from soyabean in the presence of threonine and acetoin (Besson et al. 1997). The compounds such as butyric acid, lactic acid, and diacetyl responsible for buttery flavor are found to be produced from microbial culture of *Lactococcus lactis*, *Lactobacillus* sp., *Streptococcus thermophilus*, *Leuconostoc mesenteroides* (Escamilla et al. 2000; Ibragimova et al. 1980; Longo and Sanromán 2006). Vanillin is also produced from the culture media of *Oenococcus oeni* supplemented with isoeugenol, coniferyl aldehyde, or ferulic acid (Bloem et al. 2007). Several other strains of microorganisms are also reported to

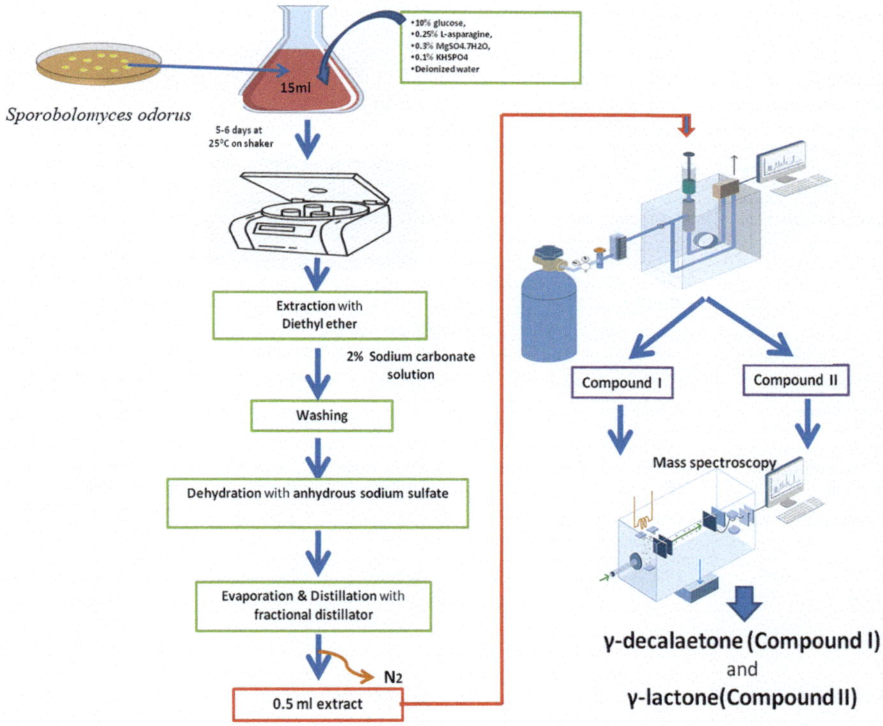


**Fig. 2.2** Steps of de novo synthesis of γ-lactone and γ-decalactone using *Sporobolomyces odorus* culture

be used in de novo synthesis of flavor compounds as mentioned in Tables 2.2 and 2.3.

#### 2.2.2.2 Flavor Production using Microbial or Enzymatic Bioconversion and Biotransformation

Specific flavor and fragrance compounds can also be produced by approaches like bioconversion or biotransformation which involves conversion of specific substrates added to the cultural medium. Several microbial enzymes are capable of converting specific substrates into flavor compounds (Janssens et al. 1989). The yield of bioconversion process is very high which leads to better opportunities towards the commercial production in comparison to de novo synthesis (Guiné et al. 2010). The process requires easily available low cost precursors and several biochemical reactions such as reduction, oxidation, dehydration, and hydrolysis for the production of different flavoring compounds. Some alternative processes like immobilized cell systems, SSF are also reported to contribute in the improvement of the yield (Feron

**Table 2.2** List of Fungal strains/Yeast involved in the de novo production of bioflavoring compounds

| Name of the organism (fungi/yeast) | Precursor | Immobilization method | Product | References |
|---|---|---|---|---|
| *Saccharomycopsis lipolytica* | Glucose, galactose | Membrane cell recycle | Citric acid | Rane and Sims (1995) |
| *Aspergillus niger* | Glucose, galactose, sucrose | Hollow fiber | Gluconic acid, citric acid | Chung and Chang (1988) |
| *Pichia pastoris* | *Benzyl alcohol* | Membrane cell recycle | Benzaldehyde | Craig and Daugulis (2014) |
| *Ceratocystis fimbriata* | Cassava bagasse, apple pomace, and soya bean | Ca-alginate beads | Esters, e.g. ethyl acetate | Soares et al. (2000) |
| *Trichoderma viride* | 2-phenylethanol (2-PE) | Liquid surface immobilization | 6-pentyl-α-pyrone | Fadel et al. (2015) |
| *Rhizopus oryzae* | Oil mill wastes | Polyurethane foam | Limonene | Guneser et al. (2017) |
| *Ischnoderma benzoinum* | Phenylalanine | Polyurethane foam | Benzaldehyde | Longo and Sanromán (2006) |

**Table 2.3** List of bacterial strains involved in the de novo production of bioflavoring compounds

| Name of the bacteria | Precursor | Immobilization technique | Product | References |
|---|---|---|---|---|
| *Escherichia coli* | Farnesyl diphosphate | Ca-alginate fibers | Geosmin | Nielsen et al. (2010) |
| *Lactococcus lactis* | Ferulic acid | Ca-alginate fibers | Vanillin | Nielsen et al. (2010) |
| *Pediococcus pentosaceus* | Peptidoglycan | Ca-alginate beads | 3-phenyl lactic acid | Kearney et al. (1990) |
| *Lactobacillus cremoris* | Lactose, whey | Membrane cell recycle | Lactic acid | Bibal et al. (1991) |
| *Lactobacillus delbrueckii* | Glucose | Hollow fiber | Lactic acid | Kulozik et al. (1992) |
| *Corynebacterium glutamicum* | α- keto glutarate | κ-Carrageenan porous glass | Glutamic acid | Lu and Chen (1988) |

et al. 1996). Both the processes (bioconversion and biotransformation) are advantageous; however, bioconversion is a multistep process, whereas biotransformation is single step process. A typical example of bioconversion and biotransformation is the production of vanillin (4-hydroxy-3-methoxybenzaldehyde) that involves natural precursor like ferulic acid from which several intermediates are produced on treatment with microbial strains like *Aspergillus niger*, *Pycnoporus cinnabarinus*,

*Phanerochaete chrysosporium*, *Pseudomonas putida*, *Streptomyces setonii* (Khoyratty, et al. 2018). The involvement of fungi such as *Pycnoporus cinnabarinus* is a well-studied example of bioconversion where agar media with selective XAD-2 resin were used for the large-scale production in a mechanically agitated and an air-lift bioreactor (Stentelaire et al. 2000). The process ultimately results into the formation of 1575 mg of vanillin per liter of fungal culture on basal medium containing maltose, diammonium tartrate, yeast extract, $KH_2P0_4$, $CaC1_2.2H_2O$, and $MgS0_4.7H_20$. However, the bacteria such as *Pseudomonas* give promising results in the production of vanillin from a non-oxidative, CoA-dependent chain-shortening mechanism from ferulic acid (Walton et al. 2003). Biochemical reactions like decarboxylation of ferulic acid, reduction of ferulic acid and coenzyme-A-independent deacetylation of ferulic acid are involved in the microbial cultures of different strains of *Pseudomonas* (*Pseudomonas mira*, *P. fluorescens*, *P. putida*), *Bacillus subtilis, Corynebacterium glutamicum,* and *Bacillus coagulans* (Priefert et al. 2001). The modern biotechnological tools are also helpful in the use of transposon induced mutant of *P. fluorescens* strain BF13, containing an insertion in the *vanA* gene that encodes the α-subunit of the vanillate O-demethylase involved in the transformation of ferulic acid to vanillic acid resulting an yield of 0.23 g vanillic acid/hr./g (wet weight) (Civolani et al. 2000).

The cheese flavor is another highly demanded flavor in the commercial market. Studies conducted by several researchers showed that the compound responsible for the development of strong cheese flavor is methyl ketone. The same can be obtained with the help of bioconversion and transformation rather than a synthetic production approach. Studies show that with the addition of medium length fatty acid as precursor can help the fungi *Penicillium roqueforti* in the production of methyl ketone via an incomplete β-oxidation of fatty acid precursor in the presence of enzyme, 3-ketoacyl CoA-thioester hydrolase (Vandamme and Soetaert 2002). The fungi *Botryodiplodia theobromae* is reported to be associated with jasmine odor generating compound, methyl(+)-7-isojasmonic acid from α-linolenic acid (Häusler and Münich 1997). The process involves esterification of jasmonic acid obtained from α-linolenic acid by fungi with the help of commercial lipases. *Hyphozyma roseoniger*, *Cryptococcus* sp. are found to convert sclareol to sclareolide which can be converted to Ambrox® (Cheetham 1997). A study shows that *Zygosaccharomyces rouxii* is used to develop the 2,5-Dimethyl-4-hydroxy-3(2H)-furanone (DMHF) from a medium supplied with D-fructose-1,6-diphosphate and glucose (Dahlen et al. 2001). Table 2.4 provides a list of few examples related to production of favoring compounds via bioconversion by microorganisms. The microbial enzymes are also found to play essential role in production of flavoring compounds. For example, the production of *l*-menthol, a key constituent of peppermint oil. *l*-menthol is produced by enantio selective hydrolysis of *dl*-menthyl acetate with esterase extracted from *Bacillus subtilis* (Zheng et al. 2009). The production of *l*-fucose from capsular exopolysaccharide (EPS) isolated from *Clavibacter* and *Klebsiella* strain by enzymatic hydrolyzation under fermentation is also a well-known example that shows the usage of microbial strain in enzymatic process of flavor production (Vanhooren and Vandamme 1999).

**Table 2.4** List of flavoring compounds produced via bioconversion by microorganisms

| Group | Name of species | Precursor | Immobilization technique | Product | References |
|---|---|---|---|---|---|
| Bacteria | *Klebsiella oxytoc* | Glucose, 1,3 propanediol | Glass wool | 2,3-Butanediol | Champluvier et al. (1989) |
| | *Rhodococcus fascians* | Limonin | κ-Carrageenan beads | Citrus juice debittering/ limonin degradation | Manjón et al. (1991) |
| | *Zymomonas mobilis* | Sucrose | Hollow fiber | Gluconic acid | Paterson et al. (1988) |
| | *Arthrobacter* sp. | D- glucose and D-xylose | κ-Carrageenan beads | Glucose isomerization | Bazaraa and Hamdy (1989) |
| | *Erwinia rhapontici* | Sucrose | Ca-alginate beads | Isomaltulose | Cheetham et al. (1982) |
| | *Brevibacterium ammoniagenes* | Fumaric acid | Polyacrylamide beads | Malic acid | Yang et al. (1992) |
| | *Pseudomonas* | Ferulic acid | Ca/Ba-alginate | Vanillic acid | Bare et al. (1994) |
| Fungi | *Saccharomyces cerevisiae* | Xylose, Oxaloacetic acid | Ca-alginate beads, agarose beads | Glucose isomerization, malic acid | Koren and Duvnjak (1992) |
| | *Candida rugosa* | Fumaric acid | Agarose beads | Malic acid | Neufeld et al. (1991) |
| | *Rhizopus delemar* | Saturated fatty acids | Polyurethane foam | Olive oil hydrolysis | Chen and McGill (1992) |
| | *Penicillium italicum* | Geraniol, nerol, or citral | Agarose beads | Methyl ketones | Lomascolo et al. (1999) |
| | *Pycnoporus cinnabarinus* | Ferulic acid | Agitated tank | Vanillin | Lomascolo et al. (1999) |

## 2.3 Conclusion

Studied performed on the production of different flavoring compounds with the help of techniques like PTC, microbial cultures or enzymes have generated greater interest in terms of industrial perspective. Bioengineering especially metabolically and genetically modified microbes are playing a major role in the production of desired flavor compound. Though de novo synthesis and biotransformation are found advantageous for enhanced production of flavoring compound, however, commercialization of such techniques is still a bigger challenge towards bioproduction. The literature stated throughout the chapter highlights the advantageous role of different microorganisms in scaling up the production of flavor compounds and fragrances; however, product inhibition, unstable characteristics of products, or formation of toxic by-products are the common challenges in the process of commercial production. Therefore, research emphasis should be given on the identification of low cost precursors, suitable substrates, enzymes, and elicitors that are more specific and efficient in terms of production of flavoring compounds. Feasible technology should be adopted for efficient product removal to enhance the productivity of desired flavor compounds within the purview of healthy natural environment leading to health benefits to the consumers.

**Acknowledgments** RR acknowledges funding from Science & Engineering Research Board (SERB), Government of India vide grant no. EEQ/2016/000501 to the laboratory during conceptualization and writing of the chapter.

**Conflict of Interest** Authors have declared no conflict of interest.

## References

Aggelopoulos T, Katsieris K, Bekatorou A, Pandey A, Banat IM, Koutinas AA (2014) Solid state fermentation of food waste mixtures for single cell protein, aroma volatiles and fat production. Food Chem 145:710–716

Akacha NB, Gargouri M (2015) Microbial and enzymatic technologies used for the production of natural aroma compounds: synthesis, recovery modeling, and bioprocesses. Food Bioprod Process 94:675–706

Bare G, Delaunois V, Rlkir R, Thonart P (1994) Bioconversion of vanillin into vanillic acid by *Pseudomonas fluorescens* strain BTP9. Appl Biochem Biotechnol 45:599–610

Bazaraa WA, Hamdy MK (1989) Fructose production by immobilized *Arthrobacter* cells. J Ind Microbiol 4:267–274

Berger RG (1995) Why novel biotechnology of aromas? In: Aroma biotechnology. Springer, Berlin, pp 35–41

Besson I, Creuly C, Gros JB, Larroche C (1997) Pyrazine production by *Bacillus subtilis* in solid-state fermentation on soybeans. Appl Microbiol Biotechnol 47:489–495

Bibal B, Vayssier Y, Goma G, Pareilleux A (1991) High-concentration cultivation of *Lactococcus cremoris* in a cell-recycle reactor. Biotechnol Bioeng 37(8):746–754

Bicas JL, Silva JC, Dionísio AP, Pastore GM (2010) Biotechnological production of bioflavors and functional sugars. Food Sci Technol 30(1):07–18

Bloem A, Bertrand A, Lonvaud-Funel A, Revel GD (2007) Vanillin production from simple phenols by wine-associated lactic acid bacteria. Lett Appl Microbiol 44(1):62–67

Borejsza-Wysocki W, Hrazdina G (1994) Biosynthesis of p-hydroxyphenylbutan-2-one in raspberry fruits and tissue cultures. Phytochemistry 35:623–628

Cabaroğlu T, Yılmaztekin T (2010) Aroma Biyoteknolojisi. In: Necla A (ed) Gıda Biyoteknolojisi. Nobel Yayın Dağıtım Tic. Ltd. Şti, Ankara, pp 375–390

Cadwallader KR, Braddock RJ, Parish ME, Higgins DP (1989) Bioconversion of (+)-Limonene by *Pseudomonas gladioli*. J Food Sci 54(5):1241–1245

Carroll AL, Desai SH, Atsumi S (2016) Microbial production of scent and flavor compounds. Curr Opin Biotechnol 37:8–15

Champluvier B, Francart B, Rouxhet PG (1989) Co-immobilization by adhesion of β-galactosidase in nonviable cells of *Kluyveromyces lactis* with *Klebsiella oxytoca*: conversion of lactose into 2, 3-butanediol. Biotechnol Bioeng 34(6):844–853

Cheetham PSJ (1997) Combining the technical push and the business pull for natural flavours. In: Berger RG et al (eds) Biotechnology of aroma compounds. Advances in biochemical engineering/biotechnology, vol 55. Springer, Berlin, pp 1–49
Cheetham PSJ, Imber CE, Isherwood J (1982) The formation of isomaltulose by immobilized *Erwinia rhapontici*. Nature 299:628–631
Chen JP, McGill SD (1992) Enzymatic hydrolysis of triglycerides by *Rhizopus delemar* immobilized on biomass support particles. Food Biotechnol 6:1–18
Christen P, Meza JC, Revah S (1997) Fruity aroma production in solid state fermentation by *Ceratocystis fimbriata*: influence of the substrate type and the presence of precursors. Mycol Res 101(8):911–919
Chung BH, Chang HN (1988) Aerobic fungal cell immobilization in a dual hollow-fiber bioreactor: continuous production of a citric acid. Biotechnol Bioeng 32(2):205–212
Civolani C, Barghini P, Roncetti AR, Ruzzi M, Schiesser A (2000) Bioconversion of ferulic acid into vanillic acid by means of a vanillate-negative mutant of *Pseudomonas fluorescens* strain BF13. Appl Environ Microbiol 66(6):2311–2317
Collins RP (1976) Terpenes and odoriferous materials from microorganisms. Lloydia 39:20–24
Craig T, Daugulis AJ (2014) Strategies for improved bioproduction of benzaldehyde by *Pichia pastoris* and the use of hytrel as tubing material for integrated product removal by *in situ* pervaporation. Biochem Eng J 82:97–104
Dahlen T, Hauck T, Wein M, Schwab W (2001) 2,5-dimethyl-4-hydroxy-3(2H)-furanone as a secondary metabolite from d-fructose-1,6-diphosphate metabolism by *Zygosaccharomyces rouxii*. J Biosci Bioeng 91(4):352–358
Damasceno S, Cereda MP, Pastore GM, Oliveira JG (2003) Production of volatile compounds by *Geotrichum fragrans* using cassava wastewater as substrate. Process Biochem 39(4):411–414
Demain AL, Jackson M, Trenner NR (1967) Thiamine-dependent accumulation of tetramethylpyrazine accompanying a mutation in the isoleucine-valine pathway. J Bacteriol 94(2):323–326
Drawert F, Barton H (1978) Biosynthesis of flavor compounds by microorganisms. 3. Production of monoterpenes by the yeast *Kluyveromyces lactis*. J Agric Food Chem 26(3):765–766
Escamilla ML, Valdés SE, Soriano J, Tomasini A (2000) Effect of some nutritional and environmental parameters on the production of diacetyl and on starch consumption by *Pediococcus pentosaceus* and *Lactobacillus acidophilus* in submerged cultures. J Appl Microbiol 88(1):142–153
Escamilla-Hurtado ML, Valdés-Martínez SE, Soriano-Santos J, Gómez-Pliego R, Verde-Calvo JR, Reyes-Dorantes A, Tomasini-Campocosio A (2005) Effect of culture conditions on production of butter flavor compounds by *Pediococcus pentosaceus* and *Lactobacillus acidophilus* in semisolid maize-based cultures. Int J Food Microbiol 105(3):305–316
Fadel HHM, Mahmoud MG, Asker MMS, Lotfy SN (2015) Characterization and evaluation of coconut aroma produced by *Trichoderma viride* EMCC-107 in solid state fermentation on sugarcane bagasse. Electron J Biotechnol 18(1):5–9
Falconnier B, Lapierre C, Lesage-Meessen L, Yonnet G, Brunerie P, Colonna-Ceccaldi B, Corrieu G, Asther M (1994) Vanillin as a product of ferulic acid biotransformation by the white-rot fungus *Pycnoporus cinnabarinus* I-937: identification of metabolic pathways. J Biotechnol 37 (2):123–132
Feron G, Bonnarme P, Durand A (1996) Prospects for the microbial production of food flavours. Trends Food Sci Technol 7(9):285–293
Fujita Y (1988) Applications of plant cell and tissue culture (Yamada Y, chairman), Ciba Foundation Symposium, vol 137. John Wiley & Sons, Chichester, pp 228–238
Funk C, Brodelius P (1990a) Influence of growth regulators and an elicitor on phenylpropanoid metabolism in suspension cultures of *Vanilla planifolia*. Phytochemistry 29(3):845–848
Funk C, Brodelius PE (1990b) Phenylpropanoid metabolism in suspension cultures of *Vanilla planifolia* Andr: II. Effects of precursor feeding and metabolic inhibitors. Plant Physiol 94 (1):95–101

Gatfield I, Sand T (1995) Process of preparing butyric acid or 2-or 3-methylbutyric acid by oxidizing the corresponding butanols with gluconobacter roseus IAM 1841 or IFO 3990. In: Google Patents

Guiné RP, Martín-Belloso O, Mínguez-Mosquera MI, Paliyath G, Pessoa FL, Le Quéré J-L, Sidhu JS, Sinha N, Stanfield P (2010) Handbook of fruit and vegetable flavors. John Wiley and Sons, Hoboken, NJ

Güneşer O, Demirkol A, Karagül Yüceer Y, Özmen Toğay S, İşleten Hoşoğlu M, Elibol M (2015) Bioflavour production from tomato and pepper pomaces by *Kluyveromyces marxianus* and *Debaryomyces hansenii*. Bioprocess Biosyst Eng 38(6):1143–1155

Guneser O, Demirkol A, Yuceer YK, Togay SO, Hosoglu MI, Elibol M (2017) Production of flavor compounds from olive mill waste by *Rhizopus oryzae* and *Candida tropicalis*. Braz J Microbiol 48(2):275–285

Gupta S, Gupta C, Garg AP, Prakash D (2015) A biotechnological approach to microbial based perfumes and flavours. J Microbiol Exp 2(1):11–18

Hallé C, Leroi F, Dousset X, Pidoux M (1994) Les kéfirs: des associations bactéries lactiques-levures. In: de Roissart H, Luquet FM (eds) Bactéries lactiques: Aspects fondamentaux et technologiques, vol 2. Lorica, Uriage, pp 169–182

Halm M, Lillie A, Sørensen AK, Jakobsen M (1993) Microbiological and aromatic characteristics of fermented maize doughs for kenkey production in Ghana. Int J Food Microbiol 19(2):135–143

Harlander S (1995) Biotechnology for the production of flavoring materials. In: Reineccius G (ed) Source book of flavors. Springer, Boston, MA, pp 155–175

Häusler A, Münich (1997) Microbial production of natural flavors. AsM News 63:551–559

Hong YCA, Huang LC, Reineccius GA, Harlander SK, Labuza TP (1990) Production of aroma compounds from strawberry cell suspension cultures by addition of precursors. Plant Cell Tissue Organ Cult 21:245–251

Hosoglu MI (2018) Aroma characterization of five microalgae species using solid-phase microextraction and gas chromatography–mass spectrometry/olfactometry. Food Chem 240:1210–1218

Hosoglu MI, Guneser O, Yuceer YK (2018) Different bioengineering approaches on production of bioflavor compounds. In: Role of materials science in food bioengineering. Elsevier, Amsterdam, pp 37–71

Hua D, Ma C, Lin S, Song L, Deng Z, Maomy Z, Zhang Z, Yu B, Xu P (2007) Biotransformation of isoeugenol to vanillin by a newly isolated Bacillus pumilus strain: identification of major metabolites. J Biotechnol 130(4):463–470

Ibragimova AZ, Yakovlev DA, Gorshkova EIV (1980) Determination of the ratio of diacetyl and acetaldehyde in streptococci cultures by gas-liquid chromatography. Molochnaya Promyshlennost 2:43–45

Janssens L, de Poorter HL, de Mey L, Vandamme EJ, Schamp N (1989) Fusel oil as a precursor for the microbial production of fruity flavours [Hansenula mrakii]. In: Mededelingen van de Faculteit Landbouwwetenschappen Rijksuniversiteit Gent (Belgium). Faculteit Landbouwwetenschappen, Ghent

Kaminski E, Stawicki S, Wasowicz E (1974) Volatile flavor compounds produced by molds of *Aspergillus*, *Penicillium*, and *Fungi imperfecti*. Appl Environ Microbiol 27(6):1001–1004

Kearney L, Upton M, McLoughlin A (1990) Meat fermentations with immobilized lactic acid bacteria. Appl Microbiol Biotechnol 33:648–651

Khoyratty S, Kodja H, Verpoorte R (2018) Vanilla flavor production methods: a review. Ind Crop Prod 125:433–442

Koren DW, Duvnjak Z (1992) Fed-batch production of concentrated fructose syrup and ethanol using *Saccharomyces cerevisiae* ATCC 36859. Acta Biotechnol 12(6):489–496

Krings U, Berger R (1998) Biotechnological production of flavours and fragrances. Appl Microbiol Biotechnol 49:1–8

Kulozik U, Hammelehle B, Pfeifer J, Kessler HG (1992) High reaction rate continuous bioconversion process in a tubular reactor with narrow residence time distributions for the production of lactic acid. J Biotechnol 22(1–2):107–116

Kumara S (1989) Interactions between Enterobacteriaceae and *Saccharomyces cerevisiae* during wort fermentation. Yeast 5:67–72

Lomascolo A, Stentelaire C, Asther M, Lesage-Meessen L (1999) Basidiomycetes as new biotechnological tools to generate natural aromatic flavours for the food industry. Trends Biotechnol 17 (7):282–289

Longo MA, Sanromán MA (2006) Production of food aroma compounds: microbial and enzymatic methodologies. Food Technol Biotechnol 44:335–353

Lu WM, Chen WC (1988) Production of L-glutamate using entrapped living cells of *Brevibacterium ammoniagenes* with calcium alginate gels. Proc Natl Sci Council, Taipai, Taiwan 6:400–406

Manjón A, Iborra JL, Martínez-Madrid C (1991) pH control of limonin debittering with entrapped *Rhodococcus fascians* cells. Appl Microbiol Biotechnol 35:176–179

Maróstica MR Jr, Pastore GM (2007) Production of R-(+)-α-terpineol by the biotransformation of limonene from orange essential oil, using cassava waste water as medium. Food Chem 101 (1):345–350

Medeiros AB, Pandey A, Freitas RJ, Christen P, Soccol CR (2000) Optimization of the production of aroma compounds by *Kluyveromyces marxianus* in solid-state fermentation using factorial design and response surface methodology. Biochem Eng J 6(1):33–39

Mi J, Becher D, Lubuta P, Dany S, Tusch K, Schewe H, Buchhaupt M, Schrader J (2014) *De novo* production of the monoterpenoid geranic acid by metabolically engineered *Pseudomonas putida*. Microb Cell Factories 13:170

Nigam PS, Pandey A (2009) Biotechnology for agro-industrial residues utilisation. Springer, Dordrecht

Neufeld R, Peleg Y, Rokem J, Pines O, Goldberg I (1991) L-malic acid formation by immobilized *Saccharomyces cerevisiae* amplified for fumarase. Enzym Microb Technol 13:991–996

Nielsen DR, Yoon SH, Yuan CJ, Prather KLJ (2010) Metabolic engineering of acetoin and *meso*-2, 3-butanediol biosynthesis in *E. coli*. Biotechnol J 5:274–284

Oliver JWK, Machado IMP, Yoneda H, Atsumi S (2013) Cyanobacterial conversion of carbon dioxide to 2, 3-butanediol. Proc Natl Acad Sci 110(4):1249–1254

Omelianski VL (1923) Aroma-producing microörganisms. J Bacteriol 8(4):393–419

Paterson SL, Fane AG, Fell CJD, Chun UH, Rogers P (1988) Sorbitol and gluconate production in a hollow fibre membrane reactor by immobilized *Zymomonas mobilis*. Biocatalysis 1(3):217–229

Pedapudi S, Chin CK, Pedersen H (2000) Production and elicitation of benzalacetone and the raspberry ketone in cell suspension cultures of *Rubus idaeus*. Biotechnol Prog 16(3):346–349

Priefert H, Rabenhorst J, Steinbüchel A (2001) Biotechnological production of vanillin. Appl Microbiol Biotechnol 56:296–314

Rane KD, Sims KA (1995) Citric acid production by *Candida lipolytica* Y 1095 in cell recycle and fed-batch fermentors. Biotechnol Bioeng 46:325–332

Rojas V, Gil JV, Piñaga F, Manzanares P (2001) Studies on acetate ester production by non-*Saccharomyces* wine yeasts. Int J Food Microbiol 70(3):283–289

Roy P, Kumar V (2019) Production of bioflavour from microbial sources and its health benefits. Indian J Biochem Biophys 56(5):352–357

Salunkhe D, Do JY, Maga JA (1976) Biogenesis of aroma constituents of fruits and vegetables. CRC Crit Rev Food Sci Nutr 8(2):161–190

Schrader J (2007) Microbial flavour production. In: Flavours and fragrances. Springer Verlag, Berlin, pp 507–574

Schrader J, Etschmann MMW, Sell D, Hilmer JM, Rabenhorst J (2004) Applied biocatalysis for the synthesis of natural flavour compounds–current industrial processes and future prospects. Biotechnol Lett 26:463–472

Schulte U, El-Shagi H, Zenk MH (1984) Optimization of 19 *Rubiaceae* species in cell culture for the production of anthraquinones. Plant Cell Rep 3:51–54

Scragg AH (2007) The production of flavours by plant cell cultures. In: Berger RG (ed) Flavours and fragrances. Springer, Berlin, pp 599–614

Shimoni E, Baasov T, Ravid U, Shoham Y (2003) Biotransformations of propenylbenzenes by an *Arthrobacter* sp. and its *t*-anethole blocked mutants. J Biotechnol 105(1–2):61–70

Soares M, Christen P, Pandey A, Raimbault M, Soccol CR (2000) A novel approach for the production of natural aroma compounds using agro-industrial residue. Bioprocess Eng 23:695–699

Speelmans G, Bijlsma A, Eggink G (1998) Limonene bioconversion to high concentrations of a single and stable product, perillic acid, by a solvent-resistant *Pseudomonas putida* strain. Appl Microbiol Biotechnol 50:538–544

Stentelaire C, Lesage-Meessen L, Oddou J, Bernard O, Bastin G, Ceccaldi BC, Asther M (2000) Design of a fungal bioprocess for vanillin production from vanillic acid at scalable level by *Pycnoporus cinnabarinus*. J Biosci Bioeng 89(3):223–230

Subramaniyam R, Vimala R (2012) Solid state and submerged fermentation for the production of bioactive substances: a comparative study. Int J Sci Nat 3:480–486

Tahara S, Fujiwara K, Ishizaka H, Mizutani J, Obata Y (1972) γ-decalactone – one of constituents of volatiles in cultured broth of *Sporobolomyces odorus*. Agric Biol Chem 36(13):2585–2587

Tamang JP (1998) Role of microorganisms in traditional fermented foods. Indian Food Ind 17 (3):162–167

Torres S, Baigorí MD, Swathy SL, Pandey A, Castro GR (2009) Enzymatic synthesis of banana flavour (isoamyl acetate) by *Bacillus licheniformis* S-86 esterase. Food Res Int 42(4):454–460

Vandamme EJ (2003) Bioflavours and fragrances via fungi and their enzymes. Fungal Divers 13:153–166

Vandamme EJ, Soetaert W (2002) Bioflavours and fragrances via fermentation and biocatalysis. J Chem Technol Biotechnol 77(12):1323–1332

Vanhooren PT, Vandamme EJ (1999) L-Fucose: occurrence, physiological role, chemical, enzymatic and microbial synthesis. J Chem Technol Biotechnol 74(6):479–497

Walton NJ, Mayer MJ, Narbad A (2003) Vanillin. Phytochemistry 63(5):505–515

Walton NJ, Narbad A, Faulds C, Williamson G (2000) Novel approaches to the biosynthesis of vanillin. Curr Opin Biotechnol 11(5):490–496

Wang CL, Shi DJ, Gong GL (2008) Microorganisms in Daqu: a starter culture of Chinese Maotai-flavor liquor. World J Microbiol Biotechnol 24:2183–2190

Willaert R, Verachtert H, van den Bremt K, Delvaux F, Derdelinckx G (2005) Bioflavouring of foods and beverages. In: Nedović V, Willaert R (eds) Applications of cell immobilisation biotechnology. Focus on biotechnology. Springer, Dordrecht

Xiao ZJ, Xie NZ, Liu PH, Hua DL, Xu P (2006) Tetramethylpyrazine production from glucose by a newly isolated *Bacillus* mutant. Appl Microbiol Biotechnol 73:512–518

Xu P, Hua D, Lin S, Song L, Ma C, Zhang Z, Du Y, Chen H, Gan L, Wei Z (2009) Streptomyces strain and the method of converting ferulic acid to vanillin by using the same. In: Google Patents

Yang LW, Wang XY, Wei S (1992) Immobilization of *Candida rugosa* cells having high fumarase activity with polyvinyl alcohol. Ann N Y Acad Sci 672(1):563–565

Zabetakis I, Holden M (1995) A study of strawberry flavour biosynthesis. In: Etievant P, Schreier P (eds) Bioflavour: analysis-precursor studies-biotechnology, vol 95, INRA edn. Paris, pp 211–216

Zenk MH (1990) Plant cell cultures: a potential in food and bio-technology. Food Biotechnol 4 (1):461–470

Zhao J, Hu Q, Guo YQ, Zhu WH (2001) Effects of stress factors, bioregulators, and synthetic precursors on indole alkaloid production in compact callus clusters cultures of *Catharanthus roseus*. Appl Microbiol Biotechnol 55:693–698

Zheng GW, Yu HL, Zhang JD, Xu JH (2009) Enzymatic production of *l*-Menthol by a high substrate concentration tolerable esterase from newly isolated *Bacillus subtilis* ECU0554. Adv Synth Catal 351(3):405–414

Zhu BF, Xu Y, Fan WL (2010) High-yield fermentative preparation of tetramethylpyrazine by *Bacillus* sp. using an endogenous precursor approach. J Ind Microbiol Biotechnol 37:179–186

Zorn H, Langhoff S, Scheibner M, Berger RG (2003) Cleavage of *β, β*-carotene to flavor compounds by fungi. Appl Microbiol Biotechnol 62:331–336

# Chapter 3
# Clinical Potential of Bacteriophage and Endolysin Based Therapeutics: A Futuristic Approach

**Vijay Singh Gondil, Fazal Mehmood Khan, Nancy Mehra, Deepak Kumar, Aastha Khullar, Tanvi Sharma, Abhishek Sharma, Rahul Mehta, and Hang Yang**

**Abstract** Antibiotic resistance is a global health challenge in the modern era. The emergence of antibiotic-resistant strains poses a serious threat to human health across the globe and compromises the arsenal of antibiotics upon which the modern healthcare system heavily relies. Antibiotic resistance diminishes the choice for effective antimicrobial agents and forces researchers to look for effective alternative agents. Bacteriophages have been established as potent antibacterial agents against most of the bacterial pathogens since the pre-antibiotic era. Additionally, the discovery and exploration of endolysins, i.e. phage coded peptidoglycan hydrolases, have further revolutionized the field of phage-based therapy. Bacteriophage and endolysin have demonstrated to be effective for clearing the infection in both in vitro and in vivo models. Nevertheless, there is a scarcity of information on the clinical potential of bacteriophage and endolysin. The present chapter will highlight the features of bacteriophage and endolysin that make them attractive and effective long-term therapeutic alternatives for the treatment of drug-resistant infections in clinical settings.

V. S. Gondil · F. M. Khan · H. Yang (✉)
CAS Key Laboratory of Special Pathogens and Biosafety, Center for Emerging Infectious Diseases, Wuhan Institute of Virology, Chinese Academy of Sciences, Wuhan, China
e-mail: yangh@wh.iov.cn

N. Mehra · A. Khullar
Post Graduate Institute of Medical Education and Research, Chandigarh, India

D. Kumar · R. Mehta (✉)
Department of Microbiology, DAV University, Jalandhar, Punjab, India

T. Sharma
Department of Biotechnology and Bioinformatics, Jaypee University of Information Technology, Waknaghat, Solan, India

A. Sharma
Department of Biotechnology, Himachal Pradesh University, Shimla, India

 
P. K. Arora (ed.), *Microbial Products for Health, Environment and Agriculture*, Microorganisms for Sustainability 31, https://doi.org/10.1007/978-981-16-1947-2_3

**Keywords** Endolysin · Lysin therapy · Bacteriophage · Phage therapy · Peptidoglycan hydrolases

## 3.1 Bacteriophage as Antimicrobial Agent

Bacteriophage, most popularly known as phage, is bacteria-infecting viruses. Phage is the most diverse, abundant, and most genetically variable biological entities on earth, with the global number estimated at $10^{30}$ to $10^{32}$ (Abedon 2008; Hemminga et al. 2010, Hatfull and Hendrix 2011). Phage plays an important role in maintaining bacterial diversity in the natural environment (Shapiro and Kushmaro 2011; Braga et al. 2018). Phage was initially discovered in 1915 and 1917 by Frederick Twort and Felix d` Terrelle, respectively. However, the term "bacteriophage" was first used by scientist Felix d` Herelle (Tammelin 1992). To date, more than 5000 classified bacteriophages are reported (Ackermann 2003). Bacteriophage, in a broader classification, is of mainly two types: lysogenic (temperate) in which bacteriophage incorporates their genome into host DNA and lytic (virulent) in which bacteriophage rapidly replicates in cells followed by burst opening the host cell to begin afresh infection cycle. Lytic bacteriophage replicates in the host bacterial cell exponentially

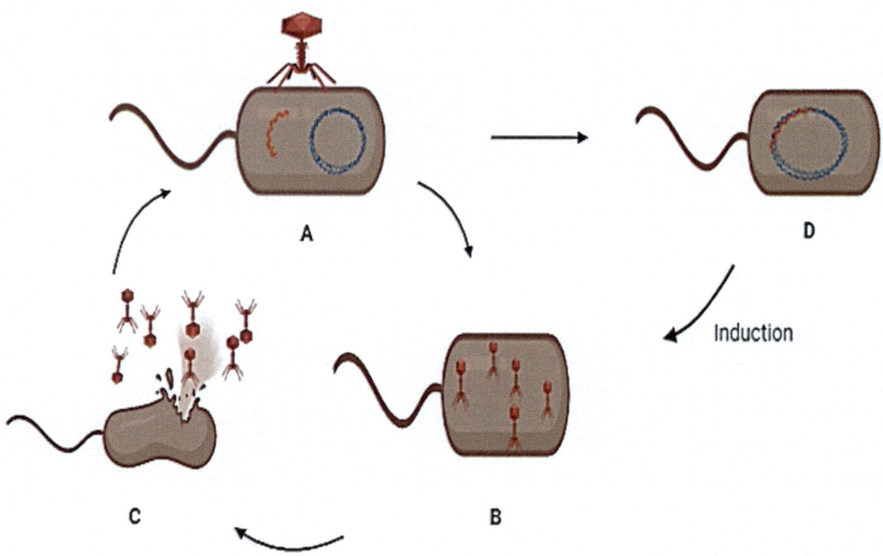


**Fig. 3.1** Steps involved in lytic and lysogenic phage replication: (**a**) Adsorption of phage to host bacterium and transfer of phage genome into the bacterial cell, (**b**) (lytic cycle) multiplication of phage in host bacterium and (**c**) lysis of host bacteria to release phage virions that could initiate infections to new host bacterial cells, (**d**) (lysogenic cycle) insertion of phage genome into the bacterial genome, which can be passed into subsequent generations or may enter into lytic cycle upon induction by environmental stressors

and is released by lysis of infected host bacterium (Gondil and Chhibber 2018). Phage is released in the environment and can potentially infect host bacteria in the near vicinity (Fig. 3.1).

Phage has been the choice of treatment for bacterial infections in the pre-antibiotic era, but progress in the development of antibiotics leads to a decline in the therapeutic application of phage (Sulakvelidze and Morris 2001). However, with the rapid emergence of antibiotic resistance, interest in utilizing bacteriophage and their products as therapeutics in controlling infections caused by *Escherichia coli, Pseudomonas, Klebsiella*, and *Proteus* strains in human and animal models has rekindled (Barrow and Soothill 1997; Brussow 2005; Gorski and Weber-Dabrowska 2005; Kumari et al. 2010). For example, as one of the phage therapy pioneers, the Institute of Immunology and Experimental Therapy (IIET) includes phage preparations that can take care of 60–80% of specific nosocomial pathogens, such as *Pseudomonas, Enterococci*, and *Acinetobacter* (Houby and Mann 2009). Generally, most of the lytic phages used as antimicrobial agents belong to the family of *Myoviridae* (having long, rigid contractile tail), *Siphoviridae* (having long flexible and noncontractile tail), and *Podoviridae* (having short noncontractile tail) (Housby and Mann 2009).

## 3.2 Advantages of Phage as a Treatment Option

Bacteriophage poses several advantages over established antibacterial drugs. Phage is highly specific to target bacterium and does not disturb the normal microflora of human hosts (Divya Ganeshan and Hosseinidoust 2019). Along with specificity, phage is self-replicating and self-limiting as long as bacteria are present. Their multiplication usually happens exponentially as long as the host bacterium is available for multiplication and subsequent infection (Ghannad and Mohammadi 2012). Phage receptors on the bacterial cell wall are primarily virulence factors, so it is believed that a bacterium will become less virulent when evolves resistance to phage (Inal 2003). In antibiotic allergic patients, options to treat bacterial infections become limited. Bacteriophage therapy in an allergic patient could be a potent alternative to conventional antibiotic treatment (Ligonenko et al. 2015). Bacteriophage does not affect eukaryotic cells, making it a clinically safe product for healthcare applications (Domingo-Calap and Delgado-Martínez 2018). Apart from infrequent and mild immune reactions, no major adverse effects have been reported upon phage therapy (Romero-Calle et al. 2019). The safety profile of bacteriophage makes it an attractive antibacterial agent to be used in combination with pre-existing antibacterial agents which include therapeutic enzymes (Nelson et al. 2006; Chhibber et al. 2018), efflux pump inhibitors (Lamers et al. 2013), phytochemicals (Moreno et al. 2013), pigments (Gondil et al. 2017), and antibacterial metal nanoparticles (Kumar et al. 2017; Gondil et al. 2019). Phage resistance to host bacterium is reported to be at a lower frequency as the rate of phage mutation is higher than that of bacteria, so if a phage-resistant bacterium emerges, the phage responds quickly (Dorval Courchesne et al. 2009). The coevolution of phage with its host entity makes it an evolving therapeutic alternative, a unique feature in all of the available therapeutic agents present to date. Considering an economical perspective, phage that can be easily scaled up with restrained resources represents an inexpensive option compared to antibiotics (Tang et al. 2019).

## 3.3 Applications of Phage Therapy in Animal Infection Models

Efficacy of phage therapy has been evaluated in in vivo antibacterial activities for more than 100 years. A variety of pathogens that include but not limited to *Staphylococcus aureus, Klebsiella pneumoniae, Pseudomonas aeruginosa*, *E. coli*, *Acinetobacter baumannii, and Salmonella enterica* that lead to fatal infections have been included in studies with numerous animal infection models. We highlight a few of the selected animal models to summarize the vast exploration of phage therapy in antibiotic sensitive as well as antibiotic-resistant infection models (Table 3.1).

## 3.4 Clinical Trials of Phage Therapy

One of the rate-limiting factors in the translation of phage therapy from the laboratory into healthcare settings is the lack of well-evidenced and validated clinical trials. A clinical trial of phage therapy is somehow different from other drug-based clinical trials due to the unique characteristics of phage (Furfaro et al. 2018). The pharmacological dose is an important factor in clinical trials as these self-replicating entities increase their number upon infection in their host bacteria (Payne and Jansen 2003). Another important factor is the nature of colonization in the site of infection, i.e. monomicrobial versus polymicrobial, which makes the situation complex. The eradication of one pathogen from the site of infection might accelerate the colonization by another pathogen, which makes monophage therapy debatable for patient care (Harper 2018). Nonetheless, these concerns can be technically addressed by phage cocktail-based strategy that covers multiple pathogens.

Humans co-exist with phage in their daily life which is a major indicator of phage safety. However, as clinical therapeutics, isolation, characterization, and scale up of phage need to follow strict guidelines to ensure uniform quality parameters (Parracho et al. 2012). In addition, for clinical outcomes and readiness of phage therapy, more intensive and multi-pronged clinical trials which include sterility, purity, and immune response mediated by sudden host lysis should also be taken into consideration (Furfaro et al. 2018). In Table 3.2, we summarize several clinical trials of phage therapy in clinical settings:

**Table 3.1** List of selected animal infection models evaluated for phage therapy

| Target pathogen | Infection model | Animal | Route of administration | References |
|---|---|---|---|---|
| *E. coli* | Systemic infection | Mice | Intramuscular injection | Smith and Huggins (1982) |
| *E. coli* | Diarrhea | Calves, piglets, lambs | Oral administration | Smith and Huggins (1982) |
| *A. baumannii, P. aeruginosa,* and *S. aureus* | Systemic infection | Mice | Intraperitoneal injection | Soothill (1992) |
| *E. coli and S. enterica* serovar Typhimurium | Systemic infection | Mice | Intraperitoneal injection | Merril et al. (1996) |
| *E. coli* | Septicaemia and meningitis | Chicken and calves | Intramuscular injection | Barrow et al. (1998) |
| Vancomycin-resistant *Enterococcus faceium* | Bacteraemia | Mice | Intraperitoneal injection | Biswas et al. (2002) |
| *S. aureus* | Bacteraemia | Mice | Intraperitoneal injection | Matsuzaki et al. (2003) |
| *E. coli* | Diarrhea | Mice | Oral administration | Chibani-Chennoufi et al. (2004) |
| Multiple drug-resistant (MDR) *K. pneumoniae* | Bacteraemia | Mice | Intraperitoneal injection | Vinodkumar et al. (2005) |
| *S. aureus* | Wound infection | Rabbit | Subcutaneous injection | Wills et al. (2005) |
| Imipenem-resistant *Pseudomonas spp.* | Bacteraemia | Mice | Intraperitoneal injection | Wang et al. (2006a) |
| β-Lactamase producing *E. coli* | Bacteraemia | Mice | Intraperitoneal injection | Wang et al. (2006b) |
| *P. aeruginosa* | Bacteraemia | Mice | Intraperitoneal injection | Watanabe et al. (2007) |
| *P. aeruginosa* | Bacteraemia | Mice | Intraperitoneal injection | Vinodkumar et al. (2008) |
| *S. aureus* | Systemic lethal infection | Mice | Intraperitoneal injection | Zimecki et al. (2010) |
| *P. aeruginosa* | Lung infection | Mice | Intranasally | Debarbieux et al. (2010) |
| *S. aureus* | Bacteraemia | Mice | Intraperitoneal injection | Sunagar et al. (2010) |
| *K. pneumoniae* | Liver Abscesses and bacteremia | Mice | Intraperitoneal injection | Hung et al. (2011) |
| *K. pneumoniae* | Burn wound infection | Mice | Tropical administration | Kumari et al. (2010) |

(continued)

**Table 3.1** (continued)

| Target pathogen | Infection model | Animal | Route of administration | References |
|---|---|---|---|---|
| *P. aeruginosa* | Lung infection | Mice | Intranasal | Morello et al. (2011) |
| *Cronobacter turicensis* | Urinary tract infection | Mice | Intraperitoneal injection | Thotova et al. (2011) |
| *P. aeruginosa* | Lung infection | Mice | Intranasal | Alemayehu et al. (2012) |
| Extended spectrum beta lactamase (ESBL) *E. coli* | Meningitis | Rat | Intraperitoneal injection | Pouillot et al. (2012) |
| *P. aeruginosa* | Keratitis | Mice | Topical | Fukuda et al. (2012) |
| Methicillin-resistant *S. aureus* | Bone infection | Rat | Medullary injection | Yilmaz et al. (2013) |
| *S. aureus* | Septicemia | Mice | Intraperitoneal injection | Takemura-Uchiyama et al. (2014) |
| *K. pneumoniae* | Lung pneumonia | Mice | Intranasal | Cao et al. (2015) |
| *E. coli* | Lung pneumonia | Mice | Intranasal | Dufour et al. (2015) |
| *A. baumannii* | Wound infection | Rat | Topical administration | Kusradze et al. (2016) |
| Methicillin-resistant *S. aureus* | Joint infection model | Mice | Coated orthopedic wires | Kaur et al. (2019) |
| *P. aeruginosa* | Cystic fibrosis | Mice | Intranasal | Pabary et al. (2016) |
| *V. Cholerae* | Diarrhea | Mice | Oral | Yen et al. (2017) |
| *E. coli, K. pneumoniae,* and *Enterobacter cloacae* | Systemic infection | Wax moth larvae (*Galleria mellonella*) | Injection | Manohar et al. (2018) |
| *E. coli* | Systemic model | Mice | Intravenous injection | Schneider et al. (2018) |
| Carbapenem-resistant *A. baumannii* | Lung infection model | Mice | Intranasal | Hua et al. (2018) |
| Multidrug-resistant *P. aeruginosa* | Joint infection model | Human case study | Local administration | Tkhilaishvili et al. (2019) |
| *S. enterica* serotype Enteritidis | Diarrhea | Mice | Oral | Dallal et al. (2019) |
| *S. aureus* | Burn wound infection | Mice | Topical administration | Kaur et al. (2019) |

**Table 3.2** List of clinical trials of phage therapy

| Target organism | Disease | Number of participants | Route | Success rate | References |
|---|---|---|---|---|---|
| *Proteus*, *Staphylococcus*, and *Streptococcus* | – | 236 | Subcutaneous/ surgical drainage | 92% | Sakandelidze and Meĭpariani (1974) |
| – | Abscess pneumonia | – | Parenteral | – | Pipiia et al. (1976) |
| *E. coli*, *Proteus*, *Staphylococcus*, and *Streptococcus* | Systemic infection | 96 | – | – | Zhukov-Verezhnikov et al. (1978) |
| *Staphylococcus* | Lung infection | 223 | – | 82% | Meladze et al. (1982) |
| *Pseudomonas* and *Staphylococcus* | Post-surgical wounds | 65 | – | 82% | Kochetkova et al. (1989) |
| *Enterococcus*, *E. coli*, *P. aeruginosa*, *Proteus*, *Staphylococcus*, and *Streptococcus* | Allergoses | 936 | | 86% | Sakandelidze et al. (1991) |
| *E. coli*, *Proteus*, and *Staphylococcus* | Acute and chronic urogenital inflammation | 46 | – | 84–92% | Perepanova et al. (1995) |
| *Proteus*, *Staphylococcus*, and *Streptococcus* | Burn wound infections | 54 | Oral tablets | – | Lazareva et al. (2001) |
| *E. coli*, *Proteus*, *Pseudomonas*, and *Staphylococcus* | Ulcers and wound infections | 96 | Topical | 70% | Markoishvili et al. (2002) |
| *P. aeruginosa* | Chronic *otitis* | 24 | Ear drops | – | Wright et al. (2009) |
| *P. aeruginosa*, *S. aureus*, and *E. coli* | Chronic venous leg ulcers | 42 | Topical | – | Rhoads et al. (2009) |
| *E. coli* | Diarrhea | 120 | Oral | – | Sarkar et al. (2016) |
| *Enterococcus*, *E. coli*, *Proteus mirabilis*, *P. aeruginosa*, *Staphylococcus*, and *Streptococcus* | Urinary tract infection | 81 | Suprapubic catheter | – | Leitner et al. (2017) |
| *S. aureus*, *E. coli*, *Streptococcus*, *P. aeruginosa*, and *P. mirabilis* | Urinary tract infection | 118 | Instillation | – | Ujmajuridze et al. (2018) |

## 3.5 Pharmacokinetics and Safety of Therapeutic Phage

Pharmacokinetics is a property of therapeutic molecule/drug to reach the specific target site and exhibit pharmacodynamic effects. The pharmacokinetic properties of the drug or any other therapeutic molecule can be classified into absorption, distribution, metabolism, and excretion of the therapeutic molecule (ADMET properties). The success of phage therapy is dependent on the concentration of phage and its multiplication at the site of host–pathogen infection. Phage can increase in situ concentration at the site of infection, thus increasing the efficiency of treatment without repetitive dosing. The increase in phage concentration over time can easily counterbalance the clearance of phage by the host immune system. Some phage delivery systems which include liposomes, transferosomes, and topical dressings not only increase the antibacterial efficiency but also increase the pharmacokinetic behavior of the loaded phage (Singla et al. 2016; Chadha et al. 2017; Gondil and Chhibber 2017; Kaur et al. 2019). These delivery systems have been well reported for in vitro as well as in vivo antibacterial efficacy in literature.

Phage is considered as safe for human because of its abundant presence in the environment and human host for a long time. Phage, like other drugs, also interacts with non-target organs and tissues, such as interactions with reticuloendothelial systems and transportation from organs to blood (Górski et al. 2006; Merril 2008). However, these interactions are not reported to lead to any adverse effects in the human system. Immunological responses to phage have been studied since long and humoral responses not only elicit immunity against viral particles but also lead to phage inactivation (Sulakvelidze et al. 2005). Phage coat protein-mediated cytokine response also has been studied, for their mammalian cell interactions (Budynek et al. 2010). The specificity of phage can be exploited as a prophylactic agent, unlike antibiotics which show broad-spectrum effects and may result in dysbiosis by killing important microflora. The specificity of phage against target bacterium can be modulated by using a single or cocktail of bacteriophage, which can increase the lytic efficiency of phage by keeping the overall activity spectrum on a narrow range. Phage may also enhance the bacterial lysis mediated toxin release, which is also mediated by a number of antibiotics. However, in the case of Gram-positive bacteria, exotoxin release is on the lower end, but in the case of Gram-negative bacteria, the presence of endotoxin in phage preparations limits their applications (Kutter 2008). Therefore, phage that is employed for systemic administration requires an additional step in preparation of removing endotoxin (Hietala et al. 2019). Gram-positive lysates also exhibit mild side effects, which may be due to the presence of bacterial pathogenicity factors (Sulakvelidze and Kutter 2005). However, these issues are of minor concern as these reactions are mild and do not complicate the treatment process, especially in localized treatment of infections. Phage demonstrates safety profile in a number of animal infection models as well as in human subjects, establishing their potential application in future clinical applications.

## 3.6 Endolysins and Their Potential in Clinical Applications

### *3.6.1 General Characteristics of Endolysin*

Endolysins (lysins) are phage derived lytic proteins, produced during the end stage of phage replication to degrade the cell wall of host bacterium and release newly synthesized phage particles. Based on their enzymatic nature and robust antibacterial activity, endolysins have been also termed as enzybiotics (Nelson et al. 2001). Endolysin exhibits several advantages over antibiotics which includes high specificity to target bacterium, rapid bacterial lysis, effectiveness in planktonic as well as biofilm cells, non-emergence of resistant mutants and potency against antibiotic sensitive as well as resistant strains (Lopez et al. 1997; Loessner et al. 2002; Schuch et al. 2002). Endolysin possesses two types of structural architecture, specifically, enzymatically active domain (EAD) and cell-wall binding domain (CBD) are present for lysins against Gram-positive bacteria, whereas in most of the endolysin against Gram-negative bacteria only EAD is present. EADs are catalytic part of endolysin responsible for bacterial peptidoglycan disruption, whereas CBDs are responsible for conferring binding and specificity function for the endolysin towards host bacterium (Bateman and Rawlings 2003; Ohnuma et al. 2008). Endolysins activity can be detected based on turbidity reduction of bacterial suspension, log killing assay, and a clear zone on bacterial or peptidoglycan lawns (Schmelcher et al. 2012). Overlay assay and zymograms are also used as variants of turbidity reduction assays for in vitro assessment of endolysin activity.

### *3.6.2 Antibacterial Spectrum and Safety of Endolysin*

To date several natural, as well as chimeric endolysins have been reported; however, lysins against Gram-positive bacteria seem to dominate the repertoire of available endolysins. The presence of an outer membrane barrier limits the peptidoglycan access to endolysins in Gram-negative pathogens, resulting in lower potency as compared to Gram-positive bacteria specified endolysins. Various endolysins against potent pathogens have been evaluated which are briefly summarized in Table 3.3.

Above described endolysins have been well characterized and established in their in vitro as well as in vivo efficacy in various animal models over time (Yang et al. 2014c; Gondil et al. 2020b). Prospective of lysin therapy has been also well-reviewed by São-José (2018). In a recent study, the chitosan-based delivery system has been also reported to augment antibacterial activity and stability of endolysins (Gondil et al. 2020a, 2021). Like phage, endolysins are considered as safe alternative therapeutics as they do not interfere with normal microflora, which is a common shortcoming with antibiotics. A lower level of cytokine response was observed in animals treated with endolysins (Witzenrath et al. 2009). Harhala et al.

**Table 3.3** List of selected endolysins against various potent pathogens

| Pathogen | Endolysin(s) | Reference(s) |
|---|---|---|
| *S. pneumoniae* | Cpl-1, pal-1, LytA, LytC, Cpl-7, Cpl-7S, Cpl-711, PL3, ClyJ, ClyJ-3, and ClyJ-3m | Loeffler et al. (2003), Rodríguez-Cerrato et al. (2007), Bustamante et al. (2010), Domenech et al. (2011), Blázquez et al. (2016), Corsini et al. (2018), Yang et al. (2019), Yang et al. (2020), Luo et al. (2020) |
| *S. suis* | Csl2 and PlySs2 | Glimer et al. (2017), Vázquez et al. (2017) |
| *S. pyogenes* | PlyC and Plypy | Cheng et al. (2005), Lood et al. (2014) |
| *S. agalactiae* | ClyR, PlyGBS, λSa2lys, and ClyV | Donovan and Foster-Frey (2008), Yang et al. (2015), Huang et al. (2020) |
| *S. aureus* | 2638A, LysK, ClyS, phi11/phi12, Ply187, Lys-phiSA012, P128, ClyF, LysGH15, PlyGRCS, CF-301, MR-10, ClyH, and ClyC | Abaev et al. (2013), Sass and Bierbaum (2007), Becker et al. (2008), Pastagia et al. (2011), Fenton et al. (2013), Singh et al. (2014), Yang et al. (2014a), Yang et al. (2014b), Chopra et al. (2015), Linden et al. (2015), Haddad Kashani et al. (2017), Schuch et al. (2017), Yang et al. (2017), Zhang et al. (2016), Channabasappa et al. (2018), Li et al. (2021) |
| *Mycobacterium spp.* | Ms6-LysB, LysA, LysB, and Bxz-2-LysB | Gil et al. (2010), Grover et al. (2014), Lai et al. (2015) |
| *A. baumannii* | LysAB2, LysAB-01, LysAB3, LysAB4, PlyE146, PlyAB1, PlyF307, ABgp46, and PD-6A3 | Lai et al. (2011), Lai et al. (2013), Huang et al. (2014), Lood et al. (2015), Thummeepak et al. (2016), Larpin et al. (2018), Zhang et al. (2018), Wu et al. (2019) |
| *P. aeruginosa* | KZ144, EL188, LysPA26, Art-175, Art-085, PlyA, and LysAB54 | Paradis-Bleau et al. (2007), Briers et al. (2014), Guo et al. (2017), Khan et al. (2021) |
| *K. pneumoniae* | K11gp3.5/K11, KP32gp15, and KP27 | Walmagh et al. (2013), Maciejewska et al. (2017), Shavrina et al. (2016) |
| *E. coli* | EndoT5, Lysep3, and Lysep3-D8 | Lv et al. (2015), Wang et al. (2017), Yan et al. (2017) |
| *S. enterica* serovar typhimurium | Lys68, SPN1S, LysBPS13, and SPN9CC | Lim et al. (2012), Park et al. (2012), Lim et al. (2014), Oliveira et al. (2014) |

(2018) also postulated the safety of Cpl-1 endolysin in cell lines and animal models in terms of modulation of gene expression and complement pathways. Fast kinetics of endolysins and their high specificity towards their target bacterium surpass the immune interactions present by the host immune system (Jado et al. 2003).

### 3.6.3 Clinical Trials of Endolysin

In lieu of alternative therapeutic exploration, some of the endolysins were exploited in clinical settings. In a clinical trial (NCT01746654), nasal carriage of *S. aureus* was studied in P128 endolysin treated 74 healthy and chronic kidney disease human subjects. The in vivo tolerability, immunogenicity, pharmacokinetics, recolonization, and safety were accessed up to 50 days. Another anti-staphylococcal phage lysin N-Rephasin® SAL200 was studied for its safety, pharmacokinetics, and thermodynamics in 36 healthy male volunteers for 50 days. Results showed that SAL200 exhibits no serious adverse effects, 1 mg/kg as the optimal dose and increased AUC and $C_{max}$ in a dose-dependent manner (Jun et al. 2017). *S. aureus* induced bloodstream bacteraemia was also studied in a placebo-controlled study to determine the safety and tolerability of CF-301 in 20 healthy subjects after a single intravenous dose. Results showed that a single dose of CF-301 is unable to elicit a significant immune response and a long-term study of 6 months also showed the absence of any hypersensitivity and antibody protection post-endolysin injection (Cassino et al. 2016). Endolysin Staphefekt SA.100 were also tested against *S. aureus* induced dermatitis in a randomized controlled double-blinded placebo trial. 100 subjects were studied for 12 weeks and the effect of endolysin on disease severity was accessed (Totté et al. 2017). Phase 2 clinical trial (NCT03089697) of N-Rephasin® SAL200 has also been started to evaluate the effect of endolysin on persistent *S. aureus* infection in 50 human subjects. Another Phase 2 clinical trial (NCT03163446) of CF-301 versus placebo was accomplished for safety, pharmacokinetics, and antibacterial efficacy against *S. aureus* bacteraemia in 121 human participants. The role of conventional antibiotics (Daptomycin, Vancomycin, and semi-synthetic Penicillin) was also studied along with CF-301 treatment up to 6 months post-endolysin treatment. Most of the completed as well ongoing clinical trials are focused on anti-staphylococcal endolysins; however, in the advent of antibiotic resistance other endolysins should also be addressed for their clinical outcomes in the near future.

## 3.7 Conclusion

Phage and endolysin are promising therapeutics in this modern era of antibiotic resistance, in which clinicians are unable to use conventional antibiotic interventions. Phage is a potent alternative agent since its discovery, the antibacterial perspectives of endolysin spur researchers to look upon these lytic molecules instead of whole phage entity. The unique characteristics of phage and endolysin also complicate their clinical outcomes, as they must also go through rigid regulatory frameworks and clinical trials. Phage has been rightly explored for their clinical potential for a long period, whereas endolysins gained the clinical validation at a higher pace as compared to phage. The upcoming time can be considered as a breakthrough for phage and endolysin for their clinical acceptance as effective, alternative therapeutic products to antibiotics. In addition to progress in clinical trials, researchers and clinicians must also look for improved delivery strategies and novel approaches of designing for phage and endolysin to fill the gap and strengthen the battery of antibacterial agents.

## References


Abaev I, Foster-Frey J, Korobova O, Shishkova N, Kiseleva N, Kopylov P, Pryamchuk S, Schmelcher M, Becker SC, Donovan DM (2013) Staphylococcal phage 2638A endolysin is lytic for Staphylococcus aureus and harbors an inter-lytic-domain secondary translational start site. Appl Microbiol Biotechnol 97(8):3449–3456

Abedon ST (ed) (2008) Bacteriophage ecology: population growth, evolution, and impact of bacterial viruses. Cambridge University Press, Cambridge, UK

Ackermann HW (2003) Bacteriophage observations and evolution. Res Microbiol 154(4):245–251

Alemayehu D, Casey PG, McAuliffe O, Guinane CM, Martin JG, Shanahan F, Coffey A, Ross RP, Hill C (2012) Bacteriophages ϕMR299-2 and ϕNH-4 can eliminate *Pseudomonas aeruginosa* in the murine lung and on cystic fibrosis lung airway cells. MBio 3(2):e00029–e00012

Barrow P, Lovell M, Berchieri A (1998) Use of lytic bacteriophage for control of experimental *Escherichia coli* septicemia and meningitis in chickens and calves. Clin Vaccine Immunol 5 (3):294–298

Barrow PA, Soothill JS (1997) Bacteriophage therapy and prophylaxis: rediscovery and renewed assessment of potential. Trends Microbiol 5(7):268–271

Bateman A, Rawlings ND (2003) The CHAP domain: a large family of amidases including GSP amidase and peptidoglycan hydrolases. Trends Biochem Sci 28:234–247

Becker SC, Foster-Frey J, Donovan DM (2008) The phage K lytic enzyme LysK and lysostaphin act synergistically to kill MRSA. FEMS Microbiol Lett 287(2):185–191

Biswas B, Adhya S, Washart P, Paul B, Trostel AN, Powell B, Carlton R, Merril CR (2002) Bacteriophage therapy rescues mice bacteraemia from a clinical isolate of vancomycin-resistant *Enterococcus faecium*. Infect Immun 70(1):204–210

Blázquez B, Fresco-Taboada A, Iglesias-Bexiga M, Menéndez M, García P (2016) PL3 amidase, a tailor-made Lysin constructed by domain shuffling with potent killing activity against pneumococci and related species. Front Microbiol 7:1156

Braga LP, Soucy SM, Amgarten DE, da Silva AM, Setubal JC (2018) Bacterial diversification in the light of the interactions with phages: the genetic symbionts and their role in ecological speciation. Front Ecol Evol 6:6

Briers Y, Walmagh M, Grymonprez B, Biebl M, Pirnay JP, Defraine V, Michiels J, Cenens W, Aertsen A, Miller S, Lavigne R (2014) Art-175 is a highly efficient antibacterial against multidrug-resistant strains and persisters of Pseudomonas aeruginosa. Antimicrob Agents Chemother 58(7):3774–3784

Brüssow H (2005) Phage therapy: the *Escherichia coli* experience. Microbiology 151 (7):2133–2140

Budynek P, Dabrowska K, Skaradziński G, Górski A (2010) Bacteriophages and cancer. Arch Microbiol 192(5):315–320

Bustamante N, Campillo NE, García E, Gallego C, Pera B, Diakun GP, Sáiz JL, García P, Díaz JF, Menéndez M (2010) Cpl-7, a lysozyme encoded by a pneumococcal bacteriophage with a novel cell wall-binding motif. J Biol Chem 285(43):33184–33196

Cao F, Wang X, Wang L, Li Z, Che J, Wang L, Li X, Cao Z, Zhang J, Jin L, Xu Y (2015) Evaluation of the efficacy of a bacteriophage in the treatment of pneumonia induced by multidrug resistance *Klebsiella pneumoniae* in mice. Biomed Res Int 2015:752930

Cassino C, Murphy M, Boyle J, Rotolo J, Wittekind M (2016) Results of the first in human study of lysin CF-301 evaluating the safety, tolerability and pharmacokinetic profile in healthy volunteers; Proceedings of the 26th European Congress of Clinical Microbiology and Infectious Diseases; Amsterdam, The Netherlands. 9–12 April 2016

Chadha P, Katare OP, Chhibber S (2017) Liposome loaded phage cocktail: enhanced therapeutic potential in resolving *Klebsiella pneumoniae* mediated burn wound infections. Burns 43 (7):1532–1543

Channabasappa S, Durgaiah M, Chikkamadaiah R, Kumar S, Joshi A, Sriram B (2018) Efficacy of novel antistaphylococcal ectolysin P128 in a rat model of methicillin-resistant Staphylococcus aureus bacteremia. Antimicrob Agents Chemother 62(2):e01358–e01317

Cheng Q, Nelson D, Zhu S, Fischetti VA (2005) Removal of group B streptococci colonizing the vagina and oropharynx of mice with a bacteriophage lytic enzyme. Antimicrob Agents Chemother 49(1):111–117

Chhibber S, Gondil VS, Kaur J (2018) Isolation, characterization, statistical optimization, and application of a novel broad-spectrum capsular depolymerase against Klebsiella pneumoniae from *Bacillus siamensis* SCVJ30. Biomed Biotechnol Res J 2(2):125

Chibani-Chennoufi S, Sidoti J, Bruttin A, Kutter E, Sarker S, Brüssow H (2004) In vitro and in vivo bacteriolytic activities of Escherichia coli phages: implications for phage therapy. Antimicrob Agents Chemother 48(7):2558–2569

Chopra S, Harjai K, Chhibber S (2015) Potential of sequential treatment with minocycline and S. aureus specific phage lysin in eradication of MRSA biofilms: an in vitro study. Appl Microbiol Biotechnol 99(7):3201–3210

Corsini B, Díez-Martínez R, Aguinagalde L, González-Camacho F, García-Fernández E, Letrado P, García P, Yuste J (2018) Chemotherapy with phage lysins reduces pneumococcal colonization of the respiratory tract. Antimicrob Agents Chemother 62(6):e02212–e02217

Dallal MM, Nikkhahi F, Alimohammadi M, Douraghi M, Rajabi Z, Foroushani AR, Azimi A, Fardsanei F (2019) Phage therapy as an approach to control *Salmonella enterica serotype* Enteritidis infection in mice. Rev Soc Bras Med Trop 52:e20190290

Debarbieux L, Leduc D, Maura D, Morello E, Criscuolo A, Grossi O, Balloy V, Touqui L (2010) Bacteriophages can treat and prevent *Pseudomonas aeruginosa* lung infections. J Infect Dis 201 (7):1096–1104

Divya Ganeshan S, Hosseinidoust Z (2019) Phage therapy with a focus on the human microbiota. Antibiotics 8(3):131

Domenech M, García E, Moscoso M (2011) In vitro destruction of *Streptococcus pneumoniae* biofilms with bacterial and phage peptidoglycan hydrolases. Antimicrob Agents Chemother 55:4144–4148

Domingo-Calap P, Delgado-Martínez J (2018) Bacteriophages: protagonists of a post-antibiotic era. Antibiotics 7(3):66

Donovan DM, Foster-Frey J (2008) LambdaSa2 prophage endolysin requires Cpl-7-binding domains and amidase-5 domain for antimicrobial lysis of streptococci. FEMS Microbiol Lett 287(1):22–33

Dorval Courchesne NM, Parisien A, Lan CQ (2009) Production and application of bacteriophage and bacteriophage-encoded lysins. Recent pat biotechnol 3(1):37–45

Dufour N, Debarbieux L, Fromentin M, Ricard JD (2015) Treatment of highly virulent extraintestinal pathogenic *Escherichia coli* pneumonia with bacteriophages. Crit Care Med 43 (6):e190–e198

Fenton M, Keary R, McAuliffe O, Ross RP, O'Mahony J, Coffey A (2013) Bacteriophage-derived peptidase eliminates and prevents staphylococcal biofilms. Int J Microbiol 2013:625341

Fukuda K, Ishida W, Uchiyama J, Rashel M, Kato SI, Morita T, Muraoka A, Sumi T, Matsuzaki S, Daibata M, Fukushima A (2012) *Pseudomonas aeruginosa* keratitis in mice: effects of topical bacteriophage KPP12 administration. PloS One 7(10):e47742

Furfaro LL, Payne MS, Chang BJ (2018) Bacteriophage therapy: clinical trials and regulatory hurdles. Front Cell Infect Microbiol 8:376

Ghannad MS, Mohammadi A (2012) Bacteriophage: time to re-evaluate the potential of phage therapy as a promising agent to control multidrug-resistant bacteria. Iran J Basic Med Sci 15 (2):693

Gil F, Grzegorzewicz AE, Catalao MJ, Vital J, McNeil MR, Pimentel M (2010) Mycobacteriophage Ms6 LysB specifically targets the outer membrane of Mycobacterium smegmatis. Microbiology 156(Pt 5):1497

Gilmer DB, Schmitz JE, Thandar M, Euler CW, Fischetti VA (2017) The phage lysin PlySs2 decolonizes Streptococcus suis from murine intranasal mucosa. PLoS One 12(1):e0169180

Gondil VS, Asif M, Bhalla TC (2017) Optimization of physicochemical parameters influencing the production of prodigiosin from *Serratia nematodiphila* RL2 and exploring its antibacterial activity. 3 Biotech 7(5):338

Gondil VS, Chhibber S (2017) Evading antibody mediated inactivation of bacteriophages using delivery systems. J Immunovirol 1:555–574

Gondil VS, Chhibber S (2018) Exploring potential of phage therapy for tuberculosis using model organism. Biomed Biotechnol Res J 2(1):9

Gondil VS, Dube T, Panda JJ, Yennamalli RM, Harjai K, Chhibber S (2020a) Comprehensive evaluation of chitosan nanoparticle-based phage lysin delivery system; a novel approach to counter S. pneumoniae infections. Int J Pharma 573:118850

Gondil VS, Harjai K, Chhibber S (2020b) Endolysins as emerging alternative therapeutic agent to counter drug resistant infections. Int J Antimicrob Agents 55(2):105844

Gondil VS, Harjai K, Chhibber S (2021) Investigating the potential of endolysin loaded chitosan nanoparticles in the treatment of pneumococcal pneumonia. J Drug Deliv Sci Technol 61:102142

Gondil VS, Kalaiyarasan T, Bharti VK, Chhibber S (2019) Antibiofilm potential of Sea buckthorn silver nanoparticles (SBT@ AgNPs) against *Pseudomonas aeruginosa*. 3 Biotech 9(11):402

Górski A, Ważna E, Dąbrowska BW, Dąbrowska K, Świtała-Jeleń K, Międzybrodzki R (2006) Bacteriophage translocation. FEMS Immunol Med Microbiol 46(3):313–319

Górski A, Weber-Dabrowska B (2005) The potential role of endogenous bacteriophages in controlling invading pathogens. Cell Mol Life Sci 62(5):511

Grover N, Paskaleva EE, Mehta KK, Dordick JS, Kane RS (2014) Growth inhibition of Mycobacterium smegmatis by mycobacteriophage-derived enzymes. Enzym Microb Technol 63:1–6

Guo M, Feng C, Ren J, Zhuang X, Zhang Y, Zhu Y, Dong K, He P, Guo X, Qin J (2017) A novel antimicrobial endolysin, LysPA26, against Pseudomonas aeruginosa. Front Microbiol 8:293

Haddad Kashani H, Fahimi H, Dasteh Goli Y, Moniri R (2017) A novel chimeric endolysin with antibacterial activity against methicillin-resistant Staphylococcus aureus. Front Cell Infect Microbiol 7:290

Harhala M, Nelson DC, Miernikiewicz P, Heselpoth RD, Brzezicka B, Majewska J, Linden SB, Shang X, Szymczak A, Lecion D, Marek-Bukowiec K (2018) Safety studies of pneumococcal endolysins Cpl-1 and pal. Viruses 10(11):638

Harper DR (2018) Criteria for selecting suitable infectious diseases for phage therapy. Viruses 10 (4):177

Hatfull GF, Hendrix RW (2011) Bacteriophages and their genomes. Curr Opin Virol 1:298–303

Hemminga MA, Vos WL, Nazarov PV, Koehorst RB, Wolfs CJ, Spruijt RB, Stopar D (2010) Viruses: incredible nanomachines. New advances with filamentous phages. Eur Biophys J 39 (4):541–550

Hietala V, Horsma-Heikkinen J, Carron A, Skurnik M, Kiljunen SJ (2019) The removal of endo- and enterotoxins from bacteriophage preparations. Front Microbiol 10:1674

Housby JN, Mann NH (2009) Phage therapy. Drug Discov Today 14:536–540

Hua Y, Luo T, Yang Y, Dong D, Wang R, Wang Y, Xu M, Guo X, Hu F, He P (2018) Phage therapy as a promising new treatment for lung infection caused by carbapenem-resistant Acinetobacter baumannii in mice. Front Microbiol 8:2659

Huang G, Shen X, Gong Y, Dong Z, Zhao X, Shen W, Wang J, Hu F, Peng Y (2014) Antibacterial properties of Acinetobacter baumannii phage Abp1 endolysin (PlyAB1). BMC Infect Dis 14 (1):681

Huang L, Luo D, Gondil VS, Gong Y, Jia M, Yan D, He J, Hu S, Yang H, Wei H (2020) Construction and characterization of a chimeric lysin ClyV with improved bactericidal activity against Streptococcus agalactiae in vitro and in vivo. Appl Microbiol Biotechnol 104:1609–1619

Hung CH, Kuo CF, Wang CH, Wu CM, Tsao N (2011) Experimental phage therapy in treating *Klebsiella pneumoniae*-mediated liver abscesses and bacteraemia in mice. Antimicrob Agents Chemother 55(4):1358–1365

Inal JM (2003) Phage therapy: a reappraisal of bacteriophages as antibiotics. Arch Immunol Ther Exp 51(4):237–244

Jado I, López R, García E, Fenoll A, Casal J, García P (2003) Phage lytic enzymes as therapy for antibiotic-resistant Streptococcus pneumoniae infection in a murine sepsis model. J Antimicrob Chemother 52(6):967–73

Jun SY, Jang IJ, Yoon S, Jang K, Yu KS, Cho JY, Seong MW, Jung GM, Yoon SJ, Kang SH (2017) Pharmacokinetics and tolerance of the phage endolysin-based candidate drug SAL200 after a single intravenous administration among healthy volunteers. Antimicrob Agents Chemother 61 (6):e02629–e02616

Kaur P, Gondil VS, Chhibber S (2019) A novel wound dressing consisting of PVA-SA hybrid hydrogel membrane for topical delivery of bacteriophages and antibiotics. Int J Pharm 572:118779

Khan FM, Gondil VS, Li C, Jiang M, Li J, Yu J, Wei H, Yang H (2021) A novel Acinetobacter baumannii bacteriophage endolysin LysAB54 with high antibacterial activity against multiple Gram-negative microbes. Front cell infect microbiol 11:70

Kochetkova VA, Mamontov AS, Moskovtseva RL, Erastova EI, Trofimov EI, Popov MI et al (1989) Phagotherapy of postoperative suppurative-inflammatory complications in patients with neoplasms. Soviet Med 6:23–26

Kumar M, Bala R, Gondil VS, Pandey SK, Chhibber S, Jain DV, Sharma RK, Wangoo N (2017) Combating food pathogens using sodium benzoate functionalized silver nanoparticles: synthesis, characterization and antimicrobial evaluation. J Mater Sci 52(14):8568–8575

Kumari S, Harjai K, Chhibber S (2010) Evidence to support the therapeutic potential of bacteriophage Kpn5 in burn wound infection caused by Klebsiella pneumoniae in BALB/c mice. J Microbiol Biotechnol 20(5):935–941

Kusradze I, Karumidze N, Rigvava S, Dvalidze T, Katsitadze M, Amiranashvili I et al (2016) Characterization and testing the efficiency of Acinetobacter baumannii phage vB-GEC_Ab-M-G7as an antibacterial agent. Front Microbiol 7:1590

Kutter E (2008) Phage therapy: bacteriophages as naturally occurring antimicrobials. In: Goldman E, Green LH (eds) Practical handbook of microbiology. CRC Press, Boca Raton, FL, pp 713–730

Lai MJ, Lin NT, Hu A, Soo PC, Chen LK, Chen LH, Chang KC (2011) Antibacterial activity of Acinetobacter baumannii phage ϕAB2 endolysin (LysAB2) against both gram-positive and gram-negative bacteria. Appl Microbiol Biotechnol 90(2):529–539

Lai MJ, Liu CC, Jiang SJ, Soo PC, Tu MH, Lee JJ, Chen YH, Chang KC (2015) Antimycobacterial activities of endolysins derived from a mycobacteriophage, BTCU-1. Molecules 20 (10):19277–19290

Lai MJ, Soo PC, Lin NT, Hu A, Chen YJ, Chen LK, Chang KC (2013) Identification and characterisation of the putative phage-related endolysins through full genome sequence analysis in Acinetobacter baumannii ATCC 17978. Int J Antimicrob Agents 42(2):141–148

Lamers RP, Cavallari JF, Burrows LL (2013) The efflux inhibitor phenylalanine-arginine beta-naphthylamide (PAβN) permeabilizes the outer membrane of gram-negative bacteria. PloS one 8(3):e60666

Larpin Y, Oechslin F, Moreillon P, Resch G, Entenza JM, Mancini S (2018) In vitro characterization of PlyE146, a novel phage lysin that targets Gram-negative bacteria. PLoS One 13(2): e0192507

Lazareva EB, Smirnov SV, Khvatov VB, Spiridonova TG, Bitkova EE, Darbeeva OS, Maĭskaia LM, Parfeniuk RL, Men'shikov DD (2001) Efficacy of bacteriophages in complex treatment of patients with burn wounds. Antibiot Chemother 46(1):10–14

Leitner L, Sybesma W, Chanishvili N, Goderdzishvili M, Chkhotua A, Ujmajuridze A, Schneider MP, Sartori A, Mehnert U, Bachmann LM, Kessler TM (2017) Bacteriophages for treating urinary tract infections in patients undergoing transurethral resection of the prostate: a randomized, placebo-controlled, double-blind clinical trial. BMC Urol 17(1):90

Li X, Wang S, Nyaruaba R, Liu H, Yang H, Wei H (2021) A highly active chimeric lysin with a calcium-enhanced bactericidal activity against staphylococcus aureus In Vitro and In Vivo. Antibiotics (Basel) 10

Ligonenko OV, Borysenko MM, Digtyar II, Ivashchenko DM, Zubakha AB, Chorna IO, Shumeyko IA, Storozhenko OV, Gorb LI, Ligonenko OO (2015) Application of bacteriophages in complex of treatment of a shot-gun wounds of soft tissues in the patients, suffering multiple allergy for antibiotics. Klin Khir 10:65–66

Lim JA, Shin H, Heu S, Ryu S (2014) Exogenous lytic activity of SPN9CC endolysin against gram-negative bacteria. J Microbiol Biotechnol 24(6):803–811

Lim JA, Shin H, Kang DH, Ryu S (2012) Characterization of endolysin from a Salmonella typhimurium-infecting bacteriophage SPN1S. Res Microbiol 163(3):233–241

Linden SB, Zhang H, Heselpoth RD, Shen Y, Schmelcher M, Eichenseher F, Nelson DC (2015) Biochemical and biophysical characterization of PlyGRCS, a bacteriophage endolysin active against methicillin-resistant Staphylococcus aureus. Appl Microbiol Biotechnol 99(2):741–752

Loeffler JM, Djurkovic S, Fischetti VA (2003) Phage lytic enzyme Cpl-1 as a novel antimicrobial for pneumococcal bacteremia. Infect Immun 71(11):6199–6204

Loessner MJ, Kramer K, Ebel F, Scherer S (2002) C-terminal domains of Listeria monocytogenes bacteriophage murein hydrolases determine specific recognition and high-affinity binding to bacterial cell wall carbohydrates. Mol Microbiol 44:335–349

Lood R, Raz A, Molina H, Euler CW, Fischetti VA (2014) A highly active and negatively charged Streptococcus pyogenes lysin with a rare D-alanyl-L-alanine endopeptidase activity protects mice against streptococcal bacteremia. Antimicrob Agents Chemother 58(6):3073–3084

Lood R, Winer BY, Pelzek AJ, Diez-Martinez R, Thandar M, Euler CW, Schuch R, Fischetti VA (2015) Novel phage lysin capable of killing the multidrug-resistant gram-negative bacterium Acinetobacter baumannii in a mouse bacteremia model. Antimicrob Agents Chemother 59 (4):1983–1991

López R, García E, García P, García JL (1997) The pneumococcal cell wall degrading enzymes: a modular design to create new lysins? Microb Drug Resist 3:199–211

Luo D, Huang L, Gondil VS, Zhou W, Yang W, Jia M, Hu S, He J, Yang H, Wei H (2020) A choline-recognizing monomeric lysin, clyJ-3m, shows elevated activity against streptococcus pneumoniae. Antimicrob Agents Chemother 64:e00311–20

Lv M, Wang S, Yan G, Sun C, Feng X, Gu J, Han W, Lei L (2015) Genome sequencing and analysis of an Escherichia coli phage vB_EcoM-ep3 with a novel lysin, Lysep3. Virus Genes 50 (3):487–497

Maciejewska B, Roszniowski B, Espaillat A, Kęsik-Szeloch A, Majkowska-Skrobek G, Kropinski AM, Briers Y, Cava F, Lavigne R, Drulis-Kawa Z (2017) Klebsiella phages representing a novel clade of viruses with an unknown DNA modification and biotechnologically interesting enzymes. Appl Microbiol Biotechnol 101(2):673–684

Manohar P, Nachimuthu R, Lopes BS (2018) The therapeutic potential of bacteriophages targeting gram-negative bacteria using *Galleria mellonella* infection model. BMC Microbiol 18(1):97

Markoishvili K, Tsitlanadze G, Katsarava R, Glenn J, Morris MD Jr, Sulakvelidze A (2002) A novel sustained-release matrix based on biodegradable poly (ester amide) s and impregnated with bacteriophages and an antibiotic show promise in management of infected venous stasis ulcers and other poorly healing wounds. Int J Dermatol 41(7):453–458

Matsuzaki S, Yasuda M, Nishikawa H, Kuroda M, Ujihara T, Shuin T, Shen Y, Jin Z, Fujimoto S, Nasimuzzaman MD, Wakiguchi H (2003) Experimental protection of mice against lethal Staphylococcus aureus infection by novel bacteriophage ϕMR11. J Infect Dis 187(4):613–624

Meladze GD, Mebuke MG, Chkhetia NS, Kiknadze NI, Koguashvili GG, Timoshuk II et al (1982) Efficacy of staphylococcal bacteriophage in treatment of purulent diseases of lungs and pleura. Grudn Khir 1:53–56

Merril CR (2008) Interaction of bacteriophages with animals. In: Abedon ST (ed) Bacteriophage ecology. Cambridge University Press, Cambridge, UK, pp 332–352

Merril CR, Biswas B, Carlton R, Jensen NC, Creed GJ, Zullo S, Adhya S (1996) Long-circulating bacteriophage as antibacterial agents. Proc Natl Acad Sci 93(8):3188–3192

Morello E, Saussereau E, Maura D, Huerre M, Touqui L, Debarbieux L (2011) Pulmonary bacteriophage therapy on *Pseudomonas aeruginosa* cystic fibrosis strains: first steps towards treatment and prevention. PloS One 6(2):e16963

Moreno H, Paulo R, da Costa-Issa F, Rajca-Ferreira AK, Pereira MA, Kaneko TM (2013) Native Brazilian plants against nosocomial infections: a critical review on their potential and the antimicrobial methodology. Curr Top Med Chem 13(24):3040–3078

Nelson D, Loomis L, Fischetti VA (2001) Prevention and elimination of upper respiratory colonization of mice by group a streptococci by using a bacteriophage lytic enzyme. Proc Natl Acad Sci 98:4107–4112

Nelson D, Schuch R, Chahales P, Zhu S, Fischetti VA (2006) PlyC: a multimeric bacteriophage lysin. Proc Natl Acad Sci 103(28):10765–10770

Ohnuma T, Onaga S, Murata K, Taira T, Katoh E (2008) LysM domains from *Pteris ryukyuensis* chitinase –a stability study and characterization of the chitin-binding site. J Biol Chem 283:5178–5187

Oliveira H, Thiagarajan V, Walmagh M, Sillankorva S, Lavigne R, Neves-Petersen MT, Kluskens LD, Azeredo J (2014) A thermostable Salmonella phage endolysin, Lys68, with broad bactericidal properties against gram-negative pathogens in presence of weak acids. PLoS One 9(10): e108376

Pabary R, Singh C, Morales S, Bush A, Alshafi K, Bilton D, Alton EW, Smithyman A, Davies JC (2016) Antipseudomonal bacteriophage reduces infective burden and inflammatory response in murine lung. Antimicrob Agents Chemother 60(2):744–751

Paradis-Bleau C, Cloutier I, Lemieux L, Sanschagrin F, Laroche J, Auger M, Garnier A, Levesque RC (2007) Peptidoglycan lytic activity of the Pseudomonas aeruginosa phage φKZ gp144 lytic transglycosylase. FEMS Microbiol Lett 266(2):201–209

Park J, Yun J, Lim JA, Kang DH, Ryu S (2012) Characterization of an endolysin, LysBPS13, from a Bacillus cereus bacteriophage. FEMS Microbiol Lett 332(1):76–83

Parracho HM, Burrowes BH, Enright MC, McConville ML, Harper DR (2012) The role of regulated clinical trials in the development of bacteriophage therapeutics. J Mol Genet Med 6:279

Pastagia M, Euler C, Chahales P, Fuentes-Duculan J, Krueger JG, Fischetti VA (2011) A novel chimeric lysin shows superiority to mupirocin for skin decolonization of methicillin-resistant and-sensitive Staphylococcus aureus strains. Antimicrob Agents Chemother 55(2):738–744

Payne RJ, Jansen VA (2003) Pharmacokinetic principles of bacteriophage therapy. Clin Pharmacokinet 42(4):315–325

Perepanova TS, Darbeeva OS, Kotliarova GA, Kondrat'eva EM, Maĭskaia LM, Malysheva VF, Baĭguzina FA, Grishkova NV (1995) The efficacy of bacteriophage preparations in treating inflammatory urologic diseases. Urol Nefrol 5:14–17

Pipiia VI, Eteriia GP, Gotua TP, Volobuev VI, Katsarava VS (1976) Experience with treating complicated forms of abscessing pneumonia in children. Vestnik Khirurgii Imeni i-i-Grekova 117:64–68

Pouillot F, Chomton M, Blois H, Courroux C, Noelig J, Bidet P, Bingen E, Bonacorsi S (2012) Efficacy of bacteriophage therapy in experimental sepsis and meningitis caused by a clone O25b: H4-ST131 *Escherichia coli* strain producing CTX-M-15. Antimicrob Agents Chemother 56(7):3568–3575

Rhoads DD, Wolcott RD, Kuskowski MA, Wolcott BM, Ward LS, Sulakvelidze A (2009) Bacteriophage therapy of venous leg ulcers in humans: results of a phase I safety trial. J Wound Care 18(6):237–243

Rodríguez-Cerrato V, García P, del Prado G, García E, Gracia M, Huelves L, Ponte C, Lopez R, Soriano F (2007) In vitro interactions of LytA, the major pneumococcal autolysin, with two bacteriophage lytic enzymes (Cpl-1 and pal), cefotaxime and moxifloxacin against antibiotic-susceptible and-resistant Streptococcus pneumoniae strains. J Antimicrob Chemother 60 (5):1159–1162

Romero-Calle D, Guimarães Benevides R, Góes-Neto A, Billington C (2019) Bacteriophages as alternatives to antibiotics in clinical care. Antibiotics 8(3):138

Sakandelidze VM (1991) The combined use of specific phages and antibiotics in different infectious allergoses. Vrach Delo 3:60–63

Sakandelidze VM, Meĭpariani AN (1974) Use of combined phages in suppurative-inflammatory diseases. Zh Mikrobiol Epidemiol Immunobiol 51(6):135–136

São-José C (2018) Engineering of phage-derived lytic enzymes: improving their potential as antimicrobials. Antibiotics 7(2):29

Sarker SA, Sultana S, Reuteler G, Moine D, Descombes P, Charton F, Bourdin G, McCallin S, Ngom-Bru C, Neville T, Akter M (2016) Oral phage therapy of acute bacterial diarrhea with two coliphage preparations: a randomized trial in children from Bangladesh. EBioMedicine 4:124–137

Sass P, Bierbaum G (2007) Lytic activity of recombinant bacteriophage φ11 and φ12 endolysins on whole cells and biofilms of Staphylococcus aureus. Appl Environ Microbiol 73(1):347–352

Schmelcher M, Donovan DM, Loessner MJ (2012) Bacteriophage endolysins as novel antimicrobials. Future Microbiol 7:1147–1171

Schneider G, Szentes N, Horváth M, Dorn Á, Cox A, Nagy G, Doffkay Z, Maróti G, Rákhely G, Kovács T (2018) Kinetics of targeted phage rescue in a mouse model of systemic *Escherichia coli* K1. Biomed Res Int 2018:7569645

Schuch R, Khan BK, Raz A, Rotolo JA, Wittekind M (2017) Bacteriophage lysin CF-301, a potent antistaphylococcal biofilm agent. Antimicrob Agents Chemother 61(7):e02666–e02616

Schuch R, Nelson D, Fischetti VA (2002) A bacteriolytic agent that detects and kills Bacillus anthracis. Nature 418:884

Shapiro OH, Kushmaro A (2011) Bacteriophage ecology in environmental biotechnology processes. Curr Opin Biotechnol 22(3):449–455

Shavrina MS, Zimin AA, Molochkov NV, Chernyshov SV, Machulin AV, Mikoulinskaia GV (2016) In vitro study of the antibacterial effect of the bacteriophage T5 thermostable endolysin on Escherichia coli cells. J Appl Microbiol 121(5):1282–1290

Singh PK, Donovan DM, Kumar A (2014) Intravitreal injection of the chimeric phage endolysin Ply187 protects mice from Staphylococcus aureus endophthalmitis. Antimicrob Agents Chemother 58(8):4621–4629

Singla S, Harjai K, Katare OP, Chhibber S (2016) Encapsulation of bacteriophage in liposome accentuates its entry in to macrophage and shields it from neutralizing antibodies. PLoS One 11 (4):e0153777

Smith HW, Huggins MB (1982) Successful treatment of experimental *Escherichia coli* infections in mice using phage: its general superiority over antibiotics. Microbiology 128(2):307–318

Soothill JS (1992) Treatment of experimental infections of mice with bacteriophages. J Med Microbiol 37(4):258–261

Sulakvelidze A (2005) Phage therapy: an attractive option for dealing with antibiotic-resistant bacterial infections. Drug discov today 12(10):807–809

Sulakvelidze A, Kutter E (2005) Bacteriophage therapy in humans. In: Kutter E, Sulakvelidze A (eds) Bacteriophages: biology and application. CRC Press, Boca Raton, FL, pp 381–436

Sulakvelidze A, Morris JG (2001) Bacteriophages as therapeutic agents. Ann Med 33(8):507–509

Sunagar R, Patil SA, Chandrakanth RK (2010) Bacteriophage therapy for *Staphylococcus aureus* bacteremia in streptozotocin-induced diabetic mice. Res Microbiol 161(10):854–860

Takemura-Uchiyama I, Uchiyama J, Osanai M, Morimoto N, Asagiri T, Ujihara T, Daibata M, Sugiura T, Matsuzaki S (2014) Experimental phage therapy against lethal lung-derived septicemia caused by *Staphylococcus aureus* in mice. Microbes Infect 16(6):512–517

Tammelin A (1992) Staphylococcus aureus surgical wound infection; possibility of preventing wound infection by use of bacteriophages. Nature 22:26–31

Tang SS, Biswas SK, Tan WS, Saha AK, Leo BF (2019) Efficacy and potential of phage therapy against multidrug resistant Shigella spp. PeerJ 7:e6225

Thummeepak R, Kitti T, Kunthalert D, Sitthisak S (2016) Enhanced antibacterial activity of Acinetobacter baumannii bacteriophage ØABP-01 endolysin (LysABP-01) in combination with colistin. Front Microbiol 7:1402

Tkhilaishvili T, Winkler T, Müller M, Perka C, Trampuz A (2019) Bacteriophages as adjuvant to antibiotics for the treatment of periprosthetic joint infection caused by multidrug-resistant *Pseudomonas aeruginosa*. Antimicrob Agents Chemother 64(1):e00924

Tóthová Ľ, Celec P, Bábíčková J, Gajdošová J, Al-Alami H, Kamodyova N, Drahovska H, Liptakova A, Turňa J, Hodosy J (2011) Phage therapy of Cronobacter-induced urinary tract infection in mice. Med Sci Monitor: Int Med J Exp Clin Res 17(7):BR173

Totté J, de Wit J, Pardo L, Schuren F, van Doorn M, Pasmans S (2017) Targeted anti-staphylococcal therapy with endolysins in atopic dermatitis and the effect on steroid use, disease severity and the microbiome: Study protocol for a randomized controlled trial (MAAS trial). Trials 18:404

Ujmajuridze A, Chanishvili N, Goderdzishvili M, Leitner L, Mehnert U, Chkhotua A, Kessler TM, Sybesma W (2018) Adapted bacteriophages for treating urinary tract infections. Front Microbiol 9:1832

Vázquez R, Domenech M, Iglesias-Bexiga M, Menéndez M, García P (2017) Csl2, a novel chimeric bacteriophage lysin to fight infections caused by Streptococcus suis, an emerging zoonotic pathogen. Sci Rep 7(1):1–3

Vinodkumar CS, Kalsurmath S, Neelagund YF (2008) Utility of lytic bacteriophage in the treatment of multidrug-resistant *Pseudomonas aeruginosa* septicemia in mice. Indian J Pathol Microbiol 51(3):360

Vinodkumar CS, Neelagund YF, Kalsurmath S (2005) Bacteriophage in the treatment of experimental septicemic mice from a clinical isolate of multidrug resistant *Klebsiella pneumoniae*. J Commun Dis 37(1):18–29

Walmagh M, Boczkowska B, Grymonprez B, Briers Y, Drulis-Kawa Z, Lavigne R (2013) Characterization of five novel endolysins from gram-negative infecting bacteriophages. Appl Microbiol Biotechnol 97(10):4369–4375

Wang J, Hu B, Xu M, Yan Q, Liu S, Zhu X, Sun Z, Reed E, Ding L, Gong J, Li QQ (2006a) Use of bacteriophage in the treatment of experimental animal bacteraemia from imipenem-resistant *Pseudomonas aeruginosa*. Int J Mol Med 17(2):309–317

Wang J, Hu B, Xu M, Yan Q, Liu S, Zhu X, Sun Z, Tao D, Ding L, Reed E, Gong J (2006b) Therapeutic effectiveness of bacteriophages in the rescue of mice with extended spectrum β-lactamase-producing *Escherichia coli* bacteraemia. Int J Mol Med 17(2):347–355

Wang S, Gu J, Lv M, Guo Z, Yan G, Yu L, Du C, Feng X, Han W, Sun C, Lei L (2017) The antibacterial activity of E. coli bacteriophage lysin lysep3 is enhanced by fusing the Bacillus amyloliquefaciens bacteriophage endolysin binding domain D8 to the C-terminal region. J Microbiol 55(5):403–408

Watanabe S, Tomizaki KY, Takahashi T, Usui K, Kajikawa K, Mihara H (2007) Interactions between peptides containing nucleobase amino acids and T7 phages displaying *S. cerevisiae* proteins. Pept Sci: Original Res Biomol 88(2):131–140

Wills QF, Kerrigan C, Soothill JS (2005) Experimental bacteriophage protection against *Staphylococcus aureus* abscesses in a rabbit model. Antimicrob Agents Chemother 49(3):1220–1221

Witzenrath M, Schmeck B, Doehn JM, Tschernig T, Zahlten J, Loeffler JM (2009) Systemic use of the endolysin Cpl-1 rescues mice with fatal pneumococcal pneumonia. Crit Care Med 37:642–649

Wright A, Hawkins CH, Änggård EE, Harper DR (2009) A controlled clinical trial of a therapeutic bacteriophage preparation in chronic otitis due to antibiotic-resistant Pseudomonas aeruginosa; a preliminary report of efficacy. Clin Otolaryngol 34(4):349–357

Wu M, Hu K, Xie Y, Liu Y, Mu D, Guo H, Zhang Z, Zhang Y, Chang D, Shi Y (2019) A novel phage PD-6A3, and its endolysin Ply6A3, with extended lytic activity against Acinetobacter baumannii. Front Microbiol 9:3302

Yan G, Liu J, Ma Q, Zhu R, Guo Z, Gao C, Wang S, Yu L, Gu J, Hu D, Han W (2017) The N-terminal and central domain of colicin a enables phage lysin to lyse Escherichia coli extracellularly. Antonie Van Leeuwenhoek 110(12):1627–1635

Yang H, Gong Y, Zhang H, Etobayeva I, Miernikiewicz P, Luo D, Li X, Zhang X, Dąbrowska K, Nelson DC, He J (2019) ClyJ is a novel pneumococcal chimeric lysin with a cysteine-and histidine-dependent amidohydrolase/peptidase catalytic domain. Antimicrob Agents Chemother 63(4):e02043–e02018

Yang H, Linden SB, Wang J, Yu J, Nelson DC, Wei H (2015) A chimeolysin with extended-spectrum streptococcal host range found by an induced lysis-based rapid screening method. Sci Rep 5:17257

Yang H, Luo D, Etobayeva I, Li X, Gong Y, Wang S, Li Q, Xu P, Yin W, He J, Nelson DC (2020) Linker editing of pneumococcal lysin ClyJ conveys improved bactericidal activity. Antimicrob Agents Chemoter 64(2):e01610

Yang H, Yu J, Wei H (2014c) Engineered bacteriophage lysins as novel anti-infectives. Front Microbiol 5:542

Yang H, Zhang H, Wang J, Yu J, Wei H (2017) A novel chimeric lysin with robust antibacterial activity against planktonic and biofilm methicillin-resistant Staphylococcus aureus. Sci Rep 7 (1):1–3

Yang H, Zhang Y, Huang Y, Yu J, Wei H (2014b) Degradation of methicillin-resistant Staphylococcus aureus biofilms using a chimeric lysin. Biofouling 30(6):667–674

Yang H, Zhang Y, Yu J, Huang Y, Zhang XE, Wei H (2014a) Novel chimeric lysin with high-level antimicrobial activity against methicillin-resistant Staphylococcus aureus in vitro and in vivo. Antimicrob Agents Chemother 58(1):536–542

Yen M, Cairns LS, Camilli A (2017) A cocktail of three virulent bacteriophages prevents Vibrio cholerae infection in animal models. Nat Commun 8(1):1–7

Yilmaz C, Colak M, Yilmaz BC, Ersoz G, Kutateladze M, Gozlugol M (2013) Bacteriophage therapy in implant-related infections: an experimental study. JBJS 95(2):117–125

Zhang J, Xu LL, Gan D, Zhang X (2018) In vitro study of bacteriophage AB3 Endolysin LysAB3 activity against Acinetobacter baumannii biofilm and biofilm-Bound a. baumannii. Clin Lab 64 (6):1021–1030

Zhang L, Li D, Li X, Hu L, Cheng M, Xia F, Gong P, Wang B, Ge J, Zhang H, Cai R (2016) LysGH15 kills Staphylococcus aureus without being affected by the humoral immune response or inducing inflammation. Sci Rep 6(1):1–9

Zhukov-Verezhnikov NN, Peremitina LD, Berillo EA, Komissarov VP, Bardymov VM (1978) Therapeutic effect of bacteriophage preparations in the complex treatment of suppurative surgical diseases. Sov Med 12:64–66

Zimecki M, Artym J, Kocięba M, Weber-Dąbrowska B, Borysowski J, Górski A (2010) Prophylactic effect of bacteriophages on mice subjected to chemotherapy-induced immunosuppression and bone marrow transplant upon infection with *Staphylococcus aureus*. Med Microbiol Immunol 199(2):71–79

# Chapter 4
# Probiotics: Origin, Products, and Regulations in India

**Amrita Narula**

**Abstract** The concept of probiotics is well known from the time of Greeks and Romans. The term 'Probiotics' is defined as the substances produced by microorganisms that help stimulate the secretion of another. Though the definition has been redefined by many, the most recent and acceptable definition is by FAO/WHO (2001) and has defined it as: 'Live microorganisms that when being administered in appropriate dose, confer the benefit of health to the receiver'. Probiotics are usually found in dairy and non-dairy products, infant formula, dietary supplements, and energy drinks. They are generally recommended for consumption after the antibiotic therapy and help to manifest a positive balance of valuable microbes in the intestine. The most often used probiotic species belongs to *Lactobacillus* and *Bifidobacterium*, apart from these some of the yeast *Saccharomyces cerevisiae* and some *Bacillus* and *E. coli* species are also used as they too demonstrated the probiotic properties. To be a probiotic, the strain needs to fulfil certain specific criteria (GRAS, nontoxic, stable, etc.) and their mechanism of probiosis includes manipulating gut microbial communities, immunomodulation, suppressing pathogens, stimulating epithelial cell proliferation and differentiation; and fortification of the intestinal barrier. The use of probiotics can restore the replenished good bacteria and overcome the adverse effect of chronic diseases. Yakult, Danone, Nestle, Amul, and Mother Dairy are the common probiotic brands in India which have made their remark of recognition in this industry along with many others (some minor brands) are too heading towards better quality pre and probiotic products. The laws governing probiotics are ambiguous due to the categorization of probiotics in functional food or drugs, therefore are regulated differently in countries as per their intended use. Regulations for the probiotic production and release in India have been framed by regulatory bodies of ICMR, DBT, and FSSAI. The outlook of researchers is looking out for commercialization of Indian probiotic strains and a new technique that holds promises to help prolong the shelf-life of probiotic products. This will create more acceptability in consumers and increase the probiotic market.

---

A. Narula (✉)
Department of Biosciences, Mody University, Lakshmangarh, Rajasthan, India

 
P. K. Arora (ed.), *Microbial Products for Health, Environment and Agriculture*, Microorganisms for Sustainability 31, https://doi.org/10.1007/978-981-16-1947-2_4

**Keywords** Probiotics · Prebiotics · Synbiotics · *Lactobacillus*

## 4.1 History of Probiotics

Scientists are studying the flora of the human intestine from long and generally bacteria have always been kindred with the disease and have made the human face much grief. Hence the concept of saddling bacteria for health benefits has a poetic ring to it (Joshi and Pandharbale 2015).

Ellie Metchnikoff has been accredited with the idea of working with food-friendly microbes without knowing the actual hero in the background was Stamen Grigorov. Grigorov was a Bulgarian physician and in 1905 he identified the starter culture (*Lactobacillus*) used in fermented Bulgarian dairy products. He got popularized as he published his work and was then invited to work in the Pasteur Institute of Paris. By the age of 27, he was able to accidentally retrieve that consumption of yoghurt is the secret behind the long life of the Bulgarian people. With this discovery, he was offered a post to work with the Pasteur Institute. But to live up to the promise that he had made to serve his people, he refused the offer and returned to Trun, Bulgaria. Therefore, the institute later authorized Metchnikoff to work on this subject. Later Metchnikoff and his assistants Coendi and Mikelson named the bacteria as *Lactobacillus bulgaricum* in recognition of Grigorov.

In 1917 German Professor, Alfred Nissle, discovered the non-pathogenic *Escherichia coli* Nissle strain during the outbreak of dysentery (shigellosis), using the faeces of two non-affected soldiers at the time of World War I. It was also discovered that *E. coli* Nissle strain 1917 played an important role in the food and medical industries before the antibiotics were discovered. His studies also revealed that probiotics not only could help treat the infectious diseases but also could be used in the medication of other ailments especially related to the GI tract (Sonnenborn and Schulze 2009).

Henry Tissier (working in Pasteur Institute) isolated the *Bifidobacterium* strain from a breast-fed infant and named the bacterium as *Bacillus bifidus communis* in 1889. He claimed that this bacteria helps displace the proteolytic bacteria causing diarrhoea and therefore can be prescribed to the infants suffering from the same. This discovery led to the conceptualization 'specific bacteria play role in maintaining health' (Soccol et al. 2010).

Isaac Carosso, a physician, treated numerous patients with gastrointestinal disorders by recommending yoghurt which helped in recovering the intestinal health. Following the conventional methods, he started producing the yoghurt, for that he procured the purified bacterial cultures from the Pasteur's institute. Thereafter the World War I, he commercialized the yoghurt production entitled 'Danino'—an outlet named after his son (Fuller 1995). Carosso then migrated to the USA from Paris, due to the onset of World War II, and in 1942, he launched 'Dannon Milk Products' which became the first American yoghurt plant.

After World War I, two Armenians named Sarkis and Rose Colombosian had emigrated to the USA and were collaterally working on yoghurt production. They sold their homemade product under the brand named 'Madzoon' meaning yoghurt in Armenian. Madzoon could not arouse the people's interest so to enhance their sales they replaced 'Madzoon' with 'Yogurt' and in 1929 and with their hard work they led to the establishment of their company 'Colombo and Sons Creamery'—became the USA's first yoghurt brand labelled as 'Colombo Yogurt'. Later in 1993 'Colombo Yogurt' was then sold to General Mills (Ozen and Dinleyici 2015).

Meanwhile in the 1930s in the far east Dr. Minoru Shirota isolated *Lactobacillus casei* strain *Shirota* in the Microbiological Laboratory of Kyoto University, Japan. The strain had the property of tolerance towards bile and gastric juices and therefore it could travel easily to the lower intestine, hence with this probiotic bacteria, Dr. Shirota developed the diary product 'Yakult' (Yakult 2014) hypothesizing that its day to day consumption might boost enteric health and extend the lifespan.

It was a turning point and from then on, people have been all about eating probiotics—good microbes—to benefit their health.

## 4.2 What Are Probiotics?

Probiotics derived from 'pro bios' the Greek term meaning 'for life'. The history of probiotics is well associated with the emergence of man; fermented milk and cheese, the concept popular among the Greeks and Romans. Fermentation was not just tasty; it was also known to be healthy and people were consuming fermented foods: beer, wine, yoghurt, cheese, kefir, etc.

*Probiotic* now not a new term, was first introduced by Lilly and Stillwell (Fuller 1989) in 1965 (antonym of the term 'antibiotics') to describe the substances produced by a microorganism that helps stimulate the secretion of another (Soccol et al. 2010). Later different interpretations for probiotics were given by the researchers considering their functioning and their health benefits for humans (Anandharaj 2020).

In 1974, Parker defined 'probiotic' as 'substances and microorganisms which contribute to intestinal microbial balance', whereas Fuller (1989) modified it to as 'viable microbial dietary supplement that beneficially affects the host through its effects in the intestinal tract'. de Vrese et al. (2001) in collaboration with the ILSI (International Life Sciences Institute) of Europe defined this term as 'a viable microbial food supplement which beneficially influences the health of the host' (Salminen et al. 1998). Lately in 2001 FAO/WHO defined probiotics as: 'Live microorganisms that when being administered in appropriate dose, confer benefit of health to the receiver'.

According to the current concurrences correlated to defining probiotics, the out-turn of probiotics is not pondered to be only restricted to microflora mediated, but, other types of mechanisms are getting investigated and familiarized too. This

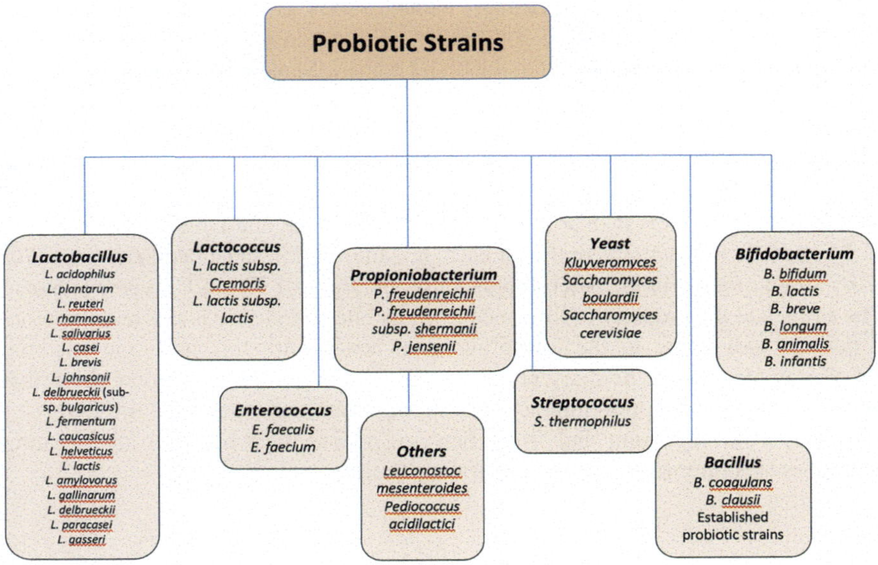


**Fig. 4.1** Probiotic strains used by industries

revelation on probiotics encouraged innovation in this field and opened the doors for a wider range of probiotic possibilities (Sanders et al. 2019).

In normal human intestines where $10^{10}$–$10^{12}$ live microorganisms per gram in the human colon have been reported (Collins and Reid 2016), and around 400 variable bacterial species coexist, making it a more nexus ecosystem. The colon only is approximate to contain above 70% of all the microorganisms in the human body. Generally, the microflora of the gut is persistent but numerous factors including age, environment, diet, stress, and medication can affect the ratio (Anandhraj 2020). The frequently used species in probiotics production belong to bacterial species of *Lactobacillus* and *Bifidobacterium* and some of *E. coli* and *Bacillus* species are recently included, apart from these some of the yeast includes *Saccharomyces cerevisiae, Saccharomyces boulardii, Kluyveromyces* are also used, as they too demonstrated the probiotic properties (Fig. 4.1).

Probiotics are naturally found in dairy and even in non-dairy products, infant formula, and dietary supplements. They are recommended for consumption after the antibiotic treatment (taken for some ailment). This generally eradicates the inhabited good microbial flora of the digestive tract in addition to the targeted harmful microbes. Therefore, probiotic products enriched with beneficial microbes is recommended for regular consumption, manifesting the positive balance of valuable microbes in the intestine.

## 4.3 Criteria's to Be a Probiotic

During recent years, almost every fermented food has been considered to possess the probiotic properties, but not all such products are probiotics. Despite different definitions given for probiotics, there have to be certain fixed criteria's to consider the microbial suspension as a probiotic. These include:

- Should be nontoxic.
- Should be genetically stable.
- Should be a lactic acid producer.
- Should possess antimicrobial activity.
- Should possess lower generation time.
- Should possess forbearance to food additives.
- Should possess stability in the food matrix.
- Should be able to avoid the effect of peristalsis.
- Should be safe for the host/non-pathogenic (GRAS).
- Should be able to evaluate its resistance to antibiotics.
- Should avoid inhibition of adhesion of pathogenic bacteria.
- Should have the power to adhere to epithelial cells and tissue.
- Should be able to produce antibacterial substances (bacteriocin).
- Should have resistance towards gastric acids and pancreatic secretions.
- Should be able to enhance the eradication rate and reduce the adverse effect when given in combination with the antibiotics.

(Pandey et al. 2015; Anandharaj 2020).

Although to fulfil all the criteria is difficult but certain properties are mandatory to be a probiotic.

## 4.4 Concept of Prebiotics and Synbiotics

### *4.4.1 Prebiotics*

The notion of Prebiotics was first defined in 1995 by Glenn Gibson and Marcel Roberfroid. According to them, prebiotics is useful in manoeuvring the microorganisms in the host to ameliorate quantifiable health outcomes. It was reframed by Gibson in 2004 as: 'A prebiotic is a selectively fermented ingredient that allows specific changes, both in composition and/or activity in the gastrointestinal microflora that confers benefits upon host wellbeing and health' (Gibson et al. 2004). 'Probiotics are live microbial feed supplements whereas prebiotics is fibre or dietary carbohydrates'. Prebiotic examples include β-fructans, lactulose, inulin, and GOS that have selective metabolism in the colon and help to escalate the numerical amount of probiotic producing bacteria like LAB (Broekaert and Walker 2006).

Prebiotics may serve as a substitute for probiotics or as ancillary support for them. However different prebiotics will help stimulate the growth of variable native enteric bacteria. Prebiotics has a lower risk of degradation and problems like allergic reactions or intolerance than probiotics due to their fibre constitution. Other beneficial effects of prebiotics include enhanced resistance to invading pathogens, improved bowel movement, lipid reduction, reduced risk of colon cancer, improved calcium, and iron utilization (Bosscher et al. 2003; Ferguson and Philpott 2007; Bruzzese et al. 2009).

### *4.4.2 Synbiotics*

Gibson and Roberfroid inaugurated the term synbiotic in 1995 and was reserved for products where the prebiotic compound(s) selectively favour the growth of the probiotic organism(s) (Cencic and Chingwaru 2010). The concept of synbiotics came into existence to overcome the toil of probiotics, as they are efficient implants in the colon and contribute to maintaining the intestinal homeostasis (Peña 2007).

The probiotic strains used in synbiotic formulations include *Bifidobacteria spp, Lactobacilli, B. coagulans, S. boulardii,* etc., whereas the prebiotics used majorly comprised of xyloseoligosaccharide (XOS), fructooligosaccharide (FOS), GOS, inulin, and even prebiotics from natural sources like yacon roots and chicory, etc., could be incorporated (Pandey et al. 2015). They have many health benefits like cholesterol reduction, antimutagenic effect, anti-hypertension, antibiotic-induced diarrhoea, boosting immunity, overcoming allergy, *Helicobacter pylori* infection, irritable and inflammatory bowel syndrome (Gupta et al. 2014).

## 4.5 Mechanism of Action of Probiotics

The process by which the probiotics deploy their biological effects is still poorly understood. Certain non-specific terminologies like competitive exclusion or colon resistance development are generally used to explain their mode of execution (Elo et al. 1991).

The concept of competitive elimination or exclusion first emerged in the early 1970s (Nurmi et al. 1992). Oelschlaeger (2010) reported three variable modes of probiotic working stating that they:

1. perhaps modulate the host's innate or acquired immune system.
2. directly affect neighbouring microbes, might be commensals or pathogenic ones.
3. could affect the microbial (like toxins) or host products (e.g. bile salts and food ingredients).

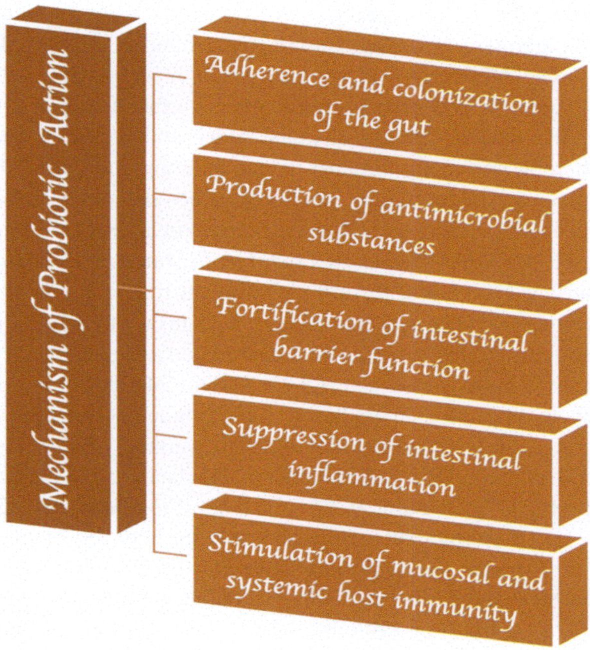


**Fig. 4.2** Mechanism of probiosis in human beings

Individually or combination of such properties in a certain specific probiotic strain could determine its action and serve as a product towards the prevention and/or treatment of a certain disease (Soccol et al. 2010).

The studies have been conducted using different tools and techniques including culturing or sequencing on diversity, composition, and function of the microbial gut flora, and hence the variability in the probiotics mode of operation has been suggested through these variable experimental models. However, the exact mechanisms remain unanswered. Though in 2004 Sartor in his research study has illustrated multiple mechanisms of probiosis, including manipulating gut flora, suppressing pathogen, immune response modulation, inducing epithelial cell proliferation and differentiation, and fortification of gastroenteric hindrances (Fig. 4.2) (Thomas and Versalovic 2010). He has stated some general rules as:

## 4.6 Adherence and Colonization of the Gut

Competition for space does exist among the native and foreign bacteria resulting in knock-back of exogenous pathogens. Certain microbes like *Lactobacilli* or *Bifidobacteria* have the potential to adhere to the mucosal membrane (Ohashi and

Ushid 2009). This enhances the probiotics intestinal endurance and limits the access of pathogens to the epithelium (Ouwehand et al. 2001; Boudeau et al. 2003).

The Glycocalyx is the outermost layer of the cells that are concocted with glycolipids and glycoproteins, plays a crucial role in protecting the epithelial cells of the intestine from mechanical injury, and also helps obstruct the invading bacteria, thereby protecting the host from contamination (Bron et al. 2012). Also during the intestinal inflammation, mucins or glycosylated proteins act as ligands for membrane receptors leading to dysbiosis (Larsson et al. 2011; Sommer et al. 2014). Commonly used probiotic species of *Lactobacillus* and *Bifidobacterium* share some common surface molecules, having a significant role in the interconnection with the mucus components (Lebeer et al. 2010). The surface molecules may include mucin-binding proteins (Mubs), surface layer associated proteins (SLAPs), and the most commonly known lipoteichoic acid (LTA). The initial adhesin is generally non-specific, driven by hydrophobic interactions, once bonded to specific cell wall components like proteinases, adherence becomes irreversible. This helps amplify hydrophobicity and consequently adhesion in some lactic acid bacteria (Radziwill-Bienkowska et al. 2017; Zhang et al. 2015; Muñoz-Provencio et al. 2012).

For example, such adherence is seen between mucins as well as the C-terminal Leu-Pro-any-Thr-Gly motif (LPxTG) of the peptidoglycan layer of the cell wall because these mucus-binding proteins contain domains of Mub and/or MucBP (MUCin-Binding Protein), which can develop this bond. Though MucBP/Mub domains are exclusively discovered in human gut isolated LAB (van Tassell and Miller 2011; Monteagudo-Mera et al. 2019; Boekhorst et al. 2006) they are even observed in the pathogenic bacteria like *Listeria monocytogenes* (Popowska et al. 2017).

Like proteins, bacterial fimbriae or pili can also promote adhesion. Such patterns have been widely characterized in case of both Gram-negative (Type IV pili) and Gram-positive bacteria. For instance, *Bifidobacterium* possesses such type of pili (Piepenbrink and Sundberg 2016; O'Connell Motherway et al. 2011) and more specific example of SpaCBA pili have been observed in some species of Lactobacillus (*Lactobacillus rhamnosus* LGG) (Toh et al. 2013; Reunanen et al. 2012). These structures could possibly be a trump card in colonizing mucosal tissues superficially (Monteagudo-Mera et al. 2019; Hospenthal et al. 2017).

A few other peripheral proteins like surface layer proteins (SLPs) and fibronectin-binding proteins (FBPs) have also been found in contributing to the phenomena of bacterial adherence to the intestinal mucosal sheath. FBPs have been recognized extracellularly in the intestine, in an insoluble form among both Gram-negative and Gram-positive bacteria. These proteins assist in intensifying the process of adhesion which is beneficial for the probiotic bacteria and excludes pathogenic strains (Monteagudo-Mera et al. 2019; Hymes et al. 2016; Lehri et al. 2015). SLP's are extracellular too but on contrary to FBPs have para-crystalline proteins covering the entire bacterial cell surface. These SLP's perform variable functions like generating virulence in pathogenic bacteria or constructing structural components, among these adhesion promoters aid in probiotics functioning. These also act as immunomodulators and assist in probiotic bacterial interaction with host's intestinal receptors (Konstantinov et al. 2008).

Such adhesion mechanisms have been studied by various researchers and have produced fine results.

## 4.7 Suppressing Growth of Pathogenic Bacteria Using Antimicrobial Substances

Antimicrobial agents or Antagonistic compounds are chemical in nature, used to demolish microbes (especially pathogenic) or to prevent their burgeon. Among the broad antimicrobial spectrum properties, this ability is of utmost importance for probiotics functionality (Fijan 2016). Probiotics are more responsive and metabolically active in (in vivo) intestinal environment (Walter et al. 2003; Bron et al. 2004) hence to restrain the epithelial invasion by the pathogens they either instigate cells of the host to produce peptides or directly liberate peptides causing interference in pathogenic activities (Gogineni et al. 2013).

Certain antimicrobial peptides like defensins (hBD protein, elafin, and SLPI), cathelicidins, hydrogen peroxide, lysozyme, nitric oxide, secretory phospholipase A2, and short-chain fatty acids (SCFA), for instance, acetic and lactic acids, expressed constitutively by the Paneth cells (specialized secretory cells of small intestine located in the intestinal crypts of Lieberkuhn). These peptides exhibit antimicrobial activity towards an array of microbes (Kelsall 2008; Furrie et al. 2005) by reducing the lumen acidity (Penner et al. 2005) and help diversify the richness of beneficial gut flora (O'Hara and Shanahan 2007).

The antimicrobial peptides act in the following manner, SCFA hampers the outer membrane of bacteria (Gram-negative) (Alakomi et al. 2000) whereas bacteriocins help create pores disrupting cells (Liévin-Le Moal and Servin 2006) and Microcins along with the structural synthesizing enzymes (of DNA/RNA) attacks the inner membrane (Duquesne et al. 2007).

## 4.8 Fortification of Intestinal Barrier Function

Intestinal epithelial cells have a role in both as a barrier and immunomodulator in the gut, as epithelial and immune cells can interact and influence each other. Microbe as a whole with its structural components or metabolites produced can stimulate the epithelial cell signalling pathways (Madsen 2012). Some of the probiotics have been advised in preserving epithelial barrier function, safeguard and reformation of the damaged mucosal sheath, incited by various factors including enteric pathogens, drugs, food antigens, or pro-inflammatory cytokines (Resta-Lenert and Barrett 2006; Rosenfeldt et al. 2004; Montalto et al. 2004; Resta-Lenert and Barrett 2003).

Probiotics help serve to combative effects which are mediated by successive mechanisms (Madsen 2012):

1. mucus secretion by goblet cells (Chichlowski et al. 2007),
2. maintaining cytoskeletal and tight junction proteins by phosphorylation (Brown 2011),
3. refurbishing chloride secretion,
4. enhancing trans-epithelial resistance (O'Hara and Shanahan 2007).

## 4.9 Suppression of Intestinal Inflammation

Researchers have presented sufficient scientific evidence supporting the role of probiotics in mucosal inflammation, particularly by restraining or restoring 'leaky' epithelial barriers (Leisched 2014). The anti-inflaming property of probiotics negates the source of pro-inflammatory stimuli and is used as therapeutic against chronic diseases like Gastroenteritis, Inflammatory bowel syndrome, Lactose Intolerance, UTI's, etc.

Combining comprehensively researched probiotic strains can assist to control and restore inflammation directly and indirectly by various modes such as:

1. arresting probable key stimulant of acute inflammation, including LPS (Claros et al. 2013),
2. simultaneously modulating multiple signalling pathways (Bermudez-Brito et al. 2012; Thomas and Versalovic 2010),
3. yielding short-chain fatty acids with anti-inflaming properties (e.g. butyrate),
4. synthesis of antimicrobial peptides (Leisched 2014),
5. synthesis of heat shock proteins (Ohland and Mac Naughton 2010; Rao and Samak 2013),
6. increased expression of mucins (Ohland and Mac Naughton 2010; Rao and Samak 2013),
7. release of metabolites and bioactive molecules (Ohland and Mac Naughton 2010; Rao and Samak 2013),
8. suppression of oxidative stress (Ohland and Mac Naughton 2010; Rao and Samak 2013),
9. interference with inflammatory pathways (Ohland and Mac Naughton 2010; Rao and Samak 2013),
10. augment levels of IgA (Ohland and Mac Naughton 2010; Rao and Samak 2013),
11. acting as a ligand for Toll-Like Receptors (TLR) influencing key pro-inflammatory pathways (Thomas and Versalovic 2010; Bermudez-Brito et al. 2012),
12. Influence development, maturation, and differentiation of dendritic and T-cells (Bermudez-Brito et al. 2012; Thomas and Versalovic 2010).
13. Influence synthesis of the important regulatory cytokines like IL-10 and TGF-β (Smits et al. 2005; Hseih et al. 2012).

## 4.10 Stimulation of Mucosal and Systemic Host Immunity

The researchers have demonstrated that on Oral administration of a fragment of probiotic bacteria only, a complex network of signals gets initiated inside the Interstitial Epithelial Cells (IECs). The commensal bacteria through pattern recognition receptors leads to (1) immune engagement and demonstrable systemic immunologic changes (McCarthy et al. 2003) associated with the tissue of GALT in the lamina propria like mucosal; (2) immune development; (3) to maintain and repair gut (Rakoff-Nahoum et al. 2004; Fukata et al. 2005); and (4) activating mainly the innate response and the cytokines released by T-cells (Galdeano and Perdigón 2004). Even the immune sensory cells (i.e. dendritic cells, M cells, and enterocytes) in the alimentary canal constantly respond to intestinal bacteria (O'Hara and Shanahan 2007).

Transmitting the antigenic information to the underlying cells is a crucial process towards immune fate: leading to activation versus suppression/tolerance. Three such mechanisms are involved which help process the antigenic material and lay before the underlying immune cells. These aforesaid mechanisms are further controlled by distinct antigen-presenting cells (APCs) (Hardy et al. 2013).

Consuming probiotic strains '*Bifidobacterium lactis* Bb-12, *Lactobacillus* GG (Rautava et al. 2006), and *Saccharomyces boulardii'* (Rodrigues et al. 2000) alleviates the production of IgA and its secretion via cytokine environs causes an alteration in the gut mucosa. These bacteria manifest epithelial cell expression of interleukins IL-6, IL-10 as well as TGFβ (transforming growth factor-β) and help favour IgA production through the medium of antibody class-switching, B-cell maturation phenomena (Shang et al. 2008; He et al. 2007). Finally augmenting polymeric Ig receptors expression into the gut lumen (Reséndiz-Albor et al. 2010).

## 4.11 Health Benefits by Probiotics

The intestinal tract harbours a complex and dynamic microbial ecosystem, capable of producing elevated concentrations of chemicals for detection and signalling particles—molecules affecting cells in the entire body. The bacteria lying in the gut produces some influential proteins which can affect the chemical and signalling molecules of the intestine either positively or negatively (Yan et al. 2007). If the ratio of good-to-bad bacteria shows disparity, initiates the activity of many of these detectors probably in negative ways, triggering a host of diseases, not just those associated with the gut but even in other body areas too (Furness et al. 1999). Fortunately, the use of probiotics can restore the replenished good bacteria and reverses the signalling which can lead to chronic diseases (Kotzamanidis et al. 2010; Ley 2010; Vyas and Ranganathan 2012; Mortaz et al. 2013). The summary for such diseases along with the probiotic use has been stated below (Table 4.1):

**Table 4.1** Use of probiotics for treatment of various diseases

| Sr. No | Disease | Cause | Clinical Symptoms | Probiotic Microbes Used in treatment | References |
|---|---|---|---|---|---|
| 1 | Chronic Diarrhoea | Small bowel bacterial overgrowth | Loose or watery stools that persist for weeks, abdominal cramps, bloating, nausea | *Lactobacillus acidophilus* and *Lactobacillus casei*<br>*Saccharomyces boulardii* | Gaon et al. (2002), Xiao et al. (2003), Le Luyer et al. (2010) |
| 2 | Inflammatory bowel disease (IBD) | Use of antibiotics caused inflamed mucosal tissue decreasing the count of lactobacilli | Diarrhoea, abdominal pain and cramping, fever, fatigue, blood in your stool, anorexia, and weight loss | VSL#3 *Bifidobacterium* strains, four lactobacilli strains, and one Streptococcus strain) | Favier et al. (1997), Gionchetti et al. (2000) |
| 3 | Irritable bowel syndrome (IBS) | Weak intestinal muscle contractions, poor coordinated signals between brain and intestines, inflammation in intestine, severe infection, and change in the gut microflora | Gas, cramps, abdominal pain and swelling, and constipation or diarrhoea, or even both | *Propionibacterium freudenreichii* ssp. *shermanii* JS, *L. rhamnosus* LC705, *Bifidobacterium breve* Bb99, and GG<br>*Lactobacillus plantarum* significantly lowered flatulence | Kajander et al. (2005), Nobaek et al. (2000) |
| 4 | Lactose intolerance (LI) | Small intestine gets deficient of lactose causing indigestion of milk sugar, also leads to calcium deficiency causing osteoporosis | Borborygmic, abdominal pain and distension, flatulence, and diarrhoea (between 30–-120 minutes of lactose ingestion) | *Lactobacillus acidophilus* W22, *Lactobacillus acidophilus* W70<br>*Lactobacillus salivarius* W24<br>*Streptococcus thermophilus* W69<br>*Bifidobacterium lactis* W52<br>*B. animalis*<br>*Streptococcus thermophilus, Lactobacillus reuteri,*<br>*B. longum, Lactobacillus bulgaricus, Lactobacillus rhamnosus, Lactobacillus acidophilus* | Besseling-van der Vaart, (2016), Monteagudo-Mera et al. (2019), Le Luyer et al. (2010), de Vrese et al. (2001), He et al. (2008), Rampengan et al. (2010), Ojetti et al. (2010), Parra and Martınez (2007), Vesa et al. (1996), Agustina et al. (2007) |

| | | | | | |
|---|---|---|---|---|---|
| 5 | Atopic dermatitis (cow's milk allergy) | Lactobacilli and Bifidobacteria is significantly reduced | Eczematous skin lesions, increased serum IgE levels, or family history.<br>Environmental factors: Microbes, irritants, and extremes of temperature, psychological stress, and food allergens | *L. rhamnosus* GG<br>*L. reuteri* and *L. rhamnosus* GG<br>*L. paracasei* and *L. rhamnosus* GG | Viljanen et al. (2005), Rosenfeldt et al. (2003), Myllyluoma et al. (2005) |
| 6 | Urinary tract infections (UTI) | Bacterial infection in kidneys, bladder, and urethra, generally occurs in young and pregnant women | Inflammation and irritation, painful and frequent urination, urinary incontinence, bad-smelling urine, haematuria, mild fever, pain or pressure in lower abdominal, and dysuria | *L. Rhamnosus and*<br>*L. fermentes*<br>*E. coli*<br>*Lactobacillus fermentum RC-14 and Lactobacillus rhamnosus GR-1*<br>*Lactobacillus rhamnosus (1 x 109 CFU per 1 billion) and a Lactobacillus reuteri*<br>*L. Acidophilus 01; Lactobacillus salivarius UCM572, and L. plantarum CLC17* | Reid et al. (2001), Morelli et al. (2004), Darouiche et al. (2001), Hull et al. (2000), Reid et al. (2001), Anukam et al. (2006), de Llano et al. (2017) |
| 7 | Crohn's disease | Inflammation of digestive tract (small intestine and colon), significant reduction in faecal bifidobacteria | Enduring diarrhoea, often turns to dysentery, weight reduction, fever, abdominal pain, tenderness, and hematochezia. | *Lactobacillus rhamnosus* GG | Favier et al. (1997) |
| 8 | Hepatic encephalopathy (HE)/ cirrhosis | Dreaded (chronic) complication of liver disease, cannot remove toxins and hence decline in brain function | Disoriented, personality changes, lack of focus, confusion, amnesia, anxiety, seizures, fatigue, and shaky hands. | *Clostridium* cluster I and *Bifidobacterium*<br>Lactobacillus acidophilus<br>*Enterococcus faecium S68/lactolus* | Xia et al. (2018), Ziada et al. (2013), Loguercio et al. (1995) |

(continued)

**Table 4.1** (continued)

| Sr. No | Disease | Cause | Clinical Symptoms | Probiotic Microbes Used in treatment | References |
|---|---|---|---|---|---|
| 9 | Acute pancreatitis (AP) | Acute abdomen inflammation of pancreatic and peri-pancreatic tissues | Gall stones, sudden immune system attack, pancreatic or gall bladder damage due to surgery/injury; excessive fat (triglycerides) in blood; alcohol abuse, cystic fibrosis, etc. | *Streptococcus Thermophilus, Lactobacillus acidophilus, and Bifidobacterium lactis* | Muftuoglu et al. (2006) |
| 10 | Dental caries/ periodontal diseases | Results from a homeostatic imbalance between the host and microbiota causing plaque formation or plaque attacks | Toothache, tooth sensitivity; tooth cavity; fluorosis staining | *Lactobacilli* spp. and *bifidobacteria* spp.<br>*Streptococcus dentisani*<br>*Lactobacillus rhamnosus* GG, *Lactobacillus reuteri,* and *Bifidobacterium*<br>*Streptococcus thermophilus*<br>*L. Paracasei SD1* | Twetman et al. (2009), Keller et al. (2011), Hasslof et al. (2010), Lopez-Lopez et al. (2017), Näse et al. (2001), Caglar et al. (2005), Cotter and Hill (2003), Arioli et al. (2010), Wattanarat et al. (2015) |
| 11 | Anaemia | Caused by blood loss, decreased or faulty RBC production or destruction of RBCs | Dizziness, fast or unusual heart pulsation, body pain, joints pain, headache, difficulty in breathing, pale skin, fatigue, cold hands and feet, growth maturation problems for children and teens | *Lactobacillus plantarum* 299v *Streptococcus thermophilus*, *L. acidophilus*, *Lactobacillus bulgaricus*, *Lactobacillus casei*, *Lactobacillus rhamnosus*, *Bifidobacterium longum*, and *Bifidobacterium breve*<br>*B. lactis* W52, *Bifidobacterium bifidum* W23, *Lactobacillus acidophilus* W37, *B. lactis* W51, *L. casei*W56, *L. brevis* W63, *Lactococcus lactis* W19, *L. salivarius* W24, and *Lc. lactis* W58 | Korčok et al. (2018), Shariaty et al. (2017), Skrypnik et al. (2019) |

| | | | | | |
|---|---|---|---|---|---|
| 12 | Diabetes | Metabolic syndrome causing increased blood sugar level.<br>Type I: Autoimmune disorder<br>Type II: Genetic and lifestyle disorder | Increased hunger and thirst, weight reduction, dysuria, blurry vision, extreme exhaustion, wound/sore healing delayed, decreased eroticism (especially in men), ED and poor muscle strength whereas women have UTI, yeast infection, dry and itchy skin | *Lactobacillus plantarum* DSM 15313 and L gasseri BNR17<br>*L. rhamnosus* GG<br>*Lactobacillus acidophilus* and *L. casei* | Andersson et al. (2010), Yun et al. (2009), Tabuchi et al. (2003), Yadav et al. (2007), Yadav et al. (2006) |
| 13 | Obesity | Consuming more calories than burn causes excessive amount of body fat deposition, no physical exercise, medication or it may be genetic, behavioural, metabolic, and hormonal influences on body weight, pregnancy, PCOs, hypothyroidism | Body mass index (BMI) is 30 or higher | *L. gasseri, Lactobacillus (L. casei strain Shirota (LAB13), L. plantarum, and L. rhamnosus, among others) and Bifidobacterium (mainly B. longum, B. infantis, and B. breve B3)*<br>*Lactobacillus curvatus HY7601 and Lactobacillus plantarumKY1032*<br>*Lactobacillus paracasei CNCM I-4270 and Lactobacillus rhamnosus CNCM I-3690*<br>*Bifidobacterium* spp. (*B. longum* SPM 1205, *B. pseudocatenulatum* SPM 1204, and *B. longum* SPM 1207 or *B. adolescentis*<br>*Pediococcus pentosaceus* LP28<br>*Bacteroides uniformis* CECT 7771<br>*Akkermansia muciniphila* | Ejtahed et al. (2019), Barengolts (2016), Chen et al. (2012), Yoo et al. (2013), Wang et al. (2015), An et al. (2011), Chen et al. (2012), Zhao et al. (2012), Gauffin et al. (2012), Everard et al. (2013) |

(continued)

**Table 4.1** (continued)

| Sr. No | Disease | Cause | Clinical Symptoms | Probiotic Microbes Used in treatment | References |
|---|---|---|---|---|---|
| 14 | Cardiovascular diseases | High blood pressure, diabetes, smoking, high cholesterol, laziness, obesity, inherited | Chest pain, tightness, and chest discomfort (angina), difficulty in breathing; loss of sensation, weakness in your legs or arms due to narrowed blood vessels; pain in the neck, jaw, throat, belly, or back. | *Lactobacillus, Lactococcus, and Bifidobacterium*<br>*L. fermentum* NCIMB 5221 and NCIMB 2797<br>*L. reuteri* NCIMB 30242 | Ishimwe et al. (2015), Kumari et al. (2011), Tomaro-Duchesneau et al. (2015), Ryan et al. (2015) |
| 15 | Cancer issues | Uncontrolled cell growth, mutations, error in DNA repair | Lethargy, thickening under the skin, unintended weight loss or gain, delayed sore healing, dyspepsia, croakiness, constant trouble in breathing or coughing, skin discolouration (yellowing, darkening, or redness), changes in existing moles, incontinence bowel or bladder habits, extreme muscle or joint pain, inexplainable fevers or night sweats and bleeding, dysphagia | *Lactobacillus acidophilus*, and *L. casei*;<br>*L. rhamnosus* GG and *L. casei*<br>*L. rhamnosus* GG and<br>*L. paracasei*IMPC2.1<br>*L. casei* ATCC 39392 and<br>*L. acidophilus* ATCC 4356<br>*L. Plantarum*<br>*L. johnsonii* BCRC17010<br>*Pichia kudriavzevii* AS-12<br>*Saccharomyces cerevisiae*<br>*Pediococcus pentosaceus* FP3 | Baldwin et al. (2010), Escamilla et al. (2012), Orlando and Refolo (2012), Soltan Dallal et al. (2015), An and Ha (2016), Chen et al. (2017), Saber et al. (2017), Sambrani et al. (2019), Thirabunyanon and Hongwittayakorn (2013) |
| 16 | Atopic dermatitis (AD/eczema) | May be related to gene variation or environmental factors, irritants, and allergens | Dry skin, itching predominantly at night, red or grey patches on hands, wrist feet, neck, bends of elbow, chest, eyelids, ankle, and knees especially in infants, face, scalp, or swollen skin may leak if scratched excessively | *Bifidobacterium longum* BB536 and *Bifidobacterium breve* M-16 V<br>*Lactobacillus fermentum, Lactobacillus salivarius*<br>*Lactobacillus reuteri* DSM 122460; *Lactobacillus rhamnosus* 19,070–2 | Enomoto et al. (2014), Huang et al. (2017), Jacobsen et al. (1999), Rosenfeldt et al. (2003) |

| | | | | | |
|---|---|---|---|---|---|
| 17 | Arthritis | Reduction in cartilage tissue, infection or injury in joints rheumatoid arthritis (RA) is autoimmune disorder-attacks synovium. | Joint pain, stiffness, swelling, redness around joint; in case of RA it causes tiredness, loss of appetite, lower RBC, and fever at times. | *Bifidobacterium bifidum, Lactobacillus casei,* and *Lactobacillus acidophilus* *Lactobacillus casei* | Zamani et al. (2016), Wescombe et al. (2009) |
| 18 | Gastritis | *Helicobacter pylori* infection, regular use of pain killers, excessive alcohol consumption | Nausea, abdominal pain and bloating, recurrent upset stomach, vomiting, dyspepsia, heartburn in stomach usually after meals, hi-cough, vomiting, anorexia, black and tarry stools | *Lactobacillus reuteri* "Folk yogurt", containing yeast and lactobacilli | Mukai et al. (2002), Oh et al. (2002) |
| 19 | Chronic fatigue syndrome (CFS) | Viral attack, weak immune system, stress, and hormonal imbalances | Chronic insomnia, other sleep disorders, loss of memory, reduced concentration, orthostatic intolerance, muscle pain, sore throat, multi-joint pain, and swollen lymph nodes | *Lactobacillus casei strain Shirota* (*LcS*) | Rao and Samak (2013) |

## 4.12 Probiotic Scenario in India

'Probiotics'—the term has created a buzz as it has entered the Indian food market since 2007. Though this merchandise in India is diminutive (only 1%) in comparison with the Western countries, at the moment it is all set to shoot up. The world's largest cattle population exists in India hence the largest producer of milk. Therefore, India is progressing to be a future prime probiotic market and can play a crucial role in the probiotic revolution as it is the rapidly expanding arena among the functional foods. Its annual growth rate of 13.56% was observed in 2019 and is expected to reach a market size of US$961.856 million in 2025 (Research and Market Reports, June 2020). Yakult, Danone, Nestle, Amul, and Mother Dairy are the common probiotic brands in India which have made their remark of recognition in this industry. Minor brands like Carbamide Forte, Wow Probiotics, Nature's Velvet, BoldFit, and many others are too heading towards better quality pre and probiotic products.

These well-known probiotic industries presently are using the outsourced strains, i.e. microbes of foreign origin. Scientists believe that the intestinal environment is home for gut flora as the members of a specific community have originated (from different food habits) and henceforth well accommodated, giving tough competition to the evolved probiotic bacteria. Though they have proven results adapting to the Indian gut easily since, isolated from a similar gut environment. But the probiotics in either form have to be taken daily as they can only transiently colonize the gut and presumed to be washed out anyway. Not all the scientists and researchers would agree, as there is no proof, this statement is merely based on the existing knowledge.

On that account, scientists of different Indian universities like NDRI (Karnal), Christian Medical College (CMC) in Vellore (Tamil Nadu), Anand Agricultural University (Gujarat), and some others are working on the optimization and production of Indian probiotic strains. For decades, Virender K Batish (Emeritus scientist) and his colleague Sunita Grover (Principal scientist) from NDRI (Karnal) have been working in this field and have generated a repository collection of 120 types of bacteria. They have planned to market *Lactobacillus fermentum*-1 (Lf-1) and *Lactobacillus plantarum*-91 (Lp-91) in near future for colitis and cholesterol reduction, respectively. These strains have shown motivating results in animal trials and studies are continuing on humans. They are even working on probiotics to overcome India's burgeoning obesity epidemic.

Not only in NDRI but other institutes like in Anand Agricultural University, Gujarat, the head of the dairy microbiology department, J B Prajapati, is working on species of Lactobacilli for the past 25 years. His team has standardized and very well incorporated these beneficial probiotic bacteria (MTCC 5463- L. helveticus, MTCC 5462- L. rhamnosus) into curd, buttermilk, and lassi (sweet buttermilk). The role of these microbes is now even tested for improving immunity especially in people between the age of 65 and 75 years. B.S. Ramakrishna (HOD Gastroenterology, CMC) is also working on the Lactobacillus genus but has studied various bacteria isolated from Indian dairy animals. His work is in the nascent stage, will take around the next 5–10 years to reach out to the market. These researchers are facing certain bottlenecks mainly funds for biosafety trials and tie-up with a commercial entity.

Companies in India have launched a range of probiotic products, Amul has introduced the concept of ice-creams and lassi enriched with probiotic. B-Activ Probiotic Curd, b-Activ Dahi, b-Activ Probiotic Lassi, and Nutrifit (Strawberry and Mango) are the Mother Dairy's range of probiotic products. Nestle launched Nesvita-India's first probiotic Dahi. Yakult, Danone Group's probiotic drink is prepared using Lactobacillus, some sugar, and fermented milk. The probiotic drug market is also emerging with companies like Ranbaxy (Binifit), Dr. Reddy's Laboratories, comprising four sub-brands: Unichem, Zydus Cadila, Glaxo SmithKline, and JB Chem. Major pharmaceutical companies have come in-action and attempting to formulate newer supplements aiming specific needs like immunodeficiency and gastrointestinal problems. These products are listed in Table 4.2.

## 4.13 Regulatory Guidelines for Probiotics in India

Industrialization based on probiotics is enhancing in India as well as in other countries, but the status for the release of any probiotic product is still ambiguous. Probiotics nowadays are being produced under variable categories of food products (functional foods), nutraceuticals, health supplements, or energy drinks (examples as stated above in Table 4.2). This being the case, are regulated diversely in countries and as per their desired use. In India, products enriched with pre and probiotics are in huge demand due to their unusual health care benefits and some claims to cure certain diseases (but not all of them are certified). Though, in India, they are recommended once in a while by doctors as part of medicament, not as a drug substitute. Henceforth, these claims by manufacturers and the absence of particular regulations have made the regulating bodies in various countries to elucidate parameters and guidelines. These should be on par with drugs to regulate their safety, efficacy, claims, and quality (Gokhale and Nadkarni 2007).

Initially, with the advent of probiotics in India, there were no specific regulations. So for evaluating the safety and to avoid popularization of probiotic products with false claims, the Indian Council of Medical Research (ICMR) in association with the Department of Biotechnology (DBT) constituted a task group, framing the regulatory guidelines for probiotic production in India (Arora and Baldi 2015).

## 4.14 The ICMR-DBT Guidelines Are Stated Under the Following Sub-Requirements as

(a) *Identification of Genus, Species, and Strain for Probiotic Use*:

**Table 4.2** A list of commercial probiotics

| Sr. No. | Probiotic Company | Incorporated strains | Consumable Amount | Product claims | No of strains used | Contains |
|---|---|---|---|---|---|---|
| **Probiotic capsules for men and women** | | | | | | |
| 1 | Cabamide forte | L. Plantarum, L. acidophilus, L. rhamnosus, L. gasseri, L. fermentum, L. casei, L. reuteri, L. paracasei, B. lactis, B. infantum, B. infantis, B. bifidum, B. longum, B. breve, S. thermophilus, and S. boulardii | 30 billion CFU/capsule | Post-workout supplement, digestion and immunity booster | 16 | 0.20 g carbohydrates, 0.59 g fat, 0.03 g protein, and 6.82Kcal of energy |
| 2 | Wow probiotics | L. Plantarum, L. casei, L. rhamnosus, L. reuteri, L. salivarius, L. fermentum, L. paracasei, L. gasseri, B. lactis, B. infantis, L. acidophilus, B. bifidum, B. breve, and S. thermophilus | 20 billion CFU/capsule | Improves immunity, effective for digestion, constipation and maximizes nutrient absorption. | 14 | 2000Kcal |
| 3 | Dr. formulated probiotic for men | L. Plantarum, L. rhamnosus, L. gasseri, L. fermentum, L. casei, L. paracasei, L. brevis, L. bulgaricus, L. salivarius, B. lactis, B. infantis, L. acidophilus, B. bifidum, B. breve, and B. longum | 50 billion CFU/capsule | Supports colon health, improve the digestive and immune system, and designed for men's daily gastrointestinal problems | 15 | |

| | | | | | | |
|---|---|---|---|---|---|---|
| 4 | Now foods probiotic | L. Plantarum, L. rhamnosus, L. salivarius, L. casei, L. paracasei, B. breve, B. lactis, L. acidophilus, B. longum, and S. thermophilus | 25 billion CFU/capsule | Boost your immunity, digestive system, and gut health, supports breakdown, delivery, and utilization of nutrients obtained from diet | 10 | |
| 5 | Cabamide forte probiotic supplement | Saccharomyces boulardii, L. acidophilus, L. rhamnosus, and B. longum | 2.75 billion CFU/capsule | Relives from unwanted gas and acidic pain, improve your immunity system, digestion power, and gut health | 4 | 4.28 kcal, 0.15 g carbohydrate, 0.02 g protein, and 0.38 g fats |
| 6 | Inlife pre and probiotic supplement | Saccharomyces boulardii, L. rhamnosus, B. bifidus, B. longum, and L. acidophilus | 2.75 billion CFU/capsule | Keeps you away from digestion and gas-related issues | 5 | 0.545 kcal, 0.062 g carbohydrate, 0.0355 g protein, and 0 g fats |
| 7 | Simply herbal probiotic | L. Plantarum, L. fermentum, L. acidophilus, L. rhamnosus L. casei, L. reuteri, L. salivarius, L. paracasei, L. gasseri, B. breve, B. infantis, B. lactis, B. bifidum, and S. thermophilus | 25 billion CFU/capsule | Easier digestion, stronger immunity and fuller nutrition, this formula can reach the large intestine making it 20% more effective | 14 | |
| 8 | Complete probiotics by Purayati | B. bifidum, L. rhamnosus B. longum, L. acidophilus, and Saccharomyces boulardii | | Friendly bacteria maintains intestinal microflora and play major role in metabolizing vitamins and collect energy from undigested carbohydrates | 5 | 0.36 kcal, 90 mg carbohydrate, 0 g protein, and 0 g fats |
| 9 | Daily probiotic by my protein | Lactobacillus salivarius, Bifidobacterium bifidum, Lactobacillus acidophilus, | | Supports digestive system and builds muscle when | 10 | |

(continued)

**Table 4.2** (continued)

| Sr. No. | Probiotic Company | Incorporated strains | Consumable Amount | Product claims | No of strains used | Contains |
|---|---|---|---|---|---|---|
| | | Lactobacillus plantarum, Lactobacillus casei, Lactobacillus rhamnosus, Bifidobacterium breve, Bifidobacterium longum, Lactobacillus bulgaricus, and Lactobacillus lactis | | taken in addition to protein sources | | |
| 10 | Nature's velvet probiotic | Bacillus coagulans | | | | |
| 11 | Hawaiian herbal probiotic plus powder | Bifidobacterium lactis, Bifidobacterium longum, Lactobacillus paracasei, lactobacillus plantarum, and Lactobacillus acidophilus | Not FDA approved | Along with an immunity booster, useful in numerous health issues including Crohn's disease, diarrhoea, inflammatory or irritable bowel syndrome, peptic ulcers, necrotizing enterocolitis (generally in premature babies), urinary tract infections, respiratory infections, milk intolerance, eczema, hypercholesteremia, prevents helicobacter pylori infections | 5 | |
| 12 | Women's probiotic | Lactobacillus rhamnosus LR-32, Lactobacillus acidophilus LA-14, Bifidobacterium bifidum BB-06, Lactobacillus casei LC-11, Lactobacillus | 5 billion CFU/capsule | | | |

| | | | | | | |
|---|---|---|---|---|---|---|
| | | plantarum LP-115, Lactobacillus salivarius LS-33, Lactobacillus paracasei LPC-37, Bifidobacterium lactis BL-04 | | | | |
| 13 | Flora⁺ probiotic | Bifidobacterium lactis BL-04, Lactobacillus rhamnosus LR-32 | 4.5 B/CFU and 1.8 B/CFU, respectively/capsule | Contributes to natural healthy gut flora, restores and maintain digestive, respiratory, and immune health, supports brain gut axis | 2 | |
| 14 | Kapiva probiotic with Amla gummies | Lactowise (Bacillus organisms/Galactomanan) | 3 billion CFU/Gummie | Combined with ripe amlas it provides a natural solution for strong gut health, in a convenient gummy form. These break down the bile in the gut and reduce the level of bad cholesterol in the gut | | 23.20 kcal, 5.80 g carbohydrate, 0 g protein, 4.10 g sugars, amla 200 mg, 0 g fats, and Lactowise 30 mg |
| 15 | Boldfit probiotics supplement | L. Acidophilus, L. casei, L. rhamnosus, L. gasseri, L. plantarum, L. fermentum, L. reuteri, B. lactis, B. infantum, B. infantis, B. bifidum, B. longum, B. breve, L. paracasei, S. thermophilus, and S. boulardii | 30 billion CFU/capsule | Balances optimal digestive health leads to healthy and light gut, improves immunity, and supports energy level and heart health | 16 | 6.82 kcal, 0.20 g carbohydrate, 0.03 g protein, and 0.59 g fats |
| 16 | Herbora probiotics for children | Bifidobacterium infantis, Bifidobacterium animalis ssp. lactis, Lactobacillus casei, Lactobacillus acidophilus, Lactobacillus reuteri | 7 x 109 CFU /g, 7 x 109 CFU /g, 7 x 109 CFU / g, 7 x 109 CFU /g, 7 x 109 CFU /g, respectively | Preventive treatment for infections and loss of intestinal flora, after an infectious process, diarrhoea, gas, digestive discomfort, etc., including colitis of the baby | 5 | |

(continued)

**Table 4.2** (continued)

| Sr. No. | Probiotic Company | Incorporated strains | Consumable Amount | Product claims | No of strains used | Contains |
|---|---|---|---|---|---|---|
| 17 | Complete pre and probiotic by Olympian lab Inc. | Bifidobacterium lactis HA-194, L. rhamnosus HA-111, L. plantarum HA-119, L. acidophilus HA-122, Bifidobacterium longum HA-135 | 25 billion CFU/capsule | The live bacteria of probiotic with the energy source of prebiotic—Helps to maintain optimal digestive growth | 5 | |
| 18 | Healthy hey nutrition probiotic | Lactobacillus sporogenes (Bacillus coagulans) | 20 billion CFU/capsule | Temperature stabilized capsules for digestion and immune health. Delayed release capsule, hence protects it from stomach acid. Helpful in constipation, diarrhoea, and overall health | 1 | |
| 19 | Health XP pre and probiotic | Lactobacillus sporogenes | 20 billion CFU/capsule | Boosts immunity and fights fatigue, provides relief from gas pain, acidity, stress, and strain of antibiotics by promoting digestion | 1 | 0.84 kcal, 0.21 g carbohydrate, 0 g protein, and 0 g fats |
| 20 | Probiotic immune by zenith nutrition | L. Acidophilus, B. bifidum, L. rhamnosus, and B. longum | 2 billion CFU/capsule | Helps maintain intestinal flora, dietary supplement | 4 | |
| 21 | Acidophilus probiotic by Nature's bounty | Lactobacillus acidophilus | 100 million CFU/capsule | Stomach friendly, supports overall well being | 1 | |
| 22 | BigLac | Bifidobacterium Bifidum, Lactobacillus acidophilus, | – | Diarrhoea due to excessive antibiotics consumption | 4 | |

| | | | | | | |
|---|---|---|---|---|---|---|
| | | Saccharomyces Boulardii, and Streptococcus Thermophilus. | | | | |
| 23 | New chapter probiotic AllFlora | Bifidobacterium bifidum, Lactobacillus plantarum LP01 Bifidobacterium breve BR03, Lactobacillus rhamnosus and Lactobacillus acidophilus, Saccharomyces boulardii | 10 billion CFU/capsule | Boosts immune defences and promotes digestive health including bowel regularity, bloating, diarrhoea, gas, and constipation, helps to reduce problematic yeast like Candida while replenishing healthy gut bacteria | 6 | |
| 24 | Meadberry probiotic | L. Plantarum, L. reuteri, L. rhamnosus L. casei, L. salivarius, L. paracasei, L. gasseri, L. acidophilus, L. fermentum, B. breve, B. infantis, B. lactis, B. bifidum, and S. thermophilus | 20 billion CFU/capsule | Dietary supplement for moderate working men and women | 14 | 2.0 kcal, 0.122 g carbohydrate, 0.086 g protein, and 0 g fats |
| 25 | GNC probiotic complex | L. Acidophilus (CUL-21), B. animalis subsp. lactis (CUL-34), B. bifidum (CUL-20), L. acidophilus (CUL-60) | 25 billion CFU/capsule | Provides digestive and immune support | 4 | |
| **Probiotic drinks for men and women** | | | | | | |
| 1 | Yakult | Lactobacillus casei strain Shirota (LcS) | 6.5 billion CFU/serving | Prevents digestive disorders such as constipation, diarrhoea, boosts immunity and reduces infection. | | 50 cal/bottle, 12 g carbohydrate, 1 g protein, and 0 g fats |
| 2 | GoodBelly | L. Plantarum (LP299V) | 12 billion CFU/serving | Delicious probiotic drink in different flavours, suitable for | | |

(continued)

**Table 4.2** (continued)

| Sr. No. | Probiotic Company | Incorporated strains | Consumable Amount | Product claims | No of strains used | Contains |
|---|---|---|---|---|---|---|
| | | | | expecting mothers and improves digestive health. | | |
| 3 | Beyond berry | | 38 CFU/serving | Excessive inflammation, helps protect cells from oxygen damage, improves potential to respond to stress, supports your adrenal glands, enhances energy, protects against radical damage, brings mental clarity and focus, cellular degeneration, and ageing, strengthens your immune system, rich in antioxidants prevents free radical damage | | Low calorie |
| 4 | DanActive by Danon | Lactobacillus casei Immunitas® (L. casei DN-114001) | 10 billion CFU/serving | Provides required nutrients like calcium, survives and remains active in intestine | | |
| 5 | Pre probiotic enhancer | | 2.5 billion CFU/serving | Variety of flavours, healthy boost to your digestion at any time of the day | | Zero calorie drink |
| 6 | Kyo Dophilus probiotics plus fibre | B. longum, L. gasseri, and B. bifidum | 1 billion CFU/serving | Fibre-rich and improves the functioning of the digestive system, it enhances immunity and promotes good health | | 15 calories, 3 g carbohydrate, 3 g dietary fibre |

| | | | | | | |
|---|---|---|---|---|---|---|
| 7 | PHD 2Go probiotic drink | | 20 billion CFU/serving | Helps to detox your digestive system, strengthens immunity, improves overall health | 6 | |
| 8 | CoCo biotic by body ecology | Lactobacillus delbrueckii and Lactobacillus acidophilus | 4 billion CFU/serving | Improves digestion, reduces sugar cravings, energy booster and improves liver cleansing | | 10 calories, 3 g carbohydrate, 2 g sugar, 10 mg sodium, and 22 mg potassium |
| **Probiotics for children** | | | | | | |
| 1 | Children's probiotic | Lactobacillus bulgaricus LB-87, Lactobacillus casei LC-11, Lactobacillus acidophilus LA-14, Lactobacillus salivarius LS-33, Streptococcus thermophilus ST-21, Bifidobacterium bifidum BB-06, Lactobacillus paracasei LPC-37, Bifidobacterium longum BL-05, Lactobacillus brevis LBR-35, Lactobacillus gasseri LG-36, Bifidobacterium breve BB-03, Lactobacillus plantarum LP-115, Bifidobacterium lactis BL-04, Lactobacillus rhamnosus LR-32 | 3 billion CFU/capsule | Neutralize the harmful bacteria that would otherwise cause occasional constipation, gas, and bloating | 14 | |
| 2 | BioGaia | Lactobacillus reuteri DSM 17938 | 100 million/5 drops | Provides colic relief that reduces excessive crying, helps build stronger immune system | 1 | |

(continued)

**Table 4.2** (continued)

| Sr. No. | Probiotic Company | Incorporated strains | Consumable Amount | Product claims | No of strains used | Contains |
|---|---|---|---|---|---|---|
| 3 | Dr. formulated probiotics | Bifidobacterium lactis, Lactobacillus gasseri, Bifidobacterium bifidum, Lactobacillus plantarum, Lactobacillus Casei, Bifidobacterium breve, Lactobacillus brevis, Lactobacillus acidophilus, Lactobacillus rhamnosus, Lactobacillus salivarius, Lactobacillus bulgaricus, Lactobacillus paracasei, Bifidobacterium infantis, Bifidobacterium longum | 5 billion CFU/capsule | Promotes immune system and digestive health | 14 | 5 calories, 1 g carbohydrate, >1 g sugar |
| 4 | Billion cheers probiotic immunity Junior's drink | Lactobacillus rhamnosus, Bifidobacterium lactis and Lactobacillus acidophilus | 5 billion CFU/serving | Probiotic drink for kids $5^+$ | 3 | 156 kcal, 0.9674 g carbohydrate |
| 5 | Good day chocolate probiotic | Bacillus coagulans MTCC 5856 | 1 billion CFU/serving | Promotes healthy digestion and keeps thing moving and growing | | 15 calories, 3 g carbohydrate, 2 g sugar, < 2 g fat, < 1 mg protein and 2 mg sodium |
| 6 | Cipla active kids Unobiotics junior | Lactobacillus rhamnosus GG | 30 billion CFU/serving | Delivers 5 times more probiotic activity than traditional uncoated freeze-dried form, maintains the health and integrity of the intestinal cells | | 4.024 kcal, 75 g carbohydrate, 0.024 g protein |

| | | | | | | |
|---|---|---|---|---|---|---|
| 7 | ProbiTec | Lactobacillus reuteri FloraActive®; Lactobacillus rhamnosus FloraActive® | 250 million CFU/serving | Boosts immune system and contributes to optimal development of teeth and bones | 2 | |
| 8 | Neo' peques probiotic | Saccharomyces boulardii, Bifidobacterium infantis SP37, Lactobacillus acidophilus LA3, Lactobacillus rhamnosus SP1, Bifidobacterium lactis W18, Lactobacillus casei BGP93 | – | Enables an optimal balance of gut microflora in infants and young children in diarrhoea | 6 | |
| 9 | Culturelle probiotics kids | Lactobacillus GG | | Helps to relieve from the tummy troubles of children | 1 | |
| 10 | GNC milestone kids probiotic | | 12.5 billion CFU/serving | Supports digestive, immune, and bone health | | |
| 11 | Jarro-Dophilus Baby's probiotic | B. infantis M-63, L. rhamnosus R0011, L. casei R0215, B. longum BB536, B. breve M-16 V | 3 billion CFU/serving | *B. infantis* capable of using unique sugars in breast milk to gain full benefits, rest support immunity and digestive health | 5 | |
| 12 | Hyperbiotics pro-kids | Lactobacillus acidophilus, Lactobacillus plantarum, B. infantis, and B. lactis | 3 billion CFU/serving | Can survive stomach acid, helps to absorb nutrients | 4 | |

The use of specific strain plays an important role in probiotic products, as their effects are dependent primarily on it. Therefore, strain identity is important and directly linked to specific health effects. Standard protocols and techniques, namely 'DNA fingerprinting, viz. Pulsed Field Gel Electrophoresis (PFGE) and ribotyping; 16S rRNA sequencing and PCR' help identify phenotypic as well as genotypic traits. The identified traits and the nomenclature should be confirmed through the scientifically validated 'International Committee on Systematics of Prokaryotes (ICPS)' (available at http://www.the-icsp.org/). These identified strains must be accumulated in internationally acclaimed repositories/culture collection centres for probiotic use. Indian repositories include: Microbial Type Culture Collection (MTCC), Chandigarh; Microbial Culture Collection (MCC), Pune; National Fungal Culture Collection of India, Pune; National Collection of Industrial Microorganisms, Pune; agriculturally important National Bureau of Microorganisms, and National Collection of Dairy Cultures, Karnal, etc. MTCC and MCC are IDA (International Depository Authority) recognized (Sharma and Shouche 2014; Source: http://www.wfcc.info/ccinfo/collection/col_by_country/i/91).

(b) *Screening of potential probiotic strains*:

For screening the putative probiotic strains, following in vitro tests are a must.

1. Withstand gastric acidity.
2. Show resistance to bile acid.
3. Antimicrobial activity towards potentially pathogenic bacteria (acid and bacteriocin production).
4. Ability to curtail pathogen adhesion to surfaces.
5. Bile salt hydrolase activity.

These are performed with specific standard methodology and are subjected to pre-clinical validation.

(c) *Safety studies in animal models* (*In Vivo*):

All the potential probiotic strains* are assessed for acute, subacute, and chronic toxicity of ingestion of exceedingly large amounts of probiotics.

(* not necessary for acclaimed strains)

(d) *Efficacy Studies in animal models* (*In Vivo*):

To justify the in vitro effects of probiotic strains, efficacy must be checked in animal models, before human trials.

(e) *Evaluation for human use*:

The probiotic strain used, must be GRAS approved and needs to be assured and characterized by the following tests:

1. Determining antibiotic resistance patterns, strain should not pose significant risk concerning transferring antibiotic resistance.
2. Evaluation of inadmissible side effects.
3. If the strain used in probiotic use belongs to species of well-known mammalian toxin producer/haemolytic potential, must be tested for both toxic and haemolytic activities consequently.

(f) *Evaluation of efficacy studies in humans*:

The results of efficacy studies should be proven with similar benefits in human trials*, including parameters like:

Statistical and clinically significant improvement in conditions, symptoms, quality of life, reducing the risk of reoccurrence of disease or faster recovery.

(*Phase-3 studies must be continued only if the probiotic claims for any specific health benefit)

If the probiotic in use has documented safe piece of evidence outside India, the data could be reviewed and sufficient enough to allow marketing within the country. While these studies are taken into account, the efficacy of abroad reports should be tested on Indian subjects.

(g) *Effective dose determination*:

The minimal effective dose along with the viable count of the strain used in terms of CFU/ml/day along with the targeted population must be indicated clearly.

(h) *Requirements for Labelling*:

The general labelling guidelines under food law have to be followed. Along with these, the display of the following information is a must:

1. Genus, species, and strain designation must be stated following the standard international nomenclature.
2. The minimum viable count of each probiotic strain should be specified both at the level at which efficacy is claimed and at the end of the shelf-life.
3. Health claim(s) should be stated clearly (only if approved based on evidences).
4. The suggested serving size to deliver the minimum effective quantity of the probiotic related to the health claim.
5. Proper storage conditions must be mentioned.

(i) *Manufacturing and Handling Practices to be followed*:

Good Manufacturing Practices should be obeyed while probiotic foods are manufactured. These practices should stick to the recommendations of the Codex General Principles of Food Hygiene and Guidelines for Application of Hazard Analysis and Critical Control Point (HACCP). These practices will ensure public protection from fraud and false manufacturing practices (Arora et al. 2013). Figure 4.3 represented the ICMR Guided Regulations for Probiotic Evaluation and Release.

ICMR also envisaged the formation of a special regulatory body 'Foods Safety and Standards Authority of India (FSSAI)' along with other sub-ordinate bodies to monitor all food relevant issues. Currently in India foods and drugs are regulated under the Prevention of Food Adulteration Act (PFA) and the Food and Drug Administration (FDA), respectively (Arora and Baldi, 2015). The guidelines in India released by the Foods Safety and Standards Authority of India (FSSAI) 2016, formed under the Food Safety and Standards Act, 2006 are as follows:

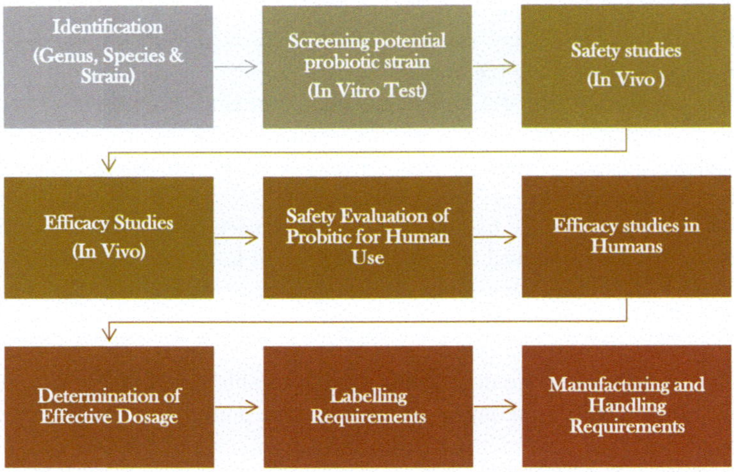


**Fig. 4.3** ICMR guided regulations for probiotic evaluation and release

## 4.15 FSSAI General Guidelines Applicable to All Food Products Under Different Categories

1. The syrups, capsules, and tablets shall meet up to the general quality requirements and standards as specified in Indian Pharmacopoeia, British Pharmacopoeia, or the United States Pharmacopoeia.
2. The food business operator may use the approved colours and additives sanctioned in *Schedule VF*, further may use natural or synthetic flavours, which have to be following FSSAI Regulations 2011. The addition of flavours must be declared on the label of the packaging.
3. The amount of nutrients incorporated into the articles of food shall not outreach the approved daily limit as specified by ICMR.
4. If the food product claims to be a health supplement, the nutrient content recommended daily for an individual shall not be less than 15% and should be greater than 30% only if the nutrient claim is higher.
5. The standard nutrients added in the food article must deliver the desired level of energy, proteins, vitamins and minerals, and other essential nutrients required for the respective age group, gender, and physiological stage following the guidelines made by ICMR.
6. The purity of the ingredients used must be covered under regulations notified by the food authorities.
7. In case such standards are not specified, the purity criteria generally accepted by pharmacopoeias, namely, Indian Pharmacopoeia, Ayurvedic Pharmacopoeia of India, relevant Bureau of Indian Standards Specifications, Quality Standards of Indian Medicinal Plants, Indian Council of Medical Research, British Pharmacopoeia, United States Pharmacopoeia, Food Chemical Codex, Joint Food, and

Agriculture Organization, or World Health Organization Expert Committee on Food Additives or CODEX Alimentarius may be adopted by food Business operators.

8. The tolerance limit for variation in the case of the food articles shall not be more than (−) 10% from the declared value of the nutrients or nutritional ingredients on the label.

## 4.16 FSSAI Guidelines Specifically Applicable to Food Products with Added Probiotic Ingredients

1. The main ingredient of the probiotic food has to be the culture of live microorganisms, the culture used may be a single strain or combination of microbes.
2. The approved strains only can be added in the probiotic products (specified in *Schedule VII*), or those microbes approved from time to time by the authority.
3. It must confer specified health benefits to the consumer.
4. It may contain added prebiotics as per FSSAI regulations.
5. The microbes must be depicted on the labelling display panel of the product.
6. These microbes must be non-GMOs.
7. The viable count of the added microbes must be $\geq 10^8$ CFU in the recommended serving size per day.
8. Probiotic food product shall not claim or mention (in labelling or even in the advertisement) to have any property of preventing, healing, or treating human disease.
9. The food authority can allow the company to mention a particular statement regarding any health claim, only if supported with scientific evidence.
10. The packaging of the probiotic product must include the following:

    a. 'PROBIOTIC FOOD' must be mentioned clearly on the label
    b. Genus and species including strain designation or culture collection number as per MTCC (if applicable) in the list of ingredients.
    c. Serving size (recommended), duration of use, storage conditions, and 'best by' date after the container is opened.
    d. Viable count at the end of the shelf-life of the probiotic strain should be stated.
    e. 'NOT FOR MEDICINAL USE'—advisory warning must be written prominently
    f. Any other warning or precaution to be taken while consuming, known side effects, contraindications, and product-drug interactions, as applicable.

11. Only additives specified in *Schedule VA* to *Schedule VF* can be used in probiotic preparations (FSSAI Regulations 2016).

## 4.17 Future Direction

Commercial availability of probiotics is a diverse range including capsules, dried powders, sprays, dietary supplements in the form of energy drinks, fortified yoghurt, probiotic enriched curd, lassi, and even ice creams. These are enriched with microbes especially Lactic Acid Bacteria (LAB) which actually could not survive for longer, may not even reach live into the gut of humans. Hence, diminishing the benefits. So, DBT Biotech Consortium is looking out for commercialization of a new technique that holds promises to help prolong the serviceable life of probiotic products. Their concept says encapsulation of probiotic strains with food grade edible strips or coatings of biopolymers (made of milk protein and plant-based waxy substances) can keep the probiotic bacteria active for 1 month, stored at 4°C. They have validated their results by building an in house lab set up, generated proofs. This concept has given promising results and now offering a license to suitable industries for commercialization. This technology has the huge market potential both in India as well as outside and can reach USD 69.3 billion markets for probiotics by 2023 reported by Dr. Bilqeesha Bhat (Jan 30, 2020; Source: https://vigyanprasar.gov.in/isw/Technology-to-increase-shelf-life-of-probiotic-products.html).

## References

Agustina R, Lukito W, Firmansyah A, Suhardjo HN, Murniati D, Bindels J (2007) The effect of early nutritional supplementation with a mixture of probiotic, prebiotic, fiber and micronutrients in infants with acute diarrhea in Indonesia. Asia Pac J Clin Nutr 16(3):435–442

Alakomi HL, Skyttä E, Saarela M, Mattila-Sandholm T, Latva-Kala K et al (2000) Lactic acid permeabilizes gram-negative bacteria by disrupting the outer membrane. Appl Environ Microbiol 66:2001–2005. https://doi.org/10.1128/aem.66.5.2001-2005.2000

An HM, Park SY, do Lee K, Kim JR, Cha MK, Lee SW et al (2011) Antiobesity and lipid-lowering effects of Bifidobacterium spp. in high fat diet-induced obese rats. Lipids Health Dis 10:116

An J, Ha E-M (2016) Combination therapy of Lactobacillus plantarum supernatant and 5-fluouracil increases chemosensitivity in colorectal cancer. J Microbiol Biotechnol 26(8):1490–1503

Anandharaj M, Sivasankari B, Rani RP (2020) Corrigendum to "Effects of probiotics, prebiotics, and synbiotics on hypercholesterolemia: a review". Chinese J Biol 2020:8236703. https://doi.org/10.1155/2020/8236703

Andersson U, Branning C, Ahrne S, Molin G, Alenfall J, Onning G, Nyman M, Holm C (2010) Probiotics lower plasma glucose in the high-fat fed C57BL/6J mouse. Benef Microbes 1:189–196

Anukam K, Osazuwa E, Ahonkhai I, Ngwu M, Osemene G, Bruce AW, Reid G (2006) Augmentation of antimicrobial metronidazole therapy of bacterial vaginosis with oral probiotic Lactobacillus rhamnosus GR-1 and Lactobacillus reuteri RC-14: randomized, double-blind, placebo-controlled trial. Microbes Infect 8(6):1450–1454

Arioli S, Ragg E, Scaglioni L, Fessas D, Signorelli M, Karp M et al (2010) Alkalizing reactions streamline cellular metabolism in acidogenic microorganisms. PLoS One 5:15520

Arora M, Baldi A (2015) Regulatory categories of probiotics across the globe: A review representing existing and recommended categorization. Indian J Med Microbiol 33:S2–S10. https://doi.org/10.4103/0255-0857.150868

Arora M, Sharma S, Baldi A (2013) Comparative insight of regulatory guidelines for probiotics in USA, India and Malaysia: a critical review. Int J Biotechnol Wellness Ind 2:51. https://doi.org/10.6000/1927-3037.2013.02.02.1

Baldwin C, Millette M, Oth D, Ruiz MT, Luquet FM, Lacroix M (2010) Probiotic Lactobacillus acidophilus and L. casei mix sensitize colorectal tumoral cells to 5-fluorouracil-induced apoptosis. Nutr Cancer 62(3):371–378

Barengolts E (2016) Gut microbiota, prebiotics, probiotics, and synbiotics in management of obesity and prediabetes: review of randomized controlled trials. Endocr Pract 22 (10):1224–1234

Bermudez-Brito M, Plaza-Diaz J, Munoz-Quezada S, Gomez-Llorente C, Gil A (2012) Probiotic mechanisms of action. Ann Nutr Metab 61:160–174

Besseling-van der Vaart I (2016) Heath, M.D.; Guagnini, F.; Kramer, M.F. in vitro evidence for efficacy in food intolerance for the multispecies probiotic formulation ecologic® tolerance (syngutTM). Benef Microbe 7:111–118

Boekhorst J, Helmer Q, Kleerebezem M, Siezen RJ (2006) Comparative analysis of proteins with a mucus-binding domain found exclusively in lactic acid bacteria. Microbiology 152:273–280. https://doi.org/10.1099/mic.0.28415-0

Bosscher D, Van Caillie-Bertrand M, Van Cauwenbergh R, Deelstra H (2003) Availabilities of calcium, iron, and zinc from dairy infant formulas is affected by soluble dietary fibres and modified starch fractions. Nutrition 19:641–645

Boudeau J, Glasser AL, Julien S, Colombel JF, Darfeuille-Michaud A (2003) Inhibitory effect of probiotic Escherichia coli strain Nissle 1917 on adhesion to and invasion of intestinal epithelial cells by adherent-invasive E. coli strains isolated from patients with Crohn's disease. Aliment Pharmacol Ther 18(1):45–56

Broekaert IJ, Walker WA (2006) Probiotics as flourishing benefactors for the human body. Gastroenterol Nurs 29:26–34

Bron PA, Marco M, Hoffer SM, Van Mullekom E, de Vos WM, Kleerebezem M (2004) Genetic characterization of the bile salt response in Lactobacillus plantarum and analysis of responsive promoters in vitro and in situ in the gastrointestinal tract. J Bacteriol 186(23):7829–7835. https://doi.org/10.1128/JB.186.23.7829-7835.2004

Bron PA, Van-Baarlen P, Kleerebezem M (2012) Emerging molecular insights into the interaction between probiotics and the host intestinal mucosa. Nat Rev Microbiol 10(1):66–78

Brown M (2011) Mode of action of probiotics: recent developments. J Anim Vet Adv 10 (14):1895–1900

Bruzzese E, Volpicelli M, Squeglia V, Bruzzese D, Salvini F, Bisceglia M, Lionetti P, Cinquetti M, Iacono G, Amarri S, Guarino A (2009) A formula containing galacto- and fructo-oligosaccharides prevents intestinal and extra-intestinal infections: an observational study. Clin Nutr 28:156–161

Caglar E, Sandalli N, Twetman S, Kavaloglu S, Ergeneli S, Selvi S (2005) Effect of yogurt with Bifidobacterium DN-173 010 on salivary mutans streptococci and lactobacilli in young adults. Acta Odontol Scand 63:317–320

Cencic A, Chingwaru W (2010) The role of functional foods, nutraceuticals, and food supplements in intestinal health. Nutrients 2(6):611–625. https://doi.org/10.3390/nu2060611

Chen J, Wang R, Li XF, Wang RL (2012) Bifidobacterium adolescentis supplementation ameliorates visceral fat accumulation and insulin sensitivity in an experimental model of the metabolic syndrome. Br J Nutr 107:1429–1434

Chen ZY, Hsieh YM, Huang CC, Tsai CC (2017) Inhibitory effects of probiotic Lactobacillus on the growth of human colonic carcinoma cell line HT-29. Molecules 22(1):107

Chichlowski M, Croom J, McBride BW, Havenstein GB, Koci MD (2007) Metabolic and physiological impact of probiotics or direct fed-microbials on poultry: a brief review of current knowledge. Int J Poult Sci 6(10):694–704. https://doi.org/10.3923/ijps.2007.694.704

Claros TG, Chang S, Gilliam EA, Maitra U, Deng H, Li L (2013) Causes and consequences of low grade endotoxemia and inflammatory diseases. Front Biosci 5:754–765

Collins S, Reid G (2016) Distant site effects of ingested prebiotics. Nutrients 8:523

Cotter PD, Hill C (2003) Surviving the acid test: responses of gram-positive bacteria to low pH. Microbiol Mol Biol Rev 67:429–453

Darouiche RO, Donovan WH, Del Terzo M, Thornby JI, Rudy DC, Hull RA (2001) Pilot trial of bacterial interference for preventing urinary tract infection. Urology 58:339

de Llano GD, Arroyo A, Cárdenas N, Rodríguez JM, Victoria M, Arribas M, Bartolomé B (2017) Strain-specific inhibition of the adherence of uropathogenic bacteria to bladder cells by probiotic Lactobacillus spp. Pathogens and Disease 75(4):ftx043. https://doi.org/10.1093/femspd/ftx043

de Vrese M, Stegelmann A, Richter B, Fenselau S, Laue C, Schre-zenmeir J (2001) Probiotics – compensation for lactase insufficiency. Am J Clin Nutr 73:421S–429S

Duquesne S, Petit V, Peduzzi J, Rebuffat S (2007) Structural and functional diversity of microcins, gene-encoded antibacterial peptides from enterobacteria. J Mol Microbiol Biotechnol 13:200–209. https://doi.org/10.1159/000104748

Ejtahed HS, Angoorani P, Soroush AR et al (2019) Probiotics supplementation for the obesity management; A systematic review of animal studies and clinical trials. J Funct Foods 52:228–242

Elo S, Saxelin M, Salminen S (1991) Attachment of *Lactobacillus casei* strain GG to human colon carcinoma cell line Caco-2: comparison with other dairy strains. Lett Appl Microbiol 13:154–156

Enomoto T, Sowa M, Nishimori K et al (2014) Effects of bifidobacterial supplementation to pregnant women and infants in the prevention of allergy development in infants and on fecal microbiota. Allergol Int 63(4):575–585. https://doi.org/10.2332/allergolint.13-OA-0683

Escamilla J, Lane MA, Maitin V (2012) Cell-free supernatants from probiotic Lactobacillus casei and Lactobacillus rhamnosus GG decrease colon cancer cell invasion in vitro. Nutr Cancer 64 (6):871–878

Everard A, Belzer C, Geurts L, Ouwerkerk JP, Druart C, Bindels LB et al (2013) Cross-talk between Akkermansia muciniphila and intestinal epithelium controls diet-induced obesity. Proc Natl Acad Sci U S A 110:9066–9071

Favier C, Neut C, Mizon C et al (1997) Fecal beta-D-galactosidase production and bifidobacteria are decreased in Crohn's disease. Dig Dis Sci 42:817–822

Ferguson LR, Philpott M (2007) Cancer prevention by dietary bioactive components that target the immune response. Curr Cancer Drug Targets 7:459–464

Fijan S (2016) Antimicrobial effect of probiotics against common pathogens probiotics and prebiotics in human nutrition and health. Intechopen, New York, pp 191–221. https://doi.org/10.5772/63141

FAO/WHO (2001) The Food and Agriculture Organization of the United Nations and the World Health Organization joint FAO/WHO expert consultation on the evaluation of health and nutritional properties of probiotics in food including powder milk with live lactic acid bacteria, FAO/WHO Report No. 10-1-2001

Food Safety and Standards Act (2006) Guidelines for evaluation of food by food safety and standard authority of India. https://www.indiacode.nic.in/bitstream/123456789/7800/1/200634_food_safety_and_standards_act%2C_2006.pdf

FSSAI Regulations (2011) Guidelines by the government of India for the food safety and standards (Food Product Standards and Food Additives) regulation. https://www.fssai.gov.in/upload/uploadfiles/files/Compendium_Food_Additives_Regulations_08_09_2020-compressed.pdf

FSSAI Regulations (2016) Amendments in the Guidelines by the government of India for the food safety and standards (Food Product Standards and Food Additives) regulation. https://old.fssai.gov.in/Portals/0/pdf/Operationalization_Caff_Beve_Blue_Tint_07_11_2016.pdf

Fukata M, Michelsen KS, Eri R, Thomas LS, Hu B, Lukasek K, Nast CC, Lechago J, Xu R, Naiki Y, Soliman A, Arditi M, Abreu MT (2005) Toll-like receptor-4 is required for intestinal response to epithelial injury and limiting bacterial translocation in a murine model of acute colitis. Am J Physiol Gastrointest Liver Physiol 288(5):G1055–G1065

Fuller LK (1995) Yogurt, yoghurt, yoghurt: an international cookbook. The Haworth Press, Binghamton, NY

Fuller R (1989) Probiotics in man and animals. J Appl Bacterial 66:365–369

Furness JB, Kunze WA, Clerc N (1999) Nutrient tasting and signaling mechanisms in the gut. II. The intestine as a sensory organ: neural, endocrine, and immune responses. Am J Phys 277 (5 Pt 1):G922–G928

Furrie E, Macfarlane S, Kennedy A et al (2005) Synbiotic therapy (Bifidobacterium longum/ synergy 1) initiates resolution of inflammation in patients with active ulcerative colitis: a randomised controlled pilot trial. Gut 54:242–249

Galdeano CM, Perdigón G (2004) Role of viability of probiotic strains in their persistence in the gut and in mucosal immune stimulation. J Appl Microbiol 97:673–681

Gaon D, Garmendia C, Murrielo NO et al (2002) Effect of Lactobacillus strains (L. casei and L. acidophilus strains cereal) on bacterial overgrowth-related chronic diarrhea. Medicina (B Aires) 62:159–163

Gauffin CP, Santacruz A, Moya A, Sanz Y (2012) Bacteroides uniformis CECT 7771 ameliorates metabolic and immunological dysfunction in mice with high-fat-diet induced obesity. PLoS One 7:e41079

Gibson GR, Peobert HM, Loo JV, Rastall RA, Roberfroid MB (2004) Dietary modulation of the human colonic microbiota: updating the concept of prebiotics. Nutr Res Rev 17:259–275

Gionchetti P, Rizzello F, Venturi A et al (2000) Oral bacteriotherapy as maintenance treatment in patients with chronic pouchitis: a double-blind, placebo-controlled therapy. Gastroenterology 119:305–309

Gogineni, et al. (2013) Probiotics: mechanisms of action and clinical applications. J Prob Health 1:1. https://doi.org/10.4172/2329-8901.1000101

Gokhale G, Nadkarni A (2007) Probiotic foods market in India soon to be regulated. India Law J 3 (1). http://www.indialawjournal.com/volume3/issue_1/article_by_aditi.html

Gupta A, Gupta P, Dhawan A (2014) Dietary supplementation of probiotics affects growth, immune response and disease resistance of Cyprinus carpio fry, fish & shellfish. Immunology 41(2):113–119. https://doi.org/10.1016/j.fsi.2014.08.023

Hardy H, Harris J, Lyon E, Beal J, Foey AD (2013) Probiotics, prebiotics and immunomodulation of gut mucosal Defences: homeostasis and immunopathology. Nutrients 5(6):1869–1912. https://doi.org/10.3390/nu5061869

Harrington LK et al (2008) A re-appraisal of lactose intolerance. Int J Clin Pract 62(10):1541–1546

Hasslof P, Hedberg M, Twetman S, Stecksen-Blicks C (2010) Growth inhibition of oral mutans streptococci and candida by commercial probiotic lactobacilli – an in vitro study. BMC Oral Health 10:18

He B, Xu W, Santini PA, Polydorides AD, Chiu A, Estrella J, Shan M, Chadburn A, Villanacci V, Plebani A et al (2007) Intestinal bacteria trigger T cell-independent immunoglobulin A2 class switching by inducing epithelial-cell secretion of the cytokine APRIL. Immunity 26:812–826. https://doi.org/10.1016/j.immuni.2007.04.014

He T, Priebe MG, Zhong Y (2008) Effects of yogurt and bifidobacteria supplementation on the colonic microbiota in lactose-intolerant subjects. J Bacteriol 104(2):595–604

Hospenthal MK, Costa TRD, Waksman G (2017) A comprehensive guide to pilus biogenesis in gram-negative bacteria. Nat Rev Microbiol 15(6):365–379. https://doi.org/10.1038/nrmicro.2017.40

Hseih PS, An Y, Tsai YC, Chen YC, Chunag CJ, Zeng CT, Want CT, An-Eri King V (2012) Potential of probiotic strains to modulate the inflammatory responses of epithelial and immune cells in vitro. New Microbiol 36(2):167–179

Huang R, Ning H, Shen M, Li J, Zhang J, Chen X (2017) Probiotics for the treatment of atopic dermatitis in children: A systematic review and meta-analysis of randomized controlled trials. Front Cell Infect Microbiol 7:392

Hull R, Rudy D, Donovan W, Svanborg C, Wieser I, Stewart C et al (2000) Urinary tract infection prophylaxis using Escherichia coli 83972 in spinal cord injured patients. J Urol 163:872–877

Hymes JP, Johnson BR, Barrangou R, Klaenhammer TR (2016) Functional analysis of an S-layer-associated fibronectin-binding protein in Lactobacillus acidophilus NCFM. Appl Environ Microbiol 82:2676–2685. https://doi.org/10.1128/AEM.00024-16

Ishimwe N, Daliri E, Lee B, Fang F, Du G (2015) The perspective on cholesterol-lowering mechanisms of probiotics. Mol Nutr Food Res 59(1):94–105

Jacobsen CN, Rosenfeldt NV, Hayford AE et al (1999) Screening of probiotic activities of forty-seven strains of Lactobacillus spp. by in vitro techniques and evaluation of the colonization ability of five selected strains in humans. Appl Environ Microbiol 65(11):4949–4956

Joshi SS, Pandharbale AA (2015) Probiotics and periodontal disease – A review. Indian J Dent Adv 7(2):136–141

Kajander K, Hatakka K, Poussa T et al (2005) A probiotic mixture alleviates symptoms in irritable bowel syndrome patients: a controlled 6-month intervention. Aliment Pharmacol Ther 22:387–394

Keller MK, Hasslof P, Stecksen-Blicks C, Twetman S (2011) Co-aggregation and growth inhibition of probiotic lactobacilli and clinical isolates of mutans streptococci: an in vitro study. Acta Odontol Scand 69:263–268

Kelsall BL (2008) Innate and adaptive mechanisms to control pathological intestinal inflammation. J Pathol 214:242–259. https://doi.org/10.1002/path.2286

Konstantinov SR, Smidt H, de Vos WM, Bruijns SCM, Singh SK, Valence F, Molle D, Lortal S, Altermann E, Klaenhammer TR, van Kooyk Y (2008) S layer protein a of Lactobacillus acidophilus NCFM regulates immature dendritic cell and T-cell functions. Proc Natl Acad Sci 105:19474–19479. https://doi.org/10.1073/pnas.0810305105

Korčok D, Nada TM, Ivanovic N, Djordjevic B (2018) Development of probiotic formulation for the treatment of Iron deficiency anemia. Chem Pharm Bull 66(4):347–352. https://doi.org/10.1248/cpb.c17-00634

Kotzamanidis C, Kourelis A, Litopoulou-Tzanetaki E, Tzanetakis N, Yiangou M (2010) Evaluation of adhesion capacity, cell surface traits and immunomodulatory activity of presumptive probiotic Lactobacillus strains. Int J Food Microbiol 140(2–3):154–163

Kumari A, Catanzaro R, Marotta F (2011) Clinical importance of lactic acid bacteria: a short review. Acta Biomed 82(3):177–180

Larsson JMH, Karlsson H, Crespo JG, Johansson MEV, Eklund L, Sjövall H, Hansson GC (2011) Altered O-glycosylation profile of MUC2 mucin occurs in active ulcerative colitis and is associated with increased inflammation. Inflamm Bowel Dis 17:2299–2307. https://doi.org/10.1002/ibd.21625

Le Luyer B, Makhoul G, Duhamel JF (2010) A multicentric study of a lactose free formula supplemented with Saccharomyces boulardii in children with acute diarrhea. Arch Pediat 17 (5):459–465

Lebeer S, Vanderleyden J, De Keersmaecker SCJ (2010) Host interactions of probiotic bacterial surface molecules: comparison with commensals and pathogens. Nat Rev Microbiol 8 (3):171–184

Lehri B, Seddon AM, Karlyshev AV (2015) Lactobacillus fermentum 3872 genome sequencing reveals plasmid and chromosomal genes potentially involved in a probiotic activity. FEMS Microbiol Lett 362:fnv068. https://doi.org/10.1093/femsle/fnv068

Leisched DV (2014) Probiotics as regulators of inflammation: a review. Funct Foods Health Dis 4 (7):299–311

Ley RE (2010) Obesity and the human microbiome. Curr Opin Gastroenterol 26(1):5–11

Liévin-Le Moal V, Servin AL (2006) The front line of enteric host defense against unwelcome intrusion of harmful microorganisms: mucins, antimicrobial peptides, and microbiota. Clin Microbiol Rev 19:315–337. https://doi.org/10.1128/CMR.19.2.315-337.2006

Loguercio C, Abbiati R, Rinaldi M, Romano A, Del Vecchio Blanco C, Coltorti M (1995) Long-term effects of enterococcus faecium SF68 versus lactulose in the treatment of patients with cirrhosis and grade 1-2 hepatic encephalopathy. J Hepatol 23(1):39–46

Lopez-Lopez A, Camelo-Castillo A, Ferrer MD, Simon-Soro A, Mira A (2017) Health- associated niche inhabitants as oral pro- biotics: the case of Streptococcus dentisani. Front Microbiol 8:379

Madsen KL (2012) Enhancement of epithelial barrier function by probiotics. J Epithelial Biol Pharmacol 5:55–59. https://doi.org/10.2174/1875044301205010055

McCarthy J, O'Mahony L, O'Callaghan L, Sheil B, Vaughan EE, Fitzsimons N, Fitzgibbon J, O'Sullivan GC, Kiely B, Collins JK, Shanahan F (2003) Double blind, placebo controlled trial of two probiotic strains in interleukin 10 knockout mice and mechanistic link with cytokine balance. Gut 52(7):975–980

Montalto M, Maggiano N, Ricci R, Curigliano V, Santoro L, Di Nicuolo F, Vecchio FM, Gasbarrini A, Gasbarrini G (2004) Lactobacillus acidophilus protects tight junctions from aspirin damage in HT-29 cells. Digestion 69:225–228. https://doi.org/10.1159/000079152

Monteagudo-Mera A, Rastall RA, Gibson GR et al (2019) Adhesion mechanisms mediated by probiotics and prebiotics and their potential impact on human health. Appl Microbiol Biotechnol 103:6463–6472. https://doi.org/10.1007/s00253-019-09978-7

Morelli L, Zonenenschain D, Del Piano M, Cognein PJ (2004) Clin Gastroenterol 38(6):S107–S110

Mortaz E, Adcock IM, Folkerts G, Barnes PJ, Paul Vos A, Garssen J (2013) Probiotics in the management of lung diseases. Mediat Inflamm 2013:751068

Muftuoglu M, Isikgor S, Tosun S et al (2006) Effects of probiotics on the severity of experimental acute pancreatitis. Eur J Clin Nutr 60:464–468. https://doi.org/10.1038/sj.ejcn.1602338

Mukai T, Asasaka T, Sato E et al (2002) Inhibition of binding of helicobacter pylori to the glycolipid receptors by probiotic Lactobacillus reuteri. FEMS Immunol Med Microbiol 32:105–110

Muñoz-Provencio D, Rodríguez-Díaz J, Collado MC, Langella P, Bermúdez-Humarán LG, Monedero V (2012) Functional analysis of the *Lactobacillus casei* BL23 sortases. Appl Environ Microbiol 78:8684–8693. https://doi.org/10.1128/AEM.02287-12

Myllyluoma E, Veijola L, Ahlroos T et al (2005) Probiotic supplementation improves tolerance to Helicobacter pylori eradication therapy— a placebo-controlled, double-blind randomized pilot study. Aliment Pharmacol Ther 21:1263–1272

Näse L, Hatakka K, Savilahti E, Saxelin M, Pönkä A, Poussa T et al (2001) Effect of long-term consumption of a probiotic bacterium, Lactobacillus rhamnosus GG, in milk on dental caries and caries risk in children. Caries Res 35:412–420

Nobaek S, Johansson ML, Molin G et al (2000) Alteration of intestinal microflora is associated with reduction in abdominal bloating and pain in patients with irritable bowel syndrome. Am J Gastroenterol 95:1231–1238

Nurmi E, Nuotio L, Schneitz C (1992) The competitive exclusion concept: development and future. Int J Food Micro Biol 15:237–240

O'Connell Motherway M, Zomer A, Leahy SC, Reunanen J, Bottacini F, Claesson MJ, O'Brien F, Flynn K, Casey PG, Moreno Munoz JA, Kearney B, Houston AM, O'Mahony C, Higgins DG, Shanahan F, Palva A, de Vos WM, Fitzgerald GF, Ventura M, O'Toole PW, van Sinderen D (2011) Functional genome analysis of Bifidobacterium breve UCC2003 reveals type IVb tight adherence (tad) pili as an essential and conserved host-colonization factor. Proc Natl Acad Sci 108:11217–11222. https://doi.org/10.1073/pnas.1105380108

O'Hara AM, Shanahan F (2007) Mechanisms of action of probiotics in intestinal diseases. Sci World J 7:31–46. https://doi.org/10.1100/tsw.2007.26

Oelschlaeger TA (2010) Mechanisms of probiotic actions – A review. Int J Med Microbiol 300:57–62

Oh Y, Osato MS, Han X et al (2002) Folk yogurt kills helicobacter pylori. J Appl Microbiol 93:1083–1088
Ohashi Y, Ushida U (2009) Health-beneficial effects of probiotics: its mode of action. Anim Sci J 80(4):361–371
Ohland CL, Mac Naughton WK (2010) Probiotic bacteria and intestinal barrier function. Am J Physiol Gastrointest Liver Physiol 298:G807–G819
Ojetti V, Gigante G, Gabrielli M (2010) The effect of oral supplementation with Lactobacillus reuteri or tilactase in lactose intolerant patients: randomized trial. Eur Rev Med Pharmacol Sci 14(3):163–170
Orlando A, Refolo MG, Messa C et al (2012) Antiproliferative and proapoptotic effects of viable or heat-killed Lactobacillus paracaseiIMPC2.1 and Lactobacillus rhamnosus GG in HGC-27 gastric and DLD-1 colon cell lines. Nutr Cancer 64(7):1103–1111
Ouwehand AC, Tuomola EM, Tölkkö S, Salminen S (2001) Assessment of adhesion properties of novel probiotic strains to human intestinal mucus. Int Food Microbiol 64(1–2):119–126. https://doi.org/10.1016/S0168-1605(00)00440-2
Ozen M, Dinleyici EC (2015) The history of probiotics: the untold story. Benef Microbes 6 (2):159–165. https://doi.org/10.3920/BM2014.0103
Pandey KR, Naik SR, Vakil BV (2015) Probiotics, prebiotics and synbiotics – a review. J Food Sci Technol 52:7577–7587. https://doi.org/10.1007/s13197-015-1921-1
Parra D, Martınez JA (2007) Amino acid uptake from a probiotic milk in lactose intolerant subjects. Br J Nutr 98(1):S101–S104
Peña AS (2007) Intestinal flora, probiotics, prebiotics, synbiotics and novel foods. Rev Esp Enferm Dig 99(11):653
Penner R, Fedorak RN, Madsen KL (2005) Probiotics and nutraceuticals: non-medicinal treatments of gastrointestinal diseases. Curr Opin Pharmacol 5(6):596–603. https://doi.org/10.1016/j.coph.2005.06.009
Piepenbrink KH, Sundberg EJ (2016) Motility and adhesion through type IV pili in gram-positive bacteria. Biochem Soc Trans 44:1659–1666. https://doi.org/10.1042/BST20160221
Popowska M, Krawczyk-Balska A, Ostrowski R, Desvaux M (2017) InlL from Listeria monocytogenes is involved in biofilm formation and adhesion to mucin. Front Microbiol 8:660
Radziwill-Bienkowska JM, Robert V, Drabot K, Chain F, Cherbuy C, Langella P, Thomas M, Bardowski JK, Mercier-Bonin M, Kowalczyk M (2017) Contribution of plasmid-encoded peptidase S8 (PrtP) to adhesion and transit in the gut of Lactococcus lactis IBB477 strain. Appl Microbiol Biotechnol 101:5709–5721. https://doi.org/10.1007/s00253-017-8334-1
Rakoff-Nahoum S, Paglino J, Eslami-Varzaneh F, Edberg S, Medzhitov R (2004) Recognition of commensal microflora by toll-like receptors is required for intestinal homeostasis. Cell 118 (2):229–241
Rampengan NH, Manoppo J, Warouw SM (2010) Comparison of efficacies between live and killed probiotics in children with lactose malabsorption. Southeast Asian J Trop Med Public Health 41 (2):474–481
Rao RK, Samak G (2013) Protection and restitution of gut barrier by probiotics: nutritional and clinical implications. Curr Nutr Food Sci 9(2):99–107
Rautava S, Arvilommi H, Isolaur E (2006) Specific probiotics in enhancing maturation of IgA responses in formula-fed infants. Pediatr Res 60:221–224. https://doi.org/10.1203/01.pdr.0000228317.72933.db
Reid G, Beuerman D, Heinemann C, Bruce AW (2001) Probiotic Lactobacillus dose required to restore and maintain a normal vaginal flora. FEMS Immunol Med Microbiol 32(1):37–41
Reséndiz-Albor AA, Reina-Garfias H, Rojas-Hernández S, Jarillo-Luna A, Rivera-Aguilar V, Miliar-García A, Campos-Rodríguez R (2010) Regionalization of pIgR expression in the mucosa of mouse small intestine. Immunol Lett 128:59–67. https://doi.org/10.1016/j.imlet.2009.11.005

Resta-Lenert S, Barrett KE (2003) Live probiotics protect intestinal epithelial cells from the effects of infection with enteroinvasive Escherichia coli (EIEC). Gut 52(7):988–997. https://doi.org/10.1136/gut.52.7.988

Resta-Lenert S, Barrett KE (2006) Probiotics and commensals reverse TNF-alpha- and IFN-gamma-induced dysfunction in human intestinal epithelial cells. Gastroenterology 130 (3):731–746. https://doi.org/10.1053/j.gastro.2005.12.015

Reunanen J, von Ossowski I, Hendrickx APA, Palva A, de Vosa WM (2012) Characterization of the SpaCBA pilus fibers in the probiotic Lactobacillus rhamnosus GG. Appl Environ Microbiol 78:2337–2344. https://doi.org/10.1128/AEM.07047-11

Rodrigues AC, Cara DC, Fretez SH, Cunha FQ, Vieira EC, Nicoli JR, Vieira LQ (2000) Saccharomyces boulardii stimulates sIgA production and the phagocytic system of gnotobiotic mice. J Appl Microbiol 89:404–414. https://doi.org/10.1046/j.1365-2672.2000.01128.x

Rosenfeldt V, Benfeldt E, Nielsen SD et al (2003) Effect of probiotic Lactobacillus strains in children with atopic dermatitis. J Allergy Clin Immunol 111:389–395

Rosenfeldt V, Benfeldt E, Valerius NH, Paerregaard A, Michaelsen KF (2004) Effect of probiotics on gastrointestinal symptoms and small intestinal permeability in children with atopic dermatitis. J Pediatr 145(5):612–616. https://doi.org/10.1016/j.jpeds.2004.06.068

Ryan P, Ross R, Fitzgerald G, Caplice N, Stanton C (2015) Functional food addressing heart health. Curr Opin Clin Nutr Metab Care 18(6):566–571

Saber B, Alipour Z, Faghfoori A, Mousavi jam, and Yari Khosroushahi, A. (2017) Secretion metabolites of probiotic yeast, Pichia kudriavzevii AS-12, induces apoptosis pathways in human colorectal cancer cell lines. Nutr Res 41:36–46

Salminen S, Bouley C, Boutron-Ruault MC et al (1998) Functional food science and gastrointestinal physiology and function. Br J Nutr 80(1):S147–S171

Sambrani R, Abdolalizadeh J, Kohan L, Jafari B (2019) Saccharomyces cerevisiae inhibits growth and metastasis and stimulates apoptosis in HT-29 colorectal cancer cell line. Comp Clin Pathol 28(4):985–995

Sanders ME, Merenstein DJ, Reid G et al (2019) Probiotics and prebiotics in intestinal health and disease: from biology to the clinic. Nat Rev Gastroenterol Hepatol 16:605–616. https://doi.org/10.1038/s41575-019-0173-3

Shang L, Fukata M, Thirunarayanan N, Martin AP, Arnaboldi P, Maussang D, Berin C, Unkeless JC, Mayer L, Abreu MT et al (2008) Toll-like receptor signaling in small intestinal epithelium promotes B-cell recruitment and IgA production in lamina propria. Gastroenterology 135:529–538. https://doi.org/10.1053/j.gastro.2008.04.020

Shariaty Z, Shan GR, Farajollahi M, Amerian M, Pour NB (2017) The effects of probiotic supplement on hemoglobin in chronic renal failure patients under hemodialysis: A randomized clinical trial. J Res Med Sci 22:74. https://doi.org/10.4103/jrms.JRMS_614_16

Sharma A, Shouche Y (2014) Microbial culture collection (MCC) and international depositary authority (IDA) at National Centre for cell science, Pune. Indian J Microbiol 54(2):129–133. https://doi.org/10.1007/s12088-014-0447-y

Skrypnik K, Bogdański P, Schmidt M et al (2019) The effect of multispecies probiotic supplementation on Iron status in rats. Biol Trace Elem Res 192:234–243. https://doi.org/10.1007/s12011-019-1658-1

Smits HH, Engering A, van der Kleij D, de Jong EC, Schipper K, van Capel TM, Zaat BA, Yazdanbakshs M, Wierenga EA, van Kooyk Y, Kapsenberg ML (2005) Selective probiotic bacteria induce IL-10-producing regulatory T cells in vitro by modulating dendritic cell function through dendritic cell-specific intercellular adhesion molecule 3-grabbing non- integrin. J Allergy Clin Immunol 115(6):1260–1267

Soccol CR et al (2010) The potential of probiotics. Food Technol Biotechnol 48(4):413–434

Soltan Dallal MM, Mojarrad M, Baghbani F, Raoofian R, Mardaneh J, Salehipour Z (2015) Effects of probiotic Lactobacillus acidophilus and Lactobacillus casei on colorectal tumor cells activity (CaCo-2). Arch Iran Med 18(3):167–172

Sommer F, Adam N, Johansson MEV, Xia L, Hansson GC, Bäckhed F (2014) Altered mucus glycosylation in core 1 O-glycan-deficient mice affects microbiota composition and intestinal architecture. PLoS One 9:e85254. https://doi.org/10.1371/journal.pone.0085254

Sonnenborn U, Schulze J (2009) The non-pathogenic Escherichia coli strain Nissle 1917 – features of a versatile probiotic. Microb Ecol Health Dis 21:122–158

Tabuchi M, Ozaki M, Tamura A, Yamada N, Ishida T, Hosoda M, Hosono A (2003) Antidiabetic effect of Lactobacillus GG in streptozotocin-induced diabetic rats. Biosci Biotechnol Biochem 67:1421–1424

Thakur N et al (2016) Probiotics: selection of criteria, safety and role in health and disease. J Innovative Biol 3(1):259–270

Thirabunyanon M, Hongwittayakorn P (2013) Potential probiotic lactic acid bacteria of human origin induce antiproliferation of colon cancer cells via synergic actions in adhesion to cancer cells and short-chain fatty acid bioproduction. Appl Biochem Biotechnol 169:511–525

Thomas C, Versalovic J (2010) Probiotics-host communication: modulation of signaling pathways in the intestine. Gut Microbes 1:148–163. https://doi.org/10.4161/gmic.1.3.11712

Toh H, Oshima K, Nakano A, Takahata M, Murakami M, Takaki T, Nishiyama H, Igimi S, Hattori M, Morita H (2013) Genomic adaptation of the Lactobacillus casei group. PLoS One 8:e75073. https://doi.org/10.1371/journal.pone.0075073

Tomaro-Duchesneau C, Saha S, Malhotra M, Jones M, Rodes L, Prakash S (2015) Lactobacillus fermentum NCIMB 5221 and NCIMB 2797 as cholesterol-lowering probiotic biotherapeutics: in vitro analysis. Benefic Microbes 6(6):861–869

Twetman L, Larsen U, Fiehn NE, Steck-sen-Blicks C, Twetman S (2009) Coaggregation between probiotic bacteria and caries- associated strains: an in vitro study. Acta Odontol Scand 67:284–288

van Tassell ML, Miller MJ (2011) Lactobacillus adhesion to mucus. Nutrients 3(5):613–636. https://doi.org/10.3390/nu3050613

Vesa TH, Korpela RA, Sahi T (1996) Tolerance to small amounts of lactose in lactose maldigesters. Am J Clin Nutr 64:197–201

Viljanen M, Pohjavuori E, Haahtela T et al (2005) Induction of inflammation as a possible mechanism of probiotic effect in atopic eczema-dermatitis syndrome. J Allergy Clin Immunol 115:1254–1259

Vyas U, Ranganathan N (2012) Probiotics, prebiotics, and synbiotics: gut and beyond. Gastroenterol Res Pract 2012:872716

Walter J, Heng NCK, Hammes WP, Loach DM, Tannock GW, Hertel C (2003) Identification of Lactobacillus reuteri genes specifically induced in the mouse gastrointestinal tract. Appl Environ Microbiol 69(4):2044–2051. https://doi.org/10.1128/aem.69.4.2044-2051.2003

Wang J, Tang H, Zhang C, Zhao Y, Derrien M, Rocher E (2015) Modulation of gut microbiota during probiotic-mediated attenuation of metabolic syndrome in high fat diet-fed mice. ISME J 9:1–15

Wattanarat O, Makeudom A, Sastraruji T et al (2015) Enhancement of salivary human neutrophil peptide 1–3 levels by probiotic supplementation. BMC Oral Health 15:19. https://doi.org/10.1186/s12903-015-0003-0

Wescombe PA, Heng NC, Burton JP, Chilcott CN, Tagg JR (2009) Streptococcal bacteriocins and the case for Streptococcus salivarius as model oral probiotics. Future Microbiol 4:81935

Xia X, Chen J, Xia J et al (2018) Role of probiotics in the treatment of minimal hepatic encephalopathy in patients with HBV-induced liver cirrhosis. J Int Med Res 46 (9):3596–3604. https://doi.org/10.1177/0300060518776064

Xiao SD, Zhang DZ, Lu H et al (2003) Multicenter, randomized, controlled trial of heat-killed Lactobacillus acidophilus LB in patients with chronic diarrhea. Adv Ther 20:253–260

Yadav H, Jain S, Sinha PR (2006) Effect of skim milk and dahi (yogurt) on blood glucose, insulin, and lipid profile in rats fed with high fructose diet. J Med Food 9:328–335

Yadav H, Jain S, Sinha PR (2007) Antidiabetic effect of probiotic dahi containing Lactobacillus acidophilus and Lactobacillus casei in high fructose fed rats. Nutrition 23:62–68

Yakult (2014) History. https://www.yakult.co.jp/english/pdf/eco2014_en.pdf
Yan F, Cao H, Cover TL, Whitehead R, Washington MK, Polk DB (2007) Soluble proteins produced by probiotic bacteria regulate intestinal epithelial cell survival and growth. Gastroenterology 132(2):562–575
Yoo SR, Kim YJ, Park DY, Jung UJ, Jeon SM, Ahn YT (2013) Probiotics L plantarum and L curvatus in combination alter hepatic lipid metabolism and suppress diet-induced obesity. Obesity 21:2571–2578
Yun SI, Park HO, Kang JH (2009) Effect of Lactobacillus gasseri BNR17 on blood glucose levels and body weight in a mouse model of type 2 diabetes. J Appl Microbiol 107:1681–1686
Zamani B, Golkar HR, Farshbaf S, Emadi-Baygi M, Tajabadi-Ebrahimi M, Jafari P, Akhavan R, Taghizadeh M, Memarzadeh MR, Asemi Z (2016) Clinical and metabolic response to probiotic supplementation in patients with rheumatoid arthritis: a randomized, double-blind, placebo-controlled trial. Int J Rheum Dis 19:869–879. https://doi.org/10.1111/1756-185X.12888
Zhang B, Zuo F, Yu R, Zeng Z, Ma H, Chen S (2015) Comparative genome-based identification of a cell wall-anchored protein from Lactobacillus plantarum increases adhesion of Lactococcus lactis to human epithelial cells. Sci Rep 5:14109. https://doi.org/10.1038/srep14109
Zhao X, Higashikawa F, Noda M, Kawamura Y, Matoba Y, Kumagai T et al (2012) The obesity and fatty liver are reduced by plant-derived Pediococcus pentosaceus LP28 in high fat diet-induced obese mice. PLoS One 7(2):e30696
Ziada DH, Soliman HH, El Yamany SA, Hamisa MF, Hasan AM (2013) Can Lactobacillus acidophilus improve minimal hepatic encephalopathy? A neurometabolite study using magnetic resonance spectroscopy. Arab J Gastroenterol 14(3):116–122. https://doi.org/10.1016/j.ajg.2013.08.002

# Chapter 5
# Fungi and Its By-Products in Food Industry: An Unexplored Area

**Ansar Alam, Komal Agrawal, and Pradeep Verma**

**Abstract** Fungi were an unsung microorganism before it was recognized for its bio-products and their utilization in daily human life. As a result of which the impact of fungi and its by-products in food and food industries is increasing day by day. They are rich sources of protein and are also utilized as animal protein replacement, e.g., mushroom, Quorn, nutritional yeast, etc. Fungal products such as amylase, cellulase, xylanase, pectinase, lipase, protease, etc. have applications in bread, brewing, milk processing, fruit juice processing, meat and fish processing. Most fermented foods, e.g., cheese, Koji rice are consumed in daily life and are produced due to the uprising practice of fungi and its enzymes. There are other fungal products as well which are used as food colouring agents. Thus, looking into the broad application of fungi in food, the present chapter will deal with the impact of fungi and its biomolecules in food, application along with its limitation and future aspects.

**Keywords** Fungi · By-products · Enzymes · Pigments · Fermented food

## 5.1 Introduction

The long search for food was an endless struggle for the survival of mankind. Our ancestors searched and ate numerous objects which resulted in the identification of few living plant and animal species as a principal source of food. The civilized agricultural habit has restricted the edible items to small number of plant and animal species indicating either the existence of small number of edible species or the lack of enthusiasm for food examination. In fact, man has used limited number of microorganisms for food and there are still numerous unverified microorganisms which can be potentially used as an alternative to satisfy the human urge for food. Matassa et al. (2016) stated that few strategies have been developed when the supply

A. Alam · K. Agrawal · P. Verma (✉)
Bioprocess and Bioenergy Laboratory, Department of Microbiology, Central University of Rajasthan, Bandarsindri, Kishangarh, Ajmer, Rajasthan, India
e-mail: pradeepverma@curaj.ac.in

 
P. K. Arora (ed.), *Microbial Products for Health, Environment and Agriculture*, Microorganisms for Sustainability 31, https://doi.org/10.1007/978-981-16-1947-2_5

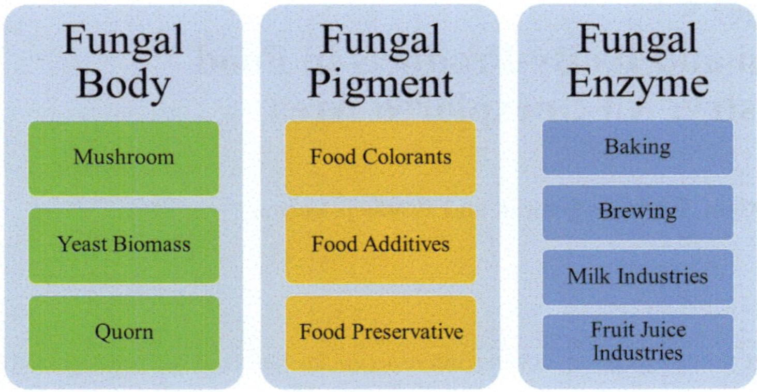


**Fig. 5.1** Diverse role of fungi in food

of food is concerned as human civilization progresses. Since centuries, man understands about use of microorganisms in preparation of better textured bread, wine and strong drink. These days man is powered with technology which allows them to point out an amazing stock of knowledge concerning microorganisms and in most of the aspect their potential as food source or application in food industries. Overgrowing population is now facing major crisis of food resulting in two-third of the population being malnourished. This has enabled the researchers to keep up the momentum or speed up the research about potential microorganism-based food (Berglund 2003) and its application at practical levels. The potential of some fungal species to contribute to the large portion of world protein supply is having great impact (Agrawal and Verma 2020; Agrawal et al. 2019; Bhardwaj et al. 2017). Though mushrooms of different type have been used by man for many years as food source, their influence may be considered negligible because they were used as condiments rather than as food staples (Fig. 5.1).

## 5.2 Contribution of Fungi in Food

Mushroom is an umbrella shaped reproductive fruiting structure that can be hypogeous or epigeous, carpophore or basidiocarp, large enough to be seen without magnification, termed as 'macrofungus' by Chang and Miles (1992). It has been used globally for many centuries for its gourmet assets, plus specific aroma and texture (Kalač 2009) and Asian countries like China, Japan, etc. are leading the foot step toward mushroom cultivation. They cultivated, *Auricularia auricularia* mushroom around 600 A.D (Kües and Liu 2000). Various mushroom varieties are valuable natural products because they are rich source of nutrients and also have biological activities (Mao et al. 2005). They contain high amount of protein and low energy content because of low level of fat molecules which makes them an admirable food for caloric diets. They provide different dietary nutrients, phenolic compounds, vitamins and their strong antioxidant capacities make them good as functional foods and as a source of bioactive compounds (Furlani and Godoy 2008; Kalač 2009; Vaz et al. 2011). It also acts as a diverse range of secondary metabolites such as phenolic compounds, polyketides, terpenes and steroids (Ishikawa et al. 1984). Various products produced via mushroom are used as a functional ingredient. Functional breads baked with mushrooms reflect valuable health benefit (Lin et al. 2008). The use of *Grifola frondosa, Hypsizygus marmoreus,* or *Pholiota nameko* mushroom (10%) with wheat flour decreases the loaf and specific volume of the bread and changes its functional properties as well (Okamura-Matsui et al. 2003).

Increasing population pressures for food increased the fungal contributions toward mankind's food supply. In present time, a notable inquisitive contradiction occurs in the human race with fungal species. At present time, in scientific and industrial sphere, research and development for fungal proteins is getting impetus (Agrawal and Verma 2021a, 2021b; Agrawal et al. 2021; Bhardwaj et al. 2020). Mushrooms have been cultured for many centuries in East Asian countries and are now becoming an important agricultural product throughout the globe (Matassa et al. 2016). China is leading in cultivation of mushroom species. More than 60 mushroom species have been cultivated around the world and around 40 different mushroom species have been cultivated only in China. The mushroom species *Lentinula*, *Pleurotus*, *Flammulina*, *Auricularia*, *Pholiata*, *Tremella*, *Agrocybe*, *Ganoderma* are cultivated on fresh wood residues, while some on slightly composted lignocellulosic materials (e.g., *Volvariella*, *Stropharia*, *Coprinus*), some on well-composted materials or animal dung (e.g., *Agaricus*), while others grow on soil and humus (e.g., *Lepiota*, *Leptista*, *Morchella*, *Gyromitra*). *Agaricus bispora*, *Lentinula edodes* and *Pleurotus* spp. are the popular species which are cultivated all around the world. Mushroom was considered for their therapeutic wealth in early civilization of the Chinese, Egyptians and Romans. Romans believed that it provides health and strength, praised as food of the gods and Chinese cherished as food of life (Miles and Chang 2004). Chang (1999) stated that Chinese were the first who cultivated *Auricularia auricula-judae* in 600 AD and *Lentinus edodes* mushroom between 1000 AD. *Agaricus bisporus* was cultivated in France in 1600 and in 1900 United States of America cultivated *Pleurotus bisporus*. Raghevendra et al. (2017) has accounted that for delicious taste, nutritional and medicinal value, mushroom has gained notable interest in recent time along with its cultivation around the globe as nutritional and protein rich agricultural product (Table 5.1).

Kim et al. (2011) stated that the mycelium of mushroom *Agaricus bisporus* was used as meat alternative. Mushroom is more preferred as meat analog over plant derived protein for their textural properties. *Agaricus bisporus* mycelium displayed better sensory assessment over soy protein (Kim et al. 2011). Because of short postharvest shelf life mushroom *Agaricus bisporus* is used as a flavouring agent. The powdered version increases the utility of mushroom as it is utilized as soup mixing and seasonings (Singh et al. 2003) agents which showed good sensory scores after 8 months of storage. Park et al. (2001) studied the production of natural food seasonings with mushroom *Pleurotus ostreatus* and *Lentinus edodes* and concluded

**Table 5.1** The reported nutritional value of various mushrooms

| S. No. | Mushroom species | Protein content | Fat content | Carbohydrate content | References |
|---|---|---|---|---|---|
| 1 | *Pleurotus Florida* | 34.56 | 2.11 | 42.83 | Alam et al. (2008) |
| 2 | *Pleurotus sajor-caju* | 36.75 | 2.22 | 31.40 | Michael et al. (2011) |
| 3 | *Pleurotus ostreatus* | 30.92 | 1.68 | 37.8 | Michael et al. (2011) |
| 4 | *Calocybe indica* | 021.4 | 4.95 | 048.5 | Alam et al. (2008) |
| 5 | *Lactarius hatsudake* | 15.3 | 1.0 | 38.2 | Yin and Zhou (2008) |
| 6 | *Lentinula edodes* | 17.1 | 1.9 | 13.4 | Zhu et al. (2007) |
| 7 | *Tricholoma matsutake* | 14.3 | 5.0 | 36.7 | Liu et al. (2010) |
| 8 | *Boletus edulis* | 26.5 | 2.8 | 65.4 | Ouzouni and Riganakos (2007) |
| 9 | *Agaricus arvensis* | 56.3 | 2.7 | 37.5 | Barros et al. (2007) |
| 10 | *Cantharellus cibarius* | 53.7 | 2.9 | 31.9 | Barros et al. (2008) |
| 11 | *Ramaria botrytis* | 39.0 | 1.4 | 50.8 | Barros et al. (2008) |

that their nucleotides are reason for savoury flavour. Food seasoned with *Pleurotus ostreatus* makes more score upon sensory assessment over *lentinus edodus*. Yoo et al. (2007) reported the *Lentinus edodes* can be used as novel functional resource for natural seasoning as it showed antioxidant activity too. Han et al. (2006) stated that *Lentinus edodus* can be used over *Agaricus bisporus* as flavouring component in brown sauce. Hong et al. (2009) reported that the use of *Tricholoma matsutake* for flavourful apple sauce can be maintained and stored for 60 days and also showed tolerable sensory scores. Leskosek-Cukalovic et al. (2010) reported that *Ganoderma lucidum* was used in traditional Chinese medicine. It contains several bioactive compounds which exhibited unique health benefits, extract of it added in brewing process led to the harvest of new type of beer with improved functionality. Its powder is added in Korean traditional rice wine called Yakju, which has best sensory score. Lin et al. (2008) reported that addition of 2–7% *Lentinus edodus* stipe with wheat flour showed high fibre content.

Various possible form of mushroom for bread making was studied (Ulziijargal et al. 2013; Okafor et al. 2012; Jeong and Shim 2004; Lee et al. 2004, 2009). Mostly the powder form is added to bread but *Ganoderma lucidum* is added as extract (Chung et al. 2004). Kim et al. (2011) conducted experiment with *Lentinus edodus* and *Pleurotus eryngii* powder in foodstuff such as muffins and cookies. Chung et al. (2004) reported that mushroom powder was added in pork patties to enhance its texture, juiciness and functionality. Cha et al. (2014) used *Tremella fuciformis* mushroom to upsurge oil holding capacity of pork patties. *Agaricus bisporus* singly accounted for more than 70% of total mushroom production globally and is the most common cultivated mushroom species. In the twentieth century, the conservative market of *Agaricus* breached by wider variety of mushroom called exotic mushroom. Fresh supply of exotic mushroom like *Lentinula, Pleurotus, Flammulina, Volvariella,* etc. species are found alongside *Agaricus* in the local markets (Moore and Chiu 2001). There is remarkable increase found to compete with *Agaricus* and is due to the substrate used for the production of these mushroom. Oyster mushroom (*Pleurotus ostreatus*, *P. cystidiosus*, *P. sajor-caju*) species grow on cotton waste, while straw mushroom *(Volvariella volvacea)* traditionally grow on rice straw but also can grow on cotton waste. Mushroom crop farming is good example of waste remediation.

Truffle is very similar to mushroom like fruiting body but grows under soil surface symbiotically with plant root to accomplish their life cycle. However, it is an Ascomycetes fruiting body which has an irregular shape and size with 2–8 cm diameter. It has a long history and praised as a gourmet food. Bokhary and Parvez (1993) stated that it has been used as food for 3000 years. The black truffle, *Tuber melanosporum* was praised as the 'diamond of French cuisine' in France. These are the highly appreciated for texture and aroma in many countries having good demand for white and black truffles. Truffles are one of the expensive delicacies ascocarp body (Kagan-Zur and Roth-Bejerano 2008; Wang and Marcone 2011). Bokhary and Parvez (1993) reported that fresh *Tuber claveryi* is rich source of protein (16%) and carbohydrate (28%). The nutritional composition of truffle estimated that the dry biomass contains 20–27% protein, 85% which is digestible by humans; 3–7.5% fat (unsaturated and saturated fatty acids); 7–13% crude fibre, approximately 60% carbohydrates; and 2–5% ascorbic acid. It also represents a wealthy source of volatile compound, regarded for traditional herbal medicine and exceptional in being consumed for its nutraceutical value (Kagan-Zur and Roth-Bejerano 2008).

## 5.3 The Fungal Mycelium as Source of Protein

The importance of production of single-cell protein is getting momentum as fungi make hope to solve the world's food shortage by industrial production of alternative meat protein and resolve the problem with fungal biotechnology (Moore and Chiu 2001). Only one prominent product named Quorn is introduced as meat substitute in the market in 1984 (Bamforth and Cook 2019). Brewer's yeast has been also used as a co-product of fermentation for food substitute of human and animals. Using brewery wastes to produce yeast product are, respectively, two most fruitful applications for fungal protein in food industry. But the successful model with fungal product available in the market is mycoprotein Quorn. This is the mycelium of filamentous soil-fungus *Fusarium.* Wilson (2001) has stated that in late 1960s a British food company Ranks Hovis McDougall (RHM) did screening for suitable mycoprotein producing fungi and results come in favour of *Fusarium* species. The species *Fusarium moniliforme* cultivated on cheap substrate produces a biomass of high nutritional value (Drouliscos et al. 1976; Macris and Kokke 1978). Another species, i.e., *Fusarium graminearum* is another strain gaining attention for the production of commercial mycoprotein where the filamentous structure is similar to the fibrous nature of meat (Seifert et al. 2003). Moore and Chiu (2001) stated that fungal biomass contains characteristic nutritional properties like low-fat, low calorie and cholesterol-free health food which allow consumers to choose Quorn as a meat alternative. Rodger (2001) has compared freshly harvested mycoprotein with egg for their nutritional value, found that it has good composition of all essential amino acids with 12% (w/w, wet weight) protein content. It contains 6% dietary fibre, 1:4 ratio of saturated and unsaturated fatty acid except cholesterol. A comparison of mycoprotein with chicken and tofu and the amount of food consumed is less than other two respective non-mycoprotein meat, it was also stated that mycoprotein has significant effect on appetite (Williamson et al. 2006). Turnbull and Ward (1995) analysed glucose and insulin level of blood has been found reduced after having mycoprotein meal. It is also a very useful low-energy food source for control of body weight and for diabetes dietary (Williamson et al. 2006). Quorn is sold as meat alternative for vegetarian people at premium price rate. Some study is also focused on cultivation of fungal species like *Trichoderma, Penicillium, Rhizopus arrhizus* and *Aspergillus oryzae* (Moore and Chiu 2001) on inexpensive cellulosic wastes.

## 5.4 Nutritional Yeast

Yeasts are single celled fungi and come under Ascomycetes and basidiomycete group. They predominantly reproduce by budding or fission and grow in vegetative state. Reed and Nagodawithana (1988) reported that since 7000 BC yeast were used for the production of fermented food. The biomass of yeast is called nutritional yeast, used as single-cell protein (SCP). The yeast biomass is rich in fats, carbohydrates, nucleic acid, vitamins and minerals with additional essential amino acids like lysine and methionine (Gonçalves et al. 2014; Nayak 2011). It is used as nutritional source which is limited in most plant and animal foods (Adedayo et al. 2011; Suman et al. 2015; Uchakalwar and Chandak 2014). Nutritional yeast cells are killed and heat deactivated for nutritional components. Burgents et al. 2004 stated that yeast SCP has inclusive amino acid range and high protein carbohydrate proportion, used for high nutrient forage substitute.

*Saccharomyces cerevisiae* and its various strains are most popular species applied for various purposes such as baking, brewing and probiotics (Fleet 2007). Yeast biomass from fruit waste and brewing process are widely used as nutrient rich food additive for human and animal feed, they can use cheap raw material, several wastes to produce biomass, protein and amino acid. According to Suman et al. (2015) both conventional and unconventional substrate are used for biomass and metabolite production where yeast can grow on wastes to help in its remediation (Adedayo et al. 2011). They need comparatively small amount of substrate to produce large amount of biomass in a short time period compared to plants and animals. Nutritional yeasts are eco-friendly because they help in waste remediation. In the developing countries, the budding pattern of population growth has resulted in food scarcity problem (Suman et al. 2015). Uchakalwar and Chandak (2014) reported that according to World Food and Agriculture Organization (FAO), 25% of the world population are suffering from protein deficiency where the use of nutritional yeast biomass as nutrient supplement can resolve the problem associated with food on earth. However, yeast protein are not commonly believed as a protein supplement and nutritional yeast production has not gained importance. For its global acceptance, one needs to find new methods by which it can be incorporated into food (Suman et al. 2015). According to Nasseri et al. (2011) drying at high temperature and under defined condition, the digestibility and shelf life of nutritional yeast-based food product enhances. Further, in developing country, nutritional yeast biomass can fulfil the diet supplement and can be used for humans and livestock (Jach et al. 2015).

## 5.5 Fungal Enzymes in Food Industry

Enzymes play an important role in food processing industries as bread making and brewing were dependent on enzymes. Enzymes play a prominent role in the food industry because the enzyme facilitates less energy requirement and high specificity for enzymatic conversation which makes it attractive workstation in the food industries. According to Vermelho and Noronha (2013) enzymes involved in the food industries are specific and break down the complex molecules into simpler ones. There are various sources to get enzymes but the fungal enzymes show advantages upon animal, vegetal origin and even bacterial. The enzyme obtained from animal and plant has advantage of their low-cost production, whereas the animal derived enzyme has limitation like ethical parameters, shortage and health of animal and their glands. Plant derived enzymes depend upon climate condition, soil properties and seed quality. These are the major factors which change the track toward microbial enzyme production (Vermelho and Noronha 2013). Advancement on genetic modification and control at every stage of production improves the attraction toward fungal enzymes to use for production at commercial scale (Fig. 5.2). The food and beverage industries are chiefly using hydrolases and transferases enzymes. These are mostly extracellular (Bhardwaj et al. 2021; Bhardwaj and Verma 2021; Chaturvedi et al. 2021), some of them are discussed as follows (Table 5.2):

### *5.5.1 Amylase*

Amylase is widely applied enzyme in the various fields of food and beverage industries. Wheat is the most important crop around the world and wheat product bread satisfies one-third protein and half of carbohydrates requirement of Europeans (Uhlig 1998). According to Mutsaers (1997) Egyptians were the first who made use

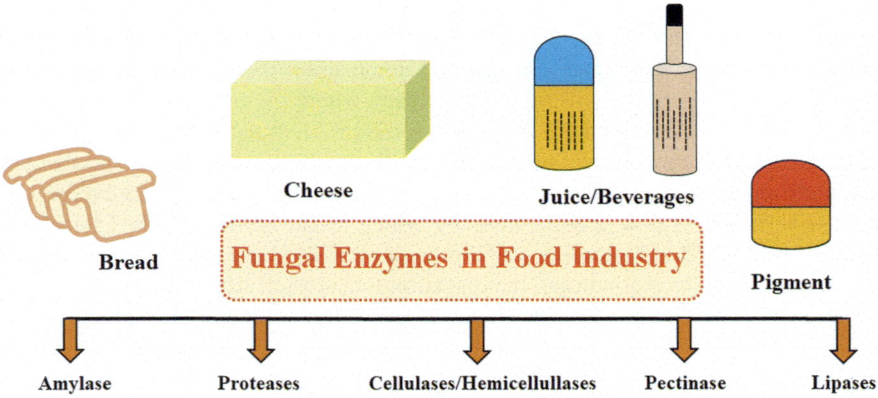


**Fig. 5.2** Utilization of various fungal enzymes in food industries

**Table 5.2** Different Enzymes and their application

| S. No. | Microorganism | Enzyme | Food Application | References |
|---|---|---|---|---|
| 1 | *Saccharomyces cerevisiae, Penicillium occitanis* | Pectinases | Mash treatment, juice clarification | Sharma et al. (2013), Maktouf et al. (2014) |
| 2 | *Penicillium roqueforti* | Proteases | Protein hydrolysis, milk clotting, low-allergenic infant food, enhanced digestibility and utilization, flavour improvement in milk and cheese, meat tenderizer, prevention of chill haze formation in brewing | Larsen et al. (1998) |
| 3 | *Aspergillus oryzae* | Amylases | Starch liquefaction and saccharification, increasing shelf life and improving quality by retaining moist, elastic and soft nature, bread softness and volume, flour adjustment, ensuring uniform yeast fermentation, juice treatment, low calorie beer | Gupta et al. (2003) |
| 4 | *Aspergillus terreus* | Cellulases | Animal feed, clarification of fruit juice | Narra et al. (2012) |
| 5 | *Aspergillus oryzae* | Xylanase | Viscosity reduction, enhanced digestibility, dough conditioning | Camacho and Aguilar (2003) |
| 6 | *Rhizomucor miehei* | Lipase | Cheese flavour development, cheddar cheese production | Jooyandeh et al. (2009) |

of enzyme in flour without knowing the exact effect. Starch is the main constituent of bread which causes hardness and unpleasant to eat. According to Singh et al. (2019) addition of amylase in dough reduces the starch crystallization and extends the shelf life of bread. It is also the first enzyme in industry which is applied in bread making flour to reduce viscosity and crumb structure of baked food. According to Raveendran et al. (2018) it is a most widely used enzyme in baking industries as anti-staling and flavour enhancement agent to improve bread quality. It converts starch of flour into smaller dextrin molecules, which is further fermented by yeast. By breaking down complex molecules into smaller molecules it generates sugar molecules in dough, which improves bread toasting quality, taste and crust colour. Couto and Sanroman (2006) reported that amylase is also used in brewing, digestive aid, fruit juice and cake production. Fungal acid amylase and amyloglucosidase are used in food industries to process fruit. Various fungi produce acid α-amylase and amyloglucosidase (Fogarty and Kelly 1980). De Souza et al. (2010) stated that these enzymes are applied to prevent post-bottling haze formation in fruit juice industries. Acid endoamylase breaks down amylose and amylopectin to dextrin, further degraded by amyloglucosidase hydrolase into glucose. Blanco et al. (2014) stated that addition of exogeneous glucoamylase to beer mash leads to additional amount of glucose in wort for fermentation but it reacts with oligosaccharide and not averts the starch haze. According to a published report, for saccharification, exo-glucoamylases of *Aspergillus niger* is used to treat starch hydrolysate for concentrate glucose syrup production. Chemical hydrolysis of starch in food processing industries is totally replaced by microbial amylases and high specificity with substrate has resulted in distinct physicochemical properties of generated products (Paloheimo et al. 2010; Amoozegar et al. 2003).

### 5.5.2 *Proteases*

Due to wide spectrum application, protease accounts for 60% of the total industrial enzymes present in the market. Proteases are most commonly used with broad-spectrum industrial application because it has several advantages over traditional chemical catalysts. Fungal proteases are generally more stable compared to animal and plant derivatives. Fungi have faster growth rate which make it more suitable for production to get abundant quantities (Rani et al. 2012). Protease is applied in various food industries like baking, brewing, meat tenderization, fruit juice pulping, etc. Wheat flour has unique property to form viscoelastic gluten matrix. This matrix gives unique baking property as it can hold gas in the dough. It is reported that specific fungal protease alters the gluten structure which improved extensibility, better structure and volume of product. Some proteases have the ability to influence the flavour and colour of the product as well. Mamo and Assefa (2018) stated that in the beer formation process, protease from *Saccharomyces fibuligera* and *Torulopsis magnolia* could inhibit the formation of hazes and acidic proteases from *Aspergillus niger* produced almost haze free beer. In the juice industries different type of proteases from species *Aspergillus niger* are used to juice clarification (Pinelo et al. 2010). Odour, colour and texture of meat are an important parameter for its merchantability. Tenderization of meat improves softness of meat, digestibility and ease to chopping. Singh et al. (2019) stated that proteases from fungus *Aspergillus Oryzae* is used for tenderization of meat. Chancharoonpong et al. (2012) has reported that at industrial production, protease enzyme from fungi *Aspergillus Oryzae* is used to reduce the salt brine fermentation time duration. Singh et al. (2019) stated that protease enzyme has great application in diary industries. Casein and whey proteins are the two most important substrates extracted from raw milk. Specific protease enzymes are used for controlled hydrolysis of protein. It is used in fortifying sport and health drinks, infant nutrition and nutritional foods. Cheese an another milk derivative is extensively used by mankind in his food habit. Ripening of cheese is a process; it is due to proteolytic breakdown of casein protein (McSweeney and Sousa 2000). Proteases as coagulants play an important role in the cheese ripening process. At lower pH protease enzyme needs to be more specific. For enhanced ripening process proteases would be added to cheese milk.

### 5.5.3 Cellulases/Hemicellulases

Cellulase and hemicellulases have ability to break down complex structure of plant biomass (Kumar et al. 2018). Cellulase is the third largest enzyme used after protease and amylase (Bajaj and Mahajan 2019). Hemicelluloses are made up of xylan, xyloglucan, glucomannan, arabinogalactan and other heteropolymers, while cellulose is homopolymer of β-1-4-linked glucose molecules. According to Juturu and Wu (2014) cellulases have been used for the extraction and clarification of fruit and vegetable juice to increase their yields. They used to macerate the fruit pulp for partial or complete liquefaction, to reduce processing time, complete liquefaction and improve extraction of valuable fruit components, these complex enzymes are used to macerate the fruit pulp. The use of these enzymes increases yield and process performance without additional capital investment (Kuhad et al. 2011, Karmakar and Ray 2011). Karmakar and Ray (2011) stated that after juice extraction, these enzymes also help to lower down juice viscosity and stability of product. Bajaj and Mahajan (2019) stated that use of the enzymes to predigest the feed makes the nutrients of the feed more absorbable; therefore, less amount of feed is required for the similar amount of weight gain of the animals. Murad and Melim Miguel et al. (2013) stated that both cellulase and xylanases are used in baking industry to increase dough softness and decrease its stickiness. These complex enzyme systems break down the polysaccharides into different simpler glucose molecules. Breakdown of these polysaccharides helps in redistributing the water and improves the overall strength and the quality of the flour (Bajaj and Mahajan 2019). These enzymes applied to frozen dough retain its properties for long duration under storage condition. Both enzymes are also used in brewing industry to help in the process of malting and macerating. These enzymes degrade the cell wall and release simple sugar, phenolic compound and aroma. They increase the yield, stability and decrease the viscosity and haze formation of wine (Sharma et al. 2014).

### 5.5.4 *Pectinase*

Pectinase enzyme can degrade high molecular weight heteropolymer polysaccharide, a major component of cell wall of plants (Khan et al. 2013). Pectinases are mixture of different enzymes which hydrolyse different part of pectin molecule (Pedrolli et al. 2009). Hoondal et al. (2002) stated that alkaline pectinase has been used in coffee and tea fermentation, while acidophilic pectinases have been extensively used in the extraction and clarification of fruit juice and wine. Fast clarification of juice resulted in a shorter process and improved quality of juice. The use of pectinases prevents post-bottling haze formation which results in smaller storage volume, better concentrate stability without spoilage. The combination with other enzymes such as hemicellulase, amylase decreases the viscosity of pulp and increases yield. Jayani et al. (2010) stated that pectinase enzyme treatment accelerates tea fermentation and also terminates the froth forming property of instant tea powders by destroying pectin. They are also used in coffee fermentation to remove mucilaginous coat from coffee beans. According to Will et al. (1992) raspberry and strawberry require enzymatic maceration and depectinization for production of clear concentrate juice. It improves juice production without disturbing any organoleptic properties of fruit.

### 5.5.5 *Lipases*

Lipases are serine hydrolases commonly used to esterification. In the food industries they are used to enhance texture and flavouring (Barbe et al. 2009) and good impact as qualitative and quantitative production. In food industry it enhances and quickens the development of flavouring aromatic substance. In cheese industries it is applied in the breakdown of milk fat in free fatty acid which puts characteristic flavour to cheese. In baking industries, it breaks down the wheat lipid into emulsifying lipid and creates flavour. It also increases bread volume, improves stability of gas cells in dough and allows it for prolonged shelf life (Carlson 1981). Phospholipase enzyme catalyses the hydrolysis of fatty acid and forms more water soluble lysophospholipid. It is also used for the production of lysolecithin which has broad range of application in food industries. Parmesan cheese is a strong flavoured product due to aldehyde and ketone compound. Fungal lipase releases long chain fatty acid which acts as substrate for oxidation. The oxidation produces flavoured compounds like aldehyde and ketone. Polyunsaturated fatty acid plays an essential part in human nutrition. Lipase enzyme have characteristic features and lipase from *Rhizomucor miehei* has good activity in low water condition. For the cocoa butter alternative, high-oleate sunflower oil and stearic acid are used with immobilized lipase enzyme.

Other enzymes such as glucose oxidase, galactosidase, transglutaminase, catalase, superoxide dismutase, lactoperoxidase and sulfhydryl oxidase have limited applications. In food industries, galactosidase enzymes have a broad-spectrum application. They are used for sugar-syrup, prebiotic production (Contesini et al. 2013). Transglutaminase enzyme has the ability to create cross link in proteinaceous substance. It catalyses the acyl transfer between glutamine and lysine amino acid. It plays a very important role in dairy food products like yoghurt production, renneting of fresh cheese making and cream whipping (Agrawal et al. 2018; Lantto et al. 2007).

The use of enzymes for food applications has increased steadily over the past two decades, not only in traditional application areas such as starch processing, brewing, fruit processing and dairies. In applications such as baking, the use of enzymes has grown even more (Godfrey and Partner 2000). The appearance of new enzyme applications is due to the increasing diversity of the enzymes available, the majority based on gene modification technology. In future new enzymes for food applications are expected to lead to major developments in the use of industrially produced enzymes (Godfrey and Partner 2000).

## 5.6 Fungal Pigments as Food Additives

Food industry and processed foods became technologically advanced due to industrial revolution. In the late nineteenth century, chemically synthesized colours replaced natural colourants. However, safety concerns arose with increasing uses of synthetic colouring agents. Consumer awareness regarding the use of synthetic colouring agent led to shift toward natural food colourants where microorganism and insects are replacing the synthetic colours. Mukherjee et al. (2017) have stated that some fungi like *Aspergillus, Fusarium, Penicillium, Trichoderma* and *Laetiporus* have been reported for production of various pigments such as quinones, anthraquinones, rubropunctatin, monascin, b-carotene, etc. accountable.

for various colours like red, purple, yellow, orange and brown. Gupta et al. (2013) have stated that due to scarcity of nutrition, fungi produce pigment known as secondary metabolites. In Asian countries, the production of pigments dates to hundreds of years (Mapari et al. 2005). Dufosse (2014) has stated that fungus *Monascus purpureus* produces red colour pigment. Mukherjee et al. (2017) have reported that fungi *Monascus* can produce various colours like yellow, red and orange. Joshi et al. (2003) stated that fungus *Blakeslea trispora* is used by European countries for the industrial production of pigments like β-carotene and lycopene. According to Garbayo et al. (2003) light influenced the mycelial growth of fungus *Gibberella fujikuroi* and presence of light induced to produce orange pigment.

Fungal pigment has been used as food colourants in the late nineteenth century and some of the fungal pigments have been approved as well (Dufosse et al. 2014). Pigments are used in cheese, candy, fruits, beverages, snacks, beer and wine wherein pigments such as riboflavin are used in beverages, ice creams and instant desserts (Chattopadhyay et al. 2008). As reported these pigments have antimicrobial and antifungal properties, different foods require different concentrations of pigment to produce the desired colour intensity which may vary the spoilage time duration of foodstuff. Cheng-Yun and Wen-Ping (2008) reported addition of 0.03 g of *Monascus* pigment per kg of sausage. El-Kholie et al. (2012) reported a high concentration of *Monascus* pigments. Rojsuntornkitti et al. (2010) noted that addition of 0.1–0.4 g of red rice powder to 100 g of Thai sausage was able to control the growth of *Salmonella, Staphylococcus aureus* and *Clostridium perfringens.*

## 5.7 Limitation and Future Prospect

Due to source, chemical nature and catalytic activity, some fungi and their associated biomolecules show threat. This exhibits allergenicity, activity related toxicity and chemical toxicity. Fungal enzymes are possible allergens and have strong effects if inhaled as dust. Quorn has many aids but it is a probable allergen too. There is low sensitization of Quorn but it might counter badly to patients who are allergic to mould. Mushroom can play an imperative part in the nourishing man and also animal without doing ill effects but some mushrooms species are known as poisonous to man only. The fungal pigment can loss their property with different pH, temperature variation, heat, etc. Food and feed processing industries have potential success using fungi and their associated products. In coming period genetic manipulation of microorganism may play a vital role in the production of enzymes and pigment. Genetically manipulated strain will be superior to the wild strain. There is need to advance the food processing technology to reduce the cost and time of processed food items. Therefore, more development and research are required in the area of commercially viable enzyme and pigment production.

## References

Adedayo MR, Ajiboye EA, Akintunde JK, Odaibo A (2011) Single cell proteins: as nutritional enhancer. Adv Appl Sci Res 2(5):396–409

Agrawal K, Verma P (2020) Production optimization of yellow laccase from *Stropharia* sp. ITCC 8422 and enzyme-mediated depolymerization and hydrolysis of lignocellulosic biomass for biorefinery application. Biomass Convers Biorefin:1–20

Agrawal K, Verma P (2021a) Applications of biomolecules of endophytic fungal origin and its future prospect. In: Fungi bio-prospects in sustainable agriculture, environment and nanotechnology. Academic Press, pp 207–230.

Agrawal K, Verma P (2021b) Fungal metabolites: a recent trend and its potential biotechnological applications. In: New and future developments in microbial biotechnology and bioengineering. Elsevier, pp 1–14

Agrawal K, Chaturvedi V, Verma P (2018) Fungal laccase discovered but yet undiscovered. Bioresour Bioprocess 5(1):1–12

Agrawal K, Bhardwaj N, Kumar B, Chaturvedi V, Verma P (2019) Process optimization, purification and characterization of alkaline stable white laccase from *Myrothecium verrucaria* ITCC-8447 and its application in delignification of agroresidues. Int J Biol Macromol 125:1042–1055

Agrawal K, Alam A, Verma P (2021) Bioprospecting and applications of fungi: a game changer in present scenario. Recent trends in mycological research: volume 2: environmental and industrial perspective, 1st edn. Springer, Cham

Alam N, Amin R, Khan A, Ara I, Shim MJ, Lee MW, Lee TS (2008) Nutritional analysis of cultivated mushrooms in Bangladesh–Pleurotus ostreatus, Pleurotus sajor-caju, *Pleurotus florida* and *Calocybe indica*. Mycobiology 36(4):228–232

Amoozegar MA, Malekzadeh F, Malik KA (2003) Production of amylase by newly isolated moderate halophile, *Halobacillus* sp. strain MA-2. J Microbiol Methods 52(3):353–359

Bajaj P, Mahajan R (2019) Cellulase and xylanase synergism in industrial biotechnology. Appl Microbiol Biotechnol 103(21–22):8711–8724

Bamforth CW, Cook DJ (2019) Food, fermentation, and micro-organisms. Wiley, Hoboken, NJ

Barbe S, Lafaquiere V, Guieysse D, Monsan P, Remaud-Siméon M, Andre I (2009) Insights into lid movements of *Burkholderia cepacia* lipase inferred from molecular dynamics simulations. Proteins Struct Funct Bioinform 77(3):509–523

Barros L, Baptista P, Correia DM, Casal S, Oliveira B, Ferreira IC (2007) Fatty acid and sugar compositions, and nutritional value of five wild edible mushrooms from Northeast Portugal. Food Chem 105(1):140–145

Barros L, Venturini BA, Baptista P, Estevinho LM, Ferreira IC (2008) Chemical composition and biological properties of Portuguese wild mushrooms: a comprehensive study. J Agric Food Chem 56(10):3856–3862

Berglund BE (2003) Human impact and climate changes—synchronous events and a causal link? Quat Int 105(1):7–12

Bhardwaj N, Verma P (2021) Microbial xylanases: a helping module for the enzyme biorefinery platform. Bioenergy research: evaluating strategies for commercialization and sustainability. Wiley

Bhardwaj N, Chanda K, Kumar B, Prasad HK, Sharma GD, Verma P (2017) Statistical optimization of nutritional and physical parameters for xylanase production from newly isolated *Aspergillus oryzae* LC1 and its application in the hydrolysis of lignocellulosic agro-residues. Bioresources 12(4):8519–8538

Bhardwaj N, Kumar B, Agrawal K, Verma P (2020) Bioconversion of rice straw by synergistic effect of in-house produced ligno-hemicellulolytic enzymes for enhanced bioethanol production. Bioresour Technol Rep 10:100352

Bhardwaj N, Agrawal K, Kumar B, Verma P (2021) Role of enzymes in deconstruction of waste biomass for sustainable generation of value-added products. In: Bioprospecting of enzymes in industry, healthcare and sustainable environment. Springer, Singapore, pp 219–250

Blanco CA, Caballero I, Barrios R, Rojas A (2014) Innovations in the brewing industry: light beer. Int J Food Sci Nutr 65(6):655–660

Bokhary HA, Parvez S (1993) Chemical composition of desert truffles *Terfezia claveryi*. J Food Compos Anal 6(3):285–293

Burgents JE, Burnett KG, Burnett LE (2004) Disease resistance of Pacific white shrimp, *Litopenaeus vannamei*, following the dietary administration of a yeast culture food supplement. Aquaculture 231(1-4):1–8

Camacho NA, Aguilar G (2003) Production, purification, and characterization of a low-molecular-mass xylanase from Aspergillus sp. and its application in baking. Appl Biochem Biotechnol 104 (3):159–171

Carlson TLG (1981) Law and order in wheat flour dough: colloidal aspects of the wheat flour dough and its lipid and protein constituents in aqueous media (doctoral dissertation, Lund: university.)

Cha MH, Heo JY, Lee C, Lo YM, Moon B (2014) Retracted: quality and sensory characterization of white jelly mushroom (*Tremella fuciformis*) as a meat substitute in pork patty formulation. J Food Process Preserv 38(3):1018–1023

Chancharoonpong C, Hsieh PC, Sheu SC (2012) Enzyme production and growth of *Aspergillus oryzae* S. on soybean koji fermentation. Apcbee Procedia 2:57–61

Chang ST (1999) World production of cultivated edible and medicinal mushrooms in 1997 with emphasis on *Lentinus edodes* (Berk.) sing, in China. Int J Med Mushrooms 1(4):291–300

Chang ST, Miles PG (1992) Mushroom biology—a new discipline. Mycologist 6(2):64–65

Chattopadhyay P, Chatterjee S, Sen SK (2008) Biotechnological potential of natural food grade biocolorants. Afr J Biotechnol 7(17):2972–2985

Chaturvedi V, Agrawal K, Verma P (2021) Chicken feathers: a treasure cove of useful metabolites and value-added products. Environ Sustain:1–13

Cheng-Yun L, Wen-Ping W (2008) Preservation and coloring of monascus color and sorghum red. Meat Res 7:46–49

Chung HC, Lee JT, Kwon OJ (2004) Bread properties utilizing extracts of Ganoderma lucidum (GL). J Korean Soc Food Sci Nutr 33(7):1201–1205

Contesini FJ, de Alencar Figueira J, Kawaguti HY, de Barros Fernandes PC, de Oliveira Carvalho P, da Graça Nascimento M, Sato HH (2013) Potential applications of carbohydrases immobilization in the food industry. Int J Mol Sci 14(1):1335–1369

Couto SR, Sanromán MA (2006) Application of solid-state fermentation to food industry—a review. J Food Eng 76(3):291–302

De Souza SGAE, Lajolo FM, Genovese MI (2010) Chemical composition and antioxidant/antidiabetic potential of Brazilian native fruits and commercial frozen pulps. J Agric Food Chem 58(8):4666–4674

Drouliscos NJ, Macris BJ, Kokke R (1976) Growth of *Fusarium moniliforme* on carob aqueous extract and nutritional evaluation of its biomass. Appl Environ Microbiol 31(5):691–694

Dufosse L, Fouillaud M, Caro Y, Mapari SA, Sutthiwong N (2014) Filamentous fungi are large-scale producers of pigments and colorants for the food industry. Curr Opin Biotechnol 26:56–61

El-Kholie EM, El Shaer MK, Abdelreheem MA, Gharib MA (2012) Detailed evaluation of a newly attained fungal pigment from *Monascus purpureus* in meat burgers. Int J Food Sci Nutr 63 (7):860–865

Fleet GH (2007) Yeasts in foods and beverages: impact on product quality and safety. Curr Opin Biotechnol 18(2):170–175

Fogarty WM, Kelly CT (1980) Amylases, amyloglucosidases and related glucaneses. In: Rose AH (ed) Microbial enzymes and bioconversions, vol 5. Academic Press, New York, pp 115–170

Furlani RPZ, Godoy HT (2008) Vitamins B1 and B2 contents in cultivated mushrooms. Food Chem 106(2):816–819

Garbayo I, Vılchez C, Nava-Saucedo JE, Barbotin JN (2003) Nitrogen, carbon and light-mediated regulation studies of carotenoid biosynthesis in immobilized mycelia of *Gibberella fujikuroi*. Enzym Microb Technol 33(5):629–634

Godfrey T, Partner ET (2000) Enzymes for grain processing-the global market. In VTT symposium, Vol. 207, VTT; 1999, pp. 141–148

Gonçalves FAG, Colen G, Takahashi JA (2014) *Yarrowia lipolytica* and its multiple applications in the biotechnological industry. Sci World J 2014:476207

Gupta C, Sharma D, Aggarwal S, Nagpal N (2013) Pigment production from *Trichoderma* spp. for dyeing of silk and wool. Int J Sci Nature 4(2):351–355

Gupta R, Gigras P, Mohapatra H, Goswami VK, Chauhan B (2003) Microbial α-amylases: a biotechnological perspective. Process Biochem 38(11):1599–1616

Han CW, Lee MY, Seong SK (2006) Quality characteristics of the brown sauce prepared with *Lentinus edodes* and *Agaricus bisporus*. J East Asian Soc Diet Life 16(3):364–370

Hong JY, Choi YJ, Kim MH, Shin SR (2009) Study on the quality of apple dressing sauce added with pine mushroom (*Tricholoma matsutake* sing) and chitosan. Korean J Food Preserv 16 (1):60–67

Hoondal G, Tiwari R, Tewari R, Dahiya NBQK, Beg Q (2002) Microbial alkaline pectinases and their industrial applications: a review. Appl Microbiol Biotechnol 59(4-5):409–418

Ishikawa Y, Morimoto K, Hamasaki T (1984) Flavoglaucin, a metabolite of *Eurotium chevalieri*, its antioxidation and synergism with tocopherol. J Am Oil Chem Soc 61(12):1864–1868

Jach ME, Serefko A, Sajnaga E, Kozak E, Poleszak E, Malm A (2015) Dietary supplements based on the yeast biomass. Curr Topics Nutraceutical Res 13(2):83–88

Jayani RS, Shukla SK, Gupta R (2010) Screening of bacterial strains for polygalacturonase activity: its production by *Bacillus sphaericus* (MTCC 7542). Enzyme Res 2010:306785

Jeong CH, Shim KH (2004) Quality characteristics of sponge cakes with addition of *Pleurotus eryngii* mushroom powders. J Korean Soc Food Sci Nutr 33:716–722

Jooyandeh H, Amarjeet K, Minhas KS (2009) Lipases in dairy industry: a review. J Food Sci Technol 46(3):181–189

Joshi VK, Attri D, Bala A, Bhushan S (2003) Microbial pigments. Int J Biotechnol 2(3):362–369

Juturu V, Wu JC (2014) Microbial cellulases: engineering, production and applications. Renew Sust Energ Rev 33:188–203

Kagan-Zur V, Roth-Bejerano N (2008) Desert truffles. Fungi 1(3):32–37

Kalač P (2009) Chemical composition and nutritional value of European species of wild growing mushrooms: a review. Food Chem 113(1):9–16

Karmakar M, Ray RR (2011) Current trends in research and application of microbial cellulases. Res J Microbiol 6(1):41–53

Khan M, Nakkeeran E, Umesh-Kumar S (2013) Potential application of pectinase in developing functional foods. Annu Rev Food Sci Technol 4:21–34

Kim K, Choi B, Lee I, Lee H, Kwon S, Oh K, Kim AY (2011) Bioproduction of mushroom mycelium of *Agaricus bisporus* by commercial submerged fermentation for the production of meat analogue. J Sci Food Agric 91(9):1561–1568

Kües U, Liu Y (2000) Fruiting body production in basidiomycetes. Appl Microbiol Biotechnol 54 (2):141–152

Kuhad RC, Gupta R, Singh A (2011) Microbial cellulases and their industrial applications. Enzyme Res 2011:280696

Kumar B, Bhardwaj N, Alam A, Agrawal K, Prasad H, Verma P (2018) Production, purification and characterization of an acid/alkali and thermo tolerant cellulase from *Schizophyllum commune* NAIMCC-F-03379 and its application in hydrolysis of lignocellulosic wastes. AMB Express 8 (1):1–16

Lantto R, Puolanne E, Katina K, Niemistö M, Buchert J, Autio K (2007) Effect of laccase and transglutaminase on the textural and water-binding properties of cooked chicken breast meat gels. Eur Food Res Technol 225(1):75–83

Larsen MD, Kristiansen KR, Hansen TK (1998) Characterization of the proteolytic activity of starter cultures of *Penicillium roqueforti* for production of blue veined cheeses. Int J Food Microbiol 43(3):215–221

Lee JY, Lee K, Kwak EJ (2009) Fermentation characteristics of bread added with *Pleurotus eryngii* powder. J Korean Soc Food Sci Nutr 38(6):757–765

Lee MJ, Kyung KH, Chang HG (2004) Effect of mushroom (*Lentinus Tuber-Regium*) powder on the bread making properties of wheat flour. Korean J Food Sci Technol 36(1):32–37

Leskosek-Cukalovic I, Despotovic S, Lakic N, Niksic M, Nedovic V, Tesevic V (2010) *Ganoderma lucidum*—medical mushroom as a raw material for beer with enhanced functional properties. Food Res Int 43(9):2262–2269

Lin LY, Tseng YH, Li RC, Mau JL (2008) Quality of shiitake stipe bread. J Food Process Preserv 32(6):1002–1015

Liu G, Wang H, Zhou B, Guo X, Hu X (2010) Compositional analysis and nutritional studies of *Tricholoma matsutake* collected from Southwest China. J Med Plants Res 4(12):1222–1227

Macris BJ, Kokke R (1978) Continuous fermentation to produce fungal protein. Effect of growth rate on the biomass yield and chemical composition of *Fusarium moniliforme*. Biotechnol Bioeng 20(7):1027–1035

Maktouf S, Neifar M, Drira SJ, Baklouti S, Fendri M, Châabouni SE (2014) Lemon juice clarification using fungal pectinolytic enzymes coupled to membrane ultrafiltration. Food Bioprod Process 92(1):14–19

Mamo J, Assefa F (2018) The role of microbial aspartic protease enzyme in food and beverage industries. J Food Qual 2018:7957269

Mao XB, Eksriwong T, Chauvatcharin S, Zhong JJ (2005) Optimization of carbon source and carbon/nitrogen ratio for cordycepin production by submerged cultivation of medicinal mushroom *Cordyceps militaris*. Process Biochem 40(5):1667–1672

Mapari SA, Nielsen KF, Larsen TO, Frisvad JC, Meyer AS, Thrane U (2005) Exploring fungal biodiversity for the production of water-soluble pigments as potential natural food colorants. Curr Opin Biotechnol 16(2):231–238

Matassa S, Boon N, Pikaar I, Verstraete W (2016) Microbial protein: future sustainable food supply route with low environmental footprint. Microb Biotechnol 9(5):568–575

McSweeney PL, Sousa MJ (2000) Biochemical pathways for the production of flavour compounds in cheeses during ripening: a review. Lait 80(3):293–324

Melim Miguel AS, Martins-Meyer TS, da Costa Figueiredo EV, Paulo Lobo BW, Dellamora–Ortiz, G.M. (2013) Enzymes in bakery. Curr Future Trends 1:287–321

Michael HW, Bultosa G, Pant LM (2011) Nutritional contents of three edible oyster mushrooms grown on two substrates at Haramaya, Ethiopia, and sensory properties of boiled mushroom and mushroom sauce. Int J Food Sci Technol 46(4):732–738

Miles PG, Chang ST (2004) Mushrooms: cultivation, nutritional value, medicinal effect, and environmental impact. CRC Press, Boca Raton, FL

Moore D, Chiu SW (2001) Fungal products as food. In: Pointing SB, Hyde KD (eds) Bio-exploitation of filamentous fungi. Fungal Diversity Research Series, vol 6. Cambridge University Press, Cambridge, UK, pp 223–251

Mukherjee G, Mishra T, Deshmukh SK (2017) Fungal pigments: an overview. In: Developments in fungal biology and applied mycology. Springer, Singapore, pp 525–541

Mutsaers JHGM (1997) The application of enzymes in baking. In proceedings of the 1st European symposium on enzymes and grain processing, pp 2–4

Narra M, Dixit G, Divecha J, Madamwar D, Shah AR (2012) Production of cellulases by solid state fermentation with *Aspergillus terreus* and enzymatic hydrolysis of mild alkali-treated rice straw. Bioresour Technol 121:355–361

Nasseri AT, Rasoul-Amini S, Morowvat MH, Ghasemi Y (2011) Single cell protein: production and process. Am J Food Technol 6(2):103–116

Nayak SK (2011) Biology of eukaryotic probiotics. In: Probiotics. Springer, Berlin, Heidelberg, pp 29–55

Okafor JNC, Okafor GI, Ozumba AU, Elemo GN (2012) Quality characteristics of bread made from wheat and Nigerian oyster mushroom (*Pleurotus plumonarius*) powder. Pak J Nutr 11(1):5–10

Okamura-Matsui T, Tomoda T, Fukuda S, Ohsugi M (2003) Discovery of alcohol dehydrogenase from mushrooms and application to alcoholic beverages. J Mol Catal B Enzym 23(2-6):133–144

Ouzouni P, Riganakos K (2007) Nutritional value and metal content profile of Greek wild edible fungi. Acta Aliment 36(1):99–110

Paloheimo M, Piironen J, Vehmaanperä J (2010) Xylanases and cellulases as feed additives. In: Enzymes in farm animal nutrition. CABI, Wallingford, UK, pp 12–53

Park MJ, Lee JS, Lee B, Lee JS (2001) Development of natural seasoning based on mushroom. J East Asian Soc Diet Life 11(3):196–203

Pedrolli DB, Monteiro AC, Gomes E, Carmona EC (2009) Pectin and pectinases: production, characterization and industrial application of microbial pectinolytic enzymes. Open Biotechnol J 3:9–18

Pinelo M, Zeuner B, Meyer AS (2010) Juice clarification by protease and pectinase treatments indicates new roles of pectin and protein in cherry juice turbidity. Food Bioprod Process 88 (2-3):259–265

Raghavendra VB, Venkitasamy C, Pan Z, Nayak C (2017) Functional foods from mushroom. In: Microbial functional foods and nutraceuticals, vol 65. Wiley, Hoboken, NJ

Rani K, Rana R, Datt S (2012) Review on latest overview of proteases. Int J Curr Life Sci 2 (1):12–18

Raveendran S, Parameswaran B, Beevi Ummalyma S, Abraham A, Kuruvilla Mathew A, Madhavan A, Pandey A (2018) Applications of microbial enzymes in food industry. Food Technol Biotechnol 56(1):16–30

Reed G, Nagodawithana TW (1988) Technology of yeast usage in winemaking. Am J Enol Vitic 39 (1):83–90

Rodger G (2001) Production and properties of mycoprotein as a meat alternative. Food Technol 55 (7):36–41

Rojsuntornkitti K, Jittrepotch N, Kongbangkerd T, Kraboun K (2010) Substitution of nitrite by Chinese red broken rice powder in Thai traditional fermented pork sausage (Nham). Int Food Res J 17:153–161

Seifert KA, McMullen CR, Yee D, Reeleder RD, Dobinson KF (2003) Molecular differentiation and detection of ginseng-adapted isolates of the root rot fungus *Cylindrocarpon destructans*. Phytopathology 93(12):1533–1542

Sharma HP, Patel H, Sharma S (2014) Enzymatic extraction and clarification of juice from various fruits—a review. Trends Post Harvest Technol 2(1):1–14

Sharma N, Rathore M, Sharma M (2013) Microbial pectinase: sources, characterization and applications. Rev Environ Sci Biotechnol 12(1):45–60

Singh R, Singh A, Sachan S (2019) Enzymes used in the food industry: friends or foes? In: Enzymes in food biotechnology. Academic Press, Cambridge, MA, pp 827–843

Singh S, Ghosh S, Patil GR (2003) Development of a mushroom-whey soup powder. Int J Food Sci Technol 38(2):217–224

Suman G, Nupur M, Anuradha S, Pradeep B (2015) Single cell protein production: a review. Int J Curr Microbiol App Sci 4(9):251–262

Turnbull WH, Ward T (1995) Mycoprotein reduces glycemia and insulinemia when taken with an oral-glucose-tolerance test. Am J Clin Nutr 61(1):135–140

Uchakalwar PR, Chandak AM (2014) Production of single cell protein from fruits waste by *Saccharomyces cerevisiae*. Int J Adv Biotechnol Res 5:770–776

Uhlig, H. (Ed.). (1998) Industrial enzymes and their applications. Wiley, Hoboken, NJ

Ulziijargal E, Yang JH, Lin LY, Chen CP, Mau JL (2013) Quality of bread supplemented with mushroom mycelia. Food Chem 138(1):70–76

Vaz JA, Barros L, Martins A, Santos-Buelga C, Vasconcelos MH, Ferreira IC (2011) Chemical composition of wild edible mushrooms and antioxidant properties of their water soluble polysaccharidic and ethanolic fractions. Food Chem 126(2):610–616

Vermelho, A.B., and Noronha, E.F. (2013). Filho EX; Ferrara MA; Bon EPS Diversity and biotechnological applications of prokaryotic enzymes The Prokaryotes, Springer, Berlin 213-240

Wang S, Marcone MF (2011) The biochemistry and biological properties of the world's most expensive underground edible mushroom: truffles. Food Res Int 44(9):2567–2581

Will F, Hasselbeck G, Dietrich H (1992) Charakterisie-rung einer gelartigen Trübung aus Himbeersaft. Flüss Obst 59:352–353

Williamson DA, Geiselman PJ, Lovejoy J, Greenway F, Volaufova J, Martin CK, Ortego L (2006) Effects of consuming mycoprotein, tofu or chicken upon subsequent eating behaviour, hunger and safety. Appetite 46(1):41–48

Wilson D (2001) Marketing mycoprotein: the Quorn foods story. Food Technol 55(7):48–50

Yin JZ, Zhou LX (2008) Analysis of nutritional components of 4 kinds of wild edible fungi in Yunnan. Food Res Develop 29:133–136

Yoo SJ, Kim SH, Choi HT, Oh HT, Choi HJ, Ham SS (2007) Antioxidative, antimutagenic and cytotoxic effects of natural seasoning using *Lentinus edodes* powder. J Korean Soc Food Sci Nutrition 36(5):515–520

Zhu XQ, Wang XJ, Xiong Z (2007) Nutrient analysis of the wild Lentinula edodes. In: Forest by-product and speciality in China, vol 2. Chinese Forestry Publishing House, Beijing, China, pp 9–11

# Chapter 6
# Biogenic Synthesis of Nanomaterials Toward Environment-Friendly Approach

**Suman Das and Dhermendra K. Tiwari**

**Abstract** Nanotechnology driven materials hosted all avenues of science for next-generation technology development. Nanomaterial synthesized by various approaches such as physical, chemical, and biological methods to achieve the defined shape, size, and morphology. Former two methods undoubtedly create high-quality nanomaterials with self-controlled and user-specific modifications in the synthesis procedure to optimize output. However, these methods are not environmentally sustainable and eco-friendly for bulk production. Biological systems created enormous scopes for eco-friendly and cheaper nanomaterial and a variety of nanomaterials has been produced in the last few years. This chapter summarizes some common biological systems such as bacteria, fungi, actinomycetes, algae, and plants used to produce various metallic and non-metallic nanomaterials and their biological applications.

**Keywords** Biopolymers · Nanomaterials · Nanoparticles · Nanotechnology

## 6.1 Introduction

Nanotechnology is one of the most emerging fields nowadays, which mainly deals with nanometer-sized particles or items (Feynman 1991). A material exhibits new property upon reduction of its size into nanoscale which is generally not observed in its macroscale or bulk form (Bogunia-Kubik and Sugisaka 2002). Nanoparticles can be categorized into three different types such as incidental nanoparticles, natural nanoparticles, and engineered nanoparticles (Buzea et al. 2007). The use of nanomaterials in biotechnology integrates biology with material science. Nanoparticles with their unique properties, such as size, surface area to volume ratio, and shapes like a rod or spherical, demonstrate a wide range of applications in

S. Das · D. K. Tiwari (✉)
Department of Biotechnology, Faculty of Life Sciences and Environment, Goa University, Taleigao Plateau, Goa, India
e-mail: dktiwari@unigoa.ac.in

 
P. K. Arora (ed.), *Microbial Products for Health, Environment and Agriculture*, Microorganisms for Sustainability 31, https://doi.org/10.1007/978-981-16-1947-2_6

the field of biology as well as other research areas (Murray et al. 2000). The usefulness and specific characteristics of nanoparticles arise from different aspects, including the equivalent size of the nanoparticles and biomolecules, such as nucleic acid and proteins (Ferrari 2005). There are numerous conventional physical and chemical methods for nanoparticle synthesis, but these pathways sometimes are given rise to many environmental challenges and are extremely pricey (Remya et al. 2017). For instance, processes like pyrolysis and chemical precipitation often result in the deposition of toxic chemical species on the surface of newly synthesized nanoparticles. The presence of such surface contaminants makes these nanoparticles undesirable for biomedical and clinical applications (Ai et al. 2011). Silver nanoparticles synthesized by chemical processes exhibited toxic effects in the human body when used for the treatment of various diseases. Owing to these facts, many researchers start to explore biological systems for nanoparticle synthesis, which are considered safe, cost-effective, and at the same time, sound eco-friendly (Thakkar et al. 2010). Nature has given ways and insight into the advanced synthesis of nanomaterials. Several studies reported that the biological systems have the potential to act as 'bio-laboratory" for the fabrication of pure metal and metal oxide particles without any surface contaminants at the nanometer scale using a biomimetic approach. The biogenic synthesis of nanoparticles involves microorganisms like bacteria, virus, fungi, algae, and plants which acts as reducing agents having the high capability of metal uptake and at the same time, the toxic substances produced during the synthesis process can easily be destroyed with the help of enzymes produced by the plants or microorganisms (Sarkar et al. 2012; Menon et al. 2017). The biological approach for the synthesis of nanoparticles provides excellent stability, polydispersity, and dimensions when compared to chemically synthesized nanoparticles (Chen et al. 2008). The biogenic methods used to synthesize nanoparticles allow the synthesis to occur at physiological pressure, pH, and temperature and thereby eliminate ruthless processing conditions. A considerable number of microorganisms have been found to possess the capability of synthesizing nanoparticles either extra or intracellularly. In recent years, nanoparticles play an important role in many fields, including environment, energy, agriculture, and healthcare due to their implausible properties (Raveendran et al. 2003). In this chapter, we are discussing the approaches and applications of nanoparticle synthesized using biological systems.

## 6.2 Biological Systems for Nanomaterial Synthesis

Our main objective is to focus on the synthesis of nanoparticles utilizing biological approaches due to its easiness of rapid synthesis, size characteristics controlling, controlled toxicity, and eco-friendly nature. A variety of natural sources exists from which nanoparticles could be synthesized including bacteria, fungi, algae, virus, actinomycetes, and plants (Fig. 6.1) (Kumar and Yadav 2009; Saifuddin et al. 2009; Balaji et al. 2009; Ali et al. 2011; Ahmad et al. 2003a; Lee et al. 2002). These

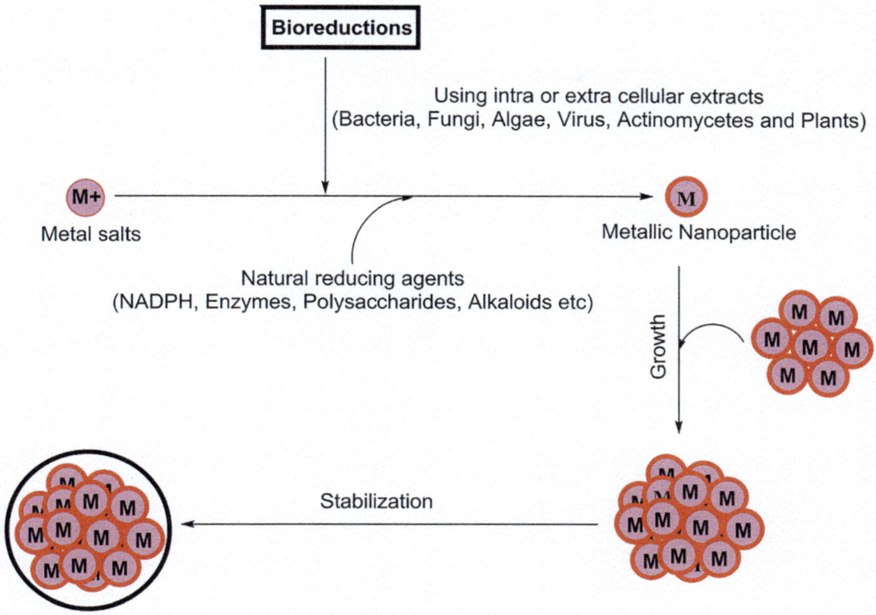


**Fig. 6.1** Schematic representation of the mechanism of biogenic nanoparticle synthesis

unicellular or multicellular organisms are capable of synthesizing inorganic nanoparticles extracellularly or intracellularly (Govindappa et al. 2016). Here we discussed all the sources which are considered as a potential source for biological nanoparticle synthesis.

## 6.3 Bacteria Mediated Nanoparticle Synthesis

In recent years, nanoparticle synthesis using bacteria is extremely increased because of its vast application. Bacterial production of nanoparticles may be intracellular or extracellular. The extracellular synthesis of nanoparticles by bacteria involves enzymatic reactions that convert metallic ions into metallic nanoparticles. In extracellular synthesis, bacteria are cultured under optimum conditions for 1–2 days and centrifuge the culture to remove the biomass. The remaining supernatant is then added to a filter-sterilized solution of a metal salt and is incubated again for nanoparticle synthesis. While in the case of an intracellular mechanism the whole biomass is used instead of supernatant and involves specific ion transportation into the negatively charged cell wall and due to the electrostatic attraction, the positively charged metal ions get diffused through the cell wall. Then the toxic metal ions converted into non-toxic metal nanoparticles by the action of enzymes present in the bacterial cell wall (Pacioni et al. 2015; Khandel and Shahi 2016). This suggests that the

**Table 6.1** Biopolymers derived from bacteria and their application

| Biopolymer | Bacteria | Application | Reference |
|---|---|---|---|
| Alginate | *Pseudomonas aeruginosa*, *Azotobacter vinelandii* | Drug delivery, tissue engineering, wound management | Pawar and Edgar (2012); Hay et al. (2013); Boekhoven et al. (2015) |
| Cyanophycin derivatives | Cyanobacteria, recombinant strains of *Escherichia coli*, *Ralstonia eutropha* and *Corynebacterium glutamicum* | Drug delivery | Sallam and Steinbüchel (2010) |
| Dextran | *Leuconostoc mesenteroides*, *Streptococcus mutans*, *Lactobacillus brevis* | Tissue engineering, drug delivery, vascular and blood applications | Vu et al. (2009); Mokhtarzadeh et al. (2016); Brøndsted et al. (1998) |
| Gellan | *Pseudomonas elodea*, *Sphingomonas paucimobilis* | Drug delivery | Morris et al. (2012); D'Arrigo et al. (2014) |
| Poly γ-glutamic acid (PGA) | *Bacillus subtilis*, *Bacillus licheniformis*, *Staphylococcus epidermidis* | Vaccine carriers, tissue engineering, biosensors, drug delivery, diagnostics, bioimaging | Ashiuchi (2013); Pereira et al. (2012); Tsai et al. (2014); Maya et al. (2014); Shu et al. (2014) |
| Polyhydroxyalkanoates | *Bacillus sp.*, *Acromonas sp.*, *Pseudomonas sp.*, *Rhodobacter sp.*, *Ralstonia sp.* | Tissue engineering, bioimaging, wound dressing in surgery, biosensors, wound management, diagnostics | Park et al. (2012); Leong et al. (2014); Wang et al. (2003); Masood et al. (2015) |
| Poly-L lysine | *Streptomyces albulus* | Biosensors, drug delivery, non-viral gene delivery vector | Pandey and Kumar (2014); Sasaki et al. (2013); Yang et al. (2014) |
| Xanthan | *Xanthomonas campestris* | Drug delivery | Rosalam and England (2006); Luo and Wang (2014) |

primary mechanism of nanoparticle synthesis by bacteria relies on enzymes (Zhang et al. 2011); for example, the enzyme called nitrate reductase has been found responsible for the synthesis of silver nanoparticles in *B. licheniformis* (Elbeshehy et al. 2015). Bacteria during fermentation produce a variety of water-soluble biopolymers which are considered as biodegradable, non-toxic, and biocompatible nanomaterials (Rodríguez-Carmona and Villaverde 2010). These biopolymers along with their source and application were summarized in Table 6.1. In recent years different species of bacteria were explored for silver, gold, and other metallic nanoparticle synthesis. Bacteria like *Magnetospirillum magnetotacticum* could be used for bio-remediation of Fe (III) metals via reduction where they took up iron and re-oxidized from low-density hydrous oxide to high-density ferrihydrite or Fe(III) oxide. In the last step, Fe(III) ions were reduced to produce magnetite within the magnetosome vesicles. Accumulation of iron within the vesicles occurs due to the presence of an intracellular protein called Ferritin, which kept it soluble and non-toxic (Pacioni et al. 2015). The thermophilic bacteria could be an exceptional tool for the extracellular synthesis of silver and gold nanoparticles (Gomathy and Sabarinathan 2010). Some *Bacillus* species showed their ability to reduce silver and synthesize extracellularly fabricated, circular nanoparticles, ranging in size from 10 to 20 nm (Sunkar and Nachiyar 2012). *Pseudomonas stutzeri AG259* isolated from silver mines and textile soil exhibit the accumulation of silver nanoparticles in their periplasmic space (Slawson et al. 1994; Rajora et al. 2016). A novel strain of *Marinobacter pelagius* was reported to synthesize monodisperse and stable gold nanoparticles (Sharma et al. 2012). Prasad and coworkers used *Lactobacillus* strains for the synthesis of titanium nanoparticles (Prasad et al. 2007). *Lactobacillus species* was also reported for silver, gold, and nanocrystals of gold-silver alloy synthesis. *Pseudomonas aeruginosa*'s cell supernatant was used for the synthesis of gold nanoparticles (Husseiny et al. 2007). It was reported that *E. coli* DH5α could synthesize gold nanoparticles on their cell surface (Du et al. 2007). In the presence of exogenous electron donor, a sulfate-reducing bacteria called *Desulfovibrio desulfuricans* NCIMB 8307 can synthesize palladium nanoparticles (Omajali et al. 2015). Studies also found that *Plectonema boryanum* UTEX485, filamentous cyanobacteria when treated with an aqueous solution of $AuCl_4^-$ and $Au(S_2O_3)_2^{-3}$ under specific conditions, they synthesize octahedral and cubic gold nanoparticles (Lengke et al. 2006). Nanoparticles synthesized from various bacterial sources together with their characteristics are listed in Table 6.2.

## 6.4 Fungi Mediated Nanoparticle Synthesis

Mycosynthesis is another approach for easy and stable biological nanoparticle synthesis using fungal biomass and their associated metabolites. Most fungi possess important metabolites with the higher capability of bioaccumulation and simple downstream processing is easy to culture for the production of low-cost and efficient nanoparticles (Alghuthaymi et al. 2015). They also have greater uptake competences and tolerances to metals, which provide an advantage for the high-yield production of nanoparticles as metal salts have a high wall-binding capability with fungal biomass (Alghuthaymi et al. 2015; Castro-Longoria et al. 2011). The process which is commonly used to synthesize nanoparticles using fungi is depicted in Fig. 6.2. Researchers proposed three possible mechanisms through which mycosynthesis of metal nanoparticles occur, including electron shuttle quinones, nitrate reductase action, or both (Alghuthaymi et al. 2015). The enzymes present in fungi such as the α-NADPH-dependent reductase and nitrate reductase have been found to have an important role in the synthesis of nanoparticle, similar to the extracellular nanoparticle synthesis mechanism found in bacteria (Kumar et al. 2007). The fungal organisms produce active biomolecules which controlled the

**Table 6.2** Bacterial-derived nanoparticles and their characteristics

| Bacterial strain | Nanoparticles | Mode of synthesis | Size | Shape | Reference |
|---|---|---|---|---|---|
| *Bacillus subtilis* | Silver and gold | Extracellular and intracellular | 5–10 | Anisotropic | Reddy et al. (2010) |
| *Bacillus licheniformis, Bacillus pumilus* and, *Bacillus persicus* | Silver | Extracellular | 77–92 | Spherical, triangular and hexagonal | Elbeshehy et al. (2015) |
| *Bacillus amyloliquefaciens* | Cadmium sulfide | Extracellular | 3–4 | Cubic/ hexagonal | Singh et al. (2011) |
| *Bacillus methylotrophicus* | Silver | Extracellular | 10–30 | Spherical | Wang et al. (2016) |
| *Bhargavaea indica* | Silver and gold | Extracellular | 30–100 | Silver anisotropic; gold, flower | Singh et al. (2015a, 2016a) |
| *Listeria monocytogenes* | Silver | – | Varied sizes | Anisotropic | Soni and Prakash (2015) |
| *Brevibacterium frigoritolerans* | Silver | Extracellular | 10–30 | Spherical | Singh et al. (2015b) |
| *Pseudomonas deceptionensis* | Silver | Extracellular | 10–30 | Spherical | Jo et al. (2016) |
| *Weissella oryzae* | Silver | Intracellular | 10–30 | Spherical | Singh et al. (2016b) |
| *Pseudomonas aeruginosa* | Gold | Extracellular | 15–30 | – | Husseiny et al. (2007) |
| *Pseudomonas stutzeri* | Silver | Intracellular | 200 | – | Klaus et al. (1999) |
| *Lactobacillus sp.* | Silver and gold | Intracellular | 60 | – | Sintubin et al. (2009) |
| *Escherichia coli* | Cadmium sulfide | Intracellular | 2–5 | Spherical | Kowshik et al. (2002a) |
| *Escherichia coli DH5α* | Gold | Intracellular | 25–33 | Spherical | Du et al. (2007) |
| *Clostridium thermoaceticum* | Cadmium sulfide | Extracellular and intracellular | 2–5 | – | Cunningham and Lundie (1993) |
| *Streptomyces albidoflavus* | Silver | Intracellular | 10–14 | – | Buddana (2012) |
| *Rhodopseudomonas capsulata* | Gold | Extracellular | 10–20 | Spherical | Syed et al. (2016) |
| *Klebsiella pneumonia* | Gold | Extracellular | 10–15 | Spherical | Prema et al. (2016) |
| *Plectonema boryanum* UTEX485 | Gold | Extracellular | 10–25 | Octahedral and cubic | Lengke et al. (2006, 2007) |
| *Marinobacter pelagius* | Gold | Extracellular | <20 | Spherical | Sharma et al. (2012) |
| *Stenotrophomonas maltophilia* | Gold | Extracellular | 40 | Spherical | Nangia et al. (2009) |

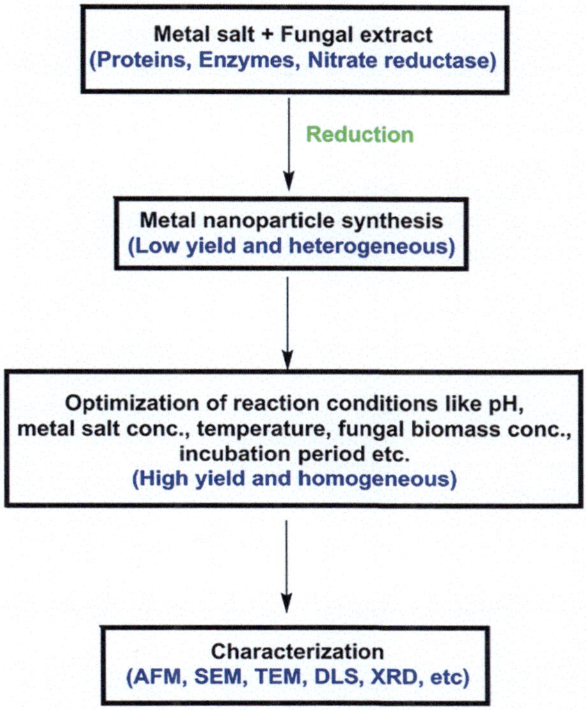


**Fig. 6.2** Mycogenic synthesis of metal nanoparticles

nanoparticle's biochemical compositions, size distribution, and shape. Fungi absorbed gold ions and subsequently produce gold nanoparticles intracellularly. The active molecules involved in the energy metabolism of fungal cells such as proteins like 3-glucan binding proteins, glyceraldehyde-3-phosphate dehydrogenase, ATPase, and reducing sugars could be used for mycosynthesis of nanoparticles. When ultrathin sections of the Au-fungal cells were studied, it was observed that gold nanoparticles were accumulated within the vacuoles of the cells (Menon et al. 2017). Some fungal strains such as *Aspergillus oryzae, Aspergillus niger,* and *Fusarium solani* were reported to have the capability of producing silver nanocrystals extracellularly (Binupriya et al. 2010a; Gade et al. 2008; Ingle et al. 2009). *Trichoderma viride* was reported to synthesize spherical nanoparticles (Thakkar et al. 2010). The silver nanoparticles produced by *Phoma glomerata* showed its antimicrobial efficiency against *P. aeruginosa, E.coli,* and *S. aureus* (Birla et al. 2009). It was found that *Trichothecium* sp. produced extracellular nanoparticles when incubated with gold ions under static conditions whereas they synthesized intracellular gold nanoparticles under shaking conditions. The possible reason behind this might be the proteins and enzymes secreted by the fungus during stationary conditions were not secreted by them during shaking conditions (Sowani et al. 2016). The *Penicillium* genus is one of the superior candidates for the synthesis of silver nanoparticles, where production occurs through the extracellular mechanism (Sadowski et al. 2008). The microscopic fungus yeast strain MKY3 under forced ecological conditions can synthesize silver nanoparticles when incubated with aqueous silver nitrate (Kowshik et al. 2002b). Dameron and coworkers reported cadmium synthesis by using *Schizosaccharomyces pombe* and *Candida glabrata* (Dameron et al. 1989). It was reported that a strain of extremophilic yeast isolated from acid mine drainage showed the capability of gold and silver nanoparticle synthesis (Mourato et al. 2011). Kowshik et al. reported the intracellular synthesis cadmium sulfide (CdS) nanocrystallites by using *Torulopsis sp.*, which exhibit quantum semiconductor properties (Kowshik et al. 2002a). It was also reported that *S. cerevisiae* biosorbes and reduces gold ions to elemental gold in the cell wall peptidoglycan layer by the aldehyde group of the reducing sugars (Khandel and Shahi 2018). The marine yeast *Rhodosporidium diobovatum* has been investigated for the stable synthesis of lead sulfide nanoparticles intracellularly (Seshadri et al. 2011). Similarly, Pichia jadinii was used to synthesize gold nanoparticles of various morphologies intracellularly (Velusamy et al. 2016). Nanoparticles synthesized from various fungal sources together with their characteristics are listed in Table 6.3.

## 6.5 Algae Mediated Nanoparticle Synthesis

Algae are photosynthetic organisms with significant ecological and economic importance. They may be unicellular such as *Chlorella* and the diatoms or multicellular such as the giant kelps (large brown alga), found in a variety of environments such as marine water, freshwater and, or moist rock surfaces (Thajuddin and Subramanian 1992, 2005; Oscar et al. 2014). They are categorized as macroscopic (macroalgae) and microscopic (microalgae). They are considered an important source for different commercial products such as biofuels and natural dyes (Lee 2018; Johansen 2012; Borowitzka 2013; Sing et al. 2013). Till now, researchers have explored various groups of algae for the biosynthesis of nanoparticles such as Cyanophyceae, Chlorophyceae, Rhodophyceae, Phaeophyceae, euglenoids, and diatoms (Sharma et al. 2016). Algae can accumulate metals and subsequently reduced metal ions into nanoparticles, which makes them an important candidate for the biosynthesis of nanoparticles (Sharma et al. 2019). They are referred to as "bionanofactories" as the dried biomasses of both dead and live algae could be used for the metallic nanoparticle synthesis (Davis et al. 1998). There are mainly three important steps for the algal synthesis of nanoparticles: first, algal extract preparation in an aqueous solvent (organic or water) by boiling or heating for a certain time, second, preparation of appropriate concentration of the ionic metallic compound and third, incubation of ionic metallic compounds and algal solutions under controlled conditions with or without continuous stirring for a certain period (Thakkar et al. 2010; Rauwel et al. 2015). The basic mechanism of nanoparticle synthesis by fungi is that the functional groups and the enzymes present in the algal cell wall react with the precursor molecules or metal ions to form complexing agents at ambient conditions causing

**Table 6.3** Fungal-derived nanoparticles and their characteristics

| Fungal strain | Nanoparticles | Mode of synthesis | Size | Shape | Reference |
|---|---|---|---|---|---|
| *Neurospora crassa* | Silver, gold, bimetallic silver and gold | Intracellular and extracellular | >100 | Quasi-spherical | Castro-Longoria et al. (2011) |
| *Fusarium oxysporum* | Silver, platinum, gold, cadmium sulfide and magnetite | Extracellular | 70–180, 20–50, 2–50 | Rectangular, triangular, spherical, needlelike and aggregates | Khandel and Shahi (2018); Durán et al. (2007) |
| *Aspergillus Niger* | Silver and gold | Extracellular | 10–20 | Spherical, elliptical, polydispersed | Gade et al. (2008); Salunke et al. (2016) |
| *Aspergillus oryzae* | Silver and gold | Mycelial surface | 5–50 for silver, 10–60 for gold | Mostly spherical | Binupriya et al. (2010a, 2010b) |
| *Rhizopus oryzae* | Gold | Cell-free filtrate | 16–25 | Spherical | Das et al. (2012) |
| *Fusarium solani* | Silver and gold | Extracellular | 5–35 for silver, 20–50 for gold | Spherical | Ingle et al. (2009); Gopinath and Arumugam (2014) |
| *Alternaria alternata* | Gold | Extracellular | 12 | Spherical, triangular, hexagonal | Sarkar et al. (2012) |
| *Amylomyces rouxii* | Silver | Extracellular | 27 | Spherical, rod shaped | Musarrat et al. (2010) |
| *Aspergillus clavatus* | Gold | Extracellular | 24.4 | Triangular, spherical and hexagonal | Verma et al. (2011) |
| *Aspergillus flavus* | Silver and magnesium oxide | Cell wall | 8.92 | Spherical | Moharrer et al. (2012); Raliya et al. (2013) |
| *Bipolaris nodulosa* | Silver | Extracellular | 12–15 | Spherical | Saha et al. (2010) |
| *Candida albicans* | Gold | Extracellular | 5 | Monodispersed spherical | Chauhan et al. (2011) |
| *Cladosporium cladosporioides* | Silver | Cell-free extract | 60–80 | Spherical | Balaji et al. (2009) |
| *Cochlibolus lunatus* | Copper and Aluminium | Extracellular | 12–15 | Quasi-spherical | Salunkhe et al. (2011) |
| *Ganoderma sp.* | Silver and gold | Extracellular | 12–22 | Monodispersed and spherical | Gurunathan et al. (2013) |

(continued)

**Table 6.3** (continued)

| Fungal strain | Nanoparticles | Mode of synthesis | Size | Shape | Reference |
|---|---|---|---|---|---|
| *Guingnardia mangiferae* | Silver | Extracellular | 32–100 | Spherical, trigonal | Balakumaran et al. (2015) |
| *Hypocrea lixii* | Copper | Extracellular | 24.50 | Spherical | Salvadori et al. (2013) |
| *Penicillium fellutanum* | Silver | Extracellular | 5–25 | Spherical, triangular and hexagonal | Kathiresan et al. (2009) |
| *Phoma glomerata* | Silver | Cell-free filtrate | 60–80 | Spherical | Birla et al. (2009) |
| *Pediococcus pentosaceus* | Silver | Extracellular | 10–40 | Spherical | Sintubin et al. (2009) |
| *Penicillium rugulosum* | Gold | Cell-free filtrate | 20–80 | Spherical | Mishra et al. (2012) |
| *Trichoderma viride* | Silver | Extracellular | 5–40 | Spherical | Thakkar et al. (2010) |
| *Trichoderma koningii* | Gold | Cell-free filtrate | 30–40 | Small spheres to polygons spheres | Maliszewska et al. (2009) |
| *Verticillium sp.* | Gold | Intracellular, cell wall and cytoplasmic membrane | 20 | Spherical | Mukherjee et al. (2001) |
| *Saccharomyces cerevisiae* | Gold | Cell wall cytoplasm | 15–20 | Spherical | Khandel and Shahi (2018) |
| *Schizosaccharomyces pombe* | Cadmium sulfide | Intracellular and extracellular | 1–1.5 | Spherical | Kowshik et al. (2002a); Williams et al. (1996) |
| *Torulopsis sp.* | Cadmium sulfide | Intracellelar | 2–5 | – | Kowshik et al. (2002a) |
| *Rhodosporidium diobovatum* | Lead sulfide | Intracellular | 2–5 | – | Seshadri et al. (2011) |
| *Pichia jadinii* | Gold | Cytoplasm | < 100 | Spherical | Velusamy et al. (2016) |
| *Extremophilic yeast* | Silver and gold | Extracellular | 20 for silver, 30–100 for gold | Irregular | Mourato et al. (2011) |

reduction and deposition of metal or metal oxide nanoparticles (Fig. 6.3) (Gade et al. 2008; Crookes-Goodson et al. 2008). The biomolecules responsible for the reduction process include pigments, polysaccharides, and peptides (Sethi 2011; Mohamed et al. 2012). A polysaccharide called fucoidan is secreted from the marine brown algae cell walls exhibited many applications in different fields like anti-cancer, anti-inflammatory, anti-coagulant, and anti-viral therapy (Lirdprapamongkol et al. 2010). Till now, different algal species were used to synthesize silver and gold nanoparticles in a cost-effective way such as *Spirulina platensis, Lyngbya majuscule,* and *Chlorella vulgaris* (Chakraborty et al. 2009; Niu and Volesky 2000; Kalabegishvili et al. 2012; Annamalai and Nallamuthu 2015). It was reported that microalgae such as *Diadesmis gallica* and *Navicula atomus* (diatoms) are capable of synthesizing silica-gold bionanocomposites and gold nanoparticles (Schröfel et al. 2011). A marine alga called *Sargassum muticum* was utilized for the extracellular synthesis of silver, gold, and gold/silver bimetallic nanoparticles (Madhiyazhagan et al. 2015; Namvar et al. 2015). Recently it was reported that green and red alga *Spirogyra insignis* and *Chondrus crispus* could be used for the synthesis of silver and gold nanoparticles, respectively (Castro et al. 2013). Senapati and coworkers used *Tetraselmis kochinensis* for intracellular synthesis of gold nanoparticles (Senapati et al. 2012). Due to these facts, biosynthesis of nanoparticles using algae should be given equal importance as other biosynthetic processes. Nanoparticles synthesized from various algal sources together with their characteristics are listed in Table 6.4.

## 6.6 Virus Mediated Nanoparticle Synthesis

Nanoparticle biosynthesis using a virus is one of the unique techniques that has been explored for the delivery of inorganic nanomaterials like cadmium sulfide (CdS), silicon dioxide ($SiO_2$), iron dioxide ($Fe_2O_3$) and Zinc sulfide (ZnS) (Yildiz et al. 2011). Viruses allowed the development of organized nanoparticle assemblies by showing great promise in interconnecting and assembling novel nano-sized components (Blum et al. 2005). They provide a useful scaffold for molecular assembly into nanoscale devices because of their size, availability of various chemical groups for modification and monodispersity (Yu et al. 2003). They have surfaces covering with capsid proteins which makes them interacting with metallic ions by creating a highly reactive surface (Yildiz et al. 2011). Tobacco mosaic virus (TMV), a plant virus on their covering surface, contains approximately 2130 molecules of capsid protein. These protein molecules can serve as the points of attachment for the scattering of nano-sized materials (Yildiz et al. 2011). A study demonstrated that upon addition of a very low concentration of TMV to gold or silver salts before adding *Nicotiana benthamiana* plant extracts, the virus-mediated solution drastically increases the nanoparticle's number as well as reducing the size of synthesized nanoparticles as compared to control. This finding suggested that TMV converts metallic ions into nanowires by acting as a bio-template. They also help in the mineralization process

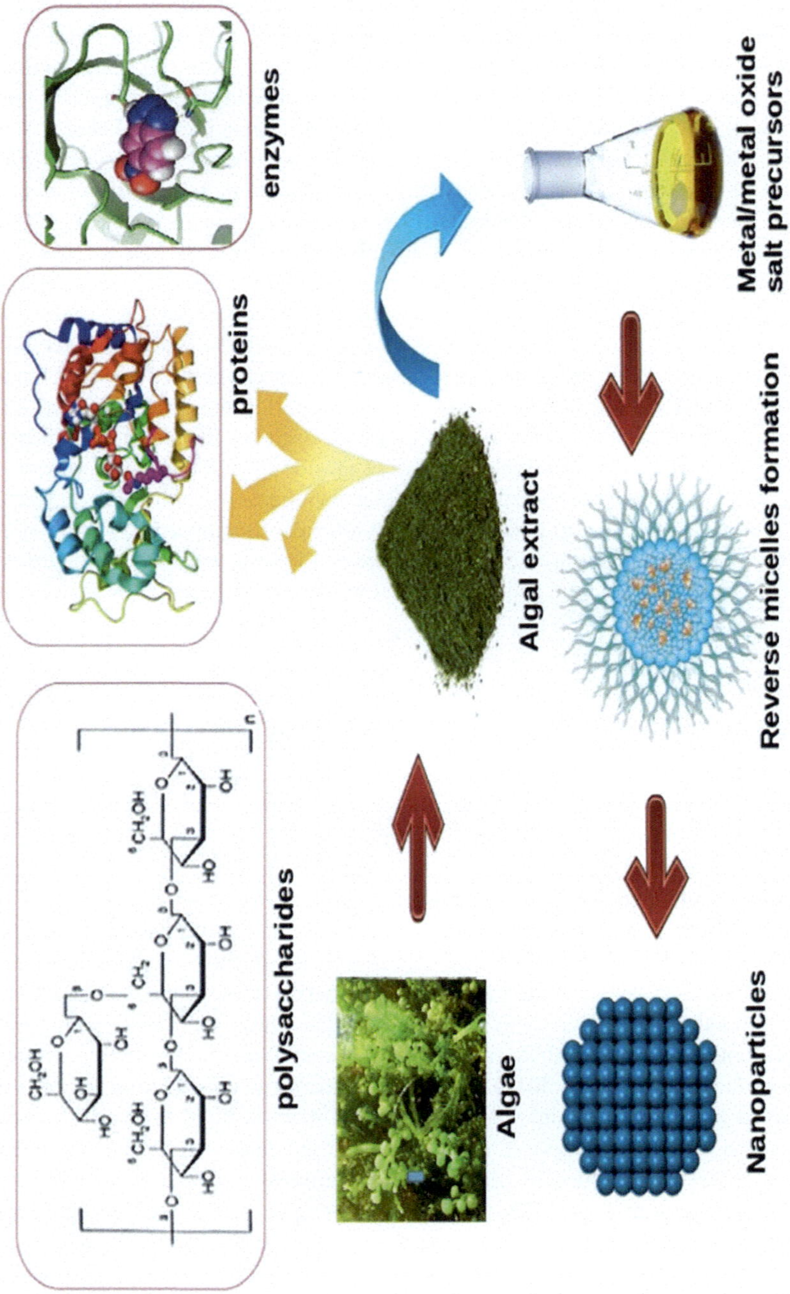


**Fig. 6.3** Mechanism of biosynthesis of nanoparticles using algae. (Image reproduce with permission from Elsevier, Ref. (Sharma et al. 2019) open access)

**Table 6.4** Algal-derived nanoparticles and their characteristics

| Algal strain | Nanoparticles | Mode of synthesis | Size | Shape | Reference |
|---|---|---|---|---|---|
| *Bifurcaria bifurcate* | Copper oxide | Intracellular | 5–45 | Spherical and elongated | Abboud et al. (2014) |
| *Chlorella vulgaris* | Gold | Extracellular | 9–20 | Spherical | Annamalai and Nallamuthu (2015) |
| *Caulerpa racemose* | Silver | Extracellular | 5–25 | Spherical and triangle | Kathiraven et al. (2015) |
| *Chlorococcum humicola* | Silver | Intracellular | 4 and 6 | Spherical | Jena et al. (2013) |
| *Spirulina platensis* | Gold | Intracellular | 20–30 | Spherical | Kalabegishvili et al. (2012) |
| *Sargassum wightii* | Silver and gold | Extracellular | 8–12 for gold, 5–22 for silver | Spherical | Singaravelu et al. (2007); Shanmugam et al. (2014) |
| *Sargassum myriocystum* | Zinc oxide and gold | Extracellular | 96–110 for zinc oxide, 15 for gold | Spherical, rectangular, triangular | Dhas et al. (2012); Nagarajan and Kuppusamy (2013) |
| *Sargassum muticum* | Gold and zinc oxide | Extracellular | 5.42 for gold and 30–57 for zinc oxide | Spherical for gold and hexagonal for zinc oxide | Madhiyazhagan et al. (2015); Namvar et al. (2015); Azizi et al. (2014) |
| *Sargassum plagiophyllum* | Silver chloride | Intracellular | 18–42 | Spherical | Dhas et al. (2014) |
| *Lyngbya majuscule* | Gold | Extracellular | <20 | Spherical | Chakraborty et al. (2009) |
| *Tetraselnis kochinensis* | Gold | Intracellular | 5–35 | – | Senapati et al. (2012) |
| *Spirogyra insignis* | Silver | – | 30 | Spherical | Castro et al. (2013) |
| *Chondrus crispus* | Gold | – | 30 | Spherical, triangular and hexagonal | Castro et al. (2013) |
| *Diadesmis gallica* | Gold | Intercellular | Various | Spherical | Schröfel et al. (2011) |
| *Navicula atomus* | Gold | Intercellular | Various | Spherical | Schröfel et al. (2011) |
| *Ecklonia cava* | Gold | Extracellular | 30 | Spherical and triangular | Venkatesan et al. (2014) |

(continued)

**Table 6.4** (continued)

| Algal strain | Nanoparticles | Mode of synthesis | Size | Shape | Reference |
|---|---|---|---|---|---|
| *Cystophora moniliformis* | Silver | Extracellular | 50–100 | Spherical | Prasad et al. (2013) |
| *Chlamydomonas reinhardtii* | Silver | Intracellular and extracellular | 5–35 | Round/ rectangular | Barwal et al. (2011) |
| *Euglena gracilis* | 2-lines ferrihydrite nanoparticles | Intracellular | 0.6–1.0 | Spherical | Brayner et al. (2012) |
| *Ulva fasciata* | Silver | Intracellular | 28–41 | Spherical | Rajesh et al. (2012) |

**Table 6.5** Virus-derived nanoparticles and their characteristics

| Viral strain | Nanoparticles | Mode of synthesis | Size | Shape | Reference |
|---|---|---|---|---|---|
| Tobacco mosaic virus (TMV) | Silicon dioxide, cadmium sulfide, Lead sulfid,e and ferric oxide | Intracellular and extracellular | 45–80 | – | Shenton et al. (1999) |
| M13 bacteriophage | Zinc sulfide and cadmium sulfide | Intracellular and extracellular | 50–100 | – | Mao et al. (2003) |

of crystalline nanowires and sulfide (Royston et al. 2009; Shenton et al. 1999). Cowpea mosaic virus (CPMV) and Cowpea chlorotic mottle virus have been utilized for the mineralization of inorganic materials (Douglas and Young 1998; Douglas et al. 2002). It was demonstrated that the surface of M13 bacteriophage contains peptides that are competent in nucleating nanocrystal growth (Mao et al. 2003). Nanoparticles synthesized from various viral sources together with their characteristics are listed in Table 6.5.

## 6.7 Actinomycetes Mediated Nanoparticle Synthesis

Nanoparticle synthesis using actinomycetes remains less explored area even though the nanoparticles synthesized by them showed good stability, monodispersity, and remarkable biocidal activity against different pathogens (Golinska et al. 2014). They can be easily modified genetically to achieve nanoparticles with better size and polydispersity (Ahmad et al. 2003a). The basic mechanism behind the actinomycetes mediated synthesis of nanoparticles is that the reductase enzyme present, reduced metal salts into metallic nanoparticles. This mechanism was concluded from the studies where zinc, silver, and copper nanoparticles were synthesized by using *Streptomyces sp.* suggested that the reductase enzyme has an important role in the

**Table 6.6** Actinomycetes-derived nanoparticles and their characteristics

| Actinomycetes strain | Nanoparticles | Mode of synthesis | Size | Shape | Reference |
|---|---|---|---|---|---|
| *Thermomonospora* sp. | Gold | Extracellular | 8 | Spherical | Ahmad et al. (2003a); Golinska et al. (2014) |
| *Streptomyces sp. LK3* | Silver | Extracellular | 5 | Spherical | Karthik et al. (2014) |
| *Streptomyces clavuligerus* | Gold | Extracellular | 8.2 | Spherical | Kumar et al. (2015) |
| *Streptomyces fulvissimus* | Gold | Extracellular | 20–50 | Spherical | Soltani et al. (2015) |
| *Streptomyces albidoflavus* | Silver | Extracellular | 10–14 | Spherical | Buddana (2012) |
| *Streptomyces hygroscopicus* | Silver | Extracellular | 20–30 | Spherical | Sadhasivam et al. (2010) |
| *Streptomyces viridogens(HM10)* | Gold | Intracellular | 18–20 | Spherical | Balagurunathan et al. (2011) |
| *Streptomyces sp. VITDDK3* | Gold | Extracellular | 90 | Hexagonal, cubical, brick and irregular | Gopal et al. (2013) |
| *Rhodococcus sp.* | Gold | Intracellular | 5–15 | – | Ahmad et al. (2003b) |
| *Gordonia amicalis HS- 11* | Silver and gold | Extracellular | 5–25 | Polycrystalline | Sowani et al. (2016) |
| *Gordonia amarae* | Gold | Extracellular and cell associated | 15–40 | Spherical | Bennur et al. (2016) |

process of metal salts reduction (Karthik et al. 2014). It was also reported that *Thermomonospora sp.*, extremophilic actinomycetes produce gold nanoparticles extracellularly with a much better polydispersity when treated with gold ions (Golinska et al. 2014). In another study, it was reported that an alkotolerant *Rhodococcus sp.* could synthesize gold nanoparticles and the metal ion concentration was found to be higher on the cell wall and cytoplasmic membrane when compared with the cytosol that might be because the enzymes catalyzed the reduction of metal ions reside on the cell wall and cytoplasmic membrane but not in the cytosol (Ahmad et al. 2003b). Nanoparticles synthesized from various actinomycetes sources together with their characteristics are listed in Table 6.6.

## 6.8 Plants Mediated Nanoparticle Synthesis

Nowadays, phytonanotechnology, which utilizes plants for the synthesis of nanomaterials, has provided new avenues for the simple, cost-effective, eco-friendly, stable, and rapid method for nanoparticle synthesis. They have the capability to synthesize various types of nanoparticles, including copper, iron, silver, zinc, gold, palladium, selenium, and platinum. There are many advantages of using phytonanotechnology, which include scalability, non-toxicity, biocompatibility, and the medical application of nanoparticles that are synthesized using water as a reducing medium (Noruzi 2015). For the synthesis of metal nanoparticles, different parts of the pants could be used including roots, stems, leaves, fruits, and their extracts (Murugan et al. 2015; Poopathi et al. 2015; Amooaghaie et al. 2015; Sadeghi et al. 2015; Zhou et al. 2014; Gogoi et al. 2015; Singh et al. 2016c). The basic protocol for phytosynthesis of nanoparticles contains the following steps: a first, collection of different plant parts and wash them thoroughly with detergent and double distilled water followed by an air dry and then cut into small pieces. Second, Boil the plant parts in an aqueous solvent (water) at a controlled temperature for a certain period to perform the extraction. Third, purify the extract by centrifugation or filtration and incubate with the respective metal salts solution at a controlled temperature for the reaction to occur, which gives a visible color change. Finally, the nanoparticles can be collected in the form of pellets after density gradient or high-speed centrifugation followed by thoroughly washing with water or other solvents (Rajeshkumar 2016; Singh et al. 2016d). Till now the exact components and mechanism behind the phytosynthesis of nanoparticles are not known but researchers proposed that organic acids, proteins, vitamins, amino acids as well as plants secondary metabolites like polyphenols, heterocyclic compounds, alkaloids, polysaccharides, terpenoids, and flavonoids significantly participate in the reduction of metal salts and consequently act as stabilizing and capping agents for nanoparticle synthesis (Pohlit et al. 2011; Doughari 2012; Duan et al. 2015). El-Kassas and coworkers depicted that the carbonyl group from proteins and the hydroxyl group from polyphenols of *Corallina officinalis* extract may help in forming and stabilizing gold nanoparticles (El-Kassas and El-Sheekh 2014). Similarly, a purgative resin called emodin with quinone compounds found to be responsible for silver nanoparticle synthesis by xerophyte plants. In the case of mesophytic plants dietchequinone, cyperoquinone and remirin were found to be useful for the synthesis of metal nanoparticles. *Cinnamomum zeylanisum* contains an important terpenoid called eugenol which was suggested to have a key role in silver and gold nanoparticles synthesis (Makarov et al. 2014). It was also suggested in reports that the mechanism of plant-mediated nanoparticle synthesis differs from species to species (Baker et al. 2013). Extracts of different plant species were explored for the synthesis of nanoparticles including *Azadirachta indica, Mirabilis jalapa,* and *Cinnamomum camphora* leaf extract, *Aloe vera* plant extracts, *Geranium* extract, etc. for silver and gold nanoparticle synthesis (Shankar et al. 2004a, 2004b; Chandran et al. 2006;

Huang et al. 2007a; Patil et al. 2012). Nanoparticles synthesized from various plant sources together with their characteristics are listed in Table 6.7.

## 6.9 Application of Biologically Derived Nanoparticles

Nanotechnology is based on modulation and the synthesis of nanoparticles demands significant modifications of the metal properties (Visweswara Rao and Hua Gan 2015). Nowadays, scientists explore the application of nanoparticles in diverse areas such as physiochemical, agriculture, biomedical and environmental areas (Visweswara Rao and Hua Gan 2015; Rai et al. 2016; Abbasi et al. 2016; Giljohann et al. 2010; Pereira et al. 2015). Gold and silver are the most commonly explored nanoparticles in the biomedical area. Gold nanoparticles were employed for specific delivery of drugs like doxorubicin, methotrexate and paclitaxel (Rai et al. 2016). They were also used for the diagnosis of a genetic disorder and genetic disease, tumor detection, photothermal therapy, detection of angiogenesis and photoimaging (Huang et al. 2007b; Khlebtsov and Dykman 2011). Based on the aggregation properties of gold nanoparticles, a susceptible, exclusive, and highly explicit system of immunoassay was developed where the gold nanoparticles coated with protein antigens became aggregated in the presence of their respective antibodies (Thanh and Rosenzweig 2002). The magnetic nanoparticle iron oxide was applied for drug delivery, tissue repair, hyperthermia, cancer therapy, magnetic resonance imaging (Mishra et al. 2012), targeting and immunoassays, magnetic responsive drug delivery, cell labeling and biological fluids detoxification process (Iv et al. 2015; Gao et al. 2008). Silver nanoparticles were reported to be used for anticancer, antimicrobial purposes, wound treatment, and anti-inflammatory applications (Ahamed et al. 2010). The endophytic fungus *Pestalotia sp.*, isolated from *Syzygium cumini* leaves able to synthesize silver nanoparticles that exhibited antibacterial activity against *S. typhi* and *S. aureus* human pathogens (Raheman et al. 2011). Titanium and zinc nanoparticles were used in various essential processing applications as well as in ultraviolet (Sharma et al. 2019)-blocking agents, biomedical and cosmetic products due to their skin-compatibility, biocompatibility, antimicrobial, self-cleansing, dermatological behaviors and non-toxic nature (Zahir et al. 2015; Ambika and Sundrarajan 2015). In electronic and optical industries, palladium and copper nanoparticles have been used in optical limiting devices, batteries, plastics plasmonic waveguides and polymers (Nasrollahzadeh and Sajadi 2015; Momeni and Nabipour 2015). Furthermore, metal nanoparticles have been applied for the visual analysis of different biomolecules with higher sensitivity and spatial resolution which include lipids, several metabolites, fatty acids, peptides, glycosphingolipids, nucleic acids, and drug molecules (Waki et al. 2015). The unique properties possess by the nanoparticles make them suitable for the development of biosensors and chemical sensors (Peng and Miller 2011). For example, researchers have developed nanosensor that may be able to detect algal toxins, mercury and mycobacteria present in drinking water (Selid et al. 2009). Scientists

**Table 6.7** Plant-derived nanoparticles and their characteristics

| Plant | Nanoparticles | Plant Tissues for Extraction | Size | Shape | Reference |
|---|---|---|---|---|---|
| *Aloe vera* | Gold | Leaves | 50–350 | Triangular | Chandran et al. (2006) |
| *Azadirachta indica* | Silver | Leaves | 41–60 | – | Poopathi et al. (2015) |
| *Anogeissus latifolia* | Silver | Gum powder | 5.5–5.9 | Spherical | Kora et al. (2012) |
| *Abutilon indicum* | Silver | Leaves | 5–25 | Spherical | Ashokkumar et al. (2015) |
| *Artocarpus gomezianus* | Zinc | Fruit | > 20 | Spherical | Suresh et al. (2015) |
| Banana | Cadmium sulfide | Peel | 1.48 | – | Zhou et al. (2014) |
| *Cocos nucifera* | Lead | Leaves | 47 | Spherical | Elango and Roopan (2015) |
| *Catharanthus roseus* | Palladium | Leaves | 40 | Spherical | Kalaiselvi et al. (2015) |
| *Citrus medica* | Copper | Fruit | 20 | – | Shende et al. (2015) |
| *Cymbopogon citratus* | Gold | Leaves | 20–50 | Spherical, hexagonal, triangular, and rod | Murugan et al. (2015) |
| *Euphorbia prostrata* | Silver and titanium dioxide ($TiO_2$) | Leaves | 10–15 for silver; 81.7–84.7 for $TiO_2$ | Spherical | Zahir et al. (2015) |
| *Ginkgo biloba* | Copper | Leaves | 15–20 | Spherical | Nasrollahzadeh and Sajadi (2015) |
| *Gardenia jasminoides* | Iron | Leaves | 32 | Rock like appearance | Naseem and Farrukh (2015) |
| *Nigella sativa* | Silver | Leaves | 15 | Spherical | Amooaghaie et al. (2015) |
| *Nyctanthes arbortristis* | Silver | Flower | 5–20 | Anisotropic | Gogoi et al. (2015) |
| *Lawsonia inermis* | Iron | Leaves | 21 | Hexagonal | Naseem and Farrukh (2015) |
| Panax ginseng | Silver and gold | Root | 10–30 for silver, 10–40 for gold | Spherical | Singh et al. (2016e) |
| *Pistacia atlantica* | Silver | Seeds | 27 | Spherical | Sadeghi et al. (2015) |

(continued)

**Table 6.7** (continued)

| Plant | Nanoparticles | Plant Tissues for Extraction | Size | Shape | Reference |
|---|---|---|---|---|---|
| *Pinus densiflora* | Silver | Cones | 30–80 | Oval and in few cases triangular | Velmurugan et al. (2015) |
| *Red ginseng* | Silver | Root | 10–30 | Spherical | Singh et al. (2016f) |
| Orange and pineapple | Silver | Fruit | 10–300 | Spherical | Hyllested et al. (2015) |

also developed nanosensors that react with auxin that helps them to study the plant's regulation of auxin (Koren et al. 2015). The nanoparticles were also used for immobilization of biomolecules, the catalyst for various electrochemical reactions, carrier for efficient electron transfer among electrode, etc. (Liu et al. 2003). Xiao et al. used gold nanoparticles to effectively immobilize horseradish peroxidase by attaching gold nanoparticles to gold electrodes which were modified with cysteamine monolayer (Xiao et al. 1999).

## 6.10 Conclusion

The diverse application of nanoparticles increases their need which in turn increases the need for industrial production with stable formulations and in eco-friendly processes. Thus, we need to focus on exploring natural resources to implement biological methods for nanoparticle synthesis as it provides a single-step process, commercially economical, environment-friendly process. The biological pathways for nanoparticle synthesis comprise many advantages compared to physical pathways, including lack of toxic contaminants, lack of complex chemical synthesis, stable production of nanoparticles with controlled shapes and sizes, and the ability for rapid synthesis using various biological resources. Nanoparticles synthesized by the biological processes can be used in the treatment of various diseases which will further open new avenues in the field of medical sciences. The nanoparticle-based biosensors have potential use in agriculture and bioremediation processes which will help the environment from getting exposed to toxic substances. Despite having benefits, the nanoparticles synthesized from biological sources still have some challenges and limitations that need to be overcome and extensive research should be done to optimize various conditions to obtain better control over shape, size, and monodispersity of synthesized nanoparticles. Additionally, the exact mechanism by which biological nanoparticles were synthesized is not clearly understood till now and hence more researches are needed to elucidate and identify the exact mechanisms and the responsible biomolecules behind the reduction and stabilization of nanoparticles.

**Acknowledgments** DKT thanks Department of Biotechnology, Govt. of India for Ramalingaswami fellowship, and University Grant Commission, India for financial support for the faculty position at Goa University.

**Conflict of Interest** The author declared no conflict of interest.

## References

Abbasi E, Milani M, Fekri Aval S, Kouhi M, Akbarzadeh A, Tayefi Nasrabadi H, Nikasa P, Joo SW, Hanifehpour Y, Nejati-Koshki K (2016) Silver nanoparticles: synthesis methods, bio-applications and properties. Crit Rev Microbiol 42(2):173–180

Abboud Y, Saffaj T, Chagraoui A, El Bouari A, Brouzi K, Tanane O, Ihssane B (2014) Biosynthesis, characterization and antimicrobial activity of copper oxide nanoparticles (CONPs) produced using brown alga extract (Bifurcaria bifurcata). Appl Nanosci 4(5):571–576

Ahamed M, AlSalhi MS, Siddiqui M (2010) Silver nanoparticle applications and human health. Clin Chim Acta 411(23–24):1841–1848

Ahmad A, Senapati S, Khan MI, Kumar R, Sastry M (2003a) Extracellular biosynthesis of monodisperse gold nanoparticles by a novel extremophilic actinomycete, Thermomonospora sp. Langmuir 19(8):3550–3553

Ahmad A, Senapati S, Khan MI, Kumar R, Ramani R, Srinivas V, Sastry M (2003b) Intracellular synthesis of gold nanoparticles by a novel alkalotolerant actinomycete, Rhodococcus species. Nanotechnology 14(7):824

Ai J, Biazar E, Jafarpour M, Montazeri M, Majdi A, Aminifard S, Zafari M, Akbari HR, Rad HG (2011) Nanotoxicology and nanoparticle safety in biomedical designs. Int J Nanomedicine 6:1117

Alghuthaymi MA, Almoammar H, Rai M, Said-Galiev E, Abd-Elsalam KA (2015) Myconanoparticles: synthesis and their role in phytopathogens management. Biotechnol Biotechnol Equip 29(2):221–236

Ali DM, Sasikala M, Gunasekaran M, Thajuddin N (2011) Biosynthesis and characterization of silver nanoparticles using marine cyanobacterium, Oscillatoria willei NTDM01. Dig J Nanomater Biostruct 6(2):385–390

Ambika S, Sundrarajan M (2015) Green biosynthesis of ZnO nanoparticles using Vitex negundo L. extract: spectroscopic investigation of interaction between ZnO nanoparticles and human serum albumin. J Photochem Photobiol B Biol 149:143–148

Amooaghaie R, Saeri MR, Azizi M (2015) Synthesis, characterization and biocompatibility of silver nanoparticles synthesized from Nigella sativa leaf extract in comparison with chemical silver nanoparticles. Ecotoxicol Environ Saf 120:400–408

Annamalai J, Nallamuthu T (2015) Characterization of biosynthesized gold nanoparticles from aqueous extract of Chlorella vulgaris and their anti-pathogenic properties. Appl Nanosci 5 (5):603–607

Ashiuchi M (2013) Microbial production and chemical transformation of poly-γ-glutamate. Microb Biotechnol 6(6):664–674

Ashokkumar S, Ravi S, Kathiravan V, Velmurugan S (2015) Retracted: synthesis of silver nanoparticles using A. indicum leaf extract and their antibacterial activity. Elsevier, Amsterdam

Azizi S, Ahmad MB, Namvar F, Mohamad R (2014) Green biosynthesis and characterization of zinc oxide nanoparticles using brown marine macroalga Sargassum muticum aqueous extract. Mater Lett 116:275–277

Baker S, Rakshith D, Kavitha KS, Santosh P, Kavitha HU, Rao Y, Satish S (2013) Plants: emerging as nanofactories towards facile route in synthesis of nanoparticles. Bioimpacts 3(3):111

Balagurunathan R, Radhakrishnan M, Rajendran RB, Velmurugan D (2011) Biosynthesis of gold nanoparticles by actinomycete Streptomyces viridogens strain HM10

Balaji D, Basavaraja S, Deshpande R, Mahesh DB, Prabhakar B, Venkataraman A (2009) Extracellular biosynthesis of functionalized silver nanoparticles by strains of Cladosporium cladosporioides fungus. Colloids Surf B: Biointerfaces 68(1):88–92

Balakumaran M, Ramachandran R, Kalaichelvan P (2015) Exploitation of endophytic fungus, Guignardia mangiferae for extracellular synthesis of silver nanoparticles and their in vitro biological activities. Microbiol Res 178:9–17

Barwal I, Ranjan P, Kateriya S, Yadav SC (2011) Cellular oxido-reductive proteins of Chlamydomonas reinhardtii control the biosynthesis of silver nanoparticles. J Nanobiotechnol 9(1):56

Bennur T, Khan Z, Kshirsagar R, Javdekar V, Zinjarde S (2016) Biogenic gold nanoparticles from the Actinomycete Gordonia amarae: application in rapid sensing of copper ions. Sensors Actuators B Chem 233:684–690

Binupriya AR, Sathishkumar M, Yun S-I (2010a) Myco-crystallization of silver ions to nanosized particles by live and dead cell filtrates of Aspergillus oryzae var. viridis and its bactericidal activity toward Staphylococcus aureus KCCM 12256. Indust Eng Chem Res 49(2):852–858

Binupriya A, Sathishkumar M, Vijayaraghavan K, Yun S-I (2010b) Bioreduction of trivalent aurum to nano-crystalline gold particles by active and inactive cells and cell-free extract of Aspergillus oryzae var. viridis. J Hazard Mater 177(1–3):539–545

Birla S, Tiwari V, Gade A, Ingle A, Yadav A, Rai M (2009) Fabrication of silver nanoparticles by Phoma glomerata and its combined effect against Escherichia coli, Pseudomonas aeruginosa and Staphylococcus aureus. Lett Appl Microbiol 48(2):173–179

Blum AS, Soto CM, Wilson CD, Brower TL, Pollack SK, Schull TL, Chatterji A, Lin T, Johnson JE, Amsinck C (2005) An engineered virus as a scaffold for three-dimensional self-assembly on the nanoscale. Small 1(7):702–706

Boekhoven J, Zha RH, Tantakitti F, Zhuang E, Zandi R, Newcomb CJ, Stupp SI (2015) Alginate–peptide amphiphile core–shell microparticles as a targeted drug delivery system. RSC Adv 5 (12):8753–8756

Bogunia-Kubik K, Sugisaka M (2002) From molecular biology to nanotechnology and nanomedicine. Biosystems 65(2–3):123–138

Borowitzka MA (2013) High-value products from microalgae—their development and commercialisation. J Appl Phycol 25(3):743–756

Brayner R, Coradin T, Beaunier P, Grenèche J-M, Djediat C, Yéprémian C, Couté A, Fiévet F (2012) Intracellular biosynthesis of superparamagnetic 2-lines ferri-hydrite nanoparticles using Euglena gracilis microalgae. Colloids Surf B: Biointerfaces 93:20–23

Brøndsted H, Andersen C, Hovgaard L (1998) Crosslinked dextran—a new capsule material for colon targeting of drugs. J Control Release 53(1–3):7–13

Buddana SK (2012) Characterization of silver nanoparticles synthesized by using marine isolate Streptomyces albidoflavus. J Microbiol Biotechnol 22(5):614–621

Buzea C, Pacheco II, Robbie K (2007) Nanomaterials and nanoparticles: sources and toxicity. Biointerphases 2(4):MR17–MR71

Castro L, Blázquez ML, Muñoz JA, González F, Ballester A (2013) Biological synthesis of metallic nanoparticles using algae. IET Nanobiotechnol 7(3):109–116

Castro-Longoria E, Vilchis-Nestor AR, Avalos-Borja M (2011) Biosynthesis of silver, gold and bimetallic nanoparticles using the filamentous fungus Neurospora crassa. Colloids Surf B: Biointerfaces 83(1):42–48

Chakraborty N, Banerjee A, Lahiri S, Panda A, Ghosh AN, Pal R (2009) Biorecovery of gold using cyanobacteria and an eukaryotic alga with special reference to nanogold formation–a novel phenomenon. J Appl Phycol 21(1):145

Chandran SP, Chaudhary M, Pasricha R, Ahmad A, Sastry M (2006) Synthesis of gold nanotriangles and silver nanoparticles using Aloevera plant extract. Biotechnol Prog 22 (2):577–583

Chauhan A, Zubair S, Tufail S, Sherwani A, Sajid M, Raman SC, Azam A, Owais M (2011) Fungus-mediated biological synthesis of gold nanoparticles: potential in detection of liver cancer. Int J Nanomedicine 6:2305

Chen H, Roco MC, Li X, Lin Y (2008) Trends in nanotechnology patents. Nat Nanotechnol 3 (3):123

Crookes-Goodson WJ, Slocik JM, Naik RR (2008) Bio-directed synthesis and assembly of nanomaterials. Chem Soc Rev 37(11):2403–2412

Cunningham DP, Lundie L (1993) Precipitation of cadmium by Clostridium thermoaceticum. Appl Environ Microbiol 59(1):7–14

D'Arrigo G, Navarro G, Di Meo C, Matricardi P, Torchilin V (2014) Gellan gum nanohydrogel containing anti-inflammatory and anti-cancer drugs: a multi-drug delivery system for a combination therapy in cancer treatment. Eur J Pharm Biopharm 87(1):208–216

Dameron C, Reese R, Mehra R, Kortan A, Carroll P, Steigerwald M, Brus L, Winge D (1989) Biosynthesis of cadmium sulphide quantum semiconductor crystallites. Nature 338 (6216):596–597

Das SK, Dickinson C, Lafir F, Brougham DF, Marsili E (2012) Synthesis, characterization and catalytic activity of gold nanoparticles biosynthesized with Rhizopus oryzae protein extract. Green Chem 14(5):1322–1334

Davis SA, Patel HM, Mayes EL, Mendelson NH, Franco G, Mann S (1998) Brittle bacteria: a biomimetic approach to the formation of fibrous composite materials. Chem Mater 10 (9):2516–2524

Dhas TS, Kumar VG, Abraham LS, Karthick V, Govindaraju K (2012) Sargassum myriocystum mediated biosynthesis of gold nanoparticles. Spectrochim Acta A Mol Biomol Spectrosc 99:97–101

Dhas TS, Kumar VG, Karthick V, Angel KJ, Govindaraju K (2014) Facile synthesis of silver chloride nanoparticles using marine alga and its antibacterial efficacy. Spectrochim Acta A Mol Biomol Spectrosc 120:416–420

Doughari JH (2012) Phytochemicals: extraction methods, basic structures and mode of action as potential chemotherapeutic agents, Phytochemicals-A global perspective of their role in nutrition and health. InTechOpen

Douglas T, Young M (1998) Host–guest encapsulation of materials by assembled virus protein cages. Nature 393(6681):152–155

Douglas T, Strable E, Willits D, Aitouchen A, Libera M, Young M (2002) Protein engineering of a viral cage for constrained nanomaterials synthesis. Adv Mater 14(6):415–418

Du L, Jiang H, Liu X, Wang E (2007) Biosynthesis of gold nanoparticles assisted by Escherichia coli DH5α and its application on direct electrochemistry of hemoglobin. Electrochem Commun 9(5):1165–1170

Duan H, Wang D, Li Y (2015) Green chemistry for nanoparticle synthesis. Chem Soc Rev 44 (16):5778–5792

Durán N, Marcato PD, De Souza GI, Alves OL, Esposito E (2007) Antibacterial effect of silver nanoparticles produced by fungal process on textile fabrics and their effluent treatment. J Biomed Nanotechnol 3(2):203–208

Elango G, Roopan SM (2015) Green synthesis, spectroscopic investigation and photocatalytic activity of lead nanoparticles. Spectrochim Acta A Mol Biomol Spectrosc 139:367–373

Elbeshehy EK, Elazzazy AM, Aggelis G (2015) Silver nanoparticles synthesis mediated by new isolates of Bacillus spp., nanoparticle characterization and their activity against Bean Yellow Mosaic Virus and human pathogens. Front Microbiol 6:453

El-Kassas HY, El-Sheekh MM (2014) Cytotoxic activity of biosynthesized gold nanoparticles with an extract of the red seaweed Corallina officinalis on the MCF-7 human breast cancer cell line. Asian Pac J Cancer Prev 15(10):4311–4317

Ferrari M (2005) Cancer nanotechnology: opportunities and challenges. Nat Rev Cancer 5 (3):161–171

Feynman R (1991) There's plenty of room at the bottom. Science 254:1300–1301

Gade A, Bonde P, Ingle A, Marcato P, Duran N, Rai M (2008) Exploitation of Aspergillus Niger for synthesis of silver nanoparticles. J Biobaased Mater Bioenergy 2(3):243–247

Gao L, Zhang D, Chen M (2008) Drug nanocrystals for the formulation of poorly soluble drugs and its application as a potential drug delivery system. J Nanopart Res 10(5):845–862

Giljohann DA, Seferos DS, Daniel WL, Massich MD, Patel PC, Mirkin CA (2010) Gold nanoparticles for biology and medicine. Angew Chem Int Ed 49(19):3280–3294

Gogoi N, Babu PJ, Mahanta C, Bora U (2015) Green synthesis and characterization of silver nanoparticles using alcoholic flower extract of Nyctanthes arbortristis and in vitro investigation of their antibacterial and cytotoxic activities. Mater Sci Eng C 46:463–469

Golinska P, Wypij M, Ingle AP, Gupta I, Dahm H, Rai M (2014) Biogenic synthesis of metal nanoparticles from actinomycetes: biomedical applications and cytotoxicity. Appl Microbiol Biotechnol 98(19):8083–8097

Gomathy M, Sabarinathan K (2010) Microbial mechanisms of heavy metal tolerance-a review. Agric Rev 31(2):133–138

Gopal JV, Thenmozhi M, Kannabiran K, Rajakumar G, Velayutham K, Rahuman AA (2013) Actinobacteria mediated synthesis of gold nanoparticles using Streptomyces sp. VITDDK3 and its antifungal activity. Mater Lett 93:360–362

Gopinath K, Arumugam A (2014) Extracellular mycosynthesis of gold nanoparticles using Fusarium solani. Appl Nanosci 4(6):657–662

Govindappa M, Farheen H, Chandrappa C, Rai RV, Raghavendra VB (2016) Mycosynthesis of silver nanoparticles using extract of endophytic fungi, Penicillium species of Glycosmis mauritiana, and its antioxidant, antimicrobial, anti-inflammatory and tyrokinase inhibitory activity. Adv Nat Sci Nanosci Nanotechnol 7(3):035014

Gurunathan S, Raman J, Abd Malek SN, John PA, Vikineswary S (2013) Green synthesis of silver nanoparticles using Ganoderma neo-japonicum Imazeki: a potential cytotoxic agent against breast cancer cells. Int J Nanomedicine 8:4399

Hay ID, Rehman ZU, Moradali MF, Wang Y, Rehm BH (2013) Microbial alginate production, modification and its applications. Microb Biotechnol 6(6):637–650

Huang J, Li Q, Sun D, Lu Y, Su Y, Yang X, Wang H, Wang Y, Shao W, He N (2007a) Biosynthesis of silver and gold nanoparticles by novel sundried Cinnamomum camphora leaf. Nanotechnology 18(10):105104

Huang X, Jain PK, El-Sayed IH, El-Sayed MA (2007b) Gold nanoparticles: interesting optical properties and recent applications in cancer diagnostics and therapy

Husseiny M, Abd El-Aziz M, Badr Y, Mahmoud M (2007) Biosynthesis of gold nanoparticles using Pseudomonas aeruginosa. Spectrochim Acta A Mol Biomol Spectrosc 67 (3–4):1003–1006

Hyllested JÆ, Palanco ME, Hagen N, Mogensen KB, Kneipp K (2015) Green preparation and spectroscopic characterization of plasmonic silver nanoparticles using fruits as reducing agents. Beilstein J Nanotechnol 6(1):293–299

Ingle A, Rai M, Gade A, Bawaskar M (2009) Fusarium solani: a novel biological agent for the extracellular synthesis of silver nanoparticles. J Nanopart Res 11(8):2079

Iv M, Telischak N, Feng D, Holdsworth SJ, Yeom KW, Daldrup-Link HE (2015) Clinical applications of iron oxide nanoparticles for magnetic resonance imaging of brain tumors. Nanomedicine 10(6):993–1018

Jena J, Pradhan N, Dash BP, Sukla LB, Panda PK (2013) Biosynthesis and characterization of silver nanoparticles using microalga Chlorococcum humicola and its antibacterial activity. Int J Nanomater Biostruct 3(1):1–8

Jo JH, Singh P, Kim YJ, Wang C, Mathiyalagan R, Jin C-G, Yang DC (2016) Pseudomonas deceptionensis DC5-mediated synthesis of extracellular silver nanoparticles. Artificial Cells Nanomed Biotechnol 44(6):1576–1581

Johansen MN (2012) Microalgae: biotechnology, microbiology, and energy. Nova Science Publisher's

Kalabegishvili T, Kirkesali E, Rcheulishvili A (2012) Synthesis of gold nanoparticles by blue-green algae Spirulina platensis. Frank Lab. of Neutron Physics

Kalaiselvi A, Roopan SM, Madhumitha G, Ramalingam C, Elango G (2015) Synthesis and characterization of palladium nanoparticles using Catharanthus roseus leaf extract and its application in the photo-catalytic degradation. Spectrochim Acta A Mol Biomol Spectrosc 135:116–119

Karthik L, Kumar G, Kirthi AV, Rahuman A, Rao KB (2014) Streptomyces sp. LK3 mediated synthesis of silver nanoparticles and its biomedical application. Bioprocess Biosyst Eng 37 (2):261–267

Kathiraven T, Sundaramanickam A, Shanmugam N, Balasubramanian T (2015) Green synthesis of silver nanoparticles using marine algae Caulerpa racemosa and their antibacterial activity against some human pathogens. Appl Nanosci 5(4):499–504

Kathiresan K, Manivannan S, Nabeel M, Dhivya B (2009) Studies on silver nanoparticles synthesized by a marine fungus, Penicillium fellutanum isolated from coastal mangrove sediment. Coll Surf B Biointerf 71(1):133–137

Khandel P, Shahi SK (2016) Microbes mediated synthesis of metal nanoparticles: current status and future prospects. Int J Nanomater Biostruct 6(1):1–24

Khandel P, Shahi SK (2018) Mycogenic nanoparticles and their bio-prospective applications: current status and future challenges. J Nanostruct Chem 8(4):369–391

Khlebtsov N, Dykman L (2011) Biodistribution and toxicity of engineered gold nanoparticles: a review of in vitro and in vivo studies. Chem Soc Rev 40(3):1647–1671

Klaus T, Joerger R, Olsson E, Granqvist C-G (1999) Silver-based crystalline nanoparticles, microbially fabricated. Proc Natl Acad Sci 96(24):13611–13614

Kora AJ, Beedu SR, Jayaraman A (2012) Size-controlled green synthesis of silver nanoparticles mediated by gum ghatti (Anogeissus latifolia) and its biological activity. Org Med Chem Lett 2 (1):1–10

Koren K, Brodersen KE, Jakobsen SL, Kühl M (2015) Optical sensor nanoparticles in artificial sediments–a new tool to visualize O2 dynamics around the rhizome and roots of seagrasses. Environ Sci Technol 49(4):2286–2292

Kowshik M, Deshmukh N, Vogel W, Urban J, Kulkarni SK, Paknikar K (2002a) Microbial synthesis of semiconductor CdS nanoparticles, their characterization, and their use in the fabrication of an ideal diode. Biotechnol Bioeng 78(5):583–588

Kowshik M, Ashtaputre S, Kharrazi S, Vogel W, Urban J, Kulkarni SK, Paknikar K (2002b) Extracellular synthesis of silver nanoparticles by a silver-tolerant yeast strain MKY3. Nanotechnology 14(1):95

Kumar V, Yadav SK (2009) Plant-mediated synthesis of silver and gold nanoparticles and their applications. J Chem Technol Biotechnol Int Res Process Environ Clean Technol 84 (2):151–157

Kumar SA, Abyaneh MK, Gosavi S, Kulkarni SK, Pasricha R, Ahmad A, Khan M (2007) Nitrate reductase-mediated synthesis of silver nanoparticles from AgNO 3. Biotechnol Lett 29 (3):439–445

Kumar CG, Poornachandra Y, Chandrasekhar C (2015) Green synthesis of bacterial mediated anti-proliferative gold nanoparticles: inducing mitotic arrest (G2/M phase) and apoptosis (intrinsic pathway). Nanoscale 7(44):18738–18750

Lee RE (2018) Phycology. Cambridge University Press, Cambridge

Lee S-W, Mao C, Flynn CE, Belcher AM (2002) Ordering of quantum dots using genetically engineered viruses. Science 296(5569):892–895

Lengke MF, Fleet ME, Southam G (2006) Morphology of gold nanoparticles synthesized by filamentous cyanobacteria from gold (I)− thiosulfate and gold (III)− chloride complexes. Langmuir 22(6):2780–2787

Lengke MF, Ravel B, Fleet ME, Wanger G, Gordon RA, Southam G (2007) Precipitation of gold by the reaction of aqueous gold (III) chloride with cyanobacteria at 25–80° C—studied by X-ray absorption spectroscopy. Can J Chem 85(10):651–659

Leong YK, Show PL, Ooi CW, Ling TC, Lan JC-W (2014) Current trends in polyhydroxyalkanoates (PHAs) biosynthesis: insights from the recombinant Escherichia coli. J Biotechnol 180:52–65

Lirdprapamongkol K, Warisnoicharoen W, Soisuwan S, Svasti J (2010) Eco-friendly synthesis of fucoidan-stabilized gold nanoparticles. Am J Appl Sci 7(8):1038

Liu S, Leech D, Ju H (2003) Application of colloidal gold in protein immobilization, electron transfer, and biosensing. Anal Lett 36(1):1–19

Luo Y, Wang Q (2014) Recent development of chitosan-based polyelectrolyte complexes with natural polysaccharides for drug delivery. Int J Biol Macromol 64:353–367

Madhiyazhagan P, Murugan K, Kumar AN, Nataraj T, Dinesh D, Panneerselvam C, Subramaniam J, Kumar PM, Suresh U, Roni M (2015) Sargassum muticum-synthesized silver nanoparticles: an effective control tool against mosquito vectors and bacterial pathogens. Parasitol Res 114(11):4305–4317

Makarov V, Love A, Sinitsyna O, Makarova S, Yaminsky I, Taliansky M, Kalinina N (2014) "Green" nanotechnologies: synthesis of metal nanoparticles using plants. Acta Nat (англоязычная версия) 6(1(20))

Maliszewska I, Aniszkiewicz L, Sadowski Z (2009) Biological synthesis of gold nanostructures using the extract of Trichoderma koningii. Acta Phys Polonica A116

Mao C, Flynn CE, Hayhurst A, Sweeney R, Qi J, Georgiou G, Iverson B, Belcher AM (2003) Viral assembly of oriented quantum dot nanowires. Proc Natl Acad Sci 100(12):6946–6951

Masood F, Yasin T, Hameed A (2015) Polyhydroxyalkanoates–what are the uses? Current challenges and perspectives. Crit Rev Biotechnol 35(4):514–521

Maya S, Sarmento B, Lakshmanan V-K, Menon D, Jayakumar R (2014) Actively targeted cetuximab conjugated γ-poly (glutamic acid)-docetaxel nanomedicines for epidermal growth factor receptor over expressing colon cancer cells. J Biomed Nanotechnol 10(8):1416–1428

Menon S, Rajeshkumar S, Kumar V (2017) A review on biogenic synthesis of gold nanoparticles, characterization, and its applications. Resour Efficient Technol 3(4):516–527

Mishra A, Tripathy SK, Yun S-I (2012) Fungus mediated synthesis of gold nanoparticles and their conjugation with genomic DNA isolated from Escherichia coli and Staphylococcus aureus. Process Biochem 47(5):701–711

Mohamed S, Hashim SN, Rahman HA (2012) Seaweeds: a sustainable functional food for complementary and alternative therapy. Trends Food Sci Technol 23(2):83–96

Moharrer S, Mohammadi B, Gharamohammadi RA, Yargoli M (2012) Biological synthesis of silver nanoparticles by Aspergillus flavus, isolated from soil of Ahar copper mine. Indian J Sci Technol 5(3):2443–2444

Mokhtarzadeh A, Alibakhshi A, Yaghoobi H, Hashemi M, Hejazi M, Ramezani M (2016) Recent advances on biocompatible and biodegradable nanoparticles as gene carriers. Expert Opin Biol Ther 16(6):771–785

Momeni S, Nabipour I (2015) A simple green synthesis of palladium nanoparticles with Sargassum alga and their electrocatalytic activities towards hydrogen peroxide. Appl Biochem Biotechnol 176(7):1937–1949

Morris ER, Nishinari K, Rinaudo M (2012) Gelation of gellan–a review. Food Hydrocoll 28 (2):373–411

Mourato A, Gadanho M, Lino AR, Tenreiro R (2011) Biosynthesis of crystalline silver and gold nanoparticles by extremophilic yeasts. Bioinorg Chem Appl 2011

Mukherjee P, Ahmad A, Mandal D, Senapati S, Sainkar SR, Khan MI, Ramani R, Parischa R, Ajayakumar P, Alam M (2001) Bioreduction of AuCl4− ions by the fungus, Verticillium

sp. and surface trapping of the gold nanoparticles formed. Angew Chem Int Ed 40 (19):3585–3588
Murray CB, Kagan AC, Bawendi M (2000) Synthesis and characterization of monodisperse nanocrystals and close-packed nanocrystal assemblies. Ann Rev Mater Sci 30(1):545–610
Murugan K, Benelli G, Panneerselvam C, Subramaniam J, Jeyalalitha T, Dinesh D, Nicoletti M, Hwang J-S, Suresh U, Madhiyazhagan P (2015) Cymbopogon citratus-synthesized gold nanoparticles boost the predation efficiency of copepod Mesocyclops aspericornis against malaria and dengue mosquitoes. Exp Parasitol 153:129–138
Musarrat J, Dwivedi S, Singh BR, Al-Khedhairy AA, Azam A, Naqvi A (2010) Production of antimicrobial silver nanoparticles in water extracts of the fungus Amylomyces rouxii strain KSU-09. Bioresour Technol 101(22):8772–8776
Nagarajan S, Kuppusamy KA (2013) Extracellular synthesis of zinc oxide nanoparticle using seaweeds of gulf of Mannar, India. J Nanobiotechnol 11(1):39
Namvar F, Azizi S, Ahmad MB, Shameli K, Mohamad R, Mahdavi M, Tahir PM (2015) Green synthesis and characterization of gold nanoparticles using the marine macroalgae Sargassum muticum. Res Chem Intermed 41(8):5723–5730
Nangia Y, Wangoo N, Goyal N, Shekhawat G, Suri CR (2009) A novel bacterial isolate Stenotrophomonas maltophilia as living factory for synthesis of gold nanoparticles. Microb Cell Factories 8(1):39
Naseem T, Farrukh MA (2015) Antibacterial activity of green synthesis of iron nanoparticles using Lawsonia inermis and Gardenia jasminoides leaves extract. J Chem 2015
Nasrollahzadeh M, Sajadi SM (2015) Green synthesis of copper nanoparticles using Ginkgo biloba L. leaf extract and their catalytic activity for the Huisgen [3+ 2] cycloaddition of azides and alkynes at room temperature. J Colloid Interface Sci 457:141–147
Niu H, Volesky B (2000) Gold-cyanide biosorption with L-cysteine. J Chem Technol Biotechnol 75 (6):436–442
Noruzi M (2015) Biosynthesis of gold nanoparticles using plant extracts. Bioprocess Biosyst Eng 38(1):1–14
Omajali JB, Mikheenko IP, Merroun ML, Wood J, Macaskie LE (2015) Characterization of intracellular palladium nanoparticles synthesized by Desulfovibrio desulfuricans and Bacillus benzeovorans. J Nanopart Res 17(6):264
Oscar FL, Bakkiyaraj D, Nithya C, Thajuddin N (2014) Deciphering the diversity of microalgal bloom in wastewater-an attempt to construct potential consortia for bioremediation. J Curr Per Appl Microbiol 2278:92
Pacioni NL, Borsarelli CD, Rey V, Veglia AV (2015) Synthetic routes for the preparation of silver nanoparticles, silver nanoparticle applications. Springer, pp 13–46
Pandey AK, Kumar A (2014) Improved microbial biosynthesis strategies and multifarious applications of the natural biopolymer epsilon-poly-l-lysine. Process Biochem 49(3):496–505
Park SJ, Kim TW, Kim MK, Lee SY, Lim S-C (2012) Advanced bacterial polyhydroxyalkanoates: towards a versatile and sustainable platform for unnatural tailor-made polyesters. Biotechnol Adv 30(6):1196–1206
Patil CD, Patil SV, Borase HP, Salunke BK, Salunkhe RB (2012) Larvicidal activity of silver nanoparticles synthesized using Plumeria rubra plant latex against Aedes aegypti and Anopheles stephensi. Parasitol Res 110(5):1815–1822
Pawar SN, Edgar KJ (2012) Alginate derivatization: a review of chemistry, properties and applications. Biomaterials 33(11):3279–3305
Peng H-I, Miller BL (2011) Recent advancements in optical DNA biosensors: exploiting the plasmonic effects of metal nanoparticles. Analyst 136(3):436–447
Pereira CL, Antunes JC, Gonçalves RM, Ferreira-da-Silva F, Barbosa MA (2012) Biosynthesis of highly pure poly-γ-glutamic acid for biomedical applications. J Mater Sci Mater Med 23 (7):1583–1591

Pereira L, Mehboob F, Stams AJ, Mota MM, Rijnaarts HH, Alves MM (2015) Metallic nanoparticles: microbial synthesis and unique properties for biotechnological applications, bioavailability and biotransformation. Crit Rev Biotechnol 35(1):114–128

Pohlit AM, Rezende AR, Baldin ELL, Lopes NP, de Andrade Neto VF (2011) Plant extracts, isolated phytochemicals, and plant-derived agents which are lethal to arthropod vectors of human tropical diseases–a review. Planta Med 77(06):618–630

Poopathi S, de Britto LJ, Praba VL, Mani C, Praveen M (2015) Synthesis of silver nanoparticles from Azadirachta indica—a most effective method for mosquito control. Environ Sci Pollut Res 22(4):2956–2963

Prasad K, Jha AK, Kulkarni A (2007) Lactobacillus assisted synthesis of titanium nanoparticles. Nanoscale Res Lett 2(5):248–250

Prasad TN, Kambala VSR, Naidu R (2013) Phyconanotechnology: synthesis of silver nanoparticles using brown marine algae Cystophora moniliformis and their characterisation. J Appl Phycol 25 (1):177–182

Prema P, Iniya P, Immanuel G (2016) Microbial mediated synthesis, characterization, antibacterial and synergistic effect of gold nanoparticles using Klebsiella pneumoniae (MTCC-4030). RSC Adv 6(6):4601–4607

Raheman F, Deshmukh S, Ingle A, Gade A, Rai M (2011) Silver nanoparticles: novel antimicrobial agent synthesized from an endophytic fungus Pestalotia sp. isolated from leaves of Syzygium cumini (L). Nano Biomed Eng 3(3):174–178

Rai M, Ingle AP, Birla S, Yadav A, Santos CAD (2016) Strategic role of selected noble metal nanoparticles in medicine. Crit Rev Microbiol 42(5):696–719

Rajesh S, Raja DP, Rathi J, Sahayaraj K (2012) Biosynthesis of silver nanoparticles using Ulva fasciata (Delile) ethyl acetate extract and its activity against Xanthomonas campestris pv. malvacearum. J Biopesticides 5:119

Rajeshkumar S (2016) Synthesis of silver nanoparticles using fresh bark of Pongamia pinnata and characterization of its antibacterial activity against gram positive and gram negative pathogens. Resour Efficient Technol 2(1):30–35

Rajora N, Kaushik S, Jyoti A, Kothari SL (2016) Rapid synthesis of silver nanoparticles by pseudomonas stutzeri isolated from textile soil under optimised conditions and evaluation of their antimicrobial and cytotoxicity properties. IET Nanobiotechnol 10(6):367–373

Raliya R, Tarafdar J, Gulecha K, Choudhary K, Rameshwar R, Prakash M, Saran R (2013) Scope of nanoscience and nanotechnology in agriculture. J Appl Biol Biotechnol 1(3):041–044

Rauwel P, Küünal S, Ferdov S, Rauwel E (2015) A review on the green synthesis of silver nanoparticles and their morphologies studied via TEM. Adv Mater Sci Eng 2015

Raveendran P, Fu J, Wallen SL (2003) Completely "green" synthesis and stabilization of metal nanoparticles. J Am Chem Soc 125(46):13940–13941

Reddy AS, Chen C-Y, Chen C-C, Jean J-S, Chen H-R, Tseng M-J, Fan C-W, Wang J-C (2010) Biological synthesis of gold and silver nanoparticles mediated by the bacteria Bacillus subtilis. J Nanosci Nanotechnol 10(10):6567–6574

Remya V, Abitha V, Rajput P, Rane A, Dutta A (2017) Silver nanoparticles green synthesis: a mini review. Chem Int 3(2):165–171

Rodríguez-Carmona E, Villaverde A (2010) Nanostructured bacterial materials for innovative medicines. Trends Microbiol 18(9):423–430

Rosalam S, England R (2006) Review of xanthan gum production from unmodified starches by Xanthomonas comprestris sp. Enzym Microb Technol 39(2):197–207

Royston ES, Brown AD, Harris MT, Culver JN (2009) Preparation of silica stabilized tobacco mosaic virus templates for the production of metal and layered nanoparticles. J Colloid Interface Sci 332(2):402–407

Sadeghi B, Rostami A, Momeni S (2015) Facile green synthesis of silver nanoparticles using seed aqueous extract of Pistacia atlantica and its antibacterial activity. Spectrochim Acta A Mol Biomol Spectrosc 134:326–332

Sadhasivam S, Shanmugam P, Yun K (2010) Biosynthesis of silver nanoparticles by Streptomyces hygroscopicus and antimicrobial activity against medically important pathogenic microorganisms. Colloids Surf B: Biointerfaces 81(1):358–362

Sadowski Z, Maliszewska I, Grochowalska B, Polowczyk I, Kozlecki T (2008) Synthesis of silver nanoparticles using microorganisms. Mater Sci-Pol 26(2):419–424

Saha S, Sarkar J, Chattopadhyay D, Patra S, Chakraborty A, Acharya K (2010) Production of silver nanoparticles by a phytopathogenic fungus Bipolaris nodulosa and its antimicrobial activity. Dig J Nanomater Biostruct 5(4):887–895

Saifuddin N, Wong C, Yasumira A (2009) Rapid biosynthesis of silver nanoparticles using culture supernatant of bacteria with microwave irradiation. J Chem 6(1):61–70

Sallam A, Steinbüchel A (2010) Dipeptides in nutrition and therapy: cyanophycin-derived dipeptides as natural alternatives and their biotechnological production. Appl Microbiol Biotechnol 87(3):815–828

Salunke BK, Sawant SS, Lee S-I, Kim BS (2016) Microorganisms as efficient biosystem for the synthesis of metal nanoparticles: current scenario and future possibilities. World J Microbiol Biotechnol 32(5):88

Salunkhe RB, Patil SV, Salunke BK, Patil CD, Sonawane AM (2011) Studies on silver accumulation and nanoparticle synthesis by Cochliobolus lunatus. Appl Biochem Biotechnol 165 (1):221–234

Salvadori MR, Lepre LF, Ando RMA, do Nascimento CAO, Corrêa B (2013) Biosynthesis and uptake of copper nanoparticles by dead biomass of Hypocrea lixii isolated from the metal mine in the Brazilian Amazon region. PLoS One 8(11):e80519

Sarkar J, Ray S, Chattopadhyay D, Laskar A, Acharya K (2012) Mycogenesis of gold nanoparticles using a phytopathogen Alternaria alternata. Bioprocess Biosyst Eng 35(4):637–643

Sasaki H, Karasawa K, Hironaka K, Tahara K, Tozuka Y, Takeuchi H (2013) Retinal drug delivery using eyedrop preparations of poly-L-lysine-modified liposomes. Eur J Pharm Biopharm 83 (3):364–369

Schröfel A, Kratošová G, Bohunická M, Dobročka E, Vávra I (2011) Biosynthesis of gold nanoparticles using diatoms—silica-gold and EPS-gold bionanocomposite formation. J Nanopart Res 13(8):3207–3216

Selid PD, Xu H, Collins EM, Striped Face-Collins M, Zhao JX (2009) Sensing mercury for biomedical and environmental monitoring. Sensors 9(7):5446–5459

Senapati S, Syed A, Moeez S, Kumar A, Ahmad A (2012) Intracellular synthesis of gold nanoparticles using alga Tetraselmis kochinensis. Mater Lett 79:116–118

Seshadri S, Saranya K, Kowshik M (2011) Green synthesis of lead sulfide nanoparticles by the lead resistant marine yeast, Rhodosporidium diobovatum. Biotechnol Progress 27(5):1464–1469

Sethi P (2011) Biochemical profile of stoechospermun marginatum. Int J Curr Pharm Res 3 (4):102–104

Shankar S, Rai A, Ahmad A, Sastry M (2004a) Biosynthesis of silver and gold nanoparticles from extracts of different parts of the geranium plant. Appl Nano Sci 1:69–77

Shankar SS, Rai A, Ahmad A, Sastry M (2004b) Rapid synthesis of Au, Ag, and bimetallic Au core–Ag shell nanoparticles using Neem (Azadirachta indica) leaf broth. J Colloid Interface Sci 275(2):496–502

Shanmugam N, Rajkamal P, Cholan S, Kannadasan N, Sathishkumar K, Viruthagiri G, Sundaramanickam A (2014) Biosynthesis of silver nanoparticles from the marine seaweed Sargassum wightii and their antibacterial activity against some human pathogens. Appl Nanosci 4(7):881–888

Sharma N, Pinnaka AK, Raje M, Ashish F, Bhattacharyya MS, Choudhury AR (2012) Exploitation of marine bacteria for production of gold nanoparticles. Microb Cell Factories 11(1):86

Sharma A, Sharma S, Sharma K, Chetri SP, Vashishtha A, Singh P, Kumar R, Rathi B, Agrawal V (2016) Algae as crucial organisms in advancing nanotechnology: a systematic review. J Appl Phycol 28(3):1759–1774

Sharma D, Kanchi S, Bisetty K (2019) Biogenic synthesis of nanoparticles: a review. Arab J Chem 12(8):3576–3600
Shende S, Ingle AP, Gade A, Rai M (2015) Green synthesis of copper nanoparticles by Citrus medica Linn.(Idilimbu) juice and its antimicrobial activity. World J Microbiol Biotechnol 31 (6):865–873
Shenton W, Douglas T, Young M, Stubbs G, Mann S (1999) Inorganic–organic nanotube composites from template mineralization of tobacco mosaic virus. Adv Mater 11(3):253–256
Shu X, Shi Q, Feng J, Xie X, Chen Y (2014) Design and in vitro evaluation of novel γ-PGA/hydroxyapatite nanocomposites for bone tissue engineering. J Mater Sci 49(22):7742–7749
Sing SF, Isdepsky A, Borowitzka MA, Moheimani NR (2013) Production of biofuels from microalgae. Mitig Adapt Strateg Glob Chang 18(1):47–72
Singaravelu G, Arockiamary J, Kumar VG, Govindaraju K (2007) A novel extracellular synthesis of monodisperse gold nanoparticles using marine alga, Sargassum wightii Greville. Coll Surf B Biointerf 57(1):97–101
Singh BR, Dwivedi S, Al-Khedhairy AA, Musarrat J (2011) Synthesis of stable cadmium sulfide nanoparticles using surfactin produced by Bacillus amyloliquifaciens strain KSU-109. Colloids Surf B: Biointerfaces 85(2):207–213
Singh P, Kim YJ, Singh H, Mathiyalagan R, Wang C, Yang DC (2015a) Biosynthesis of anisotropic silver nanoparticles by Bhargavaea indica and their synergistic effect with antibiotics against pathogenic microorganisms. J Nanomater 2015
Singh P, Kim YJ, Singh H, Wang C, Hwang KH, Farh ME-A, Yang DC (2015b) Biosynthesis, characterization, and antimicrobial applications of silver nanoparticles. Int J Nanomedicine 10:2567
Singh P, Kim YJ, Wang C, Mathiyalagan R, Yang DC (2016a) Microbial synthesis of flower-shaped gold nanoparticles. Artificial Cells Nanomed Biotechnol 44(6):1469–1474
Singh P, Kim YJ, Wang C, Mathiyalagan R, Yang DC (2016b) Weissella oryzae DC6-facilitated green synthesis of silver nanoparticles and their antimicrobial potential. Artificial Cells Nanomed Biotechnol 44(6):1569–1575
Singh J, Kaur G, Kaur P, Bajaj R, Rawat M (2016c) A review on green synthesis and characterization of silver nanoparticles and their applications: a green nanoworld. World J Pharm Pharm Sci 7:730–762
Singh P, Kim Y-J, Zhang D, Yang D-C (2016d) Biological synthesis of nanoparticles from plants and microorganisms. Trends Biotechnol 34(7):588–599
Singh P, Kim YJ, Yang DC (2016e) A strategic approach for rapid synthesis of gold and silver nanoparticles by Panax ginseng leaves. Artificial Cells Nanomed Biotechnol 44(8):1949–1957
Singh P, Kim YJ, Wang C, Mathiyalagan R, El-Agamy Farh M, Yang DC (2016f) Biogenic silver and gold nanoparticles synthesized using red ginseng root extract, and their applications. Artificial Cells Nanomed Biotechnol 44(3):811–816
Sintubin L, De Windt W, Dick J, Mast J, Van Der Ha D, Verstraete W, Boon N (2009) Lactic acid bacteria as reducing and capping agent for the fast and efficient production of silver nanoparticles. Appl Microbiol Biotechnol 84(4):741–749
Slawson R, Lohmeier-Vogel E, Lee H, Trevors J (1994) Silver resistance in pseudomonas stutzeri. Biometals 7(1):30–40
Soltani NM, Shahidi BG, Khaleghi N (2015) Biosynthesis of gold nanoparticles using streptomyces fulvissimus isolate
Soni N, Prakash S (2015) Antimicrobial and mosquitocidal activity of microbial synthesized silver nanoparticles. Parasitol Res 114(3):1023–1030
Sowani H, Mohite P, Munot H, Shouche Y, Bapat T, Kumar AR, Kulkarni M, Zinjarde S (2016) Green synthesis of gold and silver nanoparticles by an actinomycete Gordonia amicalis HS-11: mechanistic aspects and biological application. Process Biochem 51(3):374–383
Sunkar S, Nachiyar CV (2012) Biogenesis of antibacterial silver nanoparticles using the endophytic bacterium Bacillus cereus isolated from Garcinia xanthochymus. Asian Pac J Trop Biomed 2 (12):953–959

Suresh D, Shobharani R, Nethravathi P, Kumar MP, Nagabhushana H, Sharma S (2015) Artocarpus gomezianus aided green synthesis of ZnO nanoparticles: luminescence, photocatalytic and antioxidant properties. Spectrochim Acta A Mol Biomol Spectrosc 141:128–134

Syed B, Prasad NM, Satish S (2016) Endogenic mediated synthesis of gold nanoparticles bearing bactericidal activity. J Microscopy Ultrastruct 4(3):162–166

Thajuddin N, Subramanian G (1992) Survey of cyanobacterial flora of the southern east coast of India. Bot Mar 35(4):305–314

Thajuddin N, Subramanian G (2005) Cyanobacterial biodiversity and potential applications in biotechnology. Curr Sci:47–57

Thakkar KN, Mhatre SS, Parikh RY (2010) Biological synthesis of metallic nanoparticles. Nanomedicine 6(2):257–262

Thanh NTK, Rosenzweig Z (2002) Development of an aggregation-based immunoassay for anti-protein A using gold nanoparticles. Anal Chem 74(7):1624–1628

Tsai W-B, Lai H-Y, Lee J-L, Lo C-W, Chen W-S (2014) Enhancement of the cytotoxicity and selectivity of doxorubicin to hepatoma cells by synergistic combination of galactose-decorated γ-poly (glutamic acid) nanoparticles and low-intensity ultrasound. Langmuir 30(19):5510–5517

Velmurugan P, Park J-H, Lee S-M, Jang J-S, Lee K-J, Han S-S, Lee S-H, Cho M, Oh B-T (2015) Synthesis and characterization of nanosilver with antibacterial properties using Pinus densiflora young cone extract. J Photochem Photobiol B Biol 147:63–68

Velusamy P, Kumar GV, Jeyanthi V, Das J, Pachaiappan R (2016) Bio-inspired green nanoparticles: synthesis, mechanism, and antibacterial application. Toxicol Res 32(2):95–102

Venkatesan J, Manivasagan P, Kim S-K, Kirthi AV, Marimuthu S, Rahuman AA (2014) Marine algae-mediated synthesis of gold nanoparticles using a novel Ecklonia cava. Bioprocess Biosyst Eng 37(8):1591–1597

Verma VC, Singh SK, Solanki R, Prakash S (2011) Biofabrication of anisotropic gold nanotriangles using extract of endophytic Aspergillus clavatus as a dual functional reductant and stabilizer. Nanoscale Res Lett 6(1):1–7

Visweswara Rao P, Hua Gan S (2015) Recent advances in nanotechnology-based diagnosis and treatments of diabetes. Curr Drug Metab 16(5):371–375

Vu B, Chen M, Crawford RJ, Ivanova EP (2009) Bacterial extracellular polysaccharides involved in biofilm formation. Molecules 14(7):2535–2554

Waki M, Sugiyama E, Kondo T, Sano K, Setou M (2015) Nanoparticle-assisted laser desorption/ionization for metabolite imaging. In: Mass spectrometry imaging of small molecules. Springer, Berlin, pp 159–173

Wang Z, Itoh Y, Hosaka Y, Kobayashi I, Nakano Y, Maeda I, Umeda F, Yamakawa J, Nishimine M, Suenobu T (2003) Mechanism of enhancement effect of dendrimer on transdermal drug permeation through polyhydroxyalkanoate matrix. J Biosci Bioeng 96(6):537–540

Wang C, Kim YJ, Singh P, Mathiyalagan R, Jin Y, Yang DC (2016) Green synthesis of silver nanoparticles by Bacillus methylotrophicus, and their antimicrobial activity. Artificial Cells Nanomed Biotechnol 44(4):1127–1132

Williams P, Keshavarz-Moore E, Dunnill P (1996) Production of cadmium sulphide microcrystallites in batch cultivation by Schizosaccharomyces pombe. J Biotechnol 48 (3):259–267

Xiao Y, Ju H-X, Chen H-Y (1999) Hydrogen peroxide sensor based on horseradish peroxidase-labeled Au colloids immobilized on gold electrode surface by cysteamine monolayer. Anal Chim Acta 391(1):73–82

Yang B, Lv Y, Zhu J-y, Han Y-t, Jia H-z, Chen W-h, Feng J, Zhang X-z, Zhuo R-x (2014) A pH-responsive drug nanovehicle constructed by reversible attachment of cholesterol to PEGylated poly (l-lysine) via catechol–boronic acid ester formation. Acta Biomater 10 (8):3686–3695

Yildiz I, Shukla S, Steinmetz NF (2011) Applications of viral nanoparticles in medicine. Curr Opin Biotechnol 22(6):901–908

Yu L, Banerjee IA, Matsui H (2003) Direct growth of shape-controlled nanocrystals on nanotubes via biological recognition. J Am Chem Soc 125(48):14837–14840
Zahir AA, Chauhan IS, Bagavan A, Kamaraj C, Elango G, Shankar J, Arjaria N, Roopan SM, Rahuman AA, Singh N (2015) Green synthesis of silver and titanium dioxide nanoparticles using Euphorbia prostrata extract shows shift from apoptosis to G0/G1 arrest followed by necrotic cell death in Leishmania donovani. Antimicrob Agents Chemother 59(8):4782–4799
Zhang X, Yan S, Tyagi R, Surampalli R (2011) Synthesis of nanoparticles by microorganisms and their application in enhancing microbiological reaction rates. Chemosphere 82(4):489–494
Zhou GJ, Li SH, Zhang YC, Fu YZ (2014) Biosynthesis of CdS nanoparticles in banana peel extract. J Nanosci Nanotechnol 14(6):4437–4442

# Chapter 7
# Fungal Potential for the Degradation of Synthetic Dyes: An Overview of Renewable Alternatives for the Production of Lignin-Modifying Enzymes

**Clara Dourado Fernandes, Débora S. Vilar, Nádia Hortense Torres, Muhammad Bilal, Hafiz M. N. Iqbal, Ram Naresh Bharagava, Silvia Maria Egues, and Luiz Fernando Romanholo Ferreira**

**Abstract** This chapter contains a brief review of the literature on synthetic dyes, their chemical characteristics, the different types of applications, and the problems caused to the water system that receives this effluent. Possible biological treatments were addressed, such as the use of fungal ligninolytic enzymes. Some agro-industrial by-products were pointed out as substrate alternatives, in addition to being questioned about their use as a support for fungal immobilization. This approach was indicated as a prospecting methodology, which is justified by the increased induction of lignin-modifying enzymes. A brief survey on the toxicity of agro-industrial effluents was also carried out. Finally, operational optimization methodologies are presented to treat effluents through biological processes.

**Keywords** Laccase · Manganese peroxidase · Reactive Black 5 · Ligninolytic Fungi

---

C. D. Fernandes · D. S. Vilar
Graduate Program in Process Engineering, Tiradentes University, Aracaju, Sergipe, Brazil

N. H. Torres · S. M. Egues · L. F. R. Ferreira (✉)
Graduate Program in Process Engineering, Tiradentes University, Aracaju, Sergipe, Brazil

Institute of Technology and Research, Murilo Aracaju, Sergipe, Brazil

M. Bilal
School of Life Science and Food Engineering, Huaiyin Institute of Technology, Huaian, China

H. M. N. Iqbal
Tecnologico de Monterrey, School of Engineering and Sciences, Monterrey, Mexico

R. N. Bharagava
Laboratory for Bioremediation and Metagenomics Research (LBMR), Department of Microbiology (DM), Babasaheb Bhimrao Ambedkar University (A Central University), Lucknow, Uttar Pradesh, India

 
P. K. Arora (ed.), *Microbial Products for Health, Environment and Agriculture*, Microorganisms for Sustainability 31, https://doi.org/10.1007/978-981-16-1947-2_7

## 7.1 Introduction

Synthetic dyes are widely used to dye diverse products in the industry such as cosmetics, plastics leather, and mainly different fabrics (Naraian et al. 2018). According to the statistical review of world trade, the textile (SITC 65) and clothing (SITC 84) sector collected a total of US $ 315 billion and the US $ 505 billion in world exports respectively in 2018, in addition to presenting a 6% prospecting of growth in the market until 2023 (WTO (World Trade Organization) 2020).

The growing economic relevance in the textile industry raises questions about the environmental impacts caused in this segment since, in the process of processing the fibers, pigmented effluents are generated that are difficult to degrade. About 160 thousand $m^3$ of water is spent per ton of fiber in the process of textile processing in operations such as bleaching, mercerization, dyeing, and washing (Adar 2020). Due to the partial incompatibility between the dye and the fibers, it is estimated that high proportions of dyes (2–50%) are discarded after these processes (de Oliveira et al. 2018). The generated effluent has a complex variety of organic and inorganic chemicals that generate non-biodegradable hazardous waste and contributes to the imbalance of the quality standards of the receiving water body.

Dyes are synthesized molecules that have a heterocyclic or aromatic nature and can be soluble in an acidic, neutral, or basic medium (Shindy 2016). In addition to having a complex molecular structure, metals are added to its composition to increase the binding strength of the dyes to the fabric. This complexity in its chemical structure, added to its physical and chemical characteristics, makes it difficult to degrade this effluent. Its discharge without proper treatment causes a change in pH, increases the amount of total dissolved solids (STD), and total suspended solids (SST), which consequently increases the biochemical oxygen demand (BOD) and the chemical oxygen demand (COD) (Kumar and Pavithra 2019). This imbalance can cause serious environmental problems to the soil and the tributary, creating an inhibition of photosynthetic activity and favoring the anaerobic condition, conducive to eutrophication.

In this way, the complex molecular structure of the dyes associated with their loss during the industrial process raises concerns regarding human health and the environmental impacts caused. Among the diversified methods of treating this effluent, degradation by enzymatic route arouses the interest of the scientific community as it is a high-efficiency green strategy, with no toxic by-products and low energy demand. In this context, biodegradation techniques with the crude enzyme by fungal fermentation show to be a promising approach, capable of reaching 99% efficiency in removing synthetic pigments (Fernandes et al. 2020).

Fungal biodegradation occurs due to the microorganism's ability to have two types of extracellular enzyme system: the hydrolytic system, which produces hydrolases responsible for the degradation of the polysaccharide, and an oxidative ligninolytic system, which can degrade lignin-like structures in addition to opening phenyl rings (Sánchez 2009). Among the several microorganisms studied, the basidiomycete fungus *Pleurotus sajor-caju* proved to be a major producer of ligninolytic enzymes such as laccase (Lac, E.C. 1.10.3.2) and manganese peroxidase (MnP, E.C. 1.11.1.13) (Fernandes et al. 2020; Vilar et al. 2018). Such enzymes are considered extremely efficient biodegradable biological catalysts, which have high regio-selectivity and the ability to operate under mild conditions of pH, pressure, and temperature, thus enabling industrial prospecting in biorefineries. (Torres et al. 2017). Because they have high oxidative potential, MnP and Lac were cited as responsible for the degradation of synthetic contaminants such as Reactive Black 5 (Fernandes et al. 2020), azo Congo Red (Yehia and Rodriguez-Couto 2017) reactive red and green bright (Zaccaria et al. 2019).

New researches are being developed, to produce enzymes of commercial interest in a sustainable way, without using chemical substrates, which avoids the generation of new effluents. In this way, agro-industrial residues such as vinasse and pulp wash, have been re-signified as potential strategic sources of C and N for the cultivation of fungal ligninolytic enzymes (Cruz et al. 2020; Vilar et al. 2018). Another approach that has an economic and environmental impact is the immobilization of the microorganism on natural surfaces such as *Luffa cylindrica* (Sriharsha et al. 2017). This alternative facilitates operational handling and increases the useful life of the biocatalyst, allowing its recyclability (Arantes et al. 2011; Saiu et al. 2018; Sriharsha et al. 2017).

The efficiency of biotechnological processes is also associated with the interference of parameters, pH, temperature, and concentration of the substrate and contaminant (Bettin et al. 2019). For this reason, mathematical modeling that can predict the best operational condition for the degradation of synthetic dyes shows to be the approach of the future. Among the various models cited in the literature, the Response Surface Methodology (RSM) and Artificial Neural Networks (ANN) are among the most popular approaches, capable of predicting the efficiency of biodegradation processes of emerging contaminants (Mahmoodi-Babolan et al. 2019; Witek-Krowiak et al. 2014).

The implementation of agribusiness residues in the bioprocess of degradation of synthetic dyes is a sustainable alternative of high efficiency. When adding this approach with mathematical models capable of optimizing the process, one finds the approximation of the empirical to the industrial application. In these circumstances, this chapter elucidates sustainable approaches to the reuse of waste as substrates for the production of enzymes by the fungal route. The study also raises questions about the inclusion of mathematical models capable of optimizing the degradation of synthetic dyes.

## 7.2 Dye Classification

In general, dyes can be classified according to their application in the industry. However, the most specific way of classifying them is following the norms and nomenclature of the global reference classification of the Color Index (C.I.)which is biased for commercial purposes. This index has a database of dyes and pigments developed by SDC (Society of Dyers and Colorists - United Kingdom) and AATCC (American Association of Textile Chemists and Colorists). In this classification, the dyes are given a generic name determined by the application technology, followed by the C.I. number based on their chemical structure.

The classification of dyes can also be carried out according to the chemical structure. For Yesilada et al. (2018), the chemical structure determines the properties and use of dyes, in addition to providing the only rational basis for the classification of these compounds. This classification depends on the chromophore group, which is responsible for giving each dye its characteristic color, and the auxochrome which is the structure responsible for fixing the color to the fibers. The main chromophor groups are: Azo, Indigo, Anthraquinone, Nitro, Triphenylmethyl and Phthalene (Benkhaya, 2018).

The most common chromophor groups are: azo (-N=N-), methino (-CH=), carbonyl (-C=O), −nitro (-$NO_2$) and quinoidal rings while the auxochromes are: amino (-$NH_3$), carboxyl (-COOH), sulfonate (-$SO_3H$) and hydroxyl (-OH) (Bajaña et al. 2017). According to Burkinshaw and Salihu (2013), another practical way to classify dyes is through the application method, as they can be acidic, basic, direct, dispersed, mordant, metallized, solvent or reactive.

Table 7.1 summarizes the most common classes using this classification criterion and its applications in the industry.

### *7.2.1 Azo Dye*

Within the classification by functional groups, azo dyes have been overused by the industry, corresponding to 70% of global demand (Rawat et al. 2016). Its favoritism is associated with four main factors: high adaptability to application needs, simplified coupling reaction, capacity for structural variations and high molar absorbency (Said Benkhaya, 2018). These dyes are developed to resist the fading effects of exposure to light, sweat and water.

Characterized by the functional group of the azo type (-N = N-) as well as aromatic rings and sulfonic groups, these dyes form an important class of molecules with hydrophobic characteristics known as dispersed dyes, being the only one capable of dyeing polyester fibers and therefore widely used in textile processes (Almeida and Corso, 2014; Fernandes et al. 2018).

However, recent studies reveal that the azo dye has a genotoxic effect by forming cleavage products (such as aromatic amines) that cause damage to the genetic material (da Brambilla et al. 2019; Brüschweiler and Merlot 2017). These discoveries raise the concern of human exposure and bioaccumulative environmental contamination, which can lead to adverse effects due to their harmful activity to DNA (da Brambilla et al. 2019).

Among azo dyes, reactive black 5 (RB-5) is commonly used for dyeing fibers with high resistance to heat and moisture (Schubert et al. 2012). Its molecular

**Table 7.1** Classification and application of the most used dyes in the industry

| Class | Chromophore | Example | Application |
|---|---|---|---|
| Anthraquinone | 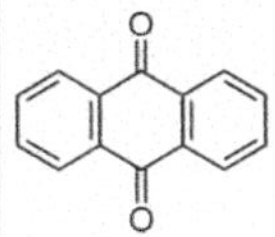 | 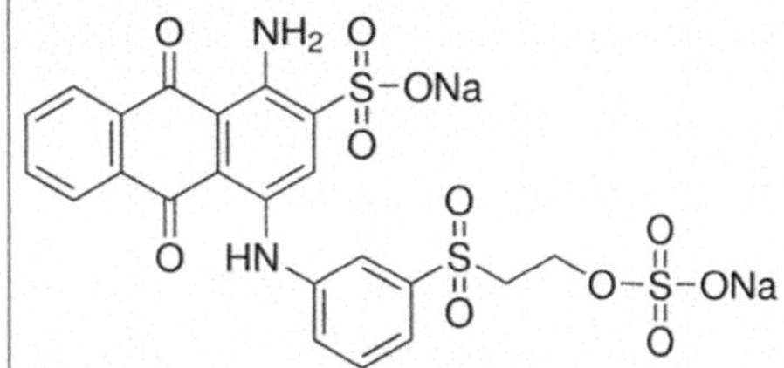 Reactive blue 19 | conventionally applied in dyeing processes in "Nylon", cotton, silk and wool (Arunagiri et al. 2014) |
| Azo | —N=N— | 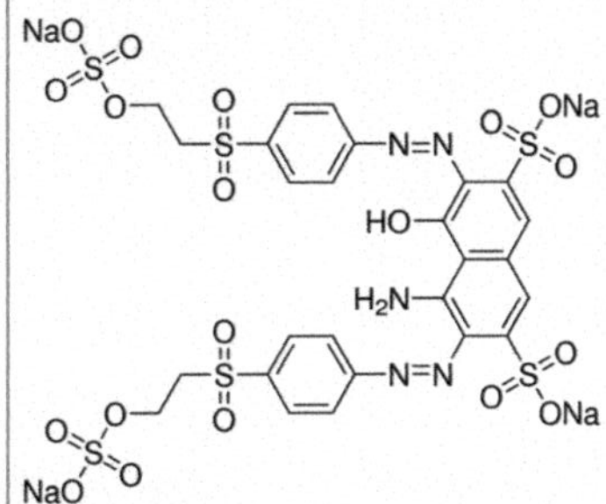 Reactive Black 5 | Azo dyes are the most common synthetic dyes preferred for coating, leather, paper, printing and textile dyeing on cotton, silk and wool (Wanyonyi et al. 2019) |

(continued)

**Table 7.1** (continued)

| Class | Chromophore | Example | Application |
|---|---|---|---|
| Quino-phthalone | 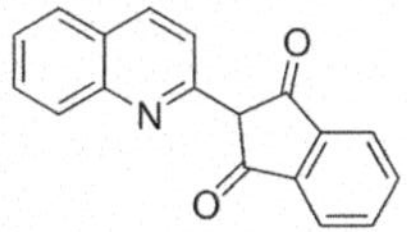 | 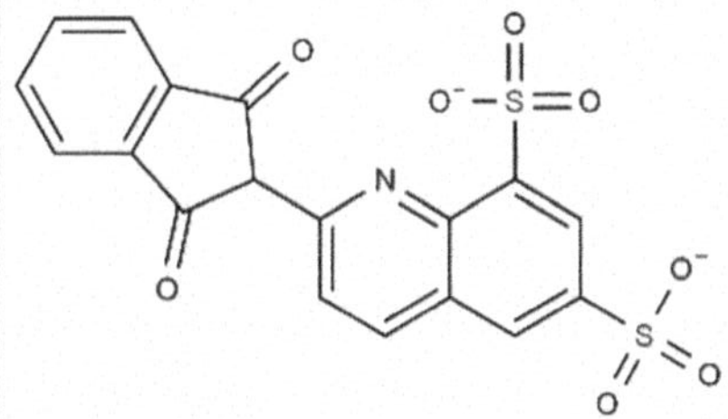 Yellow solvent 33 | Generally used as a food coloring. Quinoline yellow is anionic in nature and is used to dye wool and silk, plastic, oils, paints (Khan et al. 2019) |
| Indigo | 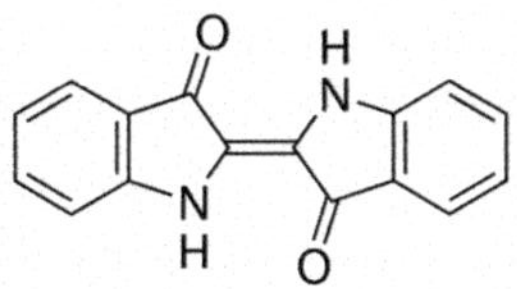 | 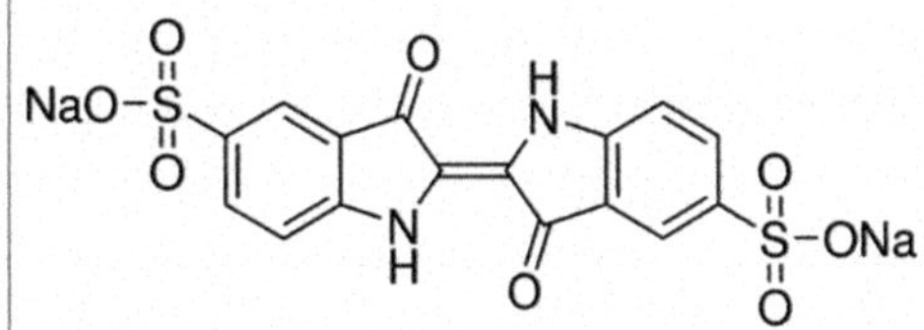 Acid blue 74 | Used to dye silk, wool, cotton, modified acrylics, but also under certain conditions on paper, leather, food and cosmetic products (Sandhya, 2010) |

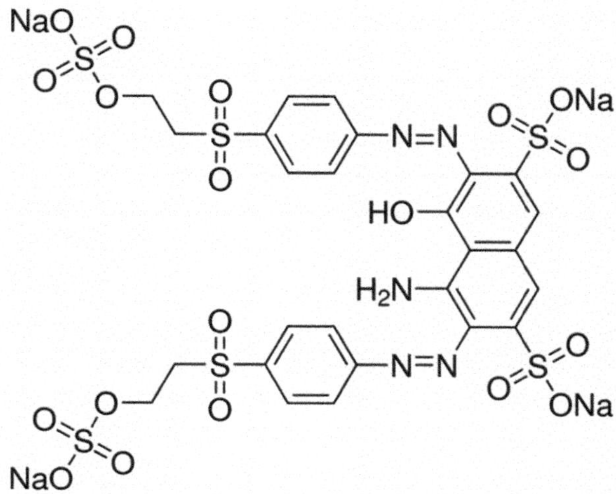


**Fig. 7.1** Molecular structure of Reactive Black 5

structure is composed of two vinyl-sulfone groups ($SO_3$ and $H_2SO_4$) that are converted into anionic sulfate in aqueous solutions (Jager et al. 2018) (Fig. 7.1).

In addition to having a complex structure, the dye can be modified to improve the application properties of the product, for example, its dispersibility, flow or resistance to flocculation. For these characteristics, RB-5 has resistance to the main oxidizing agents of conventional treatments (Fernandes et al. 2018). In this way, this azo dye raises concerns for causing environmental and human health impacts so, sustainable and economically viable approaches have been developed for its removal and degradation.

## 7.3 Treatment Methods

The effluents released by the textile industries have carcinogenic xenobiotic characteristics that are difficult degradation. Consequently, the existence of this waste in the water body is becoming a socio-environmental concern. In order to minimize the negative impacts caused to the environment and living beings, studies have been carried out in the search for methods of treating pigmented effluent that are sustainable and efficient in the long term (Katheresan et al. 2018). Among the existing methods, there are two general categories of treatment, namely physical-chemical and biological.

The physical-chemical dye removal processes have a significant percentage of efficiency that varies between 88% to 99% (Katheresan et al. 2018). This approach is characterized by using chemical methods associated with physical interaction to remove the dye, such as electrochemical oxidation (Jager et al. 2018), photo-oxidation (Traven et al. 2018) and adsorption (Jager et al. 2018; Traven et al. 2018; Vikrant et al. 2018). However, these methods are guided by economic and technical challenges, being commercially unattractive. Katheresan et al. (2018) point out in their studies specific disadvantages of these processes such as the generated sludge that is difficult to handle and dispose of, production of toxic by-products, the amount of energy and the handling of chemicals required to maneuver the process. Therefore, despite its efficiency in decolorizing the effluent, most of these processes are not considered environmentally friendly.

In contrast, biological treatments stand out in the scientific community, due to their advantages in sustainability, cost-benefit, easy handling, absence of sludge or toxic secondary waste (Vikrant et al. 2018). This approach is characterized by its flexibility in different implementation configurations, which can be installed on site (in situ) or outside (ex situ). However, the process always occurs before its disposal into the environment (Ghosh et al. 2017). Bioremediation techniques that employ fungi are called mycorremediation and can degrade recalcitrant substances in processes with or without oxygen, such as adsorption by fungal biomass and/or degradation by enzymatic route (Katheresan et al. 2018). Moreover, its practice can reconcile a diverse consortium of fungi, which can enhance its performance in decolorization (Abd El-Rahim et al. 2017; Wanyonyi et al. 2019).

Several fungi are studied for the biodegradation of dye. Nevertheless, fungi isolated from forest ecosystems, such as white rot ligninolytics, have been shown to be very efficient in decoloring the dye in submerged fermentation (Yang et al. 2016). Its performance in mycoremediation is associated with the ability of microorganisms to carry out joint actions of bioabsorption and biodegradation.

Fungal biosorption is characterized by the ability to adhere a chemical molecule to the surface or micropores of the biomass (Almeida and Corso 2014). Adhesion can occur through the fungal cell wall, which has a wide variety of functional groups (such as amino, carboxyl, thiol, lipid and phosphate groups) that form bonds with the dye molecules, enabling the subsequent removal of the system (Yang et al. 2016). In a microscopic study conducted by Singh et al. (2015), the spores of white rot fungi absorb the dye due to the hydrophobic-hydrophilic interaction between those involved, which associated with the enzymatic concentration excreted by the fungus enhances the removal of the dye.

Therefore, the processes with biosorption and biodegradation have economic, social and environmental relevance in addition to presenting a high percentage of remediation of the textile effluent. Therefore, the fusion of these methods in a single dye removal process should be considered as promising technologies to be studied.

### *7.3.1 Ligninolytic Fungi*

Mir-Tutusaus et al. (2018) reported that the term white rot fungus is not a taxonomic grouping, but rather a collection of species of fungi capable of degrade lignin. Its distinction from other microorganisms when applied to bioremediation is due to the presence of filaments called hyphae that easily pass through the substrate and reach pollutants. The hyphae cluster forms the mycelium, representing the basic units of filamentous fungi characterized by a resistant cell wall composed of polysaccharides (chitin and β-glucan) and glycoproteins (Moreno-García et al. 2018). White rot fungi are mainly basidiomycetes and some relevant species include *Phanerochaete chrysosporium*, *Aspergillus terreus*, *Pleurotus ostreatus* and *Pleurotus sajor-caju*.

The undefined multi-enzymatic system of these fungi is suitable for destabilizing molecules. Its degradation capacity is associated with the release of extracellular enzymes in the substrate that colonize them, including hydrolases and oxidoreductases (laccase (Lac), manganese-peroxidase (MnP) and lignin peroxidase (LiP)), playing a fundamental role in the removal of lignin (Mir-Tutusaus et al. 2018). Due to these characteristics, white rot microorganisms are considered efficient in the degradation of emerging pollutants, due to their high redox potential capable of recovering contaminated environments and their low specificity of the enzyme set (Ali, 2010). However, like all living organisms, these fungi can modify their environment and use the chemical compounds present in the environment as sources of energy and base for their growth and reproduction.

## 7.4 Lignin-Modifying Enzymes

### *7.4.1 Laccase*

Laccase enzymes (benzenediol: oxygen oxidoreductase) belong to the family of polyphenols oxidases, being widely found in plants, insects, bacteria, and filamentous fungi (El-Batal et al. 2015). In *Pleurotus ostreatus* laccase, this enzyme has a molecular mass ranging from 50–60 KDa, in addition to covering four copper ions (Cu) present in three binding sites, in which each ion performs a considerable function in the catalytic reaction that occurs through the oxidation of phenolic substrate, as molecular oxygen is reduced to water (Rivera-Hoyos et al. 2015).

Fungal laccases are classified according to their redox potential, as high (790 mV), medium (430–710 mV) and low (430 mV), which makes it possible to transform toxic compounds into metabolic ones, through the oxidation of these compounds (Piscitelli et al. 2011). However, these enzymes have difficulties to penetrate the substrate (biomass) and degrade lignin, due to their high molecular weight. Furthermore, they are able to degrade phenolic compounds of lignin only on the surface of the substrate. In contrast, they are unable to oxidize the non-phenolic compounds of the lignin present on the surface and have a high redox potential, as these enzymes generally have low redox potential (Widsten and Kandelbauer, 2008). With the use of chemical mediators, in which they are compounds of low molecular mass that allow oxidized radicals to react chemically with the target substrates of high redox potential, these limitations can be overcome since these mediators act as intermediate substrates of laccases (Rivera-Hoyos et al. 2015).

Most of these enzymes are inducible, so inducers such as aromatic or phenolic compounds related to lignin or lignin derivatives are responsible for increasing the production of this enzyme through fermentation processes. The same inducer can either increase the laccase production for a given species of fungus or cause no effect on another species. Therefore, the optimum inducer is not common to all fungi, and the choice of it will depend on the fungus studied (Mann et al. 2015). Laccase production is affected by the physiological differences that exist in cultivation conditions and between fungal cultures, as well as in low-cost procedures that make application of these enzymes viable (Kocyigit et al. 2012). The ideal temperature of its activity and its stability vary according to the different sources of enzymes, normally more stable in acidic pH (Majeau et al. 2010).

Laccase, according to Rivera-Hoyos et al. (2015) is a dimeric glycoprotein belonging to the family of blue copper enzymes that usually contain an active center that comprises 4 copper atoms well distributed in three groups of cupredoxin are identified: a type 1 copper (T1), responsible for the oxidation of the substrate and is covalently linked to a cysteine that provides the intense blue-green color of the enzyme, as well as having a maximum electronic absorbance at 610 nm, when copper is in the cupric state ($Cu^{2+}$) (Giardina et al. 2010; Piscitelli et al. 2011). A type 2 (T2) copper that acts as an electron acceptor, has poor absorption in the visible region, not being detected by electronic absorption, but has detectable RPE and, finally, two types 3 Cu (T3) that do not show a signal EPR, consisting of two tightly coupled copper atoms and operates as a two-electron acceptor, in addition to being responsible for the band at 330 nm (Rodríguez-Delgado et al. 2015) (Fig. 7.2).

The catalytic cycle of the enzyme Lac begins with the reduction of T1 by the substrate where Cu1 removes electrons from the substrate and transfers them to T2 and T3, through the amino acids of the polypeptide chain (His-Cys-His), and due to the strong interaction between T1 and T2 there is a reduction of one oxygen molecule to two water molecules in the active center of these coppers (Fig. 7.2). The oxidation of four substrate molecules is necessary to produce the complete reduction of molecular oxygen to water through successive monoelectronic oxidations of the substrate (variety of organic and inorganic substances) (Brijwani et al. 2010).

### 7.4.2 MnP

According to Maciel et al. (2010), manganese peroxidase (MnP) is a glycosylated extracellular enzyme with a catalytic cycle similar to that of lignin peroxidase (LiP). It has a molecular mass ranging from 40 to 52 KDa, has a heme prosthetic group and is dependent on hydrogen peroxide as a co-substrate responsible for catalysis, oxidation from $Mn^{+2}$ to $Mn^{+3}$ (Hakala et al. 2005).

The catalytic cycle of MnP is initiated by the binding of $H_2O_2$ or another organic peroxide to the enzyme's native iron, forming an iron-peroxide complex (Fig. 7.3). The subsequent breakdown of the O-O bond of the peroxide requires the transfer of two electrons from the heme group of the enzyme, which results in the formation of a complex radical $Fe^{+4}$-oxo porphyrin (MnP-I). With the breakdown of the oxygen bond, a water molecule is released, followed by a reduction that causes the formation

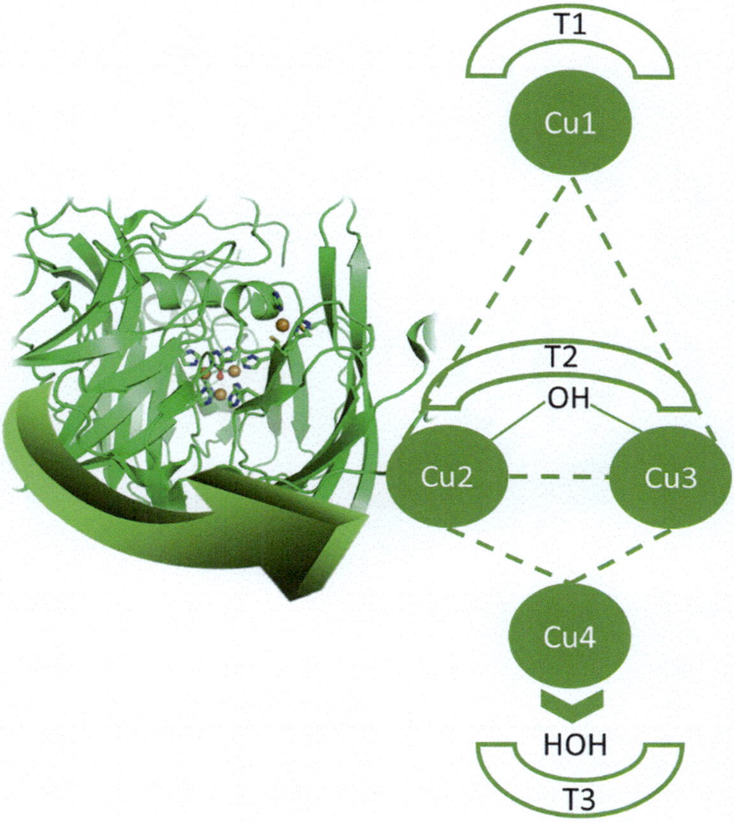


**Fig. 7.2** Protein sequence homology model of Lacase *Pleurotus ostreatu*, white rot fungus and its catalytic cycle

of non-root $Fe^{+4}$ -oxo porphyrin (MnP-II). This intermediate complex gains 1 electron from the $Mn^{+2}$ ion, thus being oxidized to $Mn^{+3}$. From $Mn^{+2}$, the reduction of MnP-II occurs, and another $Mn^{+3}$ is generated, which leads to the original formation of the enzyme and soon a second molecule of water is released (Hofriter, 2002).

During catalysis, $Mn^{+3}$ formed and stabilized by organic acids produced by the fungus itself, can displace electrons from various organic compounds, which also includes phenols, aromatic amines, sulfur compounds and unsaturated fatty acids in a non-specific way, leading to the formation of highly reactive radicals, which makes it possible to achieve recalcitrant structures with high reduction potential, although in the presence of appropriate intermediates, their performance can also be extended to non-phenolic structures (Carvalho et al. 2009).

MnP of white rot fungi is considered one of the main enzymes involved in biotechnology, due to its ability to oxidize various toxic compounds, and can also be used in the biodegradation of lignin, humic acids, synthetic dyes, polychlorinated

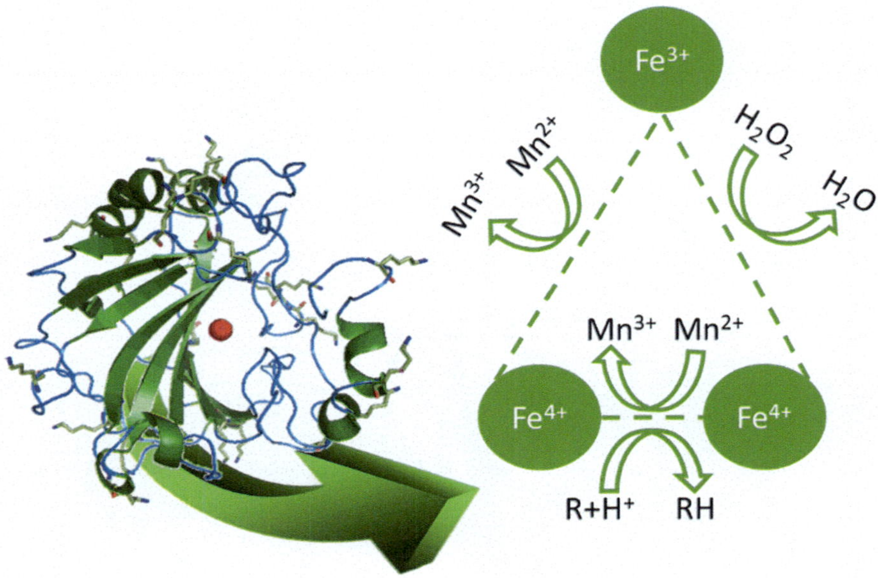


**Fig. 7.3** Homology model of MnP protein sequence *Pleurotus ostreatus* and its catalytic cycle

biphenyls (PCBs), aromatic hydrocarbons polycyclic (PHA) and aromatic xenobiotics (de Oliveira et al. 2009). Until then, no bacteria, yeast and mycorrhizal basidiomycete capable of producing manganese peroxidase have been revealed, being apparently limited to certain basidiomycete fungi. The ability to synthesize MnP is widespread among the different taxonomic groups of basidiomycetes (Kulikova et al. 2011).

In this context, according to (Fernández-Fueyo et al. 2014), the genome of *P. ostreatus*, belonging to lignicellulolytic fungi, was sequenced in a preliminary analysis in silico showed the genes that encode their manganese peroxidase ($MnP_4$). The study showed a helical structure (formed by four main helices and two to three small ones) that has a structural calcium ion, and a proximal domain composed of six other helices as illustrated in Fig. 7.3.

## 7.5 Agro-Industrial by-Products as Enzymatic Production Tool

In the course of processes conducted by microorganisms, according to (Giese, 2015), three stages can be distinguished: upstream, which involves the preparation of the inoculum and the raw material; transformation, in which the microorganism is placed in direct contact with the substrate and the formation of products arising from biochemical reactions favored by the process conditions and the downstream step, where the formed product is separated, recovered and purified when necessary. During the fermentation process, ligninolytic fungi transform the substrates present in the culture medium into one or more products of biotechnological interest. These products include ligninolytic enzymes such as Lac and MnP (Pompeu et al. 2018; Cruz et al. 2020; Vilar et al. 2018).

Among the techniques used for enzymatic production processes of fungal origin, the classification regarding the cultivation mode, usually called solid or submerged fermentation, stands out (Costa, 2016). In solid-state fermentation, filamentous fungi grow on the surface of the medium, as they usually are in nature; however, this method does not favor the downstream stage for the extraction of the generated products. On the other hand, submerged fermentation facilitates the homogenization of soluble nutrients, and demonstrates relative simplicity when the cultivation is scheduled for industrial production purposes, since the homogeneity of the medium facilitates the control of process parameters such as pH and dissolved $O_2$. Other advantages reported by Costa (2016) were reduced product degradation in enzymes with low thermo stability; control of the carbon source and ease in the extraction of the product of interest from the fermented broth, avoiding catabolic repression. Nevertheless, there are limitations in this process, such as the generation of large liquid volumes, requiring an increase in the energy demand associated with sterilization and removal of products from the environment (Costa, 2016).

In order to minimize this obstacle, several studies have been developed on submerged fermentation using agro-industrial by-products, since Brazil is one of the largest agricultural producers in the world and responsible for large amounts of lignocellulosic residues that can be used as enzyme inducers (Golveia et al. 2018).

In the process of degradation of emerging pollutants, due to the action of fungal ligninolytic enzymes capable of carrying out their degradation. These enzymes have widespread applications in several industries, being used in pharmaceutical processes (Li et al. 2015), cellulose treatment (Zhang et al. 2017a, b) and agricultural processes (Krell et al. 2018). However, despite its great versatility for industrial applications, a large amount of enzymes and with high catalytic activity are required.

Munir (2015) evaluated the production of lignin-degrading enzymes by *P. chrysosporium*, from wheat straw residues, reaching, under optimized conditions, the yield of 993.9 ± 18.4 $UI.mL^{-1}$ (MnP). On the other hand, Golveia et al. (2018) obtained a Lac production of 1,642 $UI.mL^{-1}$ during the submerged degradation of cupuaçu residues by *Pycnoporus sanguineus*. Amriani et al. (2017) reported that when using the fungus *Trametes versicolor* U80 and grown in black liquor, an activity of 80 $IU.L^{-1}$ (MnP) was achieved, which provided the decolorization of 90% of the residue. Other enzymes can also be induced, such as cellulase through sugarcane bagasse by *Trichoderma koningii* (8.2 $IU.g^{-1}$ substrate) (Salomão et al. 2019). Therefore, the application of by-products and agro-industrial residues proves to be a versatile and promising alternative for the production of diverse ligninolytic enzymes.

## 7.6 Immobilization of Fungal Mycelia

Techniques for immobilizing fungal mycelia can be defined according to the adhesion of the biomass to a solid or porous support, limiting the detached growth of the microorganism, but conserving its biological activity (García-Reys et al. 2017). This process is identified as a viable alternative to increase stability and reduce shear stress in fermentation processes, in addition to favoring the separation of biotechnological products from the fermented medium and, consequently, reducing costs in the downstream stage. As shown by Giese (2015), the support for immobilization should have a large contact surface, preferably containing the presence of functional groups that cause better cell adhesion, be easy to handle, reusable, in addition to ensuring cell viability and stability of the processes in which they are employed. Thus, in general, fungal mycelia immobilizations can be classified as fixation, in which the microorganisms adhere to the surface by chemical bonding alone; or by entrapment, which involves the retention of microorganisms in the pores of fibers or is physically trapped inside a solid or porous matrix, as exemplified in Fig. 7.4 (Moreno-García et al. 2018).

Microorganisms perform several multicellular methods of immobilization. As an example of fungal self-immobilization, Moreno-García (2018) cites the formation of cell filaments and flakes. This natural immobilization process consists of the

**Fig. 7.4** Classification of immobilization methods

aggregation of single-celled organisms in suspension to form cellular aggregates known as flakes, making their potential use in reactors viable. This technique is considered simple and economically viable, although it is easily influenced by several factors such as composition of the cell wall, medium, pH and dissolved oxygen (Moreno-García et al. 2018).

Artificial immobilizations can be obtained by trapping mycelia in microcapsules, or by an interaction surface between two immiscible fluids. Immobilization in a polymeric matrix, the microorganism binds to the support through covalent bonds where the matrix used can be preformed or synthesized in situ (Moreno-García et al. 2018). Covizzi (2007) states that the cellular matrix formed in this process, prevents the spread of the microorganism to the culture medium, without interfering with the absorption of nutrients and metabolites. Moreno-García et al. (2018), cites polysaccharide gels as alginates, agar, chitosan and polygalacturonic acid as one of the main immobilizing agents.

On the other hand, fungal adhesion on solid surfaces is the most common way for the microorganism to be present in nature, especially on rocks and minerals, being directly related to microbial activity and its survival (Costa, 2016). During this natural process, electrostatic interactions between the microorganism and the support are responsible for immobilizations such as adsorption and the formation of biofilms. In this model, the binding of fungal cells consists of a physical-chemical process resulting from the hydrophilic and / or hydrophobic properties of the cell surface, depending on the pH, ionic strength of the solution in which it is found and also on the composition of the support surface (Park et al. 2010) since, the method of immobilization by adhesion is based on the excretion of polysaccharides that helps in fixing the fungus to the support (García-Reys et al. 2017). This is a low operating cost method, easy to handle, where different types of supports can be used (Svobodová and Novotný, 2018).

In studies by Przystaś et al. (2018) specific supports were selected for immobilizing biomass from basidiomycete fungi to verify the degradation of azo dyes. Among the different substrates tested, the biomass immobilized in sawdust absorbed 82.5% of bright green. A similar study was carried out by Mazmanci and Ünyayar (2005) and Nilsson et al. (2006), both using the organic support *Luffa cylindrica*, for the adsorption of dye, reaching an efficiency of 70 and 99% of decolorization, respectively. Table 7.2 summarizes some microorganisms immobilized on different supports to adsorb on different contaminants.

Cell immobilization can also be used to complement wastewater treatments. Because it is an efficient approach to removing dangerous substances, fungal treatments can be integrated into advanced treatment processes (Ghosh Ray and Ghangrekar, 2015). In order to treat wastewater flows generated by the pulp and paper mills Zhang et al. (2017a, b) used ears of corn from agricultural residues to immobilize white rot fungi, increasing the efficiency of the treatment in terms of removing the color and COD of a secondly treated pulp and paper. Fernandes et al. (2020) also reported that the immobilization of *P. sajor-caju* on the surface of *Luffa cylindrica* brought improvements to the enzymatic activity of MnP, which could assist in the decolorization processes of the synthetic dye RB5. On the other hand,

**Table 7.2** Application of microorganisms and supports used in biotechnological processes

| Microorganism | Support | Process | Reference |
|---|---|---|---|
| *Metarhizium brunneum* CB15 | Encapsulation with: Amidated pectin, corn starch, cellulose, cellulase and yeast | Increased endophytes in potato plants | Krell et al. (2018) |
| *Aspergillus Niger* | Carbon Nano tube | Textile dye removal | Bello et al. (2017) |
| *Aspergillus niger* e, *Aspergillus terreu* | *Luffa cylindrica* | Lead removal | Sriharsha et al. (2017) |
| *Aspergillus Niger* | Alumina nanoparticles | Fluoride removal from water | Yang et al. (2017) |
| Fungal strains of white rot | Corn cob | Tertiary treatment of cellulose | Zhang et al. (2017a, b) |
| *Aspergillus Niger, Cladosporium cladosporioides e Penicillium solitum* | Rice and sodium alginate | Analysis of potential for agro-industrial use | Elizei et al. (2014) |
| *Phanerochaete chrysosporium; T. Versicolor; Pleurotus ostreatus; Pleurotus sajor-caju* | *Luffa cylindrica* | Reactive red 2 Decolorization; reactive blue 4. | Nilsson et al. (2006) |
| *Funalia trogii* | *Luffa cylindrica* | Reactive black 5 Decolorization | Mazmanci and Ünyayar (2005) |
| *P. Sajor-caju* | *Luffa cylindrica* | Reactive black 5 Decolorization | Fernandes et al. (2020) |

Sriharsha et al. (2017) stated the high efficiency of *Aspergillus niger* and *Aspergillus terreo* immobilized in *Luffa cylindrica* for the treatment of water contaminated with high concentrations of lead.

## 7.7 Effluent Toxicity

With emphasis on studies that evaluate the harmful effects of chemical substances on living organisms, ecotoxicology has been shown to be an effective approach in helping to analyze the environmental impacts caused by such contaminants, providing the validation of toxicity by bioindicators (O'Brien, 2017). The realization of ecotoxicological tests in aquatic and terrestrial organisms, allows identifying the harmful effects of several pollutants, aiming to evaluate the potential risk to the local biota (CONAMA, 2005). Bioassays are recommended by international and national organizations, which guide the use of sensitive and abundant organisms. According

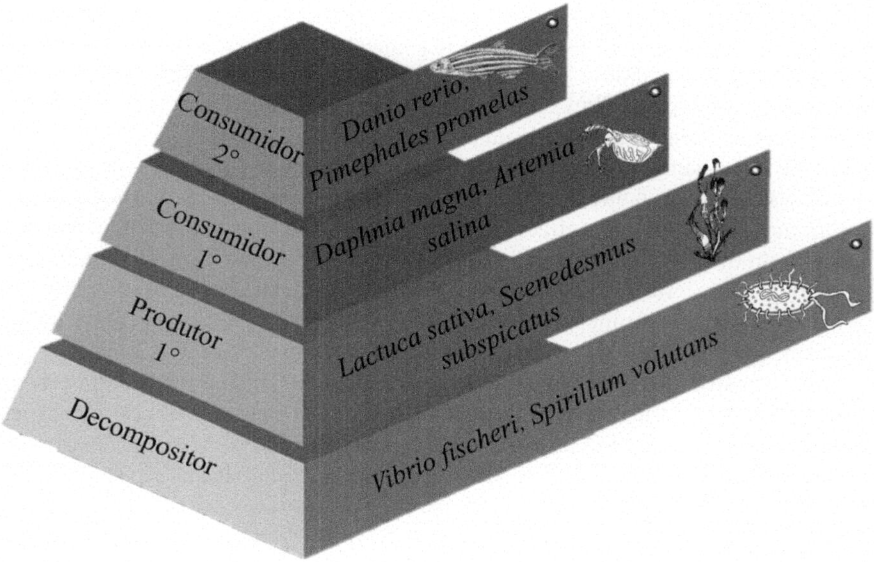


**Fig. 7.5** Classification of bioindicators by trophic level

to (Colombo et al. 2018), the analysis of ecotoxicological effects enables an understanding of the possible effects on the ecosystem surrounding the effluent-receiving water body, in addition to explaining the biological effect of solubilized compounds. In contrast, the main Brazilian legal instrument, which regulates the standards for the discharge of effluents, maintaining the quality of water bodies, is CONAMA Resolution N°. 357/05, supplemented and amended by CONAMA N°. 430/2011, where these define the relevance ecotoxicological tests, explaining methods used to detect and evaluate the ability of a toxic agent to cause harmful effects, identified by bioindicators of large groups in an ecological chain (CONAMA, 2011).

Bioindicators are particularly sensitive species, and when they are absent in the environment to which they are inserted, it suggests an adverse impact, visible by the biological responses of the species under study, which show synergistic, additive, antagonistic and potentiation effects. Depending on its concentration and property, the effluent can cause changes in the food web, which can cause biomagnification (increase) or trophic dilution (decrease) of the bioindicator (Zhou et al. 2019). The ecotoxicity tests are carried out under specific conditions and controlled in the laboratory or in situ, which maintain the veracity and experimental relevance in determining the toxicity of the contaminant. In these tests, the test organisms are exposed to different concentrations of a given sample and the toxic effects produced are observed and quantified.

To perform toxicity bioassays, different trophic levels are commonly used (Fig. 7.5). The propensity to choose these is related to some essential characteristics, such as: constant and high selectivity to contaminants, high availability, genetic stability in populations, representativeness of their trophic level, environmental relevance in the study area, commercial importance, ease of cultivation and adaptation to laboratory conditions (Sobrero and Ronco, 2004). In addition, species with clearly established physiology, genetics and behavior should be chosen, in order to promote better interpretation of results (Fernandes et al. 2018).

On the other hand, bioassays can be classified according to the time of exposure and effects caused on the test organism, thus being able to be distinguished as acute or chronic. The acute toxic effect corresponds to a short period of life of the test organism on display, culminating in its lethality or total immobility. In contrast, in the chronic toxic effect, test organisms demonstrate deleterious effects measured over a long period of time, which can affect various biological functions of the organism, such as life cycle, reproduction and behavior (CONAMA, 2005).

After exposure of the test organism to the contaminant, the test results can be analyzed for their average lethal concentration (LC50) or average effective concentration (EC50), representing the values of concentrations of toxic agents capable of causing mortality or immobility at 50% of test organisms, after the exposure period under the test conditions. In addition, the result of the bioassay can also be expressed by the Toxicity Factor - FT, to determine the lowest dilution of the contaminant, which does not cause deleterious effects during, or after exposure (CONAMA, 2011).

In order to assess the toxicity of the textile and agro-industrial effluent and prove the efficiency of the treatment process, the species *Lactuca sativa L.* belonging to the *Asteraceae* family was chosen as a bioindicator for its availability, ease of reproduction, maintenance in laboratory conditions and for its relevance environmental and economic.

### 7.7.1 *Lactuca sativa*

According to Priac et al. (2017), bioindicators used in ecotoxicological tests allow an evaluation of the residues in an integral and efficient way, before contact with the environment. Seed of *Lactuca sativa L.* (lettuce) are plant species commonly used in phytotoxicity tests and widely recommended by the United States Environmental Protection Agency (USEPA, 1996) the International Organization for Standardization (ISO, 1995) and the Organization for Economic Co-operation and Development (OECD, 2003). Colombo et al. (2018) states that phytotoxicity with *L. sativa* provides information when observing the effective or absent seed germination. This analysis is performed through the germination index (GI), where there is interference in the metabolic reactions of the organism, culminating in the harmful effects caused by the effluent, in addition to inferring the biological effect caused by the soluble compounds even in low concentrations.

IG is the phytotoxicity index commonly used to assess the toxicity of pure compounds or complex mixtures, such as industrial waste or effluents. These tests are classified as a static acute toxicity test with 120 h of exposure, where the process

**Table 7.3** Application of toxicity bioassays

| Test organism | Application | Reference |
|---|---|---|
| *L. sativa*; *A. salina* | Procion red MX-5B azo dye fungal treatments | Almeida and Corso (2014) |
| *Lactuca sativa, Cucumis sativus,* and *Lycopersicon esculentum* | Toxicity of textile dyes: DB38- azo dye; RB15- copper deftalocyanine | de Oliveira et al. (2018) |
| *Lactuca sativa, A. salina* and *Danio rerio* | Azo dye removal using Fenton-type processes | Fernandes et al. (2018) |

of germination and root development in the first days of growth is evaluated (Komilis et al. 2016). Several authors have reported acute phytotoxicity through bioassays carried out with *Lactuca sativa* on treated effluents from the textile industry (Table 7.3), (Fernandes et al. 2018; de Oliveira et al. 2018) and agro-industrial residues, such as pulp wash and vinasse (Cruz et al. 2020; Vilar et al. 2018) demonstrating the versatility of application of this bioindicator. Thus, the use of bioassays with plants represents a fast and economical method for characterizing the toxicity of environmental samples (Chan-Keb et al. 2018).

### *7.7.2 Zebrafish* (Danio rerio*)*

*Danio rerio* (popularly known as Zebrafish) belonging to the phylum *Cyprinidae*, is a diploid freshwater vertebrate used in bioassays for the toxicity of environmental pollutants that have a complex composition. This method is considered of high efficiency because it presents biological responses that show synergistic, additive, antagonistic and potentiation effects that do not appear in conventional physical-chemical analyzes (Lourenço et al. 2017).

The acute toxicity test used for this organism in its early stages of life (Fish Embryo Acute Toxicity Test (FET) is of high importance with regard to the impairment of the growth and survival phases of the organisms in polluted environments, constituting an important tool for adequate environmental monitoring. This organism has high sensitivity when exposed to different contaminants, as it has rapid absorption of compounds available in aqueous media and accumulates them in various tissues such as, in your central nervous system (Sant'Anna et al. 2011; Wang et al. 2020). Currently, zebrafish are used as an in vivo vertebrate model for ecotoxicological studies due to the high degree of genomic homology for humans (70% similarity), low cost, optical transparency, and high fertility rate (Howe et al. 2013). Thus, research with this organism in several areas is increasingly frequent to assess toxicity, the occurrence of malformations, and assess sublethal effects (Horie et al. 2017) as expressed in Table 7.4.

**Table 7.4** Applications of bioassays with zebrafish

| Applications | Reference |
|---|---|
| Toxicological analysis of uranium mining waste | Lourenço et al. (2017) |
| Toxicity of basic red 51 synthetic dye and natural erythrostominone dye | Abe et al. (2017) |
| Tannery effluent toxicity | Rocha and De Oliveira (2017) |
| Impact of vinasse on aquatic organisms | Sousa et al. (2019) |
| Toxicity by-product of the oil extraction process | Babić et al. (2019) |
| Toxicity of oil refinery waste | Kim et al. (2019) |

## 7.8 Optimization of Operational Conditions

As previously reported, efficiency in biodegradation processes is associated with the proper functioning of operating conditions, such as reaction time, pH, substrate and contaminant concentration (Bettin et al. 2019). Therefore, modeling capable of predicting and optimizing such parameters becomes an essential step to achieve the highest process yield. However, because it is a living process, extremely sensitive to external interference, biodegradation is considered a complex model to be modeled due to the non-linearities involved (Fernandes et al. 2020). According to Witek-Krowiak et al. (2014), it is possible to work around this limitation using analyzes that simulate the random relationship between the operational input parameters and the desired result. Among the methods used in recent research, the Artificial Neural Networks (ANN) and the Response Surface Methodology (RSM) are among the most efficient, being able to predict biodescoloration processes with accuracy values close to 100% (Mahmoodi-babolan et al. 2019; Fernandes et al. 2020).

With a smaller number of experimental data, RSM is the tool capable of finding the optimal region and, thus, determining the best operational conditions according to an analytical mathematical model (Shojaeimehr et al. 2014). This methodology is widely used for the design of wastewater treatment processes (Gasemloo et al. 2019). However, the RSM technique is limited by the quadratic correlation hypothesis, since it assumes second order polynomial equations (Fernandes et al. 2020).

On the other hand, Artificial Neural Networks (ANN) are considered universal approximators (Fernandes et al. 2020), In addition to being able to design empirical models simulating the process of biodecolorization, RNA has no limitations in the correlation of nonlinearities (Torregrossa and Capitanescu, 2019). In the view of Khataee et al. (2011) ANN makes an analogy with the brain of intelligent organisms, where it is possible to use artificial neurons divided into layers, and thus serve as models that design the path of complex responses. The number of neurons used in each layer is equivalent to the number of parameters (input) and responses (output). Neurons located in the middle layer are responsible for processing information. With this approach, it is also possible to analyze the representation of the contributions of each parameter under analysis for the final answer. For this, the ANN weighs the weights and bias between the neurons used (Al Aani et al. 2019). Among the existing RNA models, the multilayer perceptron is mainly used, as it allows the modeling of highly non-linear processes through different layers of neural networks (Li et al. 2019).

However, ANNs are not able to optimize the input parameters from their adjustment, it is necessary to use a global search algorithm to perform this function. Among the existing stochastic methods, Genetic Algorithms (AG) have stood out due to their efficiency. An AG is an algorithm inspired by evolutionary theory, where natural selection chooses the best individuals (possible solutions to the problem) within a population (Ghanavati Nasab et al. 2018). Another great differential is the synergy between Genetic Algorithms and neural models, making the ANN-AG technique an effective tool in the simulation and optimization of complex processes (Ghanavati Nasab et al. 2018).

More recently, RSM and ANN methods have been applied to model and optimize environmental studies (Mahmoodi-Babolan et al. 2019). In addition, bibliographic research reveals that ANN and RSM are considered effective forecasting tools for descolorization phenomena. To date, many studies have been carried out to estimate the capacity of enzymes to micro-measure wastewater using RSM. Cordova-villegas et al. (2019) performed the optimization of the decolorization of AB113 and DB38 azo dyes by peroxidase enzymes, using the response surface methodology in a Box-Behnken Designer (BBD) evaluating parameters such as pH and pollutant concentration, reaching $R^2$ value of 99, 07%. Ajmi et al. (2018) performed the optimization for color removal from wastewater using a fungal consortium reaching 99.75% efficiency by applying RSM with an $R^2$ of 90.93%. Parameters such as initial concentration of the dye, pH, reaction time, and temperature were evaluated by Torbati (2016) through the ANN to optimize the phytoremediation of the dye Green malachite. In addition, its results revealed that ANN-GA's predictions are more advantageous than those of the quadratic model provided by RSM (Baştürk and Alver, 2019; Qi et al. 2019) which shows that the predicted operational conditions lead to more high process efficiency. On the other hand, Fernandes et al. (2020) concluded that both ANN-GA and RSM were able to efficiently model the biodecolorization of the synthetic dye RB5 using crude enzyme of the fungus P. sajor-caju. The authors found an $R^2$ of 98% for RSM and 99% for ANN-GA showing that there is no statistically significant difference between the models.

## 7.9 Conclusion

Scientific progress points to the advancement of sustainable technologies capable of reframing residues on alternative highly efficient substrates. In this context, different agro-industrial residues have unique characteristics, capable of providing a good enzymatic performance of Lac and MnP. The white-rot fungus *Pleurotus sajor-caju* is considered as the promising fungus for the production of ligninolytic enzymes with the ability to degrade not only synthetic dyes but also other emerging contaminants. In this way, it was possible to present traditional biosorption techniques, as well as to raise prospects about the application of the crude enzyme as a biodecolorization agent. The search for the approximation of these techniques to industrial scale has been raising great efforts on the part of scientists and researchers. Therefore, studies that carry out mathematical models capable of predicting and optimizing operational conditions, proved to be a recent but necessary advance, to bring biotechnological processes closer to the industry. Few articles were found in this area, as there is also a lack of studies with economic viability analysis, demonstrating a gap for future works.

## References

Abd El-Rahim WM, Moawad H, Abdel Azeiz AZ, Sadowsky MJ (2017) Optimization of conditions for decolorization of azo-based textile dyes by multiple fungal species. J Biotechnol 260:11–17. https://doi.org/10.1016/J.JBIOTEC.2017.08.022

Abe FR, Mendonça JN, Moraes LAB, Oliveira GARD, Gravato C, Soares AMVM, Oliveira DPD (2017) Toxicological and behavioral responses as a tool to assess the effects of natural and synthetic dyes on zebrafish early life. Chemosphere 178:282–290. https://doi.org/10.1016/j.chemosphere.2017.03.030

Adar E (2020) Optimization of triple dye mixture removal by oxidation with Fenton. Int J Environ Sci Technol. https://doi.org/10.1007/s13762-020-02782-1

Ajmi K, Vismara E, Manai I, Haddad M, Hamdi M, Bouallagui H (2018) Polyvinyl acetate processing wastewater treatment using combined Fenton's reagent and fungal consortium: application of central composite design for conditions optimization. J Hazard Mater 358:243–255. https://doi.org/10.1016/j.jhazmat.2018.06.050

Al Aani S, Bonny T, Hasan SW, Hilal N (2019) Can machine language and artificial intelligence revolutionize process automation for water treatment and desalination? Desalination. https://doi.org/10.1016/j.desal.2019.02.005

Ali H (2010) Biodegradation of synthetic dyes - a review. Water Air Soil Pollut 213:251–273. https://doi.org/10.1007/s11270-010-0382-4

Almeida EJR, Corso CR (2014) Comparative study of toxicity of azo dye Procion red MX-5B following biosorption and biodegradation treatments with the fungi Aspergillus Niger and Aspergillus terreus. Chemosphere 112:317–322. https://doi.org/10.1016/J.CHEMOSPHERE.2014.04.060

Amriani F, Sari AA, Abimanyu H, Tachibana S (2017) Evaluation of lignin-based black liquor decolorization by Trametes versicolor U 80. In: AIP conference proceedings. AIP Publishing LLC, p 020001. https://doi.org/10.1063/1.4973128

Arantes V, Silva EM, Milagres AMF (2011) Optimal recovery process conditions for manganese-peroxidase obtained by solid-state fermentation of eucalyptus residue using Lentinula edodes. Biomass Bioenergy 35:4040–4044. https://doi.org/10.1016/j.biombioe.2011.06.042

Arunagiri A, Priya K, Kalaichelvi P, Anantharaj R (2014) Extraction of reactive Orange 107 dye from aqueous stream using triton X-114 surfactant: quantum chemical calculations and experiment. J Ind Eng Chem 20:2409–2420. https://doi.org/10.1016/J.JIEC.2013.10.021

Babić S, Malev O, Pflieger M, Lebedev AT, Mazur DM, Kužić A, Čož-Rakovac R, Trebše P (2019) Toxicity evaluation of olive oil mill wastewater and its polar fraction using multiple whole-organism bioassays. Sci Total Environ 686:903–914. https://doi.org/10.1016/j.scitotenv.2019.06.046

Bajaña FWV, Cárdenas AMV, Cedeño JKZ (2017) In: Primeira Edicón (ed.) Etudio De Remocion De Un Colorante Orgánico

Baştürk E, Alver A (2019) Modeling azo dye removal by sono-Fenton processes using response surface methodology and artificial neural network approaches. J Environ Manag 248. https://doi.org/10.1016/j.jenvman.2019.109300

Bello IA, Kabbashi NA, Alam MZ, Alkhatib MF, Murad FN (2017) Factors affecting the immobilization of fungal biomass on CNT as a biosorbent for textile dyes removal. IOP Conf Ser Mater Sci Eng, vol 217. https://doi.org/10.1088/1757-899X/217/1/012027

Benkhaya S (2018) Classifications, properties and applications of textile dyes: a review. Appl J Environ Eng Sci 3:311–320

Bettin F, Cousseau F, Martins K, Zaccaria S, Girardi V, da Silveira MM, Dillon AJP (2019) Effects of pH, Temperature and Agitation on the Decolourisation of Dyes by Laccase-Containing Enzyme Preparation from Pleurotus sajor-caju. Braz Arch Biol Technol 62. https://doi.org/10.1590/1678-4324-2019180338

Brijwani K, Rigdon A, Vadlani PV (2010) Fungal laccases: production, function, and applications in food processing. Enzyme Res. https://doi.org/10.4061/2010/149748

Brüschweiler BJ, Merlot C (2017) Azo dyes in clothing textiles can be cleaved into a series of mutagenic aromatic amines which are not regulated yet. Regul Toxicol Pharmacol 88:214–226. https://doi.org/10.1016/J.YRTPH.2017.06.012

Burkinshaw SM, Salihu G (2013) The wash-off of dyeings using interstitial water. Part 4: disperse and reactive dyes on polyester/cotton fabric. Dyes Pigments 99:548–560. https://doi.org/10.1016/j.dyepig.2013.06.006

Carvalho W, Canilha L, Ferraz A, Ferreira M, de Biotecnologia D, de Lorena EDE, Paulo UDS, Sp L (2009) Madeira: Estrutura E Composição Química 32, 2191–2195. https://doi.org/10.1109/ISMAR.2015.9

Chan-Keb CA, Agraz-Hernández CM, Perez-Balan RA, Gómez-Solano MI, Maldonado-Montiel TDNJ, Ake-Canche B, Gutiérrez-Alcántara EJ (2018) Acute toxicity of water and aqueous extract of soils from Champotón river in Lactuca sativa L. Toxicol Rep 5:593–597. https://doi.org/10.1016/j.toxrep.2018.05.009

Colombo A, Módenes AN, Trigueros DEG, de Medeiros BL, Marin P, Monte Blanco SPD, Hinterholz CL (2018) Toxicity evaluation of the landfill leachate after treatment with photo-Fenton, biological and photo-Fenton followed by biological processes. J Environ Sci Health Part A Toxic/Hazardous Substain Environ Eng:1–8. https://doi.org/10.1080/10934529.2018.1544475

CONAMA (2005) Conselho Nascional do Meio Ambiente. Resolução n° 357 de 17 de março de 2005

CONAMA (2011) Conselho Nascional do Meio Ambiente. Resolução n° 430 de 13 de maio de 2011

Cordova-Villegas LG, Cordova-Villegas AY, Taylor KE, Biswas N (2019) Response surface methodology for optimization of enzyme-catalyzed azo dye decolorization. J Environ Eng 145. https://doi.org/10.1061/(ASCE)EE.1943-7870.0001513

Costa MAL (2016) Estudo da precipitação com etanol de xilanases de complexos enzimáticos produzidos por Aspergillus niger em fermentação no estado sólido e fermentação submersa campinas 2016. UNICAMP, Faculdade de Engenharia Quimica, Campinas

Covizzi LG, Giese EC, Gomes E, Dekker RFH, da Silva R (2007) Imobilização de células microbianas e suas aplicações biotecnológicas. Semin. Ciências Exatas e Tecnológicas 28, 143. https://doi.org/10.5433/1679-0375.2007v28n2p143

Cruz YWG, Vieira YA, Vilar DS, Torres NH, Aguiar MM, Cavalcanti EB, Américo-Pinheiro JHP, Soriano RN, Bharagava RN, Lima ÁS, Ferreira LFR (2020) Pulp wash: a new source for production of ligninolytic enzymes and biomass and its toxicological evaluation after biological treatment. Environ Technol 41:1837–1847. https://doi.org/10.1080/09593330.2018.1551428

da Brambilla CC, Garcia CM, Leticia H, Silva A, da Silva FR, Taffarel SR, Grivicich I, Picada JN, Scotti A, Dalberto D, Mišík M, Knasmüller S, da Silva J (2019) Amido black 10B a widely used azo dye causes DNA damage in pro- and eukaryotic indicator cells. Chemosphere 217:430–436. https://doi.org/10.1016/j.chemosphere.2018.11.026

de Oliveira PL, Duarte MCT, Ponezi AN, Durrant LR (2009) Purification and partial characterization of manganese peroxidase from Bacillus pumilus and Paenibacillus sp. Braz J Microbiol 40:818–826. https://doi.org/10.1590/S1517-83822009000400012

de Oliveira GAR, Leme DM, de Lapuente J, Brito LB, Porredón C, de Rodrigues LB, Brull N, Serret JT, Borràs M, Disner GR, Cestari MM, de Oliveira DP (2018) A test battery for assessing the ecotoxic effects of textile dyes. Chem Biol Interact 291:171–179. https://doi.org/10.1016/J.CBI.2018.06.026

El-Batal AI, Elkenawy NM, Yassin AS, Amin MA (2015) Laccase production by *Pleurotus ostreatus* and its application in synthesis of gold nanoparticles. Biotechnol Rep 5:31–39. https://doi.org/10.1016/j.btre.2014.11.001

Elizei VG, Chalfoun SM, dos Botelho DMS, Rebelles PPR (2014) Imobilização de fungos filamentosos com potencial para uso agroindustrial. Arq Inst Biol (Sao Paulo) 81:165–172. https://doi.org/10.1590/1808-1657001032012

Fernandes NC, Brito LB, Costa GG, Taveira SF, Cunha-Filho MSS, Oliveira GAR, Marreto RN (2018) Removal of azo dye using Fenton and Fenton-like processes: evaluation of process factors by box–Behnken design and ecotoxicity tests. Chem Biol Interact 291:47–54. https://doi.org/10.1016/j.cbi.2018.06.003

Fernandes CD, Nascimento VRS, Meneses DB, Vilar DS, Torres NH, Leite MS, Vega Baudrit JR, Bilal M, Iqbal HMN, Bharagava RN, Egues SM, Romanholo Ferreira LF (2020) Fungal biosynthesis of lignin-modifying enzymes from pulp wash and Luffa cylindrica for azo dye RB5 biodecolorization using modeling by response surface methodology and artificial neural network. J Hazard Mater 399. https://doi.org/10.1016/j.jhazmat.2020.123094

Fernández-Fueyo E, Ruiz-Dueñas FJ, Martínez MJ, Romero A, Hammel KE, Medrano FJ, Martínez AT (2014) Ligninolytic peroxidase genes in the oyster mushroom genome: heterologous expression, molecular structure, catalytic and stability properties, and lignin-degrading ability. Biotechnol Biofuels 7. https://doi.org/10.1186/1754-6834-7-2

García-Reys M, Beltrán Hernández RI, Vázquez-Rodriguez GA, Coronel-Olivares C, Medina-Moreno SA, Juárez-Santillán LF, Lucho-Constantino CA (2017) Formation, morphology and biotechnological applications of filamentous fungal pellets. Mex J Chem Eng 16:703–720

Gasemloo S, Khosravi M, Sohrabi MR, Dastmalchi S, Gharbani P (2019) Response surface methodology (RSM) modeling to improve removal of Cr (VI) ions from tannery wastewater using sulfated carboxymethyl cellulose nanofilter. J Clean Prod 208:736–742. https://doi.org/10.1016/j.jclepro.2018.10.177

Ghanavati Nasab S, Semnani A, Teimouri A, Kahkesh H, Momeni Isfahani T, Habibollahi S (2018) Removal of Congo red from aqueous solution by hydroxyapatite nanoparticles loaded on Zein as an efficient and green adsorbent: response surface methodology and artificial neural network-genetic algorithm. J Polym Environ 26:3677–3697. https://doi.org/10.1007/s10924-018-1246-z

Ghosh Ray S, Ghangrekar MM (2015) Enhancing organic matter removal, biopolymer recovery and electricity generation from distillery wastewater by combining fungal fermentation and microbial fuel cell. Bioresour Technol 176:8–14. https://doi.org/10.1016/j.biortech.2014.10.158

Ghosh A, Dastidar MG, Sreekrishnan TR (2017) Bioremediation of chromium complex dyes and treatment of sludge generated during the process. Int Biodeterior Biodegrad 119:448–460. https://doi.org/10.1016/j.ibiod.2016.08.013

Giardina P, Faraco V, Pezzella C, Piscitelli A, Vanhulle S, Sannia G (2010) Laccases: a never-ending story. Cell Mol Life Sci. https://doi.org/10.1007/s00018-009-0169-1

Giese EC (2015) Série Tecnologia Ambiental do Uso de Micro-organismos

Golveia JCS, Santiago MF, Sales PTF, Sartoratto A, Ponezi AN, Thomaz DV, de Gil ES, Maria MT (2018) Cupuaçu (Theobroma grandiflorum) residue and its potential application in the bioremediation of 17-A-ethinylestradiol as a Pycnoporus sanguineus laccase inducer. Prep Biochem Biotechnol 48:541–548. https://doi.org/10.1080/10826068.2018.1466161

Hakala TK, Lundell T, Galkin S, Maijala P, Kalkkinen N, Hatakka A (2005) Manganese peroxidases, laccases and oxalic acid from the selective white-rot fungus Physisporinus rivulosus grown on spruce wood chips. Enzym Microb Technol 36:461–468. https://doi.org/10.1016/j.enzmictec.2004.10.004

Hofriter M (2002) Review: lignin conversion by manganese peroxidase (MnP). Enzym Microb Technol 30:454–466. https://doi.org/10.1016/S0141-0229(01)00528-2

Horie Y, Yamagishi T, Takahashi H, Shintaku Y, Iguchi T, Tatarazako N (2017) Assessment of the lethal and sublethal effects of 20 environmental chemicals in zebrafish embryos and larvae by using OECD TG 212. J Appl Toxicol 37:1245–1253. https://doi.org/10.1002/jat.3487

Howe K, Clark MD, Torroja CF et al (2013) The zebrafish reference genome sequence and its relationship to the human genome. Nature 496:498–503. https://doi.org/10.1038/nature12111

ISO (1995) Soil quality–determination of the effects of pollutants on Soil Flora. Part 2: effects of chemicals on the emergence of higher plants. International Organization for Standardization, Geneve

Jager D, Kupka D, Vaclavikova M, Ivanicova L, Gallios G (2018) Degradation of reactive black 5 by electrochemical oxidation. Chemosphere 190:405–416. https://doi.org/10.1016/j.chemosphere.2017.09.126

Katheresan V, Kansedo J, Lau SY (2018) Efficiency of various recent wastewater dye removal methods: a review. J Environ Chem Eng 6:4676–4697. https://doi.org/10.1016/j.jece.2018.06.060

Khan MS, Rehman MT, Bhat SA, Tabrez S, Hussain A, Husain FM, AlAjmi MF, Alamery SF, Sumbul S (2019) Food additive dye (quinoline yellow) promotes unfolding and aggregation of myoglobin: a spectroscopic and molecular docking analysis. Spectrochim Acta Part A Mol Biomol Spectrosc 214:216–226. https://doi.org/10.1016/J.SAA.2019.01.090

Khataee AR, Dehghan G, Zarei M, Ebadi E, Pourhassan M (2011) Neural network modeling of biotreatment of triphenylmethane dye solution by a green macroalgae. Chem Eng Res Des 89:172–178. https://doi.org/10.1016/j.cherd.2010.05.009

Kim K, Wang CH, Ok YS, Lee SE (2019) Heart developmental toxicity by carbon black waste generated from oil refinery on zebrafish embryos (Danio rerio): combined toxicity on heart function by nickel and vanadium. J Hazard Mater 363:127–137. https://doi.org/10.1016/j.jhazmat.2018.09.089

Kocyigit A, Pazarbasi MB, Yasa I, Ozdemir G, Karaboz I (2012) Production of laccase from Trametes trogii TEM H2: a newly isolated white-rot fungus by air sampling. J Basic Microbiol 52:661–669. https://doi.org/10.1002/jobm.201100341

Komilis D, Molina CD, Sánchez A, Young BJ, Riera NI, López VA, Barrena R, Rizzo PF, Crespo DC, Torre VD, Fernández FE (2016) Development of phytotoxicity indexes and their correlation with ecotoxicological, stability and physicochemical parameters during passive composting of poultry manure. Waste Manag 54:101–109. https://doi.org/10.1016/j.wasman.2016.05.001

Krell V, Jakobs-Schoenwandt D, Vidal S, Patel AV (2018) Cellulase enhances endophytism of encapsulated Metarhizium brunneum in potato plants. Fungal Biol 122:373–378. https://doi.org/10.1016/j.funbio.2018.03.002

Kulikova NA, Klein OI, Stepanova EV, Koroleva OV (2011) Use of basidiomycetes in industrial waste processing and utilization technologies: fundamental and applied aspects (review). Appl Biochem Microbiol 47:565–579. https://doi.org/10.1134/S000368381106007X

Kumar PS, Pavithra KG (2019) Water and textiles, water in textiles and fashion. Elsevier, Amsterdam. https://doi.org/10.1016/b978-0-08-102633-5.00002-6

Li X, de Toledo RA, Wang S, Shim H (2015) Removal of carbamazepine and naproxen by immobilized Phanerochaete chrysosporium under non-sterile condition. New Biotechnol 32:282–289. https://doi.org/10.1016/j.nbt.2015.01.003

Li S, Kazemi H, Rockaway TD (2019) Performance assessment of stormwater GI practices using artificial neural networks. Sci Total Environ 651:2811–2819. https://doi.org/10.1016/j.scitotenv.2018.10.155

Lourenço J, Marques S, Carvalho FP, Oliveira J, Malta M, Santos M, Gonçalves F, Pereira R, Mendo S (2017) Uranium mining wastes: the use of the fish embryo acute toxicity test (FET) test to evaluate toxicity and risk of environmental discharge. Sci Total Environ 605–606:391–404. https://doi.org/10.1016/j.scitotenv.2017.06.125

Maciel MJM, Castro e Silva A, Ribeiro HCT (2010) Industrial and biotechnological applications of ligninolytic enzymes of the basidiomycota: a review. Electron J Biotechnol 13:1–13. https://doi.org/10.2225/vol13-issue6-fulltext-2

Mahmoodi-Babolan N, Heydari A, Nematollahzadeh A (2019) Removal of methylene blue via bioinspired catecholamine/starch superadsorbent and the efficiency prediction by response surface methodology and artificial neural network-particle swarm optimization. Bioresour Technol 294:122084. https://doi.org/10.1016/j.biortech.2019.122084

Majeau JA, Brar SK, Tyagi RD (2010) Laccases for removal of recalcitrant and emerging pollutants. Bioresour Technol 101:2331–2350. https://doi.org/10.1016/j.biortech.2009.10.087

Mann J, Markham JL, Peiris P, Spooner-Hart RN, Holford P, Nair NG (2015) Use of olive mill wastewater as a suitable substrate for the production of laccase by Cerrena consors. Int Biodeterior Biodegrad 99:138–145. https://doi.org/10.1016/j.ibiod.2015.01.010

Mazmanci MA, Ünyayar A (2005) Decolourisation of reactive black 5 by Funalia trogii immobilised on Luffa cylindrica sponge. Process Biochem 40:337–342. https://doi.org/10.1016/j.procbio.2004.01.007

Mir-Tutusaus JA, Baccar R, Caminal G, Sarrà M (2018) Can white-rot fungi be a real wastewater treatment alternative for organic micropollutants removal? A review. Water Res 138:137–151. https://doi.org/10.1016/j.watres.2018.02.056

Moreno-García J, García-Martínez T, Mauricio JC, Moreno J (2018) Yeast immobilization systems for alcoholic wine fermentations: actual trends and future perspectives. Front Microbiol 9. https://doi.org/10.3389/fmicb.2018.00241

Munir N, Asgher M, Tahir IM, Riaz M, Bilal M, Shah SMA (2015) Utilization of agro-wastes for production of ligninolytic enzymes in liquid state fermentation by Phanerochaete chrysosporium-IBL-03. Int J Chem Biochem Sci 7:9–14

Naraian R, Kumari S, Gautam RL (2018) Biodecolorization of brilliant green carpet industry dye using three distinct *Pleurotus* spp. Environ Sustain. https://doi.org/10.1007/s42398-018-0012-4

Nilsson I, Möller A, Mattiasson B, Rubindamayugi MST, Welander U (2006) Decolorization of synthetic and real textile wastewater by the use of white-rot fungi. Enzym Microb Technol 38:94–100. https://doi.org/10.1016/j.enzmictec.2005.04.020

O'Brien AL (2017) What are the roadblocks to using population models in ecotoxicology studies? Mar Pollut Bull 124:5–8. https://doi.org/10.1016/j.marpolbul.2017.08.038

OECD (2003) Terrestrial plant test: 208—seedling emergence and seedling growth test. Organisation for Economic Cooperation and Development, Paris

Park D, Yun YS, Park JM (2010) The past, present, and future trends of biosorption. Biotechnol Bioprocess Eng 15:86–102. https://doi.org/10.1007/s12257-009-0199-4

Piscitelli A, Giardina P, Lettera V, Pezzella C, Sannia G, Faraco V (2011) Induction and transcriptional regulation of laccases in Fungi. Curr Genomics 12:104–112. https://doi.org/10.2174/138920211795564331

Pompeu GB, Pietrobon VC, Andreote CCF, Ferreira LFR, Aguiar M, Sartori SB, Cruz SH, Monteiro RTR (2018) Role of the antioxidant defense system during the production of lignocellulolytic enzymes by fungi. Int Microbiol. https://doi.org/10.1007/s10123-018-00045-1

Priac A, Badot P-M, Crini G (2017) Treated wastewater phytotoxicity assessment using Lactuca sativa: focus on germination and root elongation test parameters. C R Biol 340:188–194. https://doi.org/10.1016/j.crvi.2017.01.002

Przystaś W, Zabłocka-Godlewska E, Grabińska-Sota E (2018) Efficiency of decolorization of different dyes using fungal biomass immobilized on different solid supports. Braz J Microbiol 49:285–295. https://doi.org/10.1016/j.bjm.2017.06.010

Qi J, Zhao W, Kan Z, Meng H, Li Y (2019) Parameter optimization of double-blade normal milk processing and mixing performance based on RSM and BP-GA. Food Sci Nutr 7:3501–3512. https://doi.org/10.1002/fsn3.1198

Rawat D, Mishra V, Sharma RS (2016) Detoxification of azo dyes in the context of environmental processes. Chemosphere 155:591–605. https://doi.org/10.1016/j.chemosphere.2016.04.068

Rivera-Hoyos CM, Morales-Álvarez ED, Poveda-Cuevas SA, Reyes-Guzmán EA, Poutou-Piñales RA, Reyes-Montaño EA, Pedroza-Rodríguez AM, Rodríguez-Vázquez R, Cardozo-Bernal ÁM (2015) Computational analysis and low-scale constitutive expression of laccases synthetic genes GLLCC1 from ganoderma lucidum and POXA 1B from *Pleurotus ostreatus* in pichia pastoris. PLoS One 10:1–21. https://doi.org/10.1371/journal.pone.0116524

Rocha OP, De Oliveira DP (2017) Investigation of a Brazilian tannery effluent by means of zebra fish (Danio rerio) embryo acute toxicity (FET) test. J Toxicol Environ Health Part A Curr Issues 80:1078–1085. https://doi.org/10.1080/15287394.2017.1357356

Rodríguez-Delgado MM, Alemán-Nava GS, Rodríguez-Delgado JM, Dieck-Assad G, Martínez-Chapa SO, Barceló D, Parra R (2015) Laccase-based biosensors for detection of phenolic compounds. TrAC Trends Anal Chem. https://doi.org/10.1016/j.trac.2015.05.008

Saiu G, Tronci S, Grosso M, Cadoni E, Curreli N (2018) Pyrene and chrysene tolerance and biodegradation capability of *Pleurotus Sajor-Caju*. Open Chem Eng J 12:24–35. https://doi.org/10.2174/1874123101812010024

Salomão GSB, Agnezi JC, Paulino LB, Hencker LB, de Lira TS, Tardioli PW, Pinotti LM (2019) Production of cellulases by solid state fermentation using natural and pretreated sugarcane bagasse with different fungi. Biocatal Agric Biotechnol 17:1–6. https://doi.org/10.1016/J.BCAB.2018.10.019

Sánchez C (2009) Lignocellulosic residues: biodegradation and bioconversion by fungi. Biotechnol Adv. https://doi.org/10.1016/j.biotechadv.2008.11.001

Sandhya S (2010) Biodegradation of azo dyes under anaerobic condition: role of Azoreductase. pp 39–57. https://doi.org/10.1007/698_2009_43

Sant'Anna MCB, de Soares VM, Seibt KJ, Ghisleni G, Rico EP, Rosemberg DB, de Oliveira JR, Schröder N, Bonan CD, Bogo MR (2011) Iron exposure modifies acetylcholinesterase activity in zebrafish (Danio rerio) tissues: distinct susceptibility of tissues to iron overload. Fish Physiol Biochem 37:573–581. https://doi.org/10.1007/s10695-010-9459-7

Schubert M, Muffler A, Mourad S (2012) The use of a radial basis neural network and genetic algorithm for improving the efficiency of laccase-mediated dye decolourization. J Biotechnol 161:429–436. https://doi.org/10.1016/j.jbiotec.2012.08.003

Shindy HA (2016) Basics in colors , dyes and pigments chemistry: a review. Int Sci Org:29–36

Shojaeimehr T, Rahimpour F, Khadivi MA, Sadeghi M (2014) A modeling study by response surface methodology (RSM) and artificial neural network (ANN) on Cu2+ adsorption optimization using light expended clay aggregate (LECA). J Ind Eng Chem 20:870–880. https://doi.org/10.1016/j.jiec.2013.06.017

Singh RL, Singh PK, SINGH RP (2015) Enzymatic decolorization and degradation of azo dyes – a review. Int Biodeterior Biodegrad 104:21–31. https://doi.org/10.1016/j.ibiod.2015.04.027

Sobrero MC, Ronco A (2004) Ensayos de toxicidad aguda con semillas de lechuga 55 Ensayo de toxicidad aguda con semillas de lechuga Lactuca sativa L

Sousa RMOF, Amaral C, Fernandes JMC, Fraga I, Semitela S, Braga F, Coimbra AM, Dias AA, Bezerra RM, Sampaio A (2019) Hazardous impact of vinasse from distilled winemaking by-products in terrestrial plants and aquatic organisms. Ecotoxicol Environ Saf 183:109493. https://doi.org/10.1016/j.ecoenv.2019.109493

Sriharsha DV, Kumar RL, Savitha J (2017) Immobilized fungi on Luffa cylindrica: an effective biosorbent for the removal of lead. J Taiwan Inst Chem Eng 80:589–595. https://doi.org/10.1016/j.jtice.2017.08.032

Svobodová K, Novotný Č (2018) Bioreactors based on immobilized fungi: bioremediation under non-sterile conditions. Appl Microbiol Biotechnol 102:39–46. https://doi.org/10.1007/s00253-017-8575-z

Torbati S (2016) Artificial neural network modeling of biotreatment of malachite green by Spirodela polyrhiza: study of plant physiological responses and the dye biodegradation pathway. Process Saf Environ Prot 99:11–19. https://doi.org/10.1016/j.psep.2015.10.004

Torregrossa D, Capitanescu F (2019) Optimization models to save energy and enlarge the operational life of water pumping systems. J Clean Prod 213:89–98. https://doi.org/10.1016/j.jclepro.2018.12.124

Torres JA, Nogueira FGE, Silva MC, Lopes JH, Tavares TS, Ramalho TC, Corrêa AD (2017) Novel eco-friendly biocatalyst: soybean peroxidase immobilized onto activated carbon obtained from agricultural waste. RSC Adv 7:16460–16466. https://doi.org/10.1039/c7ra01309d

Traven VF, Cheptsov DA, Dolotov SM, Ivanov IV (2018) Control of the fluorescence of laser dyes by photooxidation of dihydrohetarenes. Dyes Pigments 158:104–113. https://doi.org/10.1016/j.dyepig.2018.05.023

USEPA (1996) Ecological effects test guidelines: 850.4200—seed germination/root elongation toxicity test. United States Environmental Protection Agency, Washington

Vikrant K, Giri BS, Raza N, Roy K, Kim KH, Rai BN, Singh RS (2018) Recent advancements in bioremediation of dye: current status and challenges. Bioresour Technol 253:355–367. https://doi.org/10.1016/j.biortech.2018.01.029

Vilar DS, Carvalho GO, Pupo MMS, Aguiar MM, Torres NH, Américo JHP, Cavalcanti EB, Eguiluz KIB, Salazar-Banda GR, Leite MS, Ferreira LFR (2018) Vinasse degradation using *Pleurotus sajor-caju* in a combined biological – electrochemical oxidation treatment. Sep Purif Technol 192:287–296. https://doi.org/10.1016/J.SEPPUR.2017.10.017

Wang G, Xiong D, Wu M, Wang L, Yang J (2020) Induction of time- and dose-dependent oxidative stress of triazophos to brain and liver in zebrafish (Danio rerio). Comp Biochem Physiol Part C Toxicol Pharmacol 228. https://doi.org/10.1016/j.cbpc.2019.108640

Wanyonyi WC, Onyari JM, Shiundu PM, Mulaa FJ (2019) Effective biotransformation of reactive black 5 dye using crude protease from Bacillus Cereus strain KM201428. Energy Proc 157:815–824. https://doi.org/10.1016/J.EGYPRO.2018.11.247

Widsten P, Kandelbauer A (2008) Laccase applications in the forest products industry: a review. Enzym Microb Technol 42:293–307. https://doi.org/10.1016/j.enzmictec.2007.12.003

Witek-Krowiak A, Chojnacka K, Podstawczyk D, Dawiec A, Pokomeda K (2014) Application of response surface methodology and artificial neural network methods in modelling and optimization of biosorption process. Bioresour Technol 160:150–160. https://doi.org/10.1016/j.biortech.2014.01.021

WTO (World Trade organization) (2020). https://www.wto.org/english/res_e/statis_e/wts2019_e/wts19_toc_e.htm. Accessed 12 Jul 2020

Yang P, Shi W, Wang H, Liu H (2016) Screening of freshwater fungi for decolorizing multiple synthetic dyes. Braz J Microbiol 47:828–834. https://doi.org/10.1016/j.bjm.2016.06.010

Yang W, Tian S, Tang Q, Chai L, Wang H (2017) Fungus hyphae-supported alumina: an efficient and reclaimable adsorbent for fluoride removal from water. J Colloid Interface Sci 496:496–504. https://doi.org/10.1016/j.jcis.2017.02.015

Yehia RS, Rodriguez-Couto S (2017) Discoloration of the azo dye Congo Red by manganese-dependent peroxidase from *Pleurotus sajor-caju*. Appl Biochem Microbiol 53:222–229. https://doi.org/10.1134/S0003683817020181

Yesilada O, Birhanli E, Geckil H (2018) Bioremediation and decolorization of textile dyes by white rot Fungi and laccase enzymes. pp 121–153. https://doi.org/10.1007/978-3-319-77386-5_5

Zaccaria S, Boff NA, Bettin F, Dillon AJP (2019) Use of micro- and ultrafiltration membranes for concentration of laccase-rich enzymatic extract of *Pleurotus sajor-caju* PS-2001 and application in dye decolorization. Chem Pap 73:3085–3094. https://doi.org/10.1007/s11696-019-00845-3
Zhang A, Wang G, Gong G, Shen J (2017a) Immobilization of white rot fungi to carbohydrate-rich corn cob as a basis for tertiary treatment of secondarily treated pulp and paper mill wastewater. Ind Crop Prod 109:538–541. https://doi.org/10.1016/j.indcrop.2017.09.006
Zhang Y-H, Lyu S-S, Liu S-T, Chen Y-X, Qin W-L, Ye Y, Chen X-G (2017b) Thermal destruction of luffa sponge under air and nitrogen. J Therm Anal Calorim 128:53–60. https://doi.org/10.1007/s10973-016-5899-3
Zhou H, Xiang N, Xie J, Diao X (2019) Ecotoxicology: the history and present direction. In: Encyclopedia of ecology. Elsevier, Amsterdam, pp 415–423. https://doi.org/10.1016/B978-0-12-409548-9.10888-7

# Chapter 8
# Industrial Scale Production of Important Therapeutic Proteins Using Bacterial Expression System

**Kunal Kerkar, Manisha Tiwari, Dhermendra K. Tiwari, and Savita Kerkar**

**Abstract** Proteins are very important for the smooth functioning of the human body since they act as structural components, enzymes, signaling molecules and cell-cell interaction mediator. Mutations in genes, might result in improper expression or maturation of proteins leading to mild or serious abnormalities. In many cases, external supply of such functionally active proteins in the body subsidizes protein related abnormalities. Some protein related abnormalities such as diabetes are common in a large section of public where insulin is the best source of protein to streamline the proper body functions. However, large productions of such proteins are required for the treatment of deficient patients. In such a scenario, natural sources are not sufficient enough to match the required demand and chemical synthesis increases the cost. However, recombinant DNA (r-DNA) technology using a host expressing system fulfilled the large-scale therapeutics requirement at a low cost. Bacterial systems when compared to other organisms are the most studied and easy to handle in generating these proteins. Apart from recombinant therapeutic protein production, bacteria naturally produce peptide antibiotics as defense metabolites which can be used to treat bacterial infections. Peptide antibiotics are thus very useful products. However, are naturally produced in minute concentrations in bacterial systems. Hence r-DNA technology is generally preferred for mass production. In this chapter review, we have discussed the major therapeutic proteins, peptide antibiotics and their industrial scale production. We also elaborated the benefits of a bacterial host system for a large-scale production of these recombinant products.

**Keywords** Baculovirus · Insulin · Lectins · Collagenase

K. Kerkar · M. Tiwari (✉) · D. K. Tiwari · S. Kerkar (✉)
Department of Biotechnology, Faculty of Life Sciences and Environment, Goa University, Taleigao Plateau, Goa, India
e-mail: manisha.tiwari@unigoa.ac.in; savita@unigoa.ac.in



P. K. Arora (ed.), *Microbial Products for Health, Environment and Agriculture*, Microorganisms for Sustainability 31, https://doi.org/10.1007/978-981-16-1947-2_8

## 8.1 Introduction

Proteins produced from bacterial systems are widely used in pharmaceutical, commercial, supplementary food, cosmetics, research, and diagnostic products (Peternel 2013). Proteins are very important for a biological system as they play important role in various biological activities such as metabolic catalyzers, structural components, cell- cell interaction and cell signaling. Natural functionality of a biological system will be affected if an individual produces mutant of functional protein, its non-functional form or deficient production of that protein. In various cases, anon-functional, mutant or deficient protein can be compensated with the administration of artificially produced protein or analogues of that protein (Gomes et al. 2018). Recombinant therapeutic proteins have gained immense importance for treating diseases that depends on administering functional proteins to restore the normal function of the affected organs or tissues. The main challenge initially was to acquire the functional protein from external sources as they were extracted from tissues of animals, seriously limiting its availability and high cost. Several biological model systems established to use as a platform for industrial scale production of variety of proteins and enzymes.

A vector used for the expression of a protein in prokaryotes, contains a selectable marker gene, promoter, regulator and a terminator. It also contains Ori C (origin of replication), that helps to determine the copy number of the vector. Promoters for high level protein expression in prokaryotes should be strong and should have the ability to produce up to 10-30% of the total cellular protein. The promoter should be highly repressible in order to avoid generation of excess protein which might lead to feedback inhibition and potentially reduce the production. Finally, the promoter should be inducible such as pET vector which uses IPTG (isopropyl thiogalactopyranoside) for induction. There are many promoters available for high level expression of protein in prokaryotes, some of them are listed below in Table 8.1 (Hannig and Makrides 1998).

Several expression vectors has been developed for recombinant protein production in bacterial system among them the pET series of vectors are the most common for expression of heterologous proteins as it contains the T7 phage promoter, which has very high affinity with T7 RNA polymerase (from T7 bacteriophage, which infect most of the *E. coli* strains) (Huang et al. 2012).

**Table 8.1** Types of promoters used in host-expression system for recombinant protein production

| Promoter | Regulation | Induction |
|---|---|---|
| T7 lac operator | LacI | IPTG |
| Lac | LacI | IPTG |
| Trp | LacI | Tryptophan starvation, indole acrylic acid, IPTG, lactose |
| recA | LexA | Nalidixic acid |
| Tac | LacI | IPTG |
| T5 lac | $LacI^{Q}$ | IPTG |
| araBAD | AraC | L-Arabinose |

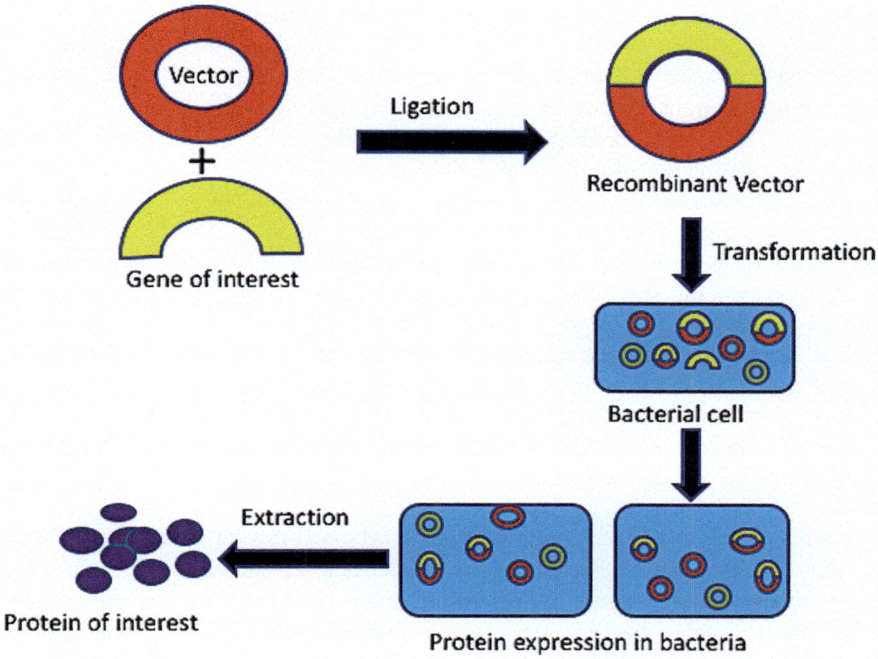


**Fig. 8.1** Production of recombinant therapeutic protein

The selection of the host cell for production is depended on the type of protein to be produced, the intellectual property rights, availability, and cost measures during production process. Bacterial system is one among the commonly used host, which is most widely used for industrial scale production of therapeutic and food supplementary proteins due to its low cost, ease of handling, faster multiplication cycle, and long term maintenance of recombinant strains (Fig. 8.1). Among them the most common and important host is the *Escherichia coli* because of its well-known genome structure, high productivity, short doubling time and minimal requirement for growth and multiplication (Graumann and Premstaller 2006; Nagesh and Srivastava 2019). However bacterial system has certain limitations such as the lack of post translational protein modifications, codon bias, endotoxin and inclusion body production which affect the high yield of recombinant protein production. To overcome these issues, certain steps such as, addition of fusion specific tag, co-expression of protein with molecular chaperones and cofactor supplementation should be enriched in growth media. BL21 is the most common *Escherichia coli* strain used for expression of proteins since it has high-level T7 bacteriophage promoter expression system, which facilitate soluble, non-toxic recombinant proteins, easy protein purification and detection. BL21 lacks the Lon protease which is responsible for degradation of the foreign proteins (Gottesman 1996). They are also deficient of OmpT protease which is an outer membrane protease which degrades the extracellularly synthesized proteins, and which may cause problems after cell lysis (Grodberg and Dunn 1988). Basically, bacterial systems are used to express non glycosylated proteins or peptides. *Lactococcus lactis* is an alternative to *Escherichia coli* as it is endotoxin free so can be used in therapeutics as well as the food industry (Yeh et al. 2009). It is a Gram-positive bacterial expression system so it can produce protein extracellularly (Morello et al. 2007).

Other than bacteria, yeast expression system is most commonly used host. *Saccharomyces cerevisiae and Pichia pastoris* are the most common candidates. Similar to *Escherichia coli*, yeast also has a short doubling time and the well understood genome (Strausberg, and National Institutes of Health Bethesda M, Strausberg, Susan L, University of Maryland, Rockville M 2000). In recent years *Pichia pastoris* has become the yeast of choice since it expresses high levels of recombinant proteins as compared to *Saccharomyces cerevisiae,* which uses methanol as its carbon source (Cregg et al. 1985). *Pichia pastoris* is used to express both intracellular as well as extracellular secretary proteins containing disulphide bonds like any other eukaryote and hence is used to express proteins containing disulphide bonds.

Baculovirus is another expression system used to express large size recombinant proteins. Baculo viral expression system has several advantages over *Escherichia coli* such as improved solubility, post-translational modifications and higher yield for secretory proteins. Baculo virus is a large double stranded DNA virus and lytic in nature. The most commonly used baculo virus for recombinant protein expression is isolated from *Autographa californica.* Insect cells used as a host for baculo virus generate disulphide bonded proteins and can produce a majority of the post translational modifications. Insect cells mostly generate N-linked glycans (Jenkins et al. 1996). This finding led to the development of glycoproteins with N-linked glycans normally found in mammalian cells (Harrison and Jarvis 2006). Finally, but not the least, mammalian cells have also been used extensively for recombinant protein expression and considered to be least effective as the strict nutrients requirements, slow growth, attachment surface requirement for most mammalian cells etc. (Figueroa et al. 2007; Wurm 2004).

Fusion proteins were originally used or constructed for large scale production of the expressed proteins to help the protein of interest to immobilize on chromatographic column and to couple the enzyme activity. Several fusion tags are added in vector either at N or C terminal of the gene of protein of interest, which facilitates easy and effective purification with high yield. Several commonly used fusion tags are mentioned in Table 8.2:

**Table 8.2** Types of fusion tags used for protein purification

| Tag | Size (kDa) | Matrix/Elution | Uses | Reference |
|---|---|---|---|---|
| Fh8 | Size of around 8 kDa and 69 amino acid long | Bind by hydrophobic interaction and EDTA for eluting | Purification, solubility and expression of proteins | Costa et al. (2014) |
| His tag | Typically, 0.86 kDa in size and 6 amino acid long. | Affinity chromatography binds to metal ions such as Ni, co, cu and Zn and eluted by lowering the pH or with imidazole | Used for detecting and purification of proteins | Porekar and Menart (2001) |
| GST | 26 kDa in size and 211 amino acids long | Glutathione, eluted using reduced glutathione | Used for increasing expression and solubility, purification and detection | Kimple et al. (2013) |
| GFP | Approximately 26.9 kDa in size and around238 long amino acid chain | – | Detection, increased expression and solubility | Hammon et al. (2009) |
| MBP | 42 kDa in size and 396 amino acid long | Amylose which is cross linked and eluted using maltose. | Purification, detection, solubility and expression | Zhao et al. (2013) |

## 8.2 Industrial Scale Production of Important Therapeutic Protein Products

### *8.2.1 Human Insulin*

Human insulin is an essential polypeptide hormone to maintain metabolic function of the body cells (Barfoed 1987). Diabetes may lead to hyperglycaemia and hence can affect the metabolism of fats, proteins and carbohydrates (Beverley and Eschwège 2003). There's a link between sugar control and complications related to diabetes as shown by the diabetes control and complications trial (Diabetes Control and Complications Trial Research Group 1993). Insulin not only acts on lowering glucose levels but also acts as a potent physiological anabolic agent (Mastick et al. 1998). Apart from this it also has many functions such as synthesis and storage of carbohydrates, lipids and proteins, prevents their degradation and releases back into circulation. Treatment of diabetes using insulin therapy was discovered by Banting and Best in 1922. Human insulin has a molecular weight of 5808 daltons and contains around 51 amino acids, produced by beta cells of pancreas and regulates the blood sugar level. Insulin is produced as pre-proinsulin which has 24 amino acid long peptide (signal peptide). Once it's in the lumen of the endoplasmic reticulum the peptide is cleaved to form proinsulin. In the endoplasmic reticulum the proinsulin is folded in a proper conformation by enzymes prohormone convertases and exoprotease carbopeptidase E to produce functional insulin (Vajo et al. 2001).

Subcutaneously injected insulin takes a long period of time to act and is very slow as compared to insulin secretion by a healthy individual (Heinemann et al. 1992). Insulin was first produced and purified from pancreas of cows and hence there were limitations in its availability until it was produced in *Escherichia coli* using recombinant DNA technology. Industrial scale production of insulin help to save a large mass of diabetic patients, who has immature or faulty insulin production in body. Large scale production mainly uses *Escherichia coli* expression system using recombinant DNA technology and is the first licensed recombinant protein based drug. Human insulin was first produced from *Escherichia coli* by Gentench, using two chain combination procedure wherein the two chains A and B encoded cDNAs were expressed in *Escherichia coli* and were purified separately and incubated together to get a functional bioactive insulin. Another approach was encoding cDNA for proinsulin and expressing it in *Escherichia coli,* followed by proteolytic digestion of the C peptide. This was a more convenient method for large scale production of insulin (Chance et al. 1999).

*Escherichia coli* has been used as a preferred microorganism for the production of insulin on industrial scale has few disadvantages such as lack of post translational activity, phosphorylation, and proteolytic processing (Jenkins 2007; Walsh and Jefferis 2006). *Campylobacter jejuni* is a bacterium having glycosylation activity has been used as an alternative host to *Escherichia coli* (Wacker et al. 2002). Initially, insulin is produced as inclusion bodies and after further multiple step for refolding and solubilisation, the functional insulin produced (Nilsson et al. 1996). In order to control the level of glucose in the blood, there was a need for insulin having faster and longer acting time. Improvement of human insulin was seriously retarded until molecular genetic techniques were introduced to create insulin analogues by changing the native structure and improving the properties of the therapeutic protein.

### *8.2.2 Lectins*

Lectins are microbial proteins that bind to carbohydrates present on the surface of the microorganisms (Procopio et al. 2017). Hence lectins are extensively studied to determine the role in pathogenicity with human host in disease development (Dias et al. 2015). Lectins have many activities such as immunomodulatory, antitumor and antifungal activities. Lectins agglutinate erythrocytes with known carbohydrate specificities. Lectins are produced by recombinant DNA technology since the yield from natural sources is limited. *Escherichia coli* (BL21-DE3) strain commonly used for industrial scale production of lectins. The yield of lectins using recombinant DNA technology has been summarized in Table 8.3 (Oliveira et al. 2013; Lam and Ng 2011).

Lectins are formed from mixtures of many isolectins (isoforms) this is due to the variability of the source from which they are extracted, structural subunits and post translation diversity (De Hoff et al. 2009). Protein mixtures which contain lectins have been studied, with respect to their functional and biological properties since it is

**Table 8.3** Lectin yield using recombinant DNA technology

| Natural sources of lectins | Yield (mg/L) | Expression system | References |
|---|---|---|---|
| Allium sativum leaf (garlic) | 5 | pET19b vector was used and cDNA was cloned into it | Upadhyay et al. (2010) |
| Galanthusnivalis | 42 | Cloning in pET vector plasmid | Luo et al. (2005) |
| Polyporussquamosus | 4-7 | Expressed BL21DE3 cells, cloned in pET vector | Tateno et al. (2004) |
| Pisum sativum | 2-5 | cDNA cloned in HindIII/BamHI/PstI restricted plasmid and expressed in Escherichia coli | Stubbs et al. (1986) |
| Nicotiana tabacumleaves | 6 | cDNA cloned in EcoRI /NotI restricted plasmid and expressed in Escherichia coli | Lannoo et al. (2007) |
| Artocarpus incise | 16 | cDNA cloned in pET25b (+) and expressed in Escherichia coli | Oliveira et al. (2009) |

difficult to separate numerous isoforms of lectins using any conventional methods. However, Isoforms may lead to complications if they have a different sugar binding affinity as it would lead to an unwanted result. For example, isoforms of *Phaseolus vulgaris* lectin (PHA-E and PHA-L) react differently with human blood cells. Both PHA-L and PHA-E have different functions, the PHA-E at low concentrations agglutinate erythrocytes whereas PHA-L agglutinate leukocytes (Raemaekers et al. 1999).

Such problems can be improved by producing and expressing recombinant lectins in a heterologous expression system using model host organisms. Since this results in lectins with high purity and having a defined amino acid sequence and the amount of lectin produced by the hosts is much higher in shorter time duration (Gemeiner et al. 2009). Apart from this, recombinant DNA technology and cDNA libraries are used to establish primary structures of lectins, studying the biosynthesis of lectins, studying the role of amino acid sequences involved in carbohydrate recognition and in obtaining novel lectins (Streicher and Sharon 2003). Many lectins produced in a variety of organisms especially plants, have been produced heterologous variants in *Escherichia coli*. Once the lectin is produced in the heterologous systems. It's functionality is tested by carrying out numerous tests such as activity assay and carbohydrate binding activity/assay. To check the toxicity of the lectins, cytotoxicity assays can also be carried out.

### 8.2.3 Human Granulocyte-Macrophage Colony Stimulating Factor (GM-CSF)

GM-CSF is a haematopoietic growth factor that stimulates development of blood immune cells such as macrophages and neutrophils. The molecular weight is 14.6 kilo Daltons and contains around 128 amino acids with disulphide linkages and N glycosylation sites. Granulocyte macrophage colony stimulating factor is a type of cytokine that helps in maturing different types of haematopoietic cells (Metcalf 2008). It is produced by fibroblasts and endothelial cells when stimulated by microbes. Since it stimulates production of haematopoietic, it has been used in immune compromised patients (Seeger 2011).

Recombinant GM-CSF is usually produced as an inclusion body in *Escherichia coli* cells. Inclusion bodies have to be resolubilized and refolded to get an active functional protein (Vallejo and Rinas 2004). So, in order to solubilise generally chaotropic agents such as 6 M guanidine hydrochloride or 8 M urea is used.

### *8.2.4 Protein Disulphide Isomerase (PDI)*

Disulphide isomerase is a protein that helps in disulphide bond formation which are important to maintain the three-dimensional structure and biological activity. It also helps in protein folding (Koivu et al. 1987; Hillson et al. 1984). It is a soluble protein that is normally found in the lumen of the endoplasmic reticulum. It has been isolated from human placenta, bovine and rat liver. It has been noticed that production of mammalian proteins in *Escherichia coli* expression system causes the production of inactive proteins due to incorrect disulphide bond formation. Hence PDI can be used to help proper folding and disulphide bond formation to get an active protein.

The proteins produced from *Escherichia coli* expression system has an additional methionine group attached to it so to get a PDI with attached alanine, *Bacillus brevis* culture was used for the production of Human PDI protein. *B. brevis* produces this protein extracellularly without much extracellular proteinase activity unlike *S. cerevisiae* and *B.subtilis* (Takagi et al. 1989).

### *8.2.5 Collagenase*

Collagen is an extracellular matrix that is most abundantly found in vertebrates. Collagen has many functions in vertebrate's right from providing scaffold to helping in growth, differentiation and survival of the cells (Theocharis et al. 2016). In order to maintain the function of the tissue, the extracellular matrix needs to be remodelled and digested (Lu et al. 2011). It can be done with the help of collagenase enzyme. Collagenase is a type of protease that degrades collagen or gelatine and non helical regions of collagen (Pseudo collagenases). Collagenases are the type of proteases that are associated with many different kinds of diseases such as bacterial infections, inflammation, rheumatoid arthritis and tumours (Watanabe 2004).

### 8.2.6 Lactase

Lactase also known as beta-galactosidase, is an enzyme that digests beta D galactosides (Gasteiger et al. 2003). It is used to digest lactose from milk products and has transgalactosylation activity. Many organisms such as plants, animals and microorganisms can produce beta-galactosidase naturally (Husain 2010; Panesar et al. 2006). Bacteria such as lactic acid bacteria and *Bifidobacterium sp* are good sources of beta-galactosidase as they are generally regarded as safe and can be used in food industry. Recombinant DNA technology has been used recently to produce large amount of beta-galactosidase, thereby increasing the economic and potential applications of beta-galactosidase. Lactase can be used as an enzyme supplement for people who are lactose intolerant.

### 8.2.7 Miscellaneous Recombinant Therapeutic Proteins

There have been many other recombinant therapeutic proteins produced from different hosts as shown in Table 8.4 (Puetz and Wurm 2019; Rao and Kroon

**Table 8.4** Therapeutic proteins and their applications

| Product | Expression system | Application |
|---|---|---|
| Eculizumab | Murine myeloma cell lines | Paroxysmal nocturnal haemoglobinuria and generalized myasthenia gravis |
| rHepatitis B surface antigen | *S. cerevisiae* | Vaccine against hepatitis B, and as a diagnostic marker |
| Rituximab | Hamster ovary | CD $20^{+}$ lymphoma's especially B cell lymphoma |
| Recombinant activated factor VIII | Hamster ovary | Prophylactic treatment in case of severe haemophilia or haemophilia A and controls and prevents bleeding |
| Protopine | *Escherichia coli* | hGH deficiency |
| Roferon A | *Escherichia coli* | Hairy cell leukaemia, hepatitis C, AIDS- related Kaposi's sarcoma and Pheladelphia chromosome positive chronic myelogenous leukemia |
| Intron A | *Escherichia coli* | Cancer, genital warts and hepatitis |
| Humatrope | *Escherichia coli* | hGH deficiency, short stature and platelet derived growth factor BB |
| Activase (Altiplase) | CHO cell line | Acute myocardial infarction and acute ischemic stroke |
| Epogen (Epoetin alpha) | CHO cell line | Anaemia and myelodysplastic syndrome, used in research of heart failure, acute kidney injury and stroke |
| Recombivax HB | *S. Cerevisiae* | Hepatitis B |
| Orthoclone OKT3 | Hybridoma cell line | Reversal of acute kidney graft rejection and rescue of steroid resistant rejection |

1993; Chien-Hung et al. 2014; Joseph et al. 2012; Mitchell et al. 2016; Dhillon 2018; Santagastino 2014), other than those mentioned above.

## 8.3 Antibiotics

Antibiotics are the substances that have the ability to inhibit the growth of other microorganisms. They have numerous applications viz. from treating infectious diseases in animals and some plants to food and biological specimen conservation (Waksman 1970). The needs of antibiotics have increased since bacterial infections are the second highest leading cause of death worldwide. Over the last decade, a large number of antibiotics have been isolated from different species and strains of the genus *Streptomyces* that has shown a broad spectrum activity against a range of Gram negative to Gram positive bacteria. (Waksman 1940). Most of the antibiotics used today have been sourced from *Streptomyces*, the most common source of antibiotics being actinomycetes (Watve et al. 2001). *Streptomyces* belong to actinomycetes and are Gram-positive (Procopio et al. 2012)., *Streptomyces sp.* produces antitumoral, antihypertensives, antiviral, antifungal and immunosuppressive compounds as secondary metabolites (Omura et al. 2001).

### *8.3.1 Peptide Antibiotics*

Antibiotics are produced as secondary metabolites in microorganisms via anabolic biosynthetic pathways (Demain et al. 1983). Antibiotics are substances having low molecular weight and are synthesised ribosomally or non-ribosomally (Nakano and Zuber 1990). Peptide antibiotics are basically derived in two ways, thus forming two classes, ribosomal synthesised peptides (natural peptides) and non-ribosomal synthesised peptides (produced from bacteria). There have been several peptide antibiotics which have been used in pharmaceutical industry such asbacilysin, gramicidin, tyrocidine, subtilin, glycopeptides, polymyxin and Bacitracin (Kleinkauf and von Dohren 1988). Non-ribosomally synthesised class of antibiotics are made by multienzyme complexes. Antibiotics such as penicillin derivatives, cephalosporin C and some glycopeptides such as vancomycin and teicoplanin fall in this category.

#### 8.3.1.1 Bacitracin

Bacitracin is a cyclic polypeptide antibiotic that is produced by *Bacillus* sp. Bacitracin is an anti-infective agent used to kill Gram positive bacteria in industrial preparations such as creams, ointments, lotions and aerosols (Yousaf 1997; Awais et al. 2008). Bacitracin producing bacillus is inoculated in nutrient broth and kept for 72 hours at 30 °C on a shaker. Once the Inoculum is ready it is transferred to a production media and incubated at 30 °C on a shaker. Samples were taken after every 24 hours and centrifuged to get a cell free supernatant. Agar diffusion method is performed to check the antimicrobial activity and the production (Sen et al. 1995). Cultures such *Staphylococcus aureus* and *Micrococcus luteus* were used as test organisms and spread plated on agar plates and wells were bored. Then 80μL of cell free supernatant was added to the wells and incubated at 37 °C and checked for inhibition zones. Paper chromatography and high-performance liquid chromatography (HPLC) can be used to further reconfirm the quality of produced peptide antibiotics (Snell et al. 1955).

#### 8.3.1.2 Polymyxins

Polymyxin E (colistin) that was isolated from a soil bacterium *Paenibacillus polymyxa* is a polypeptide antibiotic (Benedict and Langlykke 1947). Along with polymyxin B, they showed activity mostly against Gram-negative bacteria. The chemical structure is similar to other cationic antimicrobial peptides such as defensins and gramicidin (Hancock 1997). Polymyxin consists of a cyclic heptapeptides and a tripeptide side chain (Falagas et al. 2010; Li et al. 2006). It also contains a N-terminal fatty acyl segment that is acylated to the tripeptide side chain and is an important factor in the antimicrobial activity of the antibiotic (Brink et al. 2014; Gallardo-Godoy et al. 2016). Polymyxin E and Polymyxin B differ in only one amino acid residue i.e. polymyxin B has phenylalanine in place of leucine in Polymyxin E (Nation et al. 2014). Polymyxin acts on the outer lipid membrane of Gram-negative bacteria. Interaction of positively charged diamino butyric acid (Dab) on the polymyxin on the negatively charged phosphates on the membrane, destabilizes the membrane and hence causes lysis of cell (Dixon and Chopra 1986).

#### 8.3.1.3 Gramicidin

There are many types of Gramicidin based on the function and isolation. Gramicidin A is a peptide antibiotic that is isolated from *Bacillus brevis.* It acts by creating a cation permeable channel in the lipid and biological membrane. The two beta helical subunits form the channel. Gramicidin S is a peptide antibiotic that is produced from soil bacilli. It destroys the lipid bilayer barrier and interferes and displaces membrane bound proteins (Afonin et al. 2008). No intercellular structures accumulate Gramicidin S in Gramicidin producing cells and it appears that the acidic S layer can accumulate the production (Ostrovski et al. 1988). Gramicidin D is a mixture of Gramicidin A, B and C. Gramicidin D is used as eye drops to treat eye infections and is mostly effective against Gram positive bacteria like *Bacillus subtilis* and *Staphylococcus aureus* but not so effective against Gram negative bacteria. Apart from these peptide antibiotics there are many other listed in Table 8.5 (Wang et al. 2012).

**Table 8.5** Types of peptide antibiotics

| Peptide antibiotics | Applications |
|---|---|
| Boceprevir | Hepatitis C |
| Oritavancin | Bacterial infections |
| Dalbavancin | Bacterial infections |
| Enfuvirtide | HIV |
| Bacterial nisin | Stomach ulcers and oral mucositis |
| Actinomycin D | Cancer |
| Bleomycin | Tumors |

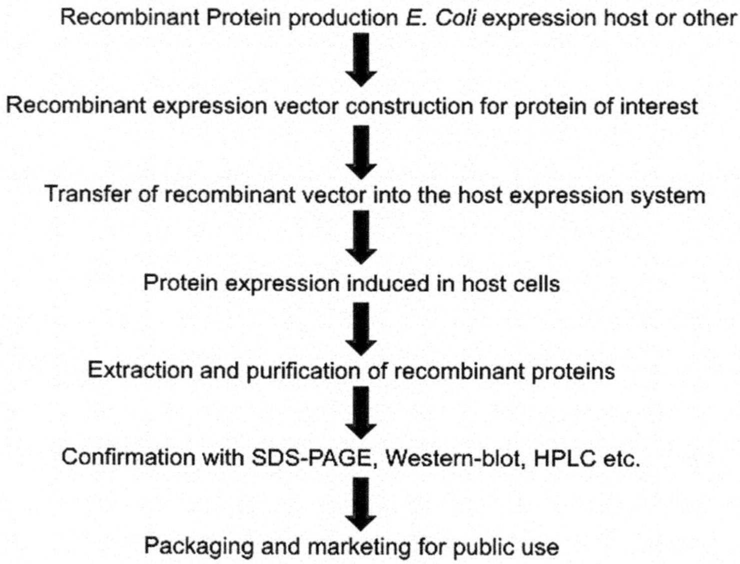


**Fig. 8.2** Production of recombinant proteins (Hirabayashi et al. 1993)

## 8.4 Production

Processes involved in the production of Recombinant therapeutic proteins from bacterial systems are depicted in Fig. 8.2. As far as peptide antibiotics are concerned, with certain exceptions such as nisin and some antibiotics produced by *Lactococcus lactis,* it is not feasible to produce peptide antibiotics from natural sources. Therefore, the two methods are in practice for production using recombinant DNA technology and protein chemistry. Chemical building of peptide antibiotics involves the use of automated peptide synthesizer which involves adding amino acids from N to C terminus. While recombinant DNA technology involves the use of genes responsible for the protein production and incorporating them into the host organism for expression in a suitable vector, which is more feasible and cost effective than the chemical synthesis.

## 8.4.1 Protein Extraction and Purification

Once the protein of interest has been produced by the host organism, the final step is extraction and purification. Most of the therapeutic proteins are intracellular. The first step involves breaking down the cell wall which can be achieved by mechanical, enzymatic or chemical method. In some cases the host expression cell can be genetically engineered so that the intracellular product is excreted out. However, making bacterial cell wall leaky using genetic manipulation is very tricky and achieved only in limited cases (Chisti and Moo-young 1986). Therefore the cell wall disintegration after successful protein expression is commonly used.

Once the cell wall has been broken, the protein should be separated from the rest of the cell contents until a purified protein is obtained. The success of cell disruption depends on the buffers, the chosen method and the presence of inhibitors. However, all the steps should be carried out carefully and always in a cold room at 4 °C for high quality protein production. Buffers are the most important factor in protein extraction. Since proteins are zwitterions, the pH of the extraction buffer should be maintained in order to get a biologically active protein (Ahmed 2004). Lysis buffer should contain phosphate or HEPES, high ionic strength i.e. approx. 300 mM NaCl for solubility and stability of protein and a reducing agent.

## 8.4.2 Protein Extraction

Bacterial cell wall can be disrupted using chemical, enzymatic or mechanical methods as discussed above. The first two are done on the small scale such as in research laboratory whereas on the industrial scale the mechanical methods are often preferred as it gives better result, cost-effective and less time consuming. The degree of disruption of the cell depends on the size of the protein of interest as small proteins need little space as compared to inclusion bodies where the mechanical stress should be more to get more pore size.

### 8.4.2.1 Chemical, Enzymatic and Mechanical Methods for Cell Disruption

Chemical lysis includes treatment of bacteria with alkali and detergent. Chemical methods are not suitable for isolation of inclusion bodies since they are soluble in detergent hence results in less protein output but if incorporated with sonication, it results in better extraction of inclusion bodies (Rodríguez-Carmona et al. 2010).

Enzymatic method depends on the enzyme lysozyme for disruption of the cells. Certain bacteria like Gram-negative *Escherichia coli* have a lipid outer membrane which makes it less susceptible to enzymatic lysis. Gram-negative bacteria are less susceptible to lysozyme since it contains the asymmetric lipid bilayer (LPS) therefore small divalent cations and polycationic small molecules have been used to permeabilize bacterial membrane. Lysozyme is positively charged as compared to inclusion bodies (IB) and cell debris which are negatively charged. Hence agglomerates of lysozyme, inclusion bodies and cell debris are formed upon lysis. When the recombinant protein is in the form of inclusion bodies, enzymatic method is not a good option since the lysozyme sticks to the inclusion bodies adding an impurity (Peternel and Komel 2010).

Mechanical methods should be as gentle as possible so that the protein of interest is not disrupted. Mechanical lysis method is preferred for large scale production since no chemicals are needed and if dealt carefully the yield is superior. Ultrasonication is a mechanical method performed on a smaller laboratory scale. It induces shear stress along with low pressure regions which finally leads to the breakage of the cell (Doulah 1977). Sonication is not applicable when working with non-classical inclusion bodies (ncIBs) as it affects the structure stability of the inclusion bodies resulting in significant loss of protein from the inclusion bodies (IBs). French press or homogenizer uses high pressure to disrupt bacterial cells and is commonly used at industrial scale (Middelberg 2000). Unlike sonication this method is suitable for extracting ncIBs since it does not affect the structure. But to get a pure IB several cycles of homogenisation is required. Hence it is of utmost importance to optimize the disruption method, being an important aspect of the biotechnological process of protein purification and mainly depends on the solubility and intended use of the protein.

### 8.4.3 Protein Purification

After bacterial disruption, the protein must be separated from the cell debris, host cell proteins and remaining cell debris. If the protein is soluble, it can be easily extracted using centrifugation to separate out the cell debris. Soluble protein once separated from the cell debris is passed through different chromatographic columns, such as ion exchange chromatography, affinity chromatography, size exclusion chromatography, hydrophobic interaction chromatography based on the tags available, to acquire purified recombinant proteins. The protein needs to be concentrated to up to 60-70 g/L which is suitable for chromatographic purification (Asenjo et al. 1989). To maximize the yield, it is important to minimize the number of steps to recover the final purified protein. The first step involves the removal of contaminants from the protein of interest and is called clean up or pre-treatment. This treatment involves, relatively inexpensive methods such as adsorption on Whatman paper, ion exchange cartridge, aqueous partitioning in two phases, hydrophobic interaction or using salts such as ammonium sulphate for precipitation.

After the primary treatment, high resolution purification is carried out which results in the recovery of protein up to 98-99%. It involves one or two ion exchange or affinity chromatography steps. After high-resolution purification, the final step is polishing which gives ultra-purity to the protein of interest. Mainly HPLC is used for

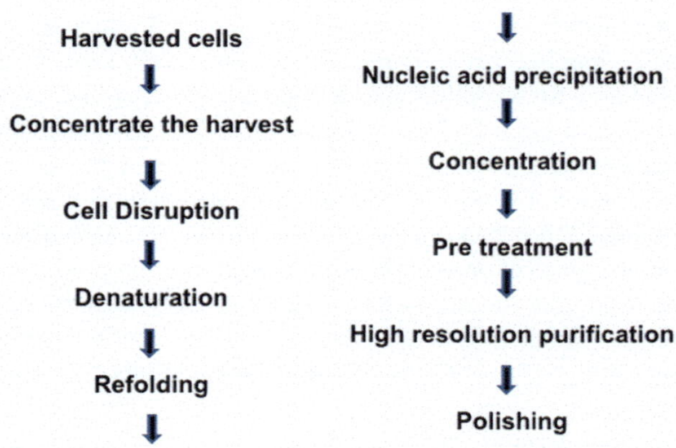


**Fig. 8.3** A flowchart of protein purification: steps involved from expression to marketing

polishing step, gel filtration can also be used to separate dimers, oligomers from monomer. Steps involved in protein extraction and purification is given in Fig. 8.3.

### *8.4.4 Storage*

Once the purification and activity of the protein is determined, it needs to be stored in proper condition so that it doesn't denature until further use. There are many methods used for storage of protein but the most commonly used is lyophilisation. The protein is lyophilised using a lyophilizer. The storage conditions vary depending on the protein characteristic and the storage period. For short term storage, proteins can be kept in a refrigerator (−20 °C). For storage at 4 °C longer than 24 h, it may be necessary to filter sterilize the protein preparation (through a 0.22μm filter) or to add a bacteriostatic agent to avoid bacterial growth. For long term storage, it becomes necessary to freeze the protein preparation either by using liquid nitrogen or a dry ice/ethanol mixture to avoid denaturation. Small aliquots are used to avoid repeated freezing and thawing which may reduce the biological activity or affect the structure. For storing the protein preparation for several months at -20 °C or -80 °C, it is necessary to add 50% glycerol to the solution to avoid freezing (Kenig et al. 2006).

## 8.5 Conclusion

With the increase in human pathologies, overcome with protein deficiency in body and treatment of several diseases related to faulty protein production, the market and potential for recombinant protein therapeutic products and peptide antibiotics is rapidly increasing day by day. In-spite of the mammalian cell lines strength, many factories still focus and use microbial cells specially *Escherichia coli* as a host organism because of its easy handling, well annotated genome and cost effectiveness, making *Escherichia coli* a potent cell factory. Since gene therapy is not available for all pathologies it has become imperative to produce an active functional protein to get rid of the abnormalities. This review chapter will help researchers to understand the importance of bacterial systems for recombinant therapeutic protein production, the types of recombinant therapeutics being produced, the way they are produced, types of bacterial host available etc. Main focus of the chapter is to bring out the types of protein products being currently produced in factories at a large scale to overcome the deficiency of gene therapy in treating human pathologies due to protein deficiency. The chapter also talks about the peptide antibiotics produced by bacterial expression system and its applications.

**Acknowledgement** We thank the Department of Biotechnology, Ministry of Science and Technology, India, for the BioCare women scientist fellowship and Goa University for providing research facilities.

## References

Afonin S, Dürr UHN, Wadhwani P et al (2008) Solid state NMR structure analysis of the antimicrobial peptide gramicidin-S in lipid membranes: and self-assembly as a β-barrel. Anion Sens:139–154. https://doi.org/10.1007/128

Ahmed H (2004) Principles and reactions of protein extraction, purification, and characterization. Princ React Protein Extr Purification, Charact, pp 1–389

Asenjo JA, Herrera L, Byrne B (1989) Development of an expert system for selection and synthesis of protein purification processes. J Biotechnol 11:275–298. https://doi.org/10.1016/0168-1656(89)90123-5

Awais M, Pervez A, Qayyum S, Saleem M (2008) Effects of glucose, incubation period and pH on the production of peptide antibiotics by *Bacillus pumilus*. Afr J Microbiol Res 2:114–119

Barfoed HC (1987) Insulin production technology. Chem Eng Prog 83(10):49–54

Benedict RG, Langlykke AF (1947) Antibiotic activity of *Bacillus polymyxa*. J Bacteriol 54(1):24

Beverley B, Eschwège E (2003) The diagnosis and classification of diabetes and impaired glucose tolerance. Textbook of diabetes (1 edn.) John C Pickup and Gareth Williams Third edition: 2.1

Brink AJ, Richards GA, Colombo G, Bortolotti F, Colombo P, Jehl F (2014) Multi component antibiotic substances produced by fermentation: implications for regulatory authorities, critically ill patients and generics. Int J Antimicrob Agents 43:1–6

Chance R, Glazer N, Wishner K (1999) Insulin Lispro (Humalog) in biopharmaceuticals, an industrial perspective. Dordrecht, pp 149–172

Chien-Hung C, Yi-Chun C, Sheng-Nan L, Chuan-Mo L, Tsung-Hui H, Chao-Hung H (2014) Serum hepatitis surface antigen level predict treatment response to nucleos(t)ide analogues. World J Gastroenterol 20(24):7686–7695

Chisti Y, Moo-young M (1986) Disruption of microbial cells for intracellular products. Butterworth, pp 1–11

Costa S, Almeida A, Castro A, Domingues L (2014) Fusion tags for protein solubility, purification, and immunogenicity in *Escherichia coli*: the novel Fh8 system. Front Microbiol 5. https://doi.org/10.3389/fmicb.2014.00063

Cregg JM, Barringer KJ, Hessler AY, Madden KR (1985) *Pichia pastoris* as a host system for transformations. Mol Cell Biol 5:3376–3385. https://doi.org/10.1128/mcb.5.12.3376

De Hoff PL, Brill LM, Hirsch AM (2009) Plant lectins: the ties that bind in root symbiosis and plant defense. Mol Genet Genomics 282:1–15. https://doi.org/10.1007/s00438-009-0460-8
Demain AL, Aharonowitz Y, Martin JF (1983) Metabolite control of secondary biosynthetic pathways. Addison-Wesley, pp 49–67
Dhillon S (2018) Eculizumab: a review in generalized myasthenia gravis. Drugs 78:367–376. https://doi.org/10.1007/s40265-018-0875-9
Diabetes Control and Complications Trial Research Group (1993) The effect of intensive treatment of diabetes on the development and the progression of long term complications in insulin-dependent diabetes mellitus. J Med 329:977–978
Dias R, Machado S, Migliolo L, Franco L (2015) Insights into animal and plant lectins with antimicrobial activities. Molecules 20:519–541
Dixon RA, Chopra I (1986) Leakage of periplasmic proteins from *Escherichia coli* mediated by polymyxin B nonapeptide. Antimicrob Agents Chemother 29:781–788. https://doi.org/10.1128/AAC.29.5.781
Doulah MS (1977) Mechanism of disintegration of biological cells in ultrasonic cavitation. Biotechnol Bioeng 19:649–660. https://doi.org/10.1002/bit.260190504
Falagas ME, Rafailidis PI, Matthaiou DK (2010) Resistance to polymyxins: mechanisms, frequency and treatment options. Drug Resist Updat 13:132–138. https://doi.org/10.1016/j.drup.2010.05.002
Figueroa B, Ailor E, Osborne D, Hardwick JM, Reff M, Betenbaugh MJ (2007) Enhanced cell culture performance using inducible anti-apoptotic genes E1B–19 K and in the production of a monoclonal antibody with Chinese hamster ovary cells. Biotechnol Bioeng 97:877–892
Gallardo-Godoy A, Muldoon C, Becker B et al (2016) Activity and predicted nephrotoxicity of synthetic antibiotics based on Polymyxin B. J Med Chem 59:1068–1077. https://doi.org/10.1021/acs.jmedchem.5b01593
Gasteiger E, Gattiker A, Hoogland C et al (2003) ExPASy: the proteomics server for in-depth protein knowledge and analysis. Nucl Acids Res 31:3784–3788. https://doi.org/10.1093/nar/gkg563
Gemeiner P, Mislovičová D, Tkáč J et al (2009) Lectinomics. II. A highway to biomedical/clinical diagnostics. Biotechnol Adv 27:1–15. https://doi.org/10.1016/j.biotechadv.2008.07.003
Gomes A, Carmo T, Carvalho L, Bahia F, Parachin N (2018) Comparison of yeasts as a host for recombinant protein production. Microorganisms 6(2):38
Gottesman S (1996) Proteases and their targets in *Escherichia coli*. Annu Rev Genet 30:465–506. https://doi.org/10.1146/annurev.genet.30.1.465
Graumann K, Premstaller A (2006) Manufacturing of recombinant therapeutic proteins in microbial systems. Biotechnol J 1:164–186
Grodberg J, Dunn JJ (1988) ompT encodes the *Escherichia coli* outer membrane protease that cleaves T7 RNA polymerase during purification. J Bacteriol 170:1245–1253. https://doi.org/10.1128/jb.170.3.1245-1253.1988
Hammon J, Palanivelu DV, Chen J et al (2009) A green fluorescent protein screen for identification of well-expressed membrane proteins from a cohort of extremophilic organisms. Protein Sci 18:121–133. https://doi.org/10.1002/pro.18
Hancock RE (1997) Peptide antibiotics. Lancet 349:418–422
Hannig G, Makrides S (1998) Strategies for optimizing heterologous protein expression in *Escherichia Coli*. Tibtech:54–61
Harrison RL, Jarvis DL (2006) Protein N-glycosylation in the Baculovirus-insect cell expression system and engineering of insect cells to produce "Mammalianized" recombinant glycoproteins. Adv Virus Res 68:159–191. https://doi.org/10.1016/S0065-3527(06)68005-6
Heinemann L, Starke AAR, Hohmann A, Berger M (1992) Timing between the subcutaneous administration of insulin and consumption of a carbohydrate rich meal. Horm Metab Res Suppl:137–139

Hillson DA, Lambert N, Freedman RB (1984) Formation and isomerization of disulfide bonds in proteins: protein disulfide-isomerase. Methods Enzymol 107:281–294. https://doi.org/10.1016/0076-6879(84)07018-X

Hirabayashi J, Sakakura Y, Kasai K (1993) Production of intact recombinant proteins in *Escherichia Coli*. Lectins Glycobiol:474–481

Huang CJ, Lin H, Yang X (2012) Industrial production of recombinant therapeutics in *Escherichia coli* and its recent advancements. J Ind Microbiol Biotechnol 39:383–399. https://doi.org/10.1007/s10295-011-1082-9

Husain Q (2010) β galactosidases and their potential applications: a review. Crit Rev Biotechnol 30:41–62. https://doi.org/10.3109/07388550903330497

Jenkins N (2007) Modifications of therapeutic proteins: challenges and prospects. Cytotechnology 53:121–125. https://doi.org/10.1007/s10616-007-9075-2

Jenkins N, Parekh RB, James DC (1996) Getting the glycosylation right: implications for the biotechnology industry. Nat Biotechnol 14:975–981. https://doi.org/10.1038/nbt0896-975

Joseph J, Jaison V, Jain K, Kakkar N, Jacob J (2012) Erythropoietin use and abuse. Indian J Endocrinol Metabol 16(2):220–227

Kenig M, Peternel Š, Gaberc-Porekar V, Menart V (2006) Influence of the protein oligomericity on final yield after affinity tag removal in purification of recombinant proteins. J Chromatogr A 1101:293–306. https://doi.org/10.1016/j.chroma.2005.09.089

Kimple ME, Brill AL, Pasker RL (2013) Overview of affinity tags for protein purification. Curr Protoc Protein Sci:1–23. https://doi.org/10.1002/0471140864.ps0909s73

Kleinkauf H, von Dohren H (1988) Peptide antibiotics, f3-lactams, and related compounds. 8

Koivu J, Myllylä R, Helaakoski T et al (1987) A single polypeptide acts both as the beta subunit of prolyl 4-hydroxylase and as a protein disulfide-isomerase. J Biol Chem 262:6447–6449

Lam SK, Ng TB (2011) Lectins: production and practical applications. Appl Microbiol Biotechnol 89:45–55. https://doi.org/10.1007/s00253-010-2892-9

Lannoo N, Vervecken W, Proost P et al (2007) Expression of the nucleocytoplasmic tobacco lectin in the yeast *Pichia pastoris*. Protein Expr Purif 53:275–282. https://doi.org/10.1016/j.pep.2007.01.007

Li J, Nation RL, Turnidge JD et al (2006) Colistin: the re-emerging antibiotic for multidrug-resistant gram-negative bacterial infections. Lancet Infect Dis 6:589–601. https://doi.org/10.1016/S1473-3099(06)70580-1

Lu P, Takai K, Weaver VM, Werb Z (2011) Extracellular matrix degradation and remodelling in development and disease. Cold Spring Harb Perspect Biol 3:12

Luo S, Zhangsun D, Tang K (2005) Functional GNA expressed in *Escherichia coli* with high efficiency and its effect on Ceratovacuna lanigera zehntner. Appl Microbiol Biotechnol 69:184–191. https://doi.org/10.1007/s00253-005-0042-6

Mastick CC, Brady MJ, Printen JA et al (1998) Spatial determinants of specificity in insulin action. Mol Cell Biochem 182:65–71. https://doi.org/10.1023/A:1006835430797

Metcalf D (2008) Hematopoietic cytokines. Blood 111:485–491

Middelberg APJ (2000) 2 Microbial cell disruption by high-pressure homogenization. 11–21. https://doi.org/10.1007/978-1-59259-027-8_2

Mitchell AC, Briquez PS, Hubbell JA, Cochran JR (2016) Engineering growth factors for regenerative medicine applications. Acta Biomater 30:1–12. https://doi.org/10.1016/j.actbio.2015.11.007

Morello E, Bermúdez-Humarán LG, Llull D et al (2007) Lactococcus lactis, an efficient cell factory for recombinant protein production and secretion. J Mol Microbiol Biotechnol 14:48–58. https://doi.org/10.1159/000106082

Nagesh T, Srivastava A (2019) Recent developments in bioprocessing of recombinant proteins: expression hosts and process development. Front Bioeng Biotecnol 7:420

Nakano MM, Zuber P (1990) Molecular biology of antibiotic production in *Bacillus*. Crit Rev Biotechnol 10:223–240

Nation RL, Li J, Cars O et al (2014) Consistent global approach on reporting of colistin doses to promote safe and effective use. Clin Infect Dis 58:139–141. https://doi.org/10.1093/cid/cit680

Nilsson J, Jonasson P, Samuelsson E et al (1996) Integrated production of human insulin and its C-peptide. J Biotechnol 48:241–250. https://doi.org/10.1016/0168-1656(96)01514-3

Oliveira C, Costa S, Teixeira JA, Domingues L (2009) cDNA cloningand functional expression of the alphaD-galactose-binding lectinfrutalin in Escherichia coli. Mol Biotechnol 43:212–220

Oliveira C, Teixeira JA, Domingues L (2013) Recombinant lectins: an array of tailor-made glycan-interaction biosynthetic tools. Crit Rev Biotechnol 33:66–80. https://doi.org/10.3109/07388551.2012.670614

Omura S, Ikeda H, Ishikawa J et al (2001) Genome sequence of an industrial microorganism Streptomyces avermitilis: deducing the ability of producing secondary metabolites. Proc Natl Acad Sci U S A 98:12215–12220. https://doi.org/10.1073/pnas.211433198

Ostrovski DN, Eremin VA, Udalova TP, Kharatian EF (1988) The cell wall of *Bacillus brevis* producing gramicidin S, a cyclodecapeptide antibiotic. Mikrobiologiia 57:1017–1023

Panesar PS, Panesar R, Singh RS et al (2006) Microbial production, immobilization and applications of $\beta$-D-galactosidase. J Chem Technol Biotechnol 81:530–543. https://doi.org/10.1002/jctb.1453

Peternel S (2013) Bacterial cell disruption: a crucial step in protein production. New Biotechnol 30:250–254. https://doi.org/10.1016/j.nbt.2011.09.005

Peternel S, Komel R (2010) Isolation of biologically active nanomaterial (inclusion bodies) from bacterial cells. Microb Cell Factories 9:1–16. https://doi.org/10.1186/1475-2859-9-66

Porekar V, Menart V (2001) Perpectives of immobilized metal affinity chromatography. Eur J Biochem 49:335–360

Procopio R, Da Silva I, Martins M, Azevedo J, Araujo J (2012) Antibiotics produced by *streptomyces*. Br J Infect Dis 16(5):466–471

Procopio T, Moura M, Albuquerque L, Gomes F, Santos N, Coelho L, Pontual E, Paiva P, Napoleao T (2017) Antibacterial lectins: action mechanisms, defensive roles and biotechnological roles. Antibacterials: Chapter 3 Nova Science Publishers, pp 70–90

Puetz J, Wurm FM (2019) Recombinant proteins for industrial versus pharmaceutical purposes: a review of process and pricing. PRO 7:476. https://doi.org/10.3390/pr7080476

Raemaekers RJM, De Muro L, Gatehouse JA, Fordham-Skelton AP (1999) Functional phytohemagglutinin (PHA) and *Galanthus nivalis* agglutinin (GNA) expressed in Pichia pastoris. Correct N-terminal processing and secretion of heterologous proteins expressed using the PHA-E signal peptide. Eur J Biochem 265:394–403. https://doi.org/10.1046/j.1432-1327.1999.00749.x

Rao P, Kroon D (1993) Orthoclone OKT3. Stability and characterization of protein and peptide drugs, pp 135–158

Rodríguez-Carmona E, Cano-Garrido O, Seras-Franzoso J et al (2010) Isolation of cell-free bacterial inclusion bodies. Microb Cell Factories 9:1–9. https://doi.org/10.1186/1475-2859-9-71

Santagastino E (2014) A new recombinant factor VIII: from genetics to clinical use. Drug Des Devel Ther 8:2507–2515

Seeger RC (2011) Immunology and immunotherapy of neuro blastoma. Semin Cancer Biol 21:229–237

Sen KS, Haque FS, Pal CS (1995) Nutrient optimization for production of broad-spectrum antibiotics by *Streptomyces antibioticus*. Microbial Hung 42:155–162

Snell N, Ijichi K, Lewis JC (1955) Paper chromatographic identification of polypeptide gram positive inhibiting antibiotics. Western Utilization Research Branch

Strausberg RL, National Institutes of Health Bethesda M, Strausberg, Susan L, University of Maryland, Rockville M (2000) Overview of protein expression in Saccharomyces cerevisiae. Curr Protoc Protein Sci:1–7

Streicher H, Sharon N (2003) Recombinant plant lectins and their mutants. Methods Enzymol 363:47–77. https://doi.org/10.1016/S0076-6879(03)01043-7

Stubbs ME, Carver JP, Dunn RJ (1986) Production of pea lectin in *Escherichia coli*. J Biol Chem 261:6141–6144

Takagi H, Kadowaki K, Udaka S (1989) Screening and characterization of protein-hyperproducing bacteria without detectable exoprotease activity. Agric Biol Chem 53:691–699. https://doi.org/10.1080/00021369.1989.10869382

Tateno H, Winter H, Goldstein I (2004) *Escherichia coli* and characterization of the recombinant Neu5Acalpha2, 6Galbeta1, 4GlcNAc-specific high-affinity lectin and its mutants from the mushroom Polyporus. Biochem J 382:667–675

Theocharis AD, Skandalis SS, Gialeli C, Karamanos NK (2016) Extracellular matrix structure. Adv Drug Deliv Rev 97:4–27. https://doi.org/10.1016/j.addr.2015.11.001

Upadhyay SK, Saurabh S, Rai P et al (2010) SUMO fusion facilitates expression and purification of garlic leaf lectin but modifies some of its properties. J Biotechnol 146:1–8. https://doi.org/10.1016/j.jbiotec.2010.01.013

Vajo Z, Fawcett J, Hamel FG, Bennett RG, Duckworth WC (2001) Insulin and analogue effects on protein degradation in different cell types. J Biol Chem 276(15):11552–11558

Vallejo LF, Rinas U (2004) Strategies for the recovery of active proteins through refolding of bacterial inclusion body proteins. Microb Cell Factories 3:11

Wacker M, Linton D, Hitchen PG et al (2002) Glycosylation *Campylobacter jejuni* N-linked. Science (80-):2649

Waksman SA (1940) Microbes in a changing world. Sci Monthly 51:422–427

Waksman A (1970) Success and failures in the search of antibiotics. Institute of Microbiology, pp 1–16

Walsh G, Jefferis R (2006) Post-translational modifications in the context of therapeutic proteins. Nat Biotechnol 24:1241–1252. https://doi.org/10.1038/nbt1252

Wang F, Qin L, Pace CJ et al (2012) Solubilized gramicidin a as potential systemic antibiotics. Chembiochem 13:51–55. https://doi.org/10.1002/cbic.201100671

Watanabe K (2004) Collagenolytic proteases from bacteria. Appl Microbiol Biotechnol 63:520–526. https://doi.org/10.1007/s00253-003-1442-0

Watve MG, Tickoo R, Jog MM, Bhole BD (2001) How many antibiotics are produced by the genus *Streptomyces*? Arch Microbiol 176:386–390. https://doi.org/10.1007/s002030100345

Wurm FM (2004) Production of recombinant protein therapeutics in cultivated mammalian cells. Nat Biotechnol 22:1393–1398. https://doi.org/10.1038/nbt1026

Yeh CM, Kao BY, Peng HJ (2009) Production of a recombinant type 1 antifreeze protein analogue by L. lactis and its applications on frozen meat and frozen dough. J Agric Food Chem 57:6216–6223. https://doi.org/10.1021/jf900924f

Yousaf M (1997) Studies on the cultural conditions for the production of antibiotic bacitracin by *B. licheniformis*. Islamia University Bahawalpur Pakistan, pp 1–15

Zhao X, Li G, Liang S (2013) Several affinity tags commonly used in chromatographic purification. J Anal Methods Chem 2013. https://doi.org/10.1155/2013/581093

# Chapter 9
# Role of Microbes and Microbial Products in Cancer Therapeutics

**Vinayak Sharma, Prakash Kumar Sinha, Jagtar Singh, and Eshu Singhal Sinha**

**Abstract** Despite all the major advancements in therapeutic research and drug synthesis, cancer still remains one of the major reasons for deaths worldwide. This calls for the need of speeding the search for new anticancer medicines. Microbes and microbial products have numerous implications in health out of which anticancer potential has been explored well. Many of these products have been screened, established, and are available in market as anticancer drugs. This chapter provides an overview of different microbes and microbial products involved in cancer therapeutics that act either directly by killing the cancer cells or indirectly by inducing immunotherapy where immune system gets activated and targets cancerous cells.

**Keywords** Cancer · Microbes · Microbial products · Anticancer agents · Enzymes

## 9.1 Introduction

Cancer is uncontrolled growth of abnormal cells in the body and is a leading cause of death worldwide. According to GLOBOCAN, 18 million new cancer cases were reported in 2018 and 9.5 million deaths were observed worldwide. Cancer incidence is estimated to increase by 5.1 million in 2030 which demands a coordinated response from public health professionals, oncologists, policy-makers, and researchers (Bray et al. 2018).

From time immemorial, natural products have played an important role in therapeutics of many human diseases including cancer. Natural products with medicinal values are available from both terrestrial and marine organisms, including microorganisms. William B. Coley, a surgeon in the Memorial Hospital in

---

Eshu Singhal Sinha and Jagtar Singh contributed equally as corresponding authors.

V. Sharma · P. K. Sinha · J. Singh (✉) · E. S. Sinha
Department of Biotechnology, Panjab University, Chandigarh, India

 
P. K. Arora (ed.), *Microbial Products for Health, Environment and Agriculture*, Microorganisms for Sustainability 31, https://doi.org/10.1007/978-981-16-1947-2_9

New York described the role of bacteria as anticancer agents for the first time in 1890 (Chakrabarty 2003; Karpiński and Adamczak 2018). He utilized a mixture composed of supernatants of *Streptococcus pyogenes* and *Serratia marcescens* for the treatment of malignancy patients. This mixture is now known as Coley's toxins. Of approximately 1200 patients, cancer regression was observed in 52 cases and a complete cure in 30 patients. The mechanism of this reaction has now been understood partially. Microbial infections activate immune cells and induce the production of inflammatory cytokines such as tumor necrosis factor-α (TNF-α) leading to cancer regression (Patyar et al. 2010). Later, in 1935 Connell used sterile filtrates from *Clostridium histolyticum* to treat advanced cancers and he observed tumor regression and explained it to be the result of enzyme production.

Besides microbes, organic compounds, e.g., pentostatin, peplomycin, and epirubicin derived from a wide range of microorganisms also have implication in the treatment of many diseases including cancer. Some of the microbial compounds can be used in their natural form and others are applicable after their synthetic modification.

Several microbes and microbial proteins/peptides have emerged as a promising group of bioactive agents that act as effective anticancer drugs. Some of these include drugs which are already in use such as actinomycin D, doxorubicin, mitomycin C, and diphtheria toxin, while other substances are either in clinical trials or are being tested in in vitro research.

## 9.2 Microbes as Anticancer Agents

The efficiency of microorganisms as antitumor agents is extremely diverse. Hypoxic core of solid tumors is resistant to major treatment methods. However, certain live, attenuated, and engineered microorganisms such as *Bifidobacterium, Bacillus, Clostridium, Salmonella, Mycobacterium,* and *Listeria* have the ability to grow in this hypoxic region of tumors (Fialho et al. 2008) and thus act as promising vectors for drug delivery for anticancer therapies (Table 9.1). Some of the microbes exhibiting the property of cancer suppression are discussed below:

### 9.2.1 *Mycobacterium bovis BCG*

A correlation between the occurrence of tuberculosis and cancer regression was observed in the beginning of the twentieth century. In 1976, Morales and his colleagues demonstrated that the use of BCG vaccine was accompanied with cancer regression through induction of both local and systemic immune response resulting in the elimination of bladder cancer cells. Subsequently, tuberculosis vaccine BCG was approved as a complementary treatment of bladder cancer (Droller 2017; Taniguchi et al. 1999). Clinical studies have proven that intra-vesical administration

**Table 9.1** Different types of microbes as anticancerous agents

| Types of organism | Strain | Cancer |
|---|---|---|
| Bacteria | | |
| *Mycobacterium bovis* | Attenuated strain Calmette-Guerin | Superficial bladder cancer |
| *Streptococcus pyogenes* | OK-432 | Lymphangioma |
| *Clostridium novyi* | Strain NT | Solid tumors |
| *Salmonella enterica* serovar typhimurium | Strain VNP20009 | Melanoma |
| *Magnetococcus marinus* | MC1 | Solid tumors and some metabolic tumors |
| Protozoan | | |
| *Toxoplasma gondii* | CPS/TLA | Pancreas, lung, and ovarian cancer and melanoma |
| *Plasmodium falciparum* | rVAR2-DT | Melanoma |

of BCG decreases the chances of cancer relapse following surgical removal of tumor or surgery and chemotherapy (Kamat et al. 2016). BCG acts on tumor cells by stimulating the immune response where $CD4^+$ and $CD8^+$ lymphocytes recognize tumor antigens along with a simultaneous enhancement of pro-inflammatory cytokines. The activation of immune system of the patient results in phagocytosis of the cancer cells (Chakrabarty 2003; Droller 2017; Felgner et al. 2016; Biot et al. 2012). Therefore, the current treatment of bladder cancer using *M. bovis* involves tumor resection and subsequent intra-vesical infusion of the microbial suspension using urethra catheters to prevent cancer relapse (Droller 2017).

### 9.2.2 *Toxoplasma gondii*

*T. gondii* is an obligatory intracellular protozoan which requires a host cell for its replication. Humans and other warm blooded animals are its major hosts and its infection causes toxoplasmosis which has a worldwide distribution. This microbe generally affects the pregnant women and individuals with compromised immunity. Despite causing several problems *Toxoplasma* also exhibits anti-tumorous activity in the host. It was reported that the protein extracts of *T. gondii* possess antitumor activity and that toxoplasmic infection exerts antitumor activity against melanoma (Pyo et al. 2014). The non-replicating *Toxoplasma* uracil auxotrophs (NRTUA) have been used in the treatment of melanoma, pancreatic cancer, lung cancer, and ovarian cancer, because these auxotrophic strains are non-pathogenic in the host organism because of the inhibition of the de novo pathway for the synthesis of UMP which is required for the synthesis of nucleic acids. Due to the absence of uracil in animals these auxotrophs are non-virulent as they do not replicate in the invaded host (Pyo et al. 2014; Sanders et al. 2016; Kim et al. 2007; Baird et al. 2013). The mechanism behind the anti-tumorigenic activity is that NRTUA leads to a rapid increase in IL-12 secretion leading to inflammation and activation of other immune cells such as $CD4^+$ and $CD8^+$ T cells. There is a significant increase in the production of T-cell mediated chemokines such as CXCL9 and CXCL10 in the tumor environment which further leads to a gradual increase in tumor antigen specific $CD8^+$ T helper cells which recognize and kill the tumor cells. NRTUA also generates hypoxic environment in the tumor site due to the inhibition of angiogenesis and lead to regression of tumor. In vivo studies reveal the development of immune memory and high titer of IgG recognizing specific tumor antigens which lead to anti-tumorigenic environment in the host (Pyo et al. 2014; Sanders et al. 2016).

### 9.2.3 *Streptococcus pyogenes OK-432*

*Streptococcus pyogenes* is a gram-positive bacterium that causes several diseases in humans such as pharyngitis, skin infections, acute rheumatic fever, scarlet fever, and toxic shock syndrome. Despite all these adverse effects on human body *S. pyogenes* also possess cytotoxic activity against different types of cancer. Dr. William Coley initially used *S. pyogenes* in the treatment of bone sarcoma. Currently, *S. pyogenes* has been used for the treatment of lymphangiomas in children (Deweerdt 2013; Olivieri et al. 2016). Lymphangiomas are tumors developed in toddlers due to excessive division of lymphatic vessel's endothelial cells. Treatment is usually done by surgical removal of cyst and it poses a high risk to the life of the toddler; therefore, alternative therapy called sclerotherapy is utilized which includes injecting *S. pyogenes* OK-432 strain into pathologically changed lymphatic vessels. Studies have proved the safety and efficacy of the strain in reducing the cyst volume by at least 50% (Olivieri et al. 2016; Ruiz Jr et al. 2004). The mechanism includes immunological activation of neutrophils and macrophages followed by increase in NK CD56 cells, TNF-α, IL-6, IL-8, IFN-γ, and VEGF (vascular endothelial growth factor). After incorporation of the strain inflammatory reactions cause swelling in the lesion but the effects are observed after few months (Olivieri et al. 2016; Ruiz Jr et al. 2004; Ohta et al. 2010a, 2010b).

### 9.2.4 *Magnetococcus marinus*

*M. marinus* MC1 is a gram-negative coccus found in the Atlantic Ocean near Rhode Island, USA. The distinctive feature includes the presence of magnetosomes which are special elements of magnetite particles ($Fe_3O_4$) surrounded by membranes which form chains in the cytosol (Bazylinski et al. 2013). The presence of magnetosomes helps the bacteria to orient with the earth's magnetic field. Another property that favors its use as an anticancer agent is the negative aerotaxis capacity where the bacteria prefers hypoxic regions (Martel 2017). Utilization of powerful magnetic fields as applied in MRI technique can direct the bacteria to tumor sites where it collects in the hypoxic core, thus increasing its capability of being used as a vector for various anticancer drugs (Felfoul et al. 2016).

### 9.2.5 *Clostridium novyi*

Presence of hypoxia in tumor core has increased the use of obligate anaerobes and facultative anaerobes in anticancer therapies. The anaerobic environment in the tumor core creates favorable environment for the growth of anaerobic bacteria (Felgner et al. 2016; Paton et al. 2012). The use of anaerobic bacteria is beneficial over chemotherapeutic drugs because they traverse through the depths of tumor, affecting only the tumor cells unlike chemotherapeutic drugs which affect both cancer and healthy cells (Paton et al. 2012; Liu et al. 2014; Staedtke et al. 2016).

In 1935, Connell firstly described the proteolytic enzymes produced by *C. histolyticum* were seen to cause regression of advanced cancers and then used this strain as an anticancer agent (Connell 1935). Since then, research on *Clostridium* for anticancerous activity began. *C. novyi* showed significant antitumor effect during experiments in mice but due to presence of lethal toxin known as α-toxin more than one-third of the mice died in a very short period of time after the spores were injected. To avoid the toxicity, *C. novyi* NT was formed by removing the α-toxin through heating. Although satisfactory results were obtained, but *C. novyi* NT is only active in the hypoxic environment of the tumor and becomes less effective at other sites. Therefore, combinatorial therapies such as conventional chemotherapy in combination with *C. novyi* NT show significant results in the treatment of tumor.

Treatment of leiomyoma has been promised in phase I and phase II of clinical trials through the attenuated strain of *C. novyi* NT (Paton et al. 2012; Liu et al. 2014; Staedtke et al. 2016). It can be utilized in active targeting where *C. novyi* is expected to produce specific enzymes, toxins, and proteins capable of conjugating to specific chemotherapeutics.

### 9.2.6 *Salmonella enterica Serovar Typhimurium*

*Salmonella enterica* serovar Typhimurium is an anaerobic rod whose attenuated strain *S. typhimurium* VNP20009 is used as anticancer agent (Bereta et al. 2007). Clinical trials on the use of this microorganism for melanoma and pancreatic cancer treatment as a vaccine started in 2002. This bacterium has a plasmid encoding expression of VEGFR2 (vascular endothelial growth factor receptor-2) which blocks the angiogenesis process (Felgner et al. 2016; Schmitz-Winnenthal et al. 2018).

### 9.2.7 *Plasmodium falciparum*

*P. falciparum* is the causative agent of malaria but it can also be used to treat cancer. It is known that *P. falciparum* expresses a malarial protein VAR2CSA in erythrocytes which binds to mucopolysaccharide-chondroitin sulfate A, present in physiological conditions on the surface of placenta cells (Salanti et al. 2015; Chishti 2015; Dimitriadis 2016). Interestingly, surface of many tumor cells also possesses chondroitin sulfate. Thus, a recombinant version of VAR2CSA, known as rVAR2 was conjugated to the appropriate part of the diphtheria toxoid and both in vitro and in vivo studies showed anticancerous effect on melanoma cells expressing high levels of chondroitin sulfate (Salanti et al. 2015; Dimitriadis 2016).

## 9.3 Purified Microbial Products as Anticancer Agents

Apart from using whole microorganism as anticancer agents and vectors, purified products of microbes can also be utilized for the same (Table 9.2). Some microbial products that exhibit anticancer activities have been discussed below:

### 9.3.1 *Antibiotics*

Antibiotics are secondary metabolites produced by certain microbes during the stationary phase of their growth. Some of these antibiotics have been demonstrated to possess anticancer activity and have been utilized as antitumor drugs. Such antibiotics include actinomycin D, doxorubicin, mitomycin C, and bleomycin.

1. *Actinomycin D*: Actinomycin D, also referred as dactinomycin is produced by *Actinomyces antibioticus*. Besides possessing antibacterial activity, it also exhibits antitumor activity. Several mechanisms explain the cytotoxic and antitumor activity of actinomycin D. These mechanisms are mainly associated with inhibition of transcription from DNA and thus indirectly lead to translation inhibition. One of the mechanisms is through intercalation of actinomycin D in DNA where it blocks RNA transcription by firmly attaching to DNA, preferably at sites with guanine residues.

   Actinomycin D results in localization of a phenoxazone ring between GpC base pair sequence in DNA. In another mechanism, actinomycin D induces a stabilization of otherwise cleavable complexes of topoisomerases I and II with DNA in which polypeptide lactone rings occupy a position in the minor groove of the DNA helix or the drug penetrates to a place in the DNA structure where topoisomerase binds with DNA (Della Latta et al. 2015). Actinomycin D is known to induce cellular p53-independent apoptosis after blockage of both DNA and RNA (Farhane et al. 2018). The slow dissociation of actinomycin D

**Table 9.2** Different microbial products possessing anticancerous activity with cancers for which they are effective

| Microbial products | | Efficacy in cancer |
|---|---|---|
| Bacteriocins | Bovicin HC5 | Human breast adenocarcinoma and liver hepatocellular carcinoma |
| | Colicin | Breast carcinoma, osteosarcoma, fibrosarcoma |
| | Laterosporulin 10 | Breast adenocarcinoma |
| | Microcin E492 | Acute T-cell leukemia, Burkitt's lymphoma, and cervical adenocarcinoma |
| Antibiotics | Actinomycin D | Wilms' cancer, neuroblastomas, and trophoblastic tumors |
| | Doxorubicin | Acute lymphoblastic tumors, ovarian carcinoma, malignant lymphoma, and gastric carcinoma |
| | Mitomycin C | Lung cancer, human colon cancer, gastric cancer |
| | Bleomycin | Hodgkin's disease, ovarian cancer, non-Hodgkin's lymphoma, and malignant pleural effusion |
| Enzymes | Arginine deiminase | Prostate cancer and hepatocellular carcinoma |
| | L-asparaginase | Myeloblastic leukemia, ovarian carcinoma, and Hodgkin's and non-Hodgkin's lymphoma |
| Non-ribosomal peptides | Arenamides | Human colon carcinoma |
| | Ariakemicins | Human lung tumor |
| | Halolitoralins | Human gastric tumor cells |
| | Heptapeptide from *P. profundus* | Human melanoma |
| Toxins | Botulinum neurotoxin type A | Benign prostatic hyperplasia, prostate cancer |
| | Diphtheria toxin | Human adrenocortical carcinoma, T-cell lymphomas |
| | Exotoxin A | Pancreatic cancer, melanoma, and head and neck squamous carcinoma |
| | Listeriolysin O | Human leukemia T-lymphocyte cells and human breast adenocarcinoma |
| Other proteins and peptides | Azurin | Breast cancer, melanoma, and oral squamous carcinoma |
| | Entap | Colorectal adenocarcinoma, gastric cancer, cervical cancer, breast cancer, and prostate cancer |
| | Pep27anal2 | Gastric cancer, breast cancer, and leukemia |

from DNA complexes, its photodynamic activity, and free radical formation also influence the biological activity of this antibiotic cum anticancer drug. The efficacy of actinomycin D has been observed in the treatment of Wilms' cancer, Ewing sarcoma, neuroblastomas, and trophoblastic tumors (Karpiński and Adamczak 2018; D'arpa and Liu 1989). Multiple variants of actinomycin D are available in the market, e.g., actinomycin D, lyovac, and cosmegen (Karpiński and Adamczak 2018). However, the clinical use of actinomycin D to inhibit malignant tumors is discouraged due to its toxic effects, thus limiting its application for experimental purposes.

2. *Doxorubicin*: Doxorubicin is one of the most effective chemotherapy drugs used against solid tumors in the treatment of several cancer types. It is an anthracycline antibiotic with antitumor activity. It acts as an amphiphilic molecule as it contains a water-insoluble aglycone and a water-soluble amino-sugar functional group (Abraham et al. 2005). Doxorubicin depicts its anticancer activity by two main mechanisms: inhibition of DNA and RNA synthesis in rapidly growing cells by blocking the replication and transcription processes (Thorn et al. 2011) and generation of iron-mediated free radicals, causing oxidative damage to cell membranes, proteins, and DNA (Cagel et al. 2017). The use of doxorubicin as anticancerous drug has been approved by Food and Drug Administration for the treatment of several cancers including acute lymphoblastic/myeloblastic leukemia, neuroblastoma, bone sarcomas, breast carcinoma, thyroid carcinoma, ovarian carcinoma, bladder carcinoma, Hodgkin's disease, malignant lymphoma, gastric carcinoma, and bronchogenic carcinoma. It has been demonstrated that combining niacin with doxorubicin can improve the treatment efficacy of skin cancers (Preet et al. 2015). Doxorubicin containing drugs include doxorubicin medac, doxorubicinum accord, doxorubicin ebewe, caelyx, adriblastine PFS, and myocet (Karpiński and Adamczak 2018).

   Resistance to chemotherapeutic agents is one of the major disadvantages of long-term anticancer treatment. Repeated doxorubicin administration leads to drug-resistant cancer cells and increased cytotoxicity. In fact, cardiotoxicity is the most common doxorubicin-induced side effect. Therefore, most of the research conducted on doxorubicin has been focused on the elimination of the anti-therapeutic effects. Potential treatment options have been developed to reduce doxorubicin-mediated cardiotoxicity, such as lowering the dosage of doxorubicin, combined therapies with cardioprotective agents (e.g., dexrazoxane), through regulation of cardiac circular RNA expression.
3. *Mitomycin C*: *Streptomyces caespitosus* strain was used to isolate mitomycin C. Mitomycin C acts as an anticancer agent as it inhibits DNA synthesis by binding to DNA during alkylation forming cross-linked double stranded DNA (Verweij and Pinedo 1990). Mitomycin C has been used for the treatment of various cancers (Bradner 2001). Mitomycin C containing drugs include mitomycin accord and mitomycin C kyowa (Karpiński and Adamczak 2018).
4. *Bleomycin*: *Streptomyces verticillus*-derived bleomycin is a mixture of glycopeptide antibiotics which have cytotoxic properties (Egger et al. 2013). Bleomycin induces oxygen- and metal ion-dependent cleavage of DNA by binding to it, releasing hydroxyl radical under the influence of molecular oxygen, thus damaging DNA. Bleomycin is used in the treatment of head and neck squamous cell carcinomas, Hodgkin's disease, non-Hodgkin's lymphoma, testicular carcinomas, ovarian cancer, and malignant pleural effusion (Segerman et al. 2013; Bayer et al. 1992). Drugs containing bleomycin include Blenoxane and Bleomycin USP (Karpiński and Adamczak 2018).

### *9.3.2 Toxins*

Toxins produced by microorganisms damage host tissues either directly at the site of microbial infection or throughout the body. Some toxins are used for therapeutic purposes (Henkel et al. 2010) and are discussed below:

1. *Botulinum Neurotoxin Type A*: Botulinum neurotoxin type A that is produced by strains of anaerobic spore forming gram-positive *Clostridium botulinum* are used for the symptomatic relief of spasticity and other movement disorders. It is also used as anticancer agent against benign prostatic hyperplasia. Since the toxin is apoptotic in nature, it reduces cell growth and proliferation of prostate cancer cell lines, PC-3 and LNCaP (Karsenty et al. 2009; Proietti et al. 2012). It is also known to initiate death by caspase-3 and -7 dependent apoptotic pathways in breast cancer cell line, T47D (Bandala et al. 2013). Injection of botulinum neurotoxin type A is also used for the treatment of post-radiation and surgical pain induced by the conventional radiotherapy with lesser side effects (Mittal and Jabbari 2020).
2. *Diphtheria Toxin*: *Corynebacterium diphtheria* produces diphtheria exotoxin which is of 60 kDa and is composed of 538 amino acids. Diphtheria toxin exhibits anticancer activity with minor side effects, so it is utilized in the antitumor therapy in combination with other agents that eliminate its side effects. Nontoxic mutant of diphtheria toxin, cross-reacting material 197 (CRM197) binds to heparin-binding epidermal growth factor and acts as inhibitor of this growth factor. CRM197 is known to inhibit angiogenesis and stimulate cell apoptosis in human adrenocortical carcinoma cells, H295R (Martarelli et al. 2009). DTAT, a diphtheria toxin-based immunotoxin is directed to cancer vascular endothelium. DTAT exhibited in vitro anticancer action in case of glioblastoma cell lines including U118MG, U373MG, U87MG (Vallera et al. 2002). A drug named Ontak is used in cutaneous T-cell lymphomas expressing CD25. It is a fusion product known as denileukin diftitox which is a DNA derived cytotoxic protein composed of enzymatically active domain of DT followed by sequences of human IL-2 and acts against the cells expressing IL-2 receptor which is highly expressed on malignant T cells (Lutz et al. 2014; Lewis et al. 2017).
3. *Exotoxin A*: *Pseudomonas aeruginosa* produces many toxins out of which exotoxin A is the main toxin which is a 66 kDa protein composed of 638 amino acids. It acts similar to diphtheria toxin and also has ADP ribosyl transferase activity. Its production is dependent on the presence of iron. Exotoxin A inhibits protein synthesis by inactivation of elongation factor-2 (EF-2). It acts as an immunotoxin with different ligands (Karpiński and Szkaradkiewicz 2013). De-immunized *Pseudomonas* exotoxin cloned with both human epidermal growth factor (EGF) and IL-4 showed anticancer activity against pancreatic cancer, PaCa-2, and prevented metastasis, selectively (Oh et al. 2012). Two exotoxin A-based immunotoxins (9.2.27PE and ABT-737) caused synergistic cytotoxicity and death of melanoma cell lines such as FEMX, Melmet-1, Melmet-5, Melmet-44, Mel RM, and MM200 by apoptosis (Risberg et al. 2011). It has also been

demonstrated that exotoxin A cloned with anti-CD133 scFv causes inhibition of head and neck squamous carcinoma (Waldron et al. 2011).

4. *Listeriolysin O*: Listeriolysin O is a pore-forming toxin produced by *Listeria monocytogenes* and is responsible for bacterial phagosomal escape. It facilitates bacterial multiplication into the cytoplasm during infection (Provoda et al. 2003). Listeriolysin O belongs to the family of cholesterol dependent cytolysins which exhibit potential cell type non-specific toxicity which is a potent source of dominant $CD4^+$ and $CD8^+$ T-cell epitopes. Therefore, the conjugated immunotoxin B3-listeriolysin O inhibits breast carcinoma cell lines MCF7 and SKBR-3 (Bergelt et al. 2009). Supernatants of *L. monocytogenes* strains showed dose-dependent cytotoxicity against human leukemia T-lymphocyte Jurkat cells (Stachowiak et al. 2012). Interestingly, listeriolysin O activity is targeted more to T cells than B cells and it may exhibit some specific anticancer activities (Stachowiak et al. 2012).

### 9.3.3 Bacteriocins

Bacteriocins are a heterogeneous group of ribosomally synthesized bacterial peptides or proteins with antimicrobial properties and some of them exhibit anticancer activity (Kaur and Kaur 2015; Mandal et al. 2016; Drider et al. 2016). Bacteriocins have been isolated from all types of bacteria. Gram-positive bacteria secrete four classes of bacteriocins.

1. *Bovicin HC5*: *Streptococcus bovis* secretes antibiotic bovicin HC5 having molecular weight of 2.4 kDa. Bovicin HC5 has structural and functional similarities to nisin and has broad spectrum antimicrobial activity against closely related species of *S. bovis* and also against Gram-positive and Gram-negative bacteria (Kaur and Kaur 2015). In vitro cytotoxic activity of bovicin HC5 has been observed against human breast adenocarcinoma (MCF-7) and human liver hepatocellular carcinoma (HepG2) with IC50 of 279.4 and 289.3 M, respectively (Paiva et al. 2012).
2. *Colicins*: Colicins are plasmid encoded antimicrobial bacteriocins having molecular weight of more than 20 kDa and are produced by the *Escherichia coli*. Basically, colicins act on *E. coli* and other closely related strains. They also possess anticancer activity against different cancers such as breast cancer, colon cancer, bone cancer, and uterus cancer.

   *E. coli* produces different colicins A, E1, and E3 and all these colicins possess cytotoxic activity against different cancers by inducing apoptosis, necrosis, and by alteration of cell cycle (Kaur and Kaur 2015). Colicins E1 and E3 are known to exhibit cytotoxic activity against BM2 cells. Colicin E3 kills cells by necrosis rather than apoptosis. However, maximum apoptotic cell death is observed by exposing cells to colicin E1 (1.25 g/mL) for 48 h (Smarda et al. 2001). Four different colicins (A, E1, U, E3) were tested in terms of their inhibitory activity against 11 cancer cell lines. Colicin E1 inhibits breast carcinoma (MCF7, ZR75, BT549, BT474, MDA-MB-231, SKBR3, and T47D), leiomyosarcoma (SKUT-

1), osteosarcoma (HOS), and fibrosarcoma (HS913T). 50% inhibition of fibrosarcoma (HS913T) and 17–40% inhibition of other cancer cells were exhibited by colicin E1 (Chumchalova and Šmarda 2003).

3. *Laterosporulin 10*: *Brevibacillus sp.* produces laterosporulin 10 (LS10), a defensin-like class IId bacteriocin which inhibits microbial pathogens. LS10 shows antibacterial activity against pathogens like *M. tuberculosis* and *Staphylococcus aureus*. Interestingly, LS10 is also known to exhibit anticancer activity. A dose-dependent cytotoxic activity is observed against different human cancer cell lines and it is observed that 10μM concentration induces highest anticancer activity against MCF-7 cells and no cytotoxicity was observed up to 15μM of concentration against normal cells. Therefore, LS10 acts as a good anticancer agent. It acts on cancer cells by mediating membrane disintegration. Notably, it induces both apoptosis and necrosis in cells at varying concentrations where apoptosis was observed at lower doses and necrosis was observed at higher doses (Baindara et al. 2017).
4. *Microcin E492*: Microcin E492 is low molecular weight bacteriocin produced by *Klebsiella pneumoniae* RYC492 strain. It shows antimicrobial activity against *Klebsiella, E. coli,* and *Salmonella.* Microcin E492 has been proven to be cytotoxic against several malignant human cell lines, including acute T-cell leukemia (Jurkat), Burkitt's lymphoma cell line, B-lymphoblastoid cell lines transformed by infection with Epstein–Barr virus (RJ2.25), and cervical adenocarcinoma (HeLa). However, human endothelial cells from human tonsils (AMG-3) and a monocyte-macrophage cell line (KG-1) were not affected by microcin E492 (Hetz et al. 2002). Similar to laterosporulin 10, microcin E492 causes apoptotic cell death at low concentrations and necrotic cell death at higher concentrations (Hetz et al. 2002). Induction of apoptosis by microcin E492 causes several biochemical and morphological changes such as cell shrinkage, DNA fragmentation and flipping of phosphatidylserine to the outer membrane. Due to apoptosis, there is an activation of caspases along with a loss of membrane potential (Hetz et al. 2002).

### 9.3.4 Non-Ribosomal Peptides

An enzyme complex present only in bacteria, cyanobacteria and fungi synthesizes non-ribosomal peptides. Besides exhibiting antimicrobial activity, some non-ribosomal peptides exhibit anticancerous activities (Agrawal et al. 2017) and are discussed below:

1. *Arenamides*: Arenamides are produced by the fermented broth of marine actinobateria, *Salinispora arenicola* which is found in the sea sediment (Great Astrolabe Reef, Kandavu Island chain, Fiji). There are three new cyclohexadepsipeptides, namely arenamides A, B, and C. Arenamides A and B exhibit anticancerous activity by blocking TNF-induced activation. Inhibition of prostaglandin E2 and nitric oxide production by arenamides also induces

moderate cytotoxic effect on human colon carcinoma, HCT-116 cells (Sieber and Marahiel 2003). The anticancer effect of these arenamides is due to suppression of NF-κB which regulates the expression of a number of genes whose effector proteins are associated with tumorigenesis (Asolkar et al. 2009).

2. *Ariakemicins*: The culture of the marine gliding bacterium of *Rapidithrix* genus (Ariake Inland Sea, Japan) produces two linear hybrid polyketidenon-ribosomal peptides (ariakemicins A and B). They are composed of threonine, two Ω amino (Ω-3) methyl carboxylic acids containing diene units and δ-isovanilloyl butyric acid. These ariakemicins antibiotics inhibit Gram-positive bacteria. Ariakemicins are also known to exhibit low cytotoxicity to human lung tumor cell line, A549, and baby hamster kidney cells (Oku et al. 2008).
3. *Halolitoralins*: Cyclic hexapeptide (halolitoralin A) and cyclic tetrapeptides (halolitoralin B and C) derived from *Halobacillus litoralis* YS3106 strain have a molecular mass 572 Da which exhibits moderate activities against human gastric tumor cells, in vitro (Yang et al. 2002).
4. 6pt?>*Heptapeptide from Paenibacillus profundus*: Glyceryl acid derived heptapeptide (glyceryl-D-leucyl-D-alanyl-D-leucyl-D-leucyl-L-valyl-D-leucyl-D-alanine) was produced from the culture of *Paenibacillus profundus*. The peptide exhibits cytotoxic activity against human melanoma cell line, SK-MEL-28 (Kalinovskaya et al. 2013).

### *9.3.5 Enzymes*

An emerging strategy in cancer therapy is to starve cancer cells through amino acid deprivation. Microbial enzymes like arginine deiminase and L-asparaginase act by amino acid deprivation in sensitive cancer cells and thus are used for the treatment of different cancers.

1. *Arginine Deiminase*: Arginine deiminase is an enzyme produced by *Mycoplasma hominis* or *Mycoplasma arginine*. It degrades arginine to citrulline in vivo, releasing ammonia (Ni et al. 2008). Thus, it acts as an anticancerous therapeutic for tumors requiring arginine.

   Interestingly, the anticancerous efficacy of pegylated arginine deiminase (ADI-PEG20) is directly correlated with the deficiency of argininosuccinate synthetase enzyme. This is because argininosuccinate synthetase enzyme is involved in the synthesis of arginine from citrulline and various tumor cells such as hepatocellular carcinoma, melanomas, pancreatic carcinomas lack the expression of this enzyme, therefore, unable to synthesize arginine. Thus, arginine deiminase inhibits the growth of cancerous cells by depleting arginine. It has been established that arginine deiminase inhibits hepatocellular carcinoma and acts as a promising drug utilizing high enzymatic deficiency of argininosuccinate synthetase in hepatocellular carcinoma. Prostate cancer cells (CWR22Rv1) are also susceptible to ADI-PEG20 in vitro. ADI-PEG20 acts on a tumor cell by inducing $G_1$ cell cycle arrest and caspase independent apoptosis (Kim et al. 2009).

2. *L-asparaginase*: *Escherichia coli* or *Erwinia* sps. produce L-asparaginase enzyme. The antitumor action of bacterial L-asparaginase enzyme is through the ability to reduce blood concentration of asparagine causing a selective inhibition of sensitive malignant cells (Avramis et al. 2002) and also by inducing apoptosis and inhibition of protein synthesis of the treated cells. Dose-dependent inhibition of brain tumor cell lines, p53 and PTEN null human glioblastomas (GBM-ES and U87) and a pediatric medulloblastoma (DAOY) has been seen after using L-asparaginase as an anticancer agent (Paiva et al. 2012). L-asparaginase has also been utilized in myeloblastic leukemia, acute lymphoblastic leukemia, Hodgkin and non-Hodgkin lymphomas, extranodal NK/T-cell lymphoma, and ovarian carcinomas (Jaccard et al. 2011; Covini et al. 2012; Yu et al. 2012).

### 9.3.6 Other Proteins and Peptides.

1. *Azurin*: *Pseudomonas aeruginosa* produces a copper containing protein known as azurin, which has molecular mass of 16 kDa. Removal of copper produces cytotoxic apo-azurin (Goto et al. 2003). Azurin is known to exhibit anticancerous properties through induction of apoptotic cell death by (1) forming complex with tumor suppressor protein p53, (2) by interfering in the receptor tyrosine kinase EphB2-mediated signaling process, (3) preventing angiogenesis through reducing VEGFR-2 tyrosine kinase activity, (4) interference with P-cadherin protein expression, and (5) inhibition of the growth of breast cancer cells (Gao et al. 2017). Azurin has a strong cytotoxic effect on breast cancer cell lines including MCF-7, MDA-MB-157, MDD2, and MDA-MB-231 (Punj et al. 2004). Azurin also exhibits anticancer activity against oral squamous carcinoma cells, YD-9 (Choi et al. 2011), and melanoma cells, UISO-Mel-2 (Yamada et al. 2002).
2. *Entap*: Clinical strains of *Enterococcus* genus produce anti-proliferative peptide (Entap). Entap is known to exhibit anti-proliferative activity against cell lines of colorectal adenocarcinoma (HT-29), human gastric adenocarcinoma (AGS), uterine cervix adenocarcinoma (HeLa), mammary gland adenocarcinoma (MDA-MB-231), and also prostatic carcinoma (22Rv1). Entap induces autophagous apoptosis in cancer cells and causes cancer cell cycle arrest in G1 phase (Karpiński and Szkaradkiewicz 2013; Karpiński et al. 2013; Karpiński 2012).
3. *Pep27anal2*: Pep27anal2 is an analogue of a signal peptide Pep27 produced by *Streptococcus pneumonia*. It is known to exhibit anticancer properties by penetrating the cell membrane and inducing caspase- and cytochrome c-independent apoptosis. Due to the membrane permeability the peptide gets hydrophobic which plays an important role against cancer cells. Pep27anal2 inhibits proliferation of cell lines of gastric cancer (SNU-601), leukemia (AML-2, HL-60, Jurkat), and breast cancer (MCF-7) (Lee et al. 2005).

### 9.3.7 Conclusion and Future Perspectives

Microbes constitute a valuable and a moderately known source of biologically active substances with significant anticancer properties. Some of these include drugs which are already in use such as actinomycin D, doxorubicin, mitomycin C, and diphtheria toxin, while other substances are either in clinical trials or are being tested in in vitro research. Majority of bacterial anticancer proteins/peptides end in the in vitro stage and only few undergo the entire procedure, from in vitro testing to clinical trials followed by registration and use as medicines. More such products should be explored for analysis at in vitro and in vivo level for production of new and efficient anticancer drugs at the earliest. The combination of different techniques such as omics with virtual and computational screening can also be utilized for the modification of the known anticancer proteins or for selection of new chemical compounds with antitumor activity.

## References

Abraham SA, Waterhouse DN, Mayer LD, Cullis PR, Madden TD, Bally MB (2005) The liposomal formulation of doxorubicin. In: Methods in enzymology. Elsevier, Amsterdam, pp 71–97

Agrawal S, Acharya D, Adholeya A, Barrow CJ, Deshmukh SK (2017) Nonribosomal peptides from marine microbes and their antimicrobial and anticancer potential. Front Pharmacol 8:828

Asolkar RN, Freel KC, Jensen PR, Fenical W, Kondratyuk TP, Park E-J et al (2009) Arenamides A-C, cytotoxic NFkappaB inhibitors from the marine actinomycete Salinispora arenicola. J Nat Prod 72(3):396–402

Avramis VI, Sencer S, Periclou AP, Sather H, Bostrom BC, Cohen LJ et al (2002) A randomized comparison of nativeEscherichia coli asparaginase and polyethylene glycol conjugated asparaginase for treatment of children with newly diagnosed standard-risk acute lymphoblastic leukemia: a Children's Cancer group study. Blood 99(6):1986–1994

Baindara P, Gautam A, Raghava GPS, Korpole S (2017) Anticancer properties of a defensin like class IId bacteriocin Laterosporulin10. Sci Rep 7:46541

Baird JR, Fox BA, Sanders KL, Lizotte PH, Cubillos-Ruiz JR, Scarlett UK et al (2013) Avirulent toxoplasma gondii generates therapeutic antitumor immunity by reversing immunosuppression in the ovarian cancer microenvironment. Cancer Res 73(13):3842–3851

Bandala C, Perez-Santos JLM, Lara-Padilla E, Delgado Lopez MG, Anaya-Ruiz M (2013) Effect of botulinum toxin a on proliferation and apoptosis in the T47D breast cancer cell line. Asian Pacific J Cancer Prev 14(2):891–894

Bayer RA, Gaynor ER, Fisher RI (1992) Bleomycin in non-Hodgkin's lymphoma. In: Seminars in oncology. Elsevier, Amsterdam, pp 46–53

Bazylinski DA, Williams TJ, Lefevre CT, Berg RJ, Zhang CL, Bowser SS et al (2013) Magnetococcus marinus gen. Nov., sp. nov., a marine, magnetotactic bacterium that represents a novel lineage (Magnetococcaceae fam. Nov., Magnetococcales Ord. Nov.) at the base of the Alphaproteobacteria. Int J Syst Evol Microbiol 63(3):801–808

Bereta M, Hayhurst A, Gajda M, Chorobik P, Targosz M, Marcinkiewicz J et al (2007) Improving tumor targeting and therapeutic potential of Salmonella VNP20009 by displaying cell surface CEA-specific antibodies. Vaccine 25(21):4183–4192

Bergelt S, Frost S, Lilie H (2009) Listeriolysin O as cytotoxic component of an immunotoxin. Protein Sci 18(6):1210–1220

Biot C, Rentsch CA, Gsponer JR, Birkhäuser FD, Jusforgues-Saklani H, Lemaître F et al (2012) Preexisting BCG-specific T cells improve intravesical immunotherapy for bladder cancer. Sci Transl Med 4(137):137ra72

Bradner WT (2001) Mitomycin C: a clinical update. Cancer Treat Rev 27(1):35–50

Bray F, Ferlay J, Soerjomataram I, Siegel RL, Torre LA, Jemal A (2018) Global cancer statistics 2018: GLOBOCAN estimates of incidence and mortality worldwide for 36 cancers in 185 countries. CA Cancer J Clin 68(6):394–424

Cagel M, Grotz E, Bernabeu E, Moretton MA, Chiappetta DA (2017) Doxorubicin: nanotechnological overviews from bench to bedside. Drug Discov Today 22(2):270–281

Chakrabarty AM (2003) Microorganisms and cancer: quest for a therapy. J Bacteriol 185 (9):2683–2686

Chishti AH (2015) Malaria selectively targets pregnancy receptors. Blood 125(2):217–218

Choi J-H, Lee M-H, Cho Y-J, Park B-S, Kim S, Kim G-C (2011) The bacterial protein azurin enhances sensitivity of oral squamous carcinoma cells to anticancer drugs. Yonsei Med J 52 (5):773–778

Chumchalova J, Šmarda J (2003) Human tumor cells are selectively inhibited by colicins. Folia Microbiol 48(1):111–115

Connell HC (1935) The study and treatment of cancer by proteolytic enzymes: preliminary report. Can Med Assoc J 33(4):364

Covini D, Tardito S, Bussolati O, Chiarelli RL, Pasquetto MV, Digilio R et al (2012) Expanding targets for a metabolic therapy of cancer: L-asparaginase. Recent Pat Anticancer Drug Discov 7 (1):4–13

D'arpa P, Liu LF (1989) Topoisomerase-targeting antitumor drugs. Biochim Biophys Acta (BBA)-reviews. Cancer 989(2):163–177

Della Latta V, Cecchettini A, Del Ry S, Morales MA (2015) Bleomycin in the setting of lung fibrosis induction: from biological mechanisms to counteractions. Pharmacol Res 97:122–130

Deweerdt S (2013) Bacteriology: a caring culture. Nature 504(7480):S4–S5

Dimitriadis E (2016) The use of malaria glycosaminoglycan to block cancers-lessons from the human placenta. Transl Cancer Res 5:S1085–S1087

Drider D, Bendali F, Naghmouchi K, Chikindas ML (2016) Bacteriocins: not only antibacterial agents. Probiot Antimicrob Proteins 8(4):177–182

Droller MJ (2017) Intracavitary bacillus Calmette-Guerin for superficial bladder tumors. J Urol 197 (2S):S146–S147

Egger C, Cannet C, Gérard C, Jarman E, Jarai G, Feige A et al (2013) Administration of bleomycin via the oropharyngeal aspiration route leads to sustained lung fibrosis in mice and rats as quantified by UTE-MRI and histology. PLoS One 8(5):e63432

Farhane Z, Bonnier F, Byrne HJ (2018) An in vitro study of the interaction of the chemotherapeutic drug Actinomycin D with lung cancer cell lines using Raman micro spectroscopy. J Biophotonics 11(1):e201700112

Felfoul O, Mohammadi M, Taherkhani S, De Lanauze D, Xu YZ, Loghin D et al (2016) Magneto-aerotactic bacteria deliver drug-containing nanoliposomes to tumour hypoxic regions. Nat Nanotechnol 11(11):941–947

Felgner S, Kocijancic D, Frahm M, Weiss S (2016) Bacteria in cancer therapy: renaissance of an old concept. Int J Microbiol 2016

Fialho AM, Das Gupta TK, Chakrabarty AM, Sleator R, Hill C (2008) Promiscuous drugs from pathogenic bacteria in the post-antibiotics era. In: Patho-Biotechnol. Landes Bioscience, Austin, TX, pp 145–162

Gao M, Zhou J, Su Z, Huang Y (2017) Bacterial cupredoxin azurin hijacks cellular signaling networks: protein–protein interactions and cancer therapy. Protein Sci 26(12):2334–2341

Goto M, Yamada T, Kimbara K, Horner J, Newcomb M, Das Gupta TK et al (2003) Induction of apoptosis in macrophages by Pseudomonas aeruginosa azurin: tumour-suppressor protein p53 and reactive oxygen species, but not redox activity, as critical elements in cytotoxicity. Mol Microbiol 47(2):549–559

Henkel JS, Baldwin MR, Barbieri JT (2010) Toxins from bacteria. In: Molecular, clinical and environmental toxicology. Springer, Cham, pp 1–29

Hetz C, Bono MR, Barros LF, Lagos R (2002) Microcin E492, a channel-forming bacteriocin from Klebsiella pneumoniae, induces apoptosis in some human cell lines. Proc Natl Acad Sci 99 (5):2696–2701

Jaccard A, Gachard N, Marin B, Rogez S, Audrain M, Suarez F et al (2011) Efficacy of L-asparaginase with methotrexate and dexamethasone (AspaMetDex regimen) in patients with refractory or relapsing extranodal NK/T-cell lymphoma, a phase 2 study. Blood 117 (6):1834–1839

Kalinovskaya NI, Romanenko LA, Kalinovsky AI, Dmitrenok PS, Dyshlovoy SA (2013) A new antimicrobial and anticancer peptide producing by the marine deep sediment strain "Paenibacillus profundus" sp. nov. Sl 79. Nat Prod Commun 8(3):1934578X1300800326

Kamat AM, Hahn NM, Efstathiou JA, Lerner SP, Malmström P-U, Choi W et al (2016) Bladder cancer. Lancet 388(10061):2796–2810

Karpiński TM (2012) New peptide (Entap) with anti-proliferative activity produced by bacteria of Enterococcus genus. Habilitation thesis. Scientific Publisher of Poznań University of Medical

Karpiński TM, Adamczak A (2018) Anticancer activity of bacterial proteins and peptides. Pharmaceutics 10(2):54

Karpiński TM, Szkaradkiewicz A, Gamian A (2013) New enterococcal anticancer peptide. In: 23rd European Congress of Clinical Microbiology and Infectious Diseases Berlin, Germany, 2013, p. 30

Karpiński TM, Szkaradkiewicz AK (2013) Anti-cancer peptides from bacteria. Bangladesh J Pharmacol 8(3):343–348

Karsenty G, Rocha J, Chevalier S, Scarlata E, Andrieu C, Zouanat FZ et al (2009) Botulinum toxin type a inhibits the growth of LNCaP human prostate cancer cells in vitro and in vivo. Prostate 69 (11):1143–1150

Kaur S, Kaur S (2015) Bacteriocins as potential anticancer agents. Front Pharmacol 6:272

Kim JO, Jung SS, Kim SY, Kim TY, Shin DW, Lee JH et al (2007) Inhibition of Lewis lung carcinoma growth by Toxoplasma gondii through induction of Th1 immune responses and inhibition of angiogenesis. J Korean Med Sci 22:S38–S46

Kim RH, Coates JM, Bowles TL, McNerney GP, Sutcliffe J, Jung JU et al (2009) Arginine deiminase as a novel therapy for prostate cancer induces autophagy and caspase-independent apoptosis. Cancer Res 69(2):700–708

Lee DG, Hahm K-S, Park Y, Kim H-Y, Lee W, Lim S-C et al (2005) Functional and structural characteristics of anticancer peptide Pep27 analogues. Cancer Cell Int 5(1):21

Lewis DJ, Dao H Jr, Nagarajan P, Duvic M (2017) Primary cutaneous anaplastic large-cell lymphoma: complete remission for 13 years after denileukin diftitox. JAAD Case Rep 3(6):501

Liu S, Xu X, Zeng X, Li L, Chen Q, Li J (2014) Tumor-targeting bacterial therapy: a potential treatment for oral cancer. Oncol Lett 8(6):2359–2366

Lutz MB, Baur AS, Schuler-Thurner B, Schuler G (2014) Immunogenic and tolerogenic effects of the chimeric IL-2-diphtheria toxin cytocidal agent Ontak® on CD25+ cells. Onco Targets Ther 3(3):e28223

Mandal SM, Pati BR, Chakraborty R, Franco OL (2016) New insights into the bioactivity of peptides from probiotics. Front Biosci 8:450–459

Martarelli D, Pompei P, Mazzoni G (2009) Inhibition of adrenocortical carcinoma by diphtheria toxin mutant CRM197. Chemotherapy 55(6):425–432

Martel S (2017) Targeting active cancer cells with smart bullets. Ther Deliv 8(5):301–312

Mittal SO, Jabbari B (2020) Botulinum neurotoxins and Cancer-a review of the literature. Toxins 12 (1):32

Ni Y, Schwaneberg U, Sun Z-H (2008) Arginine deiminase, a potential anti-tumor drug. Cancer Lett 261(1):1–11

Oh S, Todhunter DA, Panoskaltsis-Mortari A, Buchsbaum DJ, Toma S, Vallera DA (2012) A deimmunized bispecific ligand directed toxin that shows an impressive anti-pancreatic cancer effect in a systemic nude mouse orthotopic model. Pancreas 41(5):789

Ohta N, Fukase S, Suzuki Y, Ishida A, Aoyagi M (2010a) Treatments of various otolaryngological cystic diseases by OK-432 1: its indications and limitations. Laryngoscope 120(11):2193–2196

Ohta N, Fukase S, Watanabe T, Ito T, Aoyagi M (2010b) Effects and mechanism of OK-432 therapy in various neck cystic lesions. Acta Otolaryngol 130(11):1287–1292

Oku N, Adachi K, Matsuda S, Kasai H, Takatsuki A, Shizuri Y (2008) Ariakemicins a and B, novel polyketide-peptide antibiotics from a marine gliding bacterium of the genus Rapidithrix. Org Lett 10(12):2481–2484

Olivieri C, Nanni L, De Gaetano AM, Manganaro L, Pintus C (2016) Complete resolution of retroperitoneal lymphangioma with a single trial of OK-432 in an infant. Pediatr Neonatol 57 (3):240–243

Paiva AD, de Oliveira MD, de Paula SO, Baracat-Pereira MC, Breukink E, Mantovani HC (2012) Toxicity of bovicin HC5 against mammalian cell lines and the role of cholesterol in bacteriocin activity. Microbiology 158(11):2851–2858

Paton AW, Morona R, Paton JC (2012) Bioengineered microbes in disease therapy. Trends Mol Med 18(7):417–425

Patyar S, Joshi R, Byrav DSP, Prakash A, Medhi B, Das BK (2010) Bacteria in cancer therapy: a novel experimental strategy. J Biomed Sci 17(1):21

Preet S, Bharati S, Panjeta A, Tewari R, Rishi P (2015) Effect of nisin and doxorubicin on DMBA-induced skin carcinogenesis—a possible adjunct therapy. Tumor Biol 36(11):8301–8308

Proietti S, Nardicchi V, Porena M, Giannantoni A (2012) Botulinum toxin type-a Toxinactivity in prostate Cancer cell lines. Urol J 79(2):135–141

Provoda CJ, Stier EM, Lee K-D (2003) Tumor cell killing enabled by listeriolysin O-liposome-mediated delivery of the protein toxin gelonin. J Biol Chem 278(37):35102–35108

Punj V, Bhattacharyya S, Saint-Dic D, Vasu C, Cunningham EA, Graves J et al (2004) Bacterial cupredoxin azurin as an inducer of apoptosis and regression in human breast cancer. Oncogene 23(13):2367–2378

Pyo K-H, Jung B-K, Xin C-F, Lee Y-W, Chai J-Y, Shin E-H (2014) Prominent IL-12 production and tumor reduction in athymic nude mice after toxoplasma gondii lysate antigen treatment. Korean J Parasitol 52(6):605

Risberg K, Fodstad Ø, Andersson Y (2011) Synergistic anticancer effects of the 9.2. 27PE immunotoxin and ABT-737 in melanoma. PLoS One 6(9):e24012

Ruiz E Jr, Valera ET, Veríssimo F, Tone LG (2004) OK-432 therapy for lymphangioma in children. J Pediatr 80(2):154–158

Salanti A, Clausen TM, Agerbæk MØ, Al Nakouzi N, Dahlbäck M, Oo HZ et al (2015) Targeting human cancer by a glycosaminoglycan binding malaria protein. Cancer Cell 28(4):500–514

Sanders KL, Fox BA, Bzik DJ (2016) Attenuated toxoplasma gondii therapy of disseminated pancreatic cancer generates long-lasting immunity to pancreatic cancer. Onco Targets Ther 5 (4):e1104447

Schmitz-Winnenthal FH, Hohmann N, Schmidt T, Podola L, Friedrich T, Lubenau H et al (2018) A phase 1 trial extension to assess immunologic efficacy and safety of prime-boost vaccination with VXM01, an oral T cell vaccine against VEGFR2, in patients with advanced pancreatic cancer. Onco Targets Ther 7(4):e1303584

Segerman ZJ, Roy B, Hecht SM (2013) Characterization of bleomycin-mediated cleavage of a hairpin DNA library. Biochemistry 52(31):5315–5327

Sieber SA, Marahiel MA (2003) Learning from nature's drug factories: nonribosomal synthesis of macrocyclic peptides. J Bacteriol 185(24):7036–7043

Smarda J, Fialova M, Smarda J Jr (2001) Cytotoxic effects of colicins E1 and E3 on v-myb-transformed chicken monoblasts. Folia Biol 47(1):11

Stachowiak R, Lyzniak M, Budziszewska BK, Roeske K, Bielecki J, Hoser G et al (2012) Cytotoxicity of bacterial metabolic products, including listeriolysin O, on leukocyte targets. J Biomed Biotechnol 2012

Staedtke V, Roberts NJ, Bai R-Y, Zhou S (2016) Clostridium novyi-NT in cancer therapy. Genes Dis 3(2):144–152

Taniguchi K, Koga S, Nishikido M, Yamashita S, Sakuragi T, Kanetake H et al (1999) Systemic immune response after intravesical instillation of bacille Calmette-Guérin (BCG) for superficial bladder cancer. Clin Exp Immunol 115(1):131–135

Thorn CF, Oshiro C, Marsh S, Hernandez-Boussard T, McLeod H, Klein TE et al (2011) Doxorubicin pathways: pharmacodynamics and adverse effects. Pharmacogenet Genom 21 (7):440

Vallera DA, Li C, Jin N, Panoskaltsis-Mortari A, Hall WA (2002) Targeting urokinase-type plasminogen activator receptor on human glioblastoma tumors with diphtheria toxin fusion protein DTAT. J Natl Cancer Inst 94(8):597–606

Verweij J, Pinedo HM (1990) Mitomycin C: mechanism of action, usefulness and limitations. Anti-Cancer Drugs 1(1):5–13

Waldron NN, Kaufman DS, Oh S, Inde Z, Hexum MK, Ohlfest JR et al (2011) Targeting tumor-initiating cancer cells with dCD133KDEL shows impressive tumor reductions in a xenotransplant model of human head and neck cancer. Mol Cancer Ther 10(10):1829–1838

Yamada T, Goto M, Punj V, Zaborina O, Chen ML, Kimbara K et al (2002) Bacterial redox protein azurin, tumor suppressor protein p53, and regression of cancer. Proc Natl Acad Sci 99 (22):14098–14103

Yang L, Tan R, Wang Q, Huang W, Yin Y (2002) Antifungal cyclopeptides from Halobacillus litoralis YS3106 of marine origin. Tetrahedron Lett 43(37):6545–6548

Yu M, Henning R, Walker A, Kim G, Perroy A, Alessandro R et al (2012) L-asparaginase inhibits invasive and angiogenic activity and induces autophagy in ovarian cancer. J Cell Mol Med 16 (10):2369–2378

# Chapter 10
# Bacterial Cellulose: A Multifaceted Microbial Product

**Abhay Pandit and Rakesh Kumar**

**Abstract** The evolution of biotechnology revolutionized the modern era. The uses of plastic and plastic-based materials were widely accepted but due to its non-biodegradable nature the focus of research is getting shifted towards biodegradable material called biopolymer. Biopolymers are the polymers derived from biological origin with characteristic of getting degraded after its use in the normal environmental condition. Bacterial cellulose (BC) is one of the sustainable biopolymers which are biodegradable, biocompatible, and are obtained from fermentation. In this book chapter, we have discussed the two main methods to obtain BC, its biosynthesis, and its wide application with their relevant characteristic features.

**Keywords** Bacterial cellulose · Biodegradable · Agitated mode · Static mode · Mechanical properties

## 10.1 Introduction

Cellulose, one of the most enormously available polysaccharides, is being delivered by a diverse group of life forms, scaling from plants to photosynthetic eukaryotes and some prokaryotes (Brown 1886; Nobles et al. 2001; Ross et al. 1991). The annual production of cellulose ranges up to 180 billion tonnes (Amor et al. 1995; Delmer 1999). For the first time in 1886, Brown started the research related to synthesis of an extracellular gel-like mat of a linear chain pellicle produced by *Bacillus xylinum*. The produced pellicle shows chemical resemblance with the cellulose of plant origin (Bi et al. 2014; Donini et al. 2010; Hestrin and Schramm 1954; Klemm et al. 2001; Rangaswamy et al. 2015). It has been reported that, the production of bacterial cellulose (BC) is initiated through oxidative fermentation in several media by the bacterial genera *Gluconoacetobacter, Sarcina and*

A. Pandit · R. Kumar (✉)
Department of Biotechnology, Central University of South Bihar, Gaya, India
e-mail: rakeshkr@cusb.ac.in

 
P. K. Arora (ed.), *Microbial Products for Health, Environment and Agriculture*,
Microorganisms for Sustainability 31, https://doi.org/10.1007/978-981-16-1947-2_10

*Agrobacterium,* etc. (Donini et al. 2010; Esa et al. 2014; Huang et al. 2014). The glucose chains are produced from the tiny pores present on the bacterial cell membrane which forms microfibrils that further aggregates resulting in the formation of cellulosic ribbons. The aggregation of microfibrils results in cellulose ribbons, generating a 3-D structure having ample of void spaces which makes the matrix highly porous as evident from its increase in water uptake behavior (Esa et al. 2014; Shah et al. 2013). The produced membrane exhibits a property of a never dried membrane containing 99.1 wt% water of which 0.3 wt% remains associated with water and 98.5 wt% is available free water (Thompson and Hamilton 2001). BC acts as a shield against UV light or as a hurdle to some of the primitive forms of eukaryotic cells and other microorganisms but still the mechanism needs to be validated (Moon et al. 2011). BC shares one of the most dynamic behavior that is its likelihood to modify the microfibril aggregation or assembly and crystalline structure just by fine-tuning the culture conditions (Astley et al. 2003; Harris et al. 2012). BC membrane not only protects bacteria from the toxic nature of UV light rather it also serves in respiration by floating at the gas-liquid interface of the culture vessels (Reiniati et al. 2017). BC shares similar molecular formula $(C_6H_{12}O_6)_n$ as that of celluloses obtained from plant origin, but BC are devoid of other polysaccharide units like lignin, hemicellulose and pectin which make its purification much easier, simpler and a low energy process whereas purification from plant origin requires harsh chemicals (Huang et al. 2014). In future, the BC production will substantially cut the utilization of big trees to harvest white cellulose pulp from it for the manufacture of paper and related goods (Keshk and Sameshima 2006). Also, the 3-D structure of BC results in a higher degree of microfibril polymerization with improved mechanical properties (Iguchi et al. 2000; Paximada et al. 2016; Tsouko et al. 2015). The higher aspect ratio of the BC microfibrils imparts a higher specific area to the resulting cellulose ribbons, compared to plant cellulose (Sulaeva et al. 2015). BC based membrane shows increased drying time because of the tight bonding of water molecules with the free surface hydroxyl groups of the cellulose chains which enhances its water uptake ability (Gelin et al. 2007; Meftahi et al. 2010). The presence of functional groups in abundance within the BC membrane renders it an appropriate material for the introduction of functionalities for producing a diverse range of products (Siró and Plackett 2010). The high aspect ratio of BC renders it suitable for interaction with molecules like antimicrobial and active compounds like antioxidants (Shah et al. 2013). It has found application in the biomedical industry such as in tissue repair, wound bandage, artificial tubes, and as a carrier for drugs (Rajwade et al. 2015; Tsouko et al. 2015). It shows film-forming properties due to the presence of -OH chemical group in its structure, which improves the -H bonding networks (De Olyveira et al. 2016). Under the stationary condition, the BC in form of ribbons is uniaxially oriented whereas, in agitated conditions, disordered, curved, and overlapping BC ribbons are produced. Cellulose microfibrils produced from both culture condition differs between them, one from agitated culture shows thinner microfibrils unlike that produced from stationary condition (Czaja et al. 2004). Also BC is polyfunctional as well as multi-chiral in nature (Keshk and Sameshima 2006; Yoshinaga et al. 1997).

## 10.2 Biosynthesis of BC

Biosynthesis of BC remains a highly regulated mechanism with multi-step reactions. BC biosynthesis remains a highly synchronized mechanism coupled together with several intermediary steps involving different enzymes and proteins. However, the process can be divided into two distinct transitional steps. First is the establishment of intracellular 1,4-β-glucan chains and second is the aggregation and crystallization of cellulose microfibrils. In the second stage only, subfibrils containing polymerized glucose units get extruded out from tiny pores present in the cell membrane of bacteria which aggregates extracellularly and gets self-assembled to result in fibrils/ cellulose ribbons as demonstrated in **Fig. 10.1**. However, the exact mechanism regarding elongation of 1,4-β-glucan chains and its crystallization has not been confirmed till date (Lee et al. 2014; Reiniati et al. 2017).

### *10.2.1 Types of Cultivation*

Two types of cultivation modes are used to produce BC i.e., static cultivation and agitated cultivation. In stationary cultivation, a gel-like cellulose matrix is formed on the air/liquid interface of the medium while suspension pellets of different sizes (Chao et al. 2000, 2001) and solid spheres of BC are produced under agitated cultivation (Gu and Catchmark 2012).

#### 10.2.1.1 Static Mode

In static cultivation, BC pellicles are produced and keeps on floating at the liquid-air interface of the medium because of the entrapment of $CO_2$ gas that is produced during the aerobic respiration by bacterium responsible for producing BC (Lin et al. 2013). The shape and dimension of the BC produced in continuous form depend on the type of vessels used for fermentation. Cellulose synthesis under static condition is highly synchronized by the aeration rate at the interface of the medium, whose net productivity also rests on the concentration of carbon source (Budhiono et al. 1999). By increasing the cultivation period, the BC formation increases due to the hydrogen

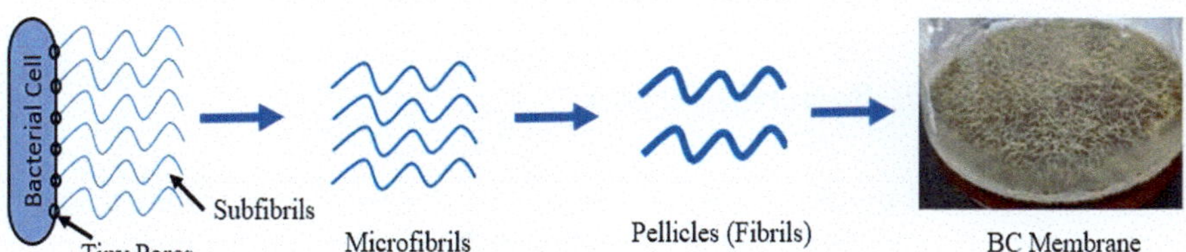


**Fig. 10.1** Mechanism of BC biosynthesis

and C-H bonding within the reaction volume (Sheykhnazari et al. 2011). When the bacterial cells get entrapped into the cellulose membrane it becomes inactive due to the insufficient oxygen and, the BC production gets limited or decreased (Borzani and de Souza 1995). The oxygen supply also gets limited due to the entrapment of bacterial cells in the produced cellulose membrane, and the nutrients of the media get continuously consumed leading to decrease in the contents of nutrients with time which finally effect the net productivity of BC (Esa et al. 2014). However, due to the time-consuming process with low productivity, it faces challenges in industrial applications (Lin et al. 2013).

#### 10.2.1.2 Agitated Mode

Under agitated cultivation, BC yield is enhanced as compared to static cultivation, which contributes to cost reduction due to the continuous mixing of oxygen into the medium (Ul-Islam et al. 2015). The stirred cultivation process favours the formation of cellulose having pellet or sphere morphology and sometimes it exists as fibrous suspensions (Esa et al. 2014), which is governed by varying the agitation rate (Ul-Islam et al. 2015). Theoretically, it is assumed that under agitation cultivation, the net BC yield should be higher due to the continuous mixing of oxygen with the media but data from literatures state that practically it is not true. The elevation in shear rate not only encourages the turbulence force of medium but also governs the maximum likelihood for the cellulose producing bacterial strains to acquire the mutations resulting in cellulose negative strains designated as $Cel^-$ (Kim et al. 2007; Park et al. 2004). However, the rate of agitation directly influences the net BC yield via the formation or transformation of cellulose producing ($Cel^+$) strains into $Cel^-$ mutants. A study conducted by Jung et al. reported BC pellet acts as an effective barrier towards elevated agitation rate (i.e., >500 rpm) created at the tip of the impeller while mutated strains i.e., $Cel^-$ get killed due to outer protector by the strong shear stress. Thus it is put forward that an optimum impeller speed is required for BC production without living $Cel^-$ mutants in agitated cultivation. In conclusion, it can be stated that an optimal agitation rate is essential for cellulose production without transforming the native strains of cellulose producer (Jung et al. 2005). Additionally the produced BC can be shaped in different forms like a membrane, multi-shaped pulps, solid spheres, tubes, and whiskers which could endow BC in industrial applications (Shi et al. 2014).

### *10.2.2 Different Bacterial Strains*

The BC is produced by the bacteria of family *Acetobacteracea* which are aerobic, able to convert ethanol into acetic acid due to which the cells are capable to grow at low pH values (Kersters et al. 2006; Yeo et al. 2004). The cellulose biofilm produced by these family, positions the cells at the surface of high oxygen tension, with zero interference in the supply of nutrients through diffusion (Iguchi et al. 2000). Few unique features of BC producing strains are discussed below. Under agitated cultivation bacterial strain like *Gluconacetobacter entanii*ACCC10215, *Komagataeibacter nataicola*Y19 produces flocks of star-shaped BC with many projections and solid ball-like BC in the presence of HS medium (Bi et al. 2014). *Komagataeibacter* genus is the major producers of BC owing to its ability to metabolize extensive carbon or nitrogen based nutrients (Islam et al. 2017; Lee et al. 2014). To increase BC production under static conditions, researchers have proposed genetically engineered strains which have have been developed so as produce in low oxygen atmosphere (Liu et al. 2018). Iguchi et al. 2000 found that *Acetobacter xylinum* produces two distinct forms of cellulose units namely cellulose I (ribbon-like polymer), and another is cellulose II (stable amorphous polymer). *Gluconacetobacterxylinum* produces solid spheres of BC under agitation culture with carbohydrate based media containing xyloglucan, xylan, arabinogalactan, and pectin (Gu and Catchmark 2012). *Gluconacetobacter persimmonis* GH-2 uses various carbon sources (2% w/v) for growth and production of cellulose (Embuscado et al. 1994). The strain utilizes effectively glucose, fructose, sucrose, mannitol, and inositol as nutrient media for the production of BC (Hungund et al. 2013). The dimensions of the BC fibers are significantly lower which ranges in the order of a few nanometers, due to which it has led to its use in the biomedical industry to the electronics world (AydIn and Aksoy 2014). However, the choice of bacterial strain for BC production depends solely on the final application of the material.

## 10.3 Properties of BC

### *10.3.1 Morphology*

Since BC ranges in the scale of few nanometers, so it's scanning electron microscopy (SEM) have been used to exactly determine its size and structure of the produced cellulose material. BC membrane shows unique porous three-dimensional cellulosic nanofibril structure which are randomly arranged (He et al. 2020). By varying the culture methods, the BC morphology of microfibril changes. BC possessing asterisk morphology has higher pore intensity when compared to BC with morphology of solid spherical masses (Bi et al. 2014). The cellulose microfibrils produced during agitation mode show twisted and curled morphology that may be due to the generation of high turbulence force inside the culture broth during agitation (Yan et al. 2008; Yun et al. 2011). Cellulose microfibrils of Komagataeibacter sp. Nov. CGMCC 17276 show marked differences in its microfibril arrangement with dense network (Fig. 10.2a) in stationary culture and looser and more porous (Fig. 10.2b) microfibrillar aggregation under agitated cultivation (Lu et al. 2020).

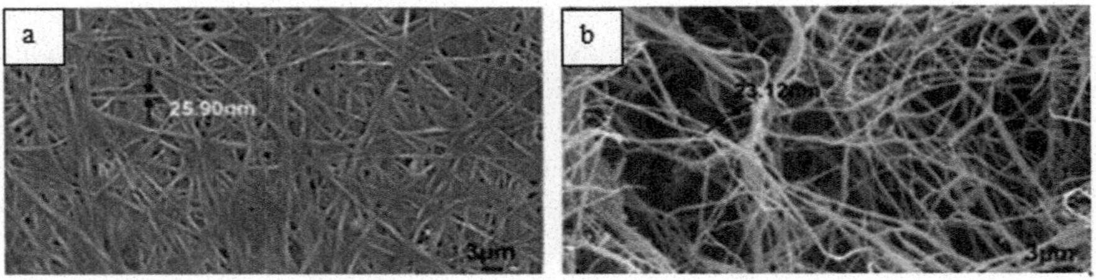


**Fig. 10.2** The microfibril arrangement of BC produced by Komagataeibacter sp. nov. CGMCC 17276. Morphology of BC produced under static (**a**) and agitated (**b**) culture conditions. (Scale bar = 3μm.) (Reproduced with the permission from Lu et al. (2020). Characterization and optimization of production of bacterial cellulosefrom strain CGMCC 17276 based on whole-genome analysis. Carbohydr Polym 232:1–14, Copyright (2020) Elsevier))

### *10.3.2 Water Holding Capacity*

BC shows hydrophilic network structures due to which it retains more percentage of moisture than their dry weight. The hydrophilic nature of BC pellicles makes it a suitable material for the manufacture of its hydrogel structures. Regardless of high water uptake (~99%), the BC membrane imparts satisfactory mechanical properties because of its unique 3-D microfibrillar arrangement and that can also be tuned chemically (Astley et al. 2003; Chanliaud and Gidley 1999; Millon and Wan 2006; Whitney et al. 1999). BC hydrogel structure can be used as a scaffold material in cardiovascular grafts, wound (as a bandage) and in tissue repair because of its biocompatibility with the living body (Gatenholm and Klemm 2010; Klemm et al. 2001, 2005, 2006; Schumann et al. 2009). The BC membrane rehydrates to ~6% which is similar to the rehydration results as obtained for the cellulose of plants. BC treated via the freeze-dried method, can reabsorb nearly 70% of water (Klemm et al. 2005). BC fibrils are about 100 times thinner as compared to cellulose of plant origin, making it an exceptionally permeable matrix letting the translocation of antibiotics or other drugs at the injury site, while simultaneously fulfilling the role of a competent physical hindrance against any foreign infectious agents. The bacterial origin cellulose membrane resembles the same properties as the hydrogels derived from polymers. BC shows effective absorption and adsorption (i.e., sorption) of fluids and also it is non-hypersensitive to immune and are sterilized without any change in its qualities (Gayathry and Gopalaswamy 2014).

### *10.3.3 Mechanical*

The excellent mechanical properties of BC are due to the crystalline nature of produced nanofibril and microfibril, making it well-suited reinforcing material in paper and textile industry. BC exhibits elevated modulus of elasticity in conjugation with a huge internal loss factor which makes it a likewise prevalent material for earphones and amplifier films (Klemm et al. 2005). BC obtained from static culture also shows great stretchability which can be used as a reinforcement material in the field of medicine such as wound healing, artificial blood vessels, and tissue engineering (Gao et al. 2020).

### *10.3.4 X-Ray Diffraction*

Cellulose is an identical polycrystalline macromolecular composite, composed of crystalline (ordered) and amorphous (less ordered) phase both. Six diverse crystalline allomorphs are reported that includes celluloses I (natural cellulose), II (hydrated cellulose), $III_I$, $III_{II}$, $IV_I$, $IV_{II}$ (Park et al. 2010). Cellulose I remains the most bottomless structure noticed in the natural world and its assembly is thermodynamically stable for a longer period. Cellulose I can be converted into any one of kind cellulose II or cellulose III regeneration and mercerization. (Klemm et al. 2005; Ross et al. 1991; O'sullivan 1997). Cellulose I consists of two polymorphs: $I_\alpha$ representing triclinic assembly and $I_\beta$ with monoclinic assembly and the existence of both form depend on the cellulose source (Nishiyama 2009; O'sullivan 1997). $I_\alpha$ is the most prominent polymorph found in most algae and bacteria, while $I_\beta$ is the most prevailing polymorph found in the case of higher plants and tunicates (Belton et al. 1989; Yamamoto and Horii 1993; Yamamoto and Horn 1994). Both $I_\alpha$ and $I_\beta$ unit cells of cellulose chains are arranged in parallel up configuration. This means that the cellulose chains are arranged such that the 1→4 link points in the same direction (Moon et al. 2011; Nishiyama 2009). Cellulose II comprises of monoclinic configuration and are being utilized cellophane (transparent films) are Rayon and Tencel™ (synthetic textile fibers) synthesis. Liquid ammonia treatments of cellulose I and cellulose II yield cellulose III, while subsequent treatment of it results in cellulose IV (Moon et al. 2011). The distinction in a bacterial strain, discussed in the above section, not only impacts the morphology of BC but also affects the microstructure of resulting cellulose such as crystallinity index (CI), crystalline units, size, $I_\alpha$ unit cell content, and mechanical characters (Yan et al. 2008). Another significant parameter i.e., $C_I$ is often utilized to portray the relative abundance of crystalline phase in cellulose (Park et al. 2010; Yang et al. 2013). The literatures stated that BC comprises ~70% $I_\alpha$ polymorph, which is the highest for any cellulosic material (Carrillo and Dobrynin 2007; Yamamoto et al. 1996). By changing the cultivation conditions, the $I_\alpha/I_\beta$ proportion can be modified by noticing a variation in the dimensions of the resulting microfibrils (Tokoh et al. 1998; Yamamoto and Horn 1994). The presence of additives disturb the normal association of the subfibril leading to BC microfibrils with a cross-section of ~6–10 nm, primarily loaded with $I_\beta$ crystal structure (Brown and Laborie 2007; Tokoh et al. 1998; Yamamoto and Horn 1994). In this manner, the openness of BC becomes greater (Park et al. 2010). The intramolecular and intermolecular H-bonding in sub elementary cellulose fibrils can be influenced by the use of different strains that can modify their fibrillar arrangement and its crystallization (Bi et al. 2014). BC of agitated culture shares a

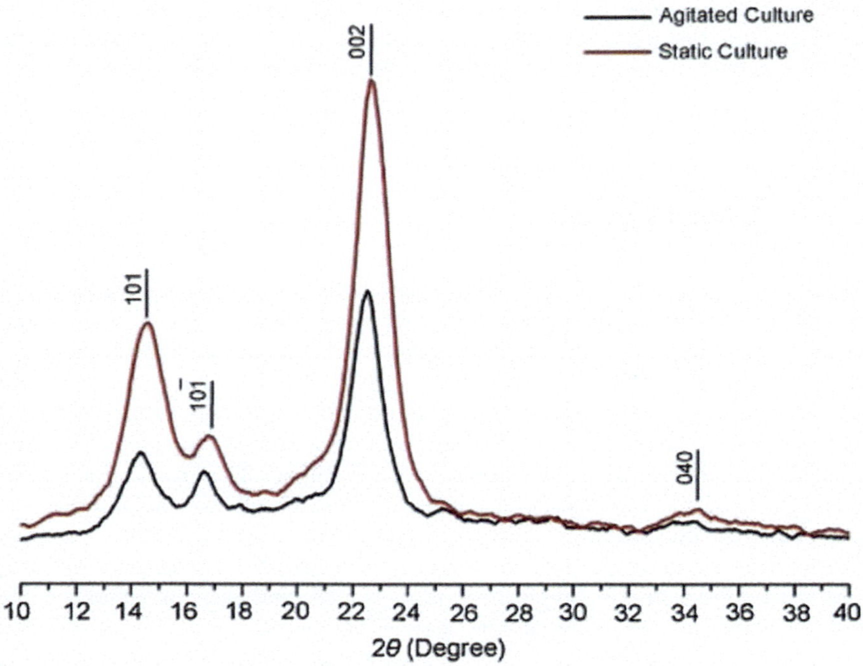


**Fig. 10.3** XRD spectra of BC under static and agitated culture conditions. (Reproduced with the permission from Lu et al. (2020). Characterization and optimization of production of bacterial cellulosefrom strain CGMCC 17276 based on whole-genome analysis. Carbohydr Polym 232:1–14, Copyright (2020) Elsevier))

reduction in $C_I$ to that of the BC of stationary culture which is influenced by the generation of high turbulence force inside the fermentation broth (Singhsa et al. 2018) (Fig. 10.3).

## 10.4 Applications

### *10.4.1 Food Industry*

BC has found its application in the food industry and it is a much explorative field due to its consumption on this planet. It can be exploited as intact membranes, disassembled BC, and BC nanocrystals. The extraordinary characteristics of BC such as porous, high mechanical strength, non-toxicity, and malleability enforce its use in the food industry. Most of the time, BC is used as a reinforcement material, accomplished by several processes such as impregnation, disassembly, acid hydrolysis, and in-situ nanocomposites. Food and drug administration (FDA) has

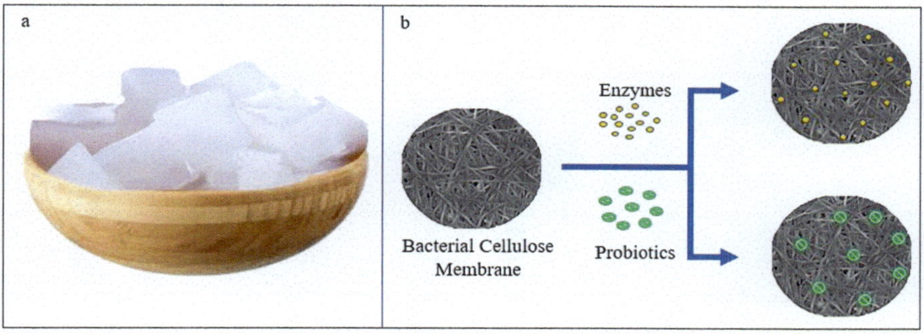


**Fig. 10.4** Application of (**a**) BC as Nata-de-coco, (**b**) BC membrane for immobilization of enzymes and probiotics

approved it to be used as a dietary fiber as a food supplement (Shi et al. 2014). For a long time, BC is being utilized as a primary source for production of Nata-de-coco, a type of food from the Philippines, produced from enriched coconut water (Iguchi et al. 2000) (Fig. 10.4a). BC acts as a fat replacer in food products containing fats to reduce the calorific value thereby avoiding the fat-related health issues. To obtain the matching shear viscosity (yield stress) of conventional commercial thickener such as 0.7% xanthan gum and 1% locust bean gum only 0.1% of BC concentration is required. Hence, BC demonstrates a good alternative thickener in food processing. BC has also been explored as an immobilizer for probiotics and enzymes (Fig. 10.4b). Some enzymes that are available as immobilized forms for regulated release from BC are lipase, laccase, and lysozyme (Azeredo et al. 2019).

## 10.4.2 Facial Mask

Cosmetic products related market is blooming in this era with launching of several beauty products. The cosmetic active substances which are responsible for improving the beauty are combined with a carrier, aiding their passage to the target i.e., skin (Manayi and Saeidnia 2014; Russell 2012). Some of conventional vehicles are creams, gels, emulsions, and lotions. Furthermore, carriers for the active ingredients must be biologically compatible, non-allergic, and non-toxic. The facial masks are the traditional vehicle for cosmetics transport and are utilized for the most part to give skin a quick, bottomless conditioning, epidermal restitution, and regulation on sebum secretion due to its effectivity, fast application and easy to use (Fig. 10.5). An appropriate substitute for a face pack may be BC, due to its film-like property and appearance as well as high mechanical strength. 3D structure of entangled nanofiber i.e., BC has profound mechanical strength, malleability, and effortless handling.

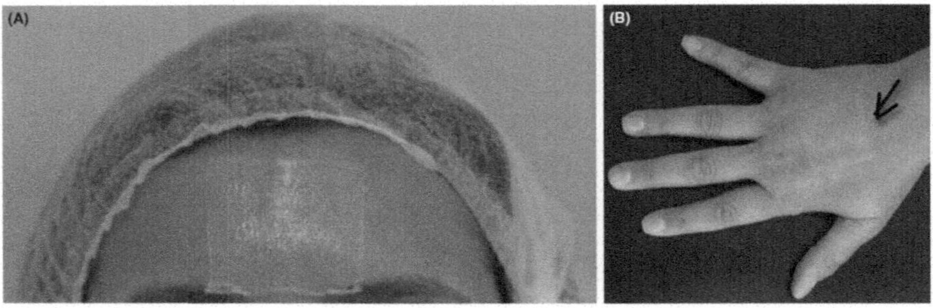


**Fig. 10.5** Pictures of BC mask adhesion to the forehead (**a**) and hand (**b**, indicated by arrow) (Reproduced with the permission from Pacheco et al. (2018). Bacterial cellulose skin masks—Properties and sensory tests. J Cosmet Dermatol 17(5):840–847, Copyright (2018) John Wiley and Sons))

Since BC has a membrane-like shape so it can be exploited as a potential material for mask manufacturing (Pacheco et al. 2018).

### *10.4.3 Cartilage Scaffold*

Tissue engineering is a branch of biomedical engineering which facilitates living tissue or cells to go through repair. Tissue engineering involves knowledge of developmental biology, tissue culture, cell differentiation and survival biology, mechano-transduction, and nanofabrication technology (Ahn et al. 2013; Melrose et al. 2008; Nasseri and Vacanti 2002). Cells, for example cartilage, tendon, ligament, bone, and skin are very much prone to breakage because of accident injury, illness, or may be surgical interventions (Gea et al. 2018). Chondrocytes, which comprise the significant segment of cartilage, regulate the development of new cartilage once any injury happens. Chondrocytes cannot reproduce themselves due to the absence of blood vessels of cartilage. In the well known process, cells are seeded over the synthetic matrix that functions as a scaffold, made from artificial polymer foam such polyglycolic acid, polylactic acid, poly(vinyl alcohol), polyhydroxyethylmethacrylate or poly N-isopropyl acrylamide (Lee and Mooney 2001; Sachlos et al. 2003). The foamy character of the matrix gives porous character. An important step in cartilage tissue engineering is the selection of the scaffold polymeric matrix so that the growth of cells or organs with desired shape and function can take place easily (Gea et al. 2018; Hutmacher 2000). However, the utilization of scaffolds made from synthetic polymers are limited because they lack biocompatibility, also comprises chemical that is toxic to the seeded cells. BC is a promising biopolymer that satisfies all necessities so it is projected scaffold candidate (cartilage, blood vessels) the same as collagen-mimicking in skeletons. (Gea et al. 2018). It has been reported that the tendency of BC to be molded into 3-D assemblies during production is facilitated due to its hydrophilic character

(Helenius et al. 2006). The optical microscopy observations, revealed that the cells grow reasonably well in the BC scaffold so this biopolymeric matrix has good future in tissue engineering (Gea et al. 2018).

### *10.4.4 Skin Tissue Repair*

Important characteristics for a material to be used in skin tissue repair is that it must be capable enough in locking exudate throughout bandage, accompanied by their expulsion after the recuperation of injury. Cellulose derived from bacteria has expanded commercially/consistently because of its potential for application in medication care products (Fig. 10.6). Its capacity lies in the high mechanical quality of the gel-like BC film, high fluid receptiveness, biocompatible and sterile nature. The low solubility of BC compared to plant-derived cellulose may also prove one of the important criteria for its use as skin tissue material. Apart from the solubility of BC compared to plant cellulose, the BC pellicles are comprised of delicate web-like nanofibrils, analogous to collagen of extracellular matrix which may prove one of the advantageous features for skin tissue repair material (Lina et al. 2011). Experiment results stated that when the co-cultivation of BC pellicles, fibroblasts, and chondrocytes compositions were implanted into nude mice, it revealed well integration of BC co-culture compositions into the skin of nude mice (Wang et al. 2009).

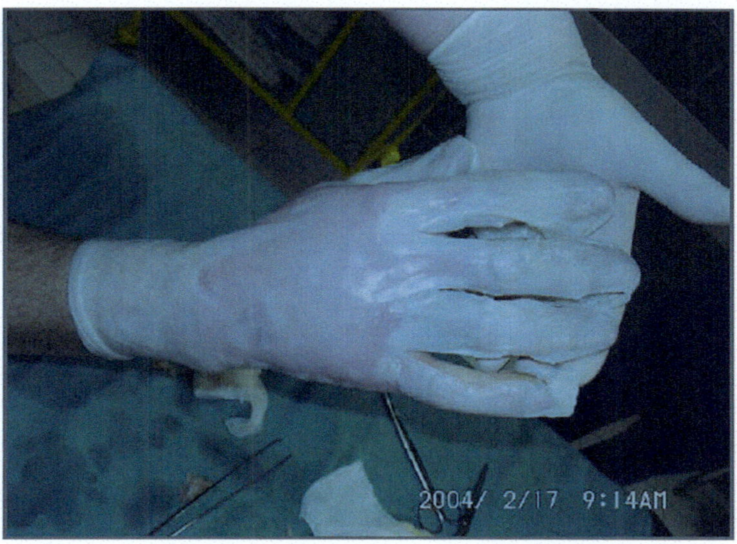


**Fig. 10.6** Bacterial cellulose dressing applied on a wounded hand. (Reproduced with permission from Czaja et al. (2006). Microbial cellulose — the natural power to heal wounds, Biomaterials, 27:149, Copyright (2006) Elsevier))

### 10.4.5 Bone Tissue Regeneration

Bone tissue engineering seeks to reestablish function concerned with physical and biological properties, focusing on healing processes parallel to natural bone (Coelho et al. 2019). Bone regeneration comprises necrotic tissue and clot reabsorption after trauma. Trauma encourages the discharge of immune cells which ultimately favors cellular differentiation that leads to subsequent growth of new bone tissue (Hing 2004). In most of the cases, bone tissue regeneration proceeds without any serious pathological issues (Einhorn 1996). However, prolonged exposure to defects may require latest technology for bone tissue regeneration. In this category, blend of bone tissue alternates for example biopolymers with live cells or tissue can be suitable approach (Kneser et al. 2006). The compatible biopolymers should have high mechanical properties, liable to chemical modifications and also it should display a delayed rate of in-vivo degradation. In comparison to other synthetic membranes, the BC membrane displays several of the above mentioned properties that makes it a suitable material in combination with hydroxyapatite (HA) for bone regeneration (Geyer et al. 1994). The addition of HA with BC exhibits similarity to bone tissue and improves the osteoconductive properties of the resulting biomaterial (Barud et al. 2015). In periodontal lesions, BC membranes are utilized for guided bone regeneration (GBR) in bone defects of significant and insignificant extent, and as a resorbable membrane impending fibroblastic cells and fibrous connective tissue into bone defects. Moreover being a low-cost treatment, the BC membrane promotes effective and quick bone formation at the site of lesions (Batista et al. 1999; dos Anjos et al. 1998; Simonpietri-C et al. 2000). Fang et al. 2009 reported the biocompatibility of developed BC-HA nanocomposite scaffolds that could stimulate in-vitro cell proliferation and differentiation using stromal cells derived from human bone marrow (hBMSC). Histological studies conclude BC-HA membrane effectivity with bone regeneration in bone lesions of rat tibiae thereby accelerating the rate of new bone formation (Saska et al. 2011). Also, solid spherical BC demonstrated reliable cell viability assay when carried on human osteoblast cells, making it advantageous for cultivating biomedical kit like bone-type tissue scaffolds. For biomimetic calcium phosphate deposition, the modification of cellulose via chemical alteration of the hydroxyl group of cellulose chains, seems to be the prerequisite step for the mineralization process. Bioactivity to the resulting material should also be incorporated in the biomimetic calcium phosphate (de Olyveira et al. 2014).

### 10.4.6 Artificial Blood Vessels

Cardiovascular disease is major cause of death due to unhealthy life style. To counter the risk of cardiovascular disease investigation of artificial blood vessels which can coordinate with individual immune system re-establishing normal biological activity can be a better option. Vascular graft and heart valves as artificial blood vessels can

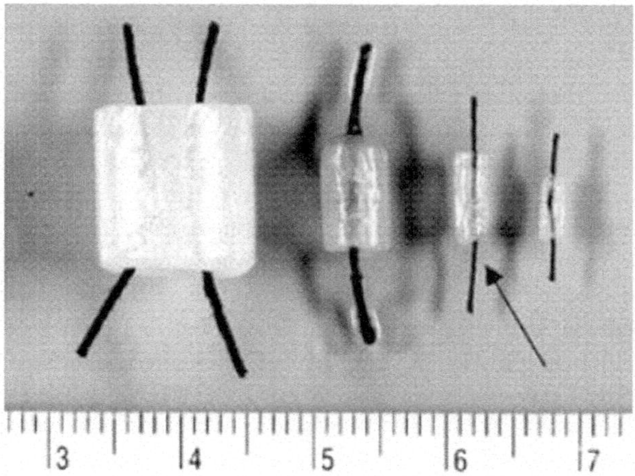


**Fig. 10.7** BASYC® tubes with different inside diameter, different wall thickness, and different lengths. Black lines symbolize the bloodstream. (Klemm et al. (2005) Angew Chemie - Int Ed 44 (22):3358–3393, (Open access))

be fabricated from polytetrafluoroethylene (PTFE), expanded-PTFE, polyurethanes and polyethylene terephthalate (PET) (Dacron®), synthetic fluorocarbon polymer (Lee and Park 2017). PET comprises the ester functional group and its hydrophobic nature that makes it suitable for the synthesis of woven large-diameter artificial blood vessels (Lessim et al. 2015). The synthetic vascular grafts made from such polymers were shown to be acceptable for an application that requires grafts >6 mm of diameter in high stream and low-obstruction flow. However, cross-sectional diameter of less than having 6 mm shows insignificant clinical advantages (Ravi and Chaikof 2010). Vascular grafts fabricated from ultrafine fibers of BC are non-toxic, homogenously pure and exhibit enhanced tensile properties. The BC derived vascular grafts have a diameter less than 6 mm (Zang et al. 2015). Gatenholom et al. demonstrated that by altering the fermentation environment and techniques, the aggregation and assembly of BC tubes can be regulated (Bodin et al. 2007). Investigations on bacterial synthesized cellulose (BASYC®) prosthesis for rodent blood vessels have demonstrated that the embedded BASYC® tube gets integrated with the carotid artery and connective tissue (Wurdinger et al. 2000) (Fig. 10.7). In-vivo biocompatibility examination of BC disclosed no macroscopic signs of inflammation along with total endothelialization including the zone of confluent endothelial layer around the implants (Zang et al. 2015). Scherner et al. (2014) demonstrated, BC grafts as a suitable scaffold for cell proliferation along with neoformation of a 3-layered vascular wall, validating it as an encouraging material for small diameter vascular grafts (Scherner et al. 2014). Inflammation examination of BC vascular graft revealed non-toxic and non-immunoreactive correlation with the leukocytes (Kim et al. 2015).

### 10.4.7 Drug Delivery

The discipline of medicine is emerging due to several number of drug delivery options. Effective drug delivery relies on several variables, including the appropriate choice of materials for research and engineering of new drug delivery systems. BC is one such biopolymer that can be considered as a drug delivery vehicle (Abeer et al. 2014). The peculiar nanofibril structure of BC represents a reasonable macromolecular support for the incorporation of drugs and the development of specific controlled release systems (Almeida et al. 2014). BC membranes are proposed for topical or transdermal drug delivery due to its ability to modulate the release and bioavailability of model drugs for percutaneous administration (Trovatti et al. 2011, 2012). The relative abundance of free surface hydroxyl groups within BC structure enabled it for macromolecular drug delivery and hydrogels (Silvestre et al. 2014). However, BC has a neutral superficial charge due to which it faces limitations for drug delivery applications over polysaccharide (Chandra and Rustgi 1998; Meyers et al. 2008). There are several studies on unmodified BC that have been designed for oral drug delivery in the form of matrices and capsule shells (Lenselink and Andriessen 2011; Scuro et al. 2004). These products revealed immediate release tendency regardless of the drug's aqueous solubility and dose, due to profound permeable structure resulting in free movement of gases, solvents, and small molecules (Lenselink and Andriessen 2011). Chemical modifications of BC is required for regulated drug delivery from BC based matrices and capsule shells (Amnuaikit et al. 2011; Lenselink and Andriessen 2011). Chemical modifications of BC can be carried out by esterification, oxidation, etherification, carbamation, and amidation of BC (Lodén and Wessman 2001). These process generate reactive functional and charged groups superficially and that is due to utilization of free hydroxyl group (Fluhr et al. 1999). Surface modification in combination with freeze-drying can be effective method to alter BC matrices that may be suitable for regulated drug release. Effective surface modification is ongoing research area and drug sustaining effect will depend on the effectiveness of the surface area of BC (Badshah et al. 2018).

### 10.4.8 Electronic Media

Today we live in the world of technological revolution which we see in all forms around us. The types of displays we use include computer screens, TV screens, and the most well-known showcase medium, the printed page on paper made up of cellulose. The display technologies that we use for digital media are derived from liquid crystal displays (LCDs), cathode ray tube displays, organic light-emitting diode (LED) displays, and plasma screens. Besides the availability of such digital medium individuals prefer to print articles instead of reading on a computer screen. The properties which made preference of paper medium are their high reflectivity, contrast, flexibility, lightweight, wide viewing angles. Above mentioned characters lead to the discovery of electronic paper also referred to as *e-paper* using microbial cellulose. *e-Paper* in general contains clear liquid containing minuscule particles and electrical charges. The dimensional stability, paper-like appearance, and unique microfibrillar nanostructure of BC can be explored to manufacture *e-paper*. Transparent probes are placed over and beneath capsules and also positive or negative electric fields are applied resulting in display of specific color onto the surface of the *e-paper*. Also to be noted that *e-Paper* consumes zero power (Shah and Brown 2005). BC has been also used to manufacture high-quality acoustic devices due to its unmatching mechanical properties. The speaker diaphragm prepared of sodium hypochlorite treated BC membranes exhibits 23.5 GPa Young's modulus merged with acoustic absorption property of 0.02 ($\tan\delta$), the sound velocity of 4522.67 m/s and density between 1–1.5 g/cm$^3$ (Indrati et al. 1998). The diaphragm prepared from BC by hot-pressing and coated with gold-sputtered electrodes showed the optimum acoustic properties had permittivity value ($\varepsilon$) of 10 and Young's modulus of 10 GPa (Markiewicz et al. 2004). Due to these properties of BC, many speaker manufacturers manufacture high-quality headphones.

## 10.5 Conclusion

BC is the purest form of cellulose produced via both static and agitated conditions utilizing oxidative fermentation in both natural and synthetic medium by the several bacterial strains. The produced BC membrane represents a 3D web-like network structure along with the presence of numerous reactive groups, imparting tailor-made properties. Under agitated culture, an ideal impeller rate favors BC assembly without altering the nature of wild type bacterial cells. Properties such as morphology, crystallinity index, mechanical, and water holding capacity can be modified by varying the cultivation environment of oxidative fermentation and the bacterial strain too. The biodegradability, biocompatibility, non-toxicity of BC led to increment in its demand ranging from the biological world to the electronic world. Still, BC is one of the many explorative biopolymers in material science.

## References

Abeer MM, Mohd Amin MCI, Martin C (2014) A review of bacterial cellulose-based drug delivery systems: their biochemistry, current approaches and future prospects. J Pharm Pharmacol 66 (8):1047–1061

Ahn S, Lee H, Kim G (2013) Functional cell-laden alginate scaffolds consisting of core/shell struts for tissue regeneration. Carbohydr Polym 98(1):936–942

Almeida IF, Pereira T, Silva NHCS, Gomes FP, Silvestre AJD, Freire CSR, Sousa Lobo JM, Costa PC (2014) Bacterial cellulose membranes as drug delivery systems: an in vivo skin compatibility study. Eur J Pharm Biopharm 86(3):332–336

Amnuaikit T, Chusuit T, Raknam P, Boonme P (2011) Effects of a cellulose mask synthesized by a bacterium on facial skin characteristics and user satisfaction. Med Devices Evid Res 4(1):77–81

Amor Y, Haigler CH, Johnson S, Wainscott M, Delmer DP (1995) A membrane-associated form of sucrose synthase and its potential role in synthesis of cellulose and callose in plants. Proc Natl Acad Sci U S A 92(20):9353–9357

Astley OM, Chanliaud E, Donald AM, Gidley MJ (2003) Tensile deformation of bacterial cellulose composites. Int J Biol Macromol 32(1–2):28–35

AydIn YA, Aksoy ND (2014) Isolation and characterization of an efficient bacterial cellulose producer strain in agitated culture: *Gluconacetobacter hansenii* P2A. Appl Microbiol Biotechnol 98(3):1065–1075

Azeredo HMC, Barud H, Farinas CS, Vasconcellos VM, Claro AM (2019) Bacterial cellulose as a raw material for food and food packaging applications. Front Sustain Food Syst 3:1–14

Badshah M, Ullah H, Khan AR, Khan S, Park JK, Khan T (2018) Surface modification and evaluation of bacterial cellulose for drug delivery. Int J Biol Macromol 113:526–533

Barud HGO, Barud HS, Cavicchioli M, Amaral TS, Junior OBO, Santos DM, Petersen ALOA, Celes F, Borges VM, Oliveira CI, Oliveira PF, Furtado RA, Tavares DC, Ribeiro SJL (2015) Preparation and characterization of a bacterial cellulose/silk fibroin sponge scaffold for tissue regeneration. Carbohydr Polym 128:41–51

Batista EL, Novaes AB, Simonpietri CJJ, Batista FC (1999) Use of bovine-derived anorganic bone associated with guided tissue regeneration in intrabony defects. Six-month evaluation at re-entry. J Periodontol 70(9):1000–1007

Belton PS, Tanner SF, Cartier N, Chanzy H (1989) High-resolution solid-state 13C nuclear magnetic resonance spectroscopy of tunicin, an animal cellulose. Macromolecules 22 (4):1615–1617

Bi JC, Liu SX, Li CF, Li J, Liu LX, Deng J, Yang YC (2014) Morphology and structure characterization of bacterial celluloses produced by different strains in agitated culture. J Appl Microbiol 117(5):1305–1311

Bodin A, Bäckdahl H, Fink H, Gustafsson L, Risberg B, Gatenholm P (2007) Influence of cultivation conditions on mechanical and morphological properties of bacterial cellulose tubes. Biotechnol Bioeng 97(2):425–434

Borzani W, de Souza SJ (1995) Mechanism of the film thickness increasing during the bacterial production of cellulose on non-agitaded liquid media. Biotechnol Lett 17(11):1271–1272

Brown AJ (1886) Brown on acetic ferment. J Chem Soc Trans 49(432):432–439

Brown EE, Laborie MPG (2007) Bioengineering bacterial cellulose/poly (ethylene oxide) nanocomposites. Biomacromolecules 8(10):3074–3081

Budhiono A, Rosidi B, Taher H, Iguchi M (1999) Kinetic aspects of bacterial cellulose formation in nata-de-coco culture system. Carbohydr Polym 40(2):137–143

Carrillo JMY, Dobrynin AV (2007) Molecular dynamics simulations of polyelectrolyte adsorption. Langmuir 23(5):2472–2482

Chandra R, Rustgi R (1998) Biodegradable polymers. Prog Polym Sci 23(7):1273–1335

Chanliaud E, Gidley MJ (1999) In vitro synthesis and properties of pectin/Acetobacter xylinus cellulose composites. Plant J 20(1):25–35

Chao Y, Ishida T, Sugano Y, Shoda M (2000) Bacterial cellulose production by Acetobacter xylinum in a 50-L internal-loop airlift reactor. Biotechnol Bioeng 68(3):345–352

Chao Y, Sugano Y, Shoda M (2001) Bacterial cellulose production under oxygen-enriched air at different fructose concentrations in a 50-liter, internal-loop airlift reactor. Appl Microbiol Biotechnol 55(6):673–679

Coelho F, Cavicchioli M, Specian SS, Scarel-Caminaga RM, de Aquino PL, de Medeiros AI, de Lima Ribeiro SJ, de Oliveira Capote TS (2019) Bacterial cellulose membrane functionalized with hydroxiapatite and anti-bone morphogenetic protein 2: a promising material for bone regeneration. PLoS One 14(8):1–23

Czaja W, Romanovicz D, Malcolm Brown R (2004) Structural investigations of microbial cellulose produced in stationary and agitated culture. Cellulose 11(3–4):403–411

Czaja W, Krystynowicz A, Bielecki S, Brown RM Jr (2006) Microbial cellulose—the natural power to heal wounds. Biomaterials 27(2):145–151

De Olyveira GM, dos Santos ML, Costa LMM, Daltro PB, Basmaji P, de Cerqueira DG, Guastaldi AC (2014) Bacterial cellulose biocomposites for guided tissue regeneration. Sci Adv Mater 6 (12):2673–2678

de Olyveira GM, Costa LMM, Riccardi CDS, dos Santos ML, Daltro PB, Basmaji P, Daltro GDC, Guastaldi AC (2016) Bacterial cellulose for advanced medical materials. In: Nanobiomaterials soft tissue eng appl of nanobiomaterials. Elsevier, Amsterdam, pp 57–82

Delmer DP (1999) Cellulose biosynthesis: exciting times for a difficult field of study. Annu Rev Plant Physiol Plant Mol Biol 50(1):245–276

Donini ÍAN, De Salvi DTB, Fukumoto FK, Lustri WR, Barud HS, Marchetto R, Messaddeq Y, Ribeiro SJL (2010) Biossíntese e recentes avanços na produção de celulose bacteriana. Eclet Quim 35(4):165–178

dos Anjos B, Novaes AB, Meffert R, Barboza EP (1998) Clinical comparison of cellulose and expanded polytetrafluoroethylene membranes in the treatment of class ii furcations in mandibular molars with 6-month re-entry. J Periodontol 69(4):454–459

Einhorn TA (1996) Enhancement of fracture healing. Instr Course Lect 45:401–416

Embuscado ME, Marks JS, BeMiller JN (1994) Bacterial cellulose. I. Factors affecting the production of cellulose by Acetobacter xylinum. Top Catal 8(5):407–418

Esa F, Tasirin SM, Rahman NA (2014) Overview of bacterial cellulose production and application. Agric Agric Sci Proc 2:113–119

Fang B, Wan YZ, Tang TT, Gao C, Dai KR (2009) Proliferation and osteoblastic differentiation of human bone marrow stromal cells on hydroxyapatite/bacterial cellulose nanocomposite scaffolds. Tissue Eng Part A 15(5):1091–1098

Fluhr JW, Gloor M, Lehmann L, Lazzerini S, Distante F, Berardesca E (1999) Glycerol accelerates recovery of barrier function in vivo. Acta Derm Venereol 79(6):418–421

Gao H, Sun Q, Han Z, Li J, Liao B, Hu L, Huang J, Zou C, Jia C, Huang J, Chang Z, Jiang D, Jin M (2020) Comparison of bacterial nanocellulose produced by different strains under static and agitated culture conditions. Carbohydr Polym 227(August):1–8

Gatenholm P, Klemm D (2010) Bacterial nanocellulose as a renewable material for biomedical applications. MRS Bull 35(3):208–213

Gayathry G, Gopalaswamy G (2014) Production and characterisation of microbial cellulosic fibre from Acetobacter xylinum. Indian J Fibre Text Res 39(1):93–96

Gea S, Sari RM, Piliang AF, Indrawan DP, Hutapea YA (2018) Study of bacterial cellulose as scaffold on cartilage tissue engineering. AIP Conf Proc 2049(December):1–9

Gelin K, Bodin A, Gatenholm P, Mihranyan A, Edwards K, Strømme M (2007) Characterization of water in bacterial cellulose using dielectric spectroscopy and electron microscopy. Polymer 48 (26):7623–7631

Geyer U, Heinze T, Stein A, Klemm D, Marsch S, Schumann D, Schmauder HP (1994) Formation, derivatization and applications of bacterial cellulose. Int J Biol Macromol 16(6):343–347

Gu J, Catchmark JM (2012) Impact of hemicelluloses and pectin on sphere-like bacterial cellulose assembly. Carbohydr Polym 88(2):547–557

Harris DM, Corbin K, Wang T, Gutierrez R, Bertolo AL, Petti C, Smilgies DM, Estevez JM, Bonetta D, Urbanowicz BR, Ehrhardt DW, Somerville CR, Rose JKC, Hong M, DeBolt S (2012) Cellulose microfibril crystallinity is reduced by mutating C-terminal transmembrane region residues CESA1 A903V and CESA3 T942I of cellulose synthase. Proc Natl Acad Sci U S A 109(11):4098–4103

He W, Zhang Z, Zheng Y, Qiao S, Xie Y, Sun Y, Qiao K, Feng Z, Wang X, Wang J (2020) Preparation of aminoalkyl-grafted bacterial cellulose membranes with improved antimicrobial properties for biomedical applications. J Biomed Mater Res Part A 108(5):1086–1098

Helenius G, Bäckdahl H, Bodin A, Nannmark U, Gatenholm P, Risberg B (2006) In vivo biocompatibility of bacterial cellulose. J Biomed Mater Res Part A 76(2):431–438

Hestrin S, Schramm M (1954) Synthesis of cellulose by Acetobacter xylinum. II. Preparation of freeze-dried cells capable of polymerizing glucose to cellulose. Biochem J 58(2):345–352
Hing KA (2004) Bone repair in the twenty-first century: biology, chemistry or engineering? Philos Trans R Soc A Math Phys Eng Sci 362(1825):2821–2850
Huang Y, Zhu C, Yang J, Nie Y, Chen C, Sun D (2014) Recent advances in bacterial cellulose. Cellulose 21(1):1–30
Hungund B, Prabhu S, Shetty C, Acharya S, Prabhu V, Gupta SG (2013) Production of bacterial cellulose from gluconacetobacter persimmonis GH-2 using dual and cheaper carbon sources. J Microb Biochem Technol 5(2):31–33
Hutmacher DW (2000) Scaffolds in tissue engineering bone and cartilage. Biomaterials 21 (24):2529–2543
Iguchi M, Yamanaka S, Budhiono A (2000) Bacterial cellulose—a masterpiece of nature's arts. J Mater Sci 35(2):261–270
Indrati L, Yudianti R, Amurwabumi K, Yuli N (1998) Application of biocellulose as an acoustic membrane. Indones J Biotechnol 3(6):180–184
Islam MU, Ullah MW, Khan S, Shah N, Park JK (2017) Strategies for cost-effective and enhanced production of bacterial cellulose. Int J Biol Macromol 102:1166–1173
Jung JY, Park JK, Chang HN (2005) Bacterial cellulose production by Gluconacetobacter hansenii in an agitated culture without living non-cellulose producing cells. Enzym Microb Technol 37 (3):347–354
Kersters K, Lisdiyanti P, Komagata K, Swings J (2006) The family acetobacteraceae: the genera acetobacter, acidomonas, asaia, gluconacetobacter, gluconobacter, and kozakia. The Prokaryotes 5:163–200
Keshk S, Sameshima K (2006) The utilization of sugar cane molasses with/without the presence of lignosulfonate for the production of bacterial cellulose. Appl Microbiol Biotechnol 72 (2):291–296
Kim YJ, Kim JN, Wee YJ, Park DH, Ryu HW (2007) Bacterial cellulose production by gluconacetobacter sp. RKY5 in a rotary biofilm contactor. Appl Biochem Biotechnol 136 (4):529–537
Kim GD, Lee SE, Yang H, Park HR, Son GW, Park CS, Park YS (2015) β2 integrins (CD11/18) are essential for the chemosensory adhesion and migration of polymorphonuclear leukocytes on bacterial cellulose. J Biomed Mater Res Part A 103(5):1809–1817
Klemm D, Schumann D, Udhardt U, Marsch S (2001) Bacterial synthesized cellulose - artificial blood vessels for microsurgery. Prog Polym Sci 26(9):1561–1603
Klemm D, Heublein B, Fink HP, Bohn A (2005) Cellulose: fascinating biopolymer and sustainable raw material. Angew Chem Int Ed 44(22):3358–3393
Klemm D, Schumann D, Kramer F, Heßler N, Hornung M, Schmauder HP, Marsch S (2006) Nanocelluloses as innovative polymers in research and application. Adv Polym Sci 205 (1):49–96
Kneser U, Schaefer DJ, Polykandriotis E, Horch RE (2006) Tissue engineering of bone: the reconstructive surgeon's point of view. J Cell Mol Med 10(1):7–19
Lee KY, Mooney DJ (2001) Hydrogels for tissue engineering. Chem Rev 101(7):1869–1879
Lee SE, Park YS (2017) The role of bacterial cellulose in artificial blood vessels. Mol Cell Toxicol 13(3):257–261
Lee KY, Buldum G, Mantalaris A, Bismarck A (2014) More than meets the eye in bacterial cellulose: biosynthesis, bioprocessing, and applications in advanced fiber composites. Macromol Biosci 14(1):10–32
Lenselink E, Andriessen A (2011) A cohort study on the efficacy of a polyhexanide-containing biocellulose dressing in the treatment of biofilms in wounds. J Wound Care 20(11):534–539
Lessim S, Oughlis S, Lataillade JJ, Migonney V, Changotade S, Lutomski D, Poirier F (2015) Protein selective adsorption properties of a polyethylene terephtalate artificial ligament grafted with poly(sodium styrene sulfonate) (polyNaSS): correlation with physicochemical parameters of proteins. Biomed Mater 10(6):1–10

Lin SP, Calvar IL, Catchmark JM, Liu JR, Demirci A, Cheng KC (2013) Biosynthesis, production and applications of bacterial cellulose. Cellulose 20(5):2191–2219
Lina F, Yue Z, Jin Z, Guang Y (2011) Bacterial cellulose for skin repair materials. Biomed Eng Front Challenges:249–274
Liu M, Li S, Xie Y, Jia S, Hou Y, Zou Y, Zhong C (2018) Enhanced bacterial cellulose production by Gluconacetobacter xylinus via expression of Vitreoscilla hemoglobin and oxygen tension regulation. Appl Microbiol Biotechnol 102(3):1155–1165
Lodén M, Wessman C (2001) The influence of a cream containing 20% glycerin and its vehicle on skin barrier properties. Int J Cosmet Sci 23(2):115–119
Lu T, Gao H, Liao B, Wu J, Zhang W, Huang J, Liu M, Huang J, Chang Z, Jin M, Yi Z, Jiang D (2020) Characterization and optimization of production of bacterial cellulosefrom strain CGMCC 17276 based on whole-genome analysis. Carbohydr Polym 232:1–14
Manayi A, Saeidnia S (2014) Cosmetics and personal care products. Encycl Toxicol Third Ed:1043–1049
Markiewicz E, Hilczer B, Pawlaczyk C (2004) Dielectric and acoustic response of biocellulose. Ferroelectrics 304(1):39–42
Meftahi A, Khajavi R, Rashidi A, Sattari M, Yazdanshenas ME, Torabi M (2010) The effects of cotton gauze coating with microbial cellulose. Cellulose 17(1):199–204
Melrose J, Chuang C, Whitelock J (2008) Tissue engineering of cartilages using biomatrices. J Chem Technol Biotechnol 83(4):444–463
Meyers MA, Chen PY, Lin AYM, Seki Y (2008) Biological materials: structure and mechanical properties. Prog Mater Sci 53(1):1–206
Millon LE, Wan WK (2006) The polyvinyl alcohol-bacterial cellulose system as a new nanocomposite for biomedical applications. J Biomed Mater Res Part B Appl Biomater 79 (2):245–253
Moon RJ, Martini A, Nairn J, Simonsen J, Youngblood J (2011) Cellulose nanomaterials review: structure, properties and nanocomposites. Chem Soc Rev 40(7):3941–3994
Nasseri BA, Vacanti JP (2002) Tissue engineering in the 21st century. Surg Technol Int 10:25–37
Nishiyama Y (2009) Structure and properties of the cellulose microfibril. J Wood Sci 55 (4):241–249
Nobles DR, Romanovicz DK, Brown RM (2001) Cellulose in cyanobacteria. Origin of vascular plant cellulose synthase? Plant Physiol 127(2):529–542
O'sullivan AC (1997) Cellulose: the structure slowly unravels. Cellulose 4(3):173–207
Pacheco G, de Mello CV, Chiari-Andréo BG, Isaac VLB, Ribeiro SJL, Pecoraro É, Trovatti E (2018) Bacterial cellulose skin masks—properties and sensory tests. J Cosmet Dermatol 17 (5):840–847
Park JK, Hyun SH, Jung JY (2004) Conversion of G. hansenii PJK into non-cellulose-producing mutants according to the culture condition. Biotechnol Bioprocess Eng 9(5):383–388
Park S, Baker JO, Himmel ME, Parilla PA, Johnson DK (2010) Cellulose crystallinity index: measurement techniques and their impact on interpreting cellulase performance. Biotechnol Biofuels 3:1–10
Paximada P, Tsouko E, Kopsahelis N, Koutinas AA, Mandala I (2016) Bacterial cellulose as stabilizer of o/w emulsions. Food Hydrocoll 53:225–232
Rajwade JM, Paknikar KM, Kumbhar JV (2015) Applications of bacterial cellulose and its composites in biomedicine. Appl Microbiol Biotechnol 99(6):2491–2511
Rangaswamy BE, Vanitha KP, Hungund BS (2015) Microbial cellulose production from bacteria isolated from rotten fruit. Int J Polym Sci:1–8
Ravi S, Chaikof EL (2010) Biomaterials for vascular tissue engineering. Regen Med 5(1):107–120
Reiniati I, Hrymak AN, Margaritis A (2017) Recent developments in the production and applications of bacterial cellulose fibers and nanocrystals. Crit Rev Biotechnol 37(4):510–524
Ross P, Mayer R, Benziman ANDM (1991) Cellulose biosynthesis and function in bacteria positive control: diguanylate cyclase. Microbiology 55(1):35–58

Russell R (2012) Cosmetics use: psychological perspectives. In: Encyclopedia of body image and human appearance, vol 1

Sachlos E, Czernuszka JT, Gogolewski S, Dalby M (2003) Making tissue engineering scaffolds work. Review on the application ofsolid freeform fabrication technology to the production of tissue engineeringscaffolds. Eur Cells Mater 5(29):39–40

Saska S, Barud HS, Gaspar AMM, Marchetto R, Ribeiro SJL, Messaddeq Y (2011) Bacterial cellulose-hydroxyapatite nanocomposites for bone regeneration. Int J Biomater:1–9

Scherner M, Reutter S, Klemm D, Sterner-Kock A, Guschlbauer M, Richter T, Langebartels G, Madershahian N, Wahlers T, Wippermann J (2014) In vivo application of tissue-engineered blood vessels of bacterial cellulose as small arterial substitutes: proof of concept? J Surg Res 189 (2):340–347

Schumann DA, Wippermann J, Klemm DO, Kramer F, Koth D, Kosmehl H, Wahlers T, Salehi-Gelani S (2009) Artificial vascular implants from bacterial cellulose: preliminary results of small arterial substitutes. Cellulose 16(5):877–885

Scuro LS, Simioni PU, Grabriel DL, Saviani EE, Modolo LV, Tamashiro WMSC, Salgado I (2004) Suppression of nitric oxide production in mouse macrophages by soybean flavonoids accumulated in response to nitroprusside and fungal elicitation. BMC Biochem 5(1):1–8

Shah J, Brown RM (2005) Towards electronic paper displays made from microbial cellulose. Appl Microbiol Biotechnol 66(4):352–355

Shah N, Ul-Islam M, Khattak WA, Park JK (2013) Overview of bacterial cellulose composites: a multipurpose advanced material. Carbohydr Polym 98(2):1585–1598

Sheykhnazari S, Tabarsa T, Ashori A, Shakeri A, Golalipour M (2011) Bacterialsynthesized cellulose nanofibers: effects of growth times and culture mediums on the structural characteristics. Carbohydr Polym 86(3):1187–1191

Shi Z, Zhang Y, Phillips GO, Yang G (2014) Utilization of bacterial cellulose in food. Food Hydrocoll 35:539–545

Silvestre AJD, Freire CSR, Neto CP (2014) Do bacterial cellulose membranes have potential in drug-delivery systems? Expert Opin Drug Deliv 11(7):1113–1124

Simonpietri-C JJ, Novaes AB, Batista EL, Feres Filho EJ (2000) Guided tissue regeneration associated with bovine-derived anorganic bone in mandibular class ii furcation defects. 6-month results at re-entry. J Periodontol 71(6):904–911

Singhsa P, Narain R, Manuspiya H (2018) Physical structure variations of bacterial cellulose produced by different Komagataeibacter xylinus strains and carbon sources in static and agitated conditions. Cellulose 25(3):1571–1581

Siró I, Plackett D (2010) Microfibrillated cellulose and new nanocomposite materials: a review. Cellulose 17(3):459–494

Sulaeva I, Henniges U, Rosenau T, Potthast A (2015) Bacterial cellulose as a material for wound treatment: properties and modifications: a review. Biotechnol Adv 33(8):1547–1571

Thompson DN, Hamilton MA (2001) Production of bacterial cellulose from alternate feedstocks. Appl Biochem Biotechnol 91(1-9): Article No. 503

Tokoh C, Takabe K, Fujita M, Saiki H (1998) Cellulose synthesized by Acetobacter xylinum in the presence of acetyl glucomannan. Cellulose 5(4):249–261

Trovatti E, Silva NHCS, Duarte IF, Rosado CF, Almeida IF, Costa P, Freire CSR, Silvestre AJD, Neto CP (2011) Biocellulose membranes as supports for dermal release of lidocaine. Biomacromolecules 12(11):4162–4168

Trovatti E, Freire CSR, Pinto PC, Almeida IF, Costa P, Silvestre AJD, Neto CP, Rosado C (2012) Bacterial cellulose membranes applied in topical and transdermal delivery of lidocaine hydrochloride and ibuprofen: in vitro diffusion studies. Int J Pharm 435(1):83–87

Tsouko E, Kourmentza C, Ladakis D, Kopsahelis N, Mandala I, Papanikolaou S, Paloukis F, Alves V, Koutinas A (2015) Bacterial cellulose production from industrial waste and by-product streams. Int J Mol Sci 16(7):14832–14849

Ul-Islam M, Khan S, Ullah MW, Park JK (2015) Bacterial cellulose composites: synthetic strategies and multiple applications in bio-medical and electro-conductive fields. Biotechnol J10 (12):1847–1861

Wang ZL, Jia YY, Shi Y, Cong DL, Chen YY, Jia SR, Zhou YL (2009) Research on characterization and biocompatibility of nano-bacterial cellulose membrane. Chem J Chin U Chin 30 (8):1553–1558

Whitney SEC, Gothard MGE, Mitchell JT, Gidley MJ (1999) Roles of cellulose and xyloglucan in determining the mechanical properties of primary plant cell walls. Plant Physiol 121(2):657–663

Wurdinger J, Marsch S, Udhardt U, Schumann HD (2000) BASYC (bacterial synthesized cellulose)—the vitalization of a microvessel-prosthesis in the rat. Microsurgery 20:268

Yamamoto H, Horii F (1993) CPMAS carbon-13 NMR analysis of the crystal transformation induced for Valonia cellulose by annealing at high temperatures. Macromolecules 26 (6):1313–1317

Yamamoto H, Horn F (1994) In situ crystallization of bacterial cellulose I. influences of polymeric additives, stirring and temperature on the formation celluloses Iα and Iβ as revealed by cross polarization/magic angle spinning (CP/MAS)13C NMR spectroscopy. Cellulose 1(1):57–66

Yamamoto H, Horii F, Hirai A (1996) In situ crystallization of bacterial cellulose II. Influences of different polymeric additives on the formation of celluloses Iα and Iβ at the early stage of incubation. Cellulose 3(4):229–242

Yan Z, Chen S, Wang H, Wang B, Jiang J (2008) Biosynthesis of bacterial cellulose/multi-walled carbon nanotubes in agitated culture. Carbohydr Polym 74(3):659–665

Yang Y, Jia J, Xing J, Chen J, Lu S (2013) Isolation and characteristics analysis of a novel high bacterial cellulose producing strain Gluconacetobacter intermedius CIs26. Carbohydr Polym 92 (2):2012–2017

Yeo SH, Lee OS, Lee IS, Kim HS, Yu TS, Jeong YJ (2004) Gluconacetobacter persimmonis sp. nov., isolated from Korean traditional persimmon vinegar. J Microbiol Biotechnol 14 (2):276–283

Yoshinaga F, Tonouchi N, Watanabe K (1997) Research progress in production of bacterial cellulose by aeration and agitation culture and its application as a new industrial material. Biosci Biotechnol Biochem 61(2):219–224

Yun YS, Bak H, Cho SY, Jin HJ (2011) Morphology and pore characteristics of bacterial cellulose/ multiwalled carbon nanotube composite cryogels. J Nanosci Nanotechnol 11(1):806–809

Zang S, Zhang R, Chen H, Lu Y, Zhou J, Chang X, Qiu G, Wu Z, Yang G (2015) Investigation on artificial blood vessels prepared from bacterial cellulose. Mater Sci Eng C 46:111–117

# Chapter 11
# Bioremediation: Going the "Nano" Way

**Abdul M. Kader, Karan K. Pahuja, Soma Mukherjee, Madhusudan Reddy, and Debarati Paul**

**Abstract** Owing to the persistence of chemical pollutants and increasing number of environmental disasters, the public has become sensitive toward environmental issues and policy makers are striving toward developing clean-up technologies locally, as well as globally. Bioremediation is preferred because of its efficiency and environment-friendliness above other methods of environmental cleanup. The concept of utilizing microbial biofilm for bioremediation of contaminated environments has been accepted widely for its sensitivity, efficiency, and economy. Nanomaterials have been evaluated for developing efficient matrices for immobilizing the microbes capable of bioremediation, since they possess exceptional properties, e.g. high porosity and small pore size that are compatible with the dimensions and needs of microbes. Surface morphology and biocompatibility also allowed faster colonization of the nano fiber surface by microorganisms, thereby reducing the cost. Various nanomaterials have been studied for their use in waste water treatment and bioremediation but their toxicity and high costs have proven to be a barrier. Advanced nanomaterials are being developed by combining them with natural substances to improve their surface properties for efficient binding of cells. Therefore, nanobioremediation is an emerging technology for use in the area of environmental cleanup.

**Keywords** Biofilms · Bioremediation · Nanomaterials · Combination · Microbes

A. M. Kader · K. K. Pahuja · D. Paul (✉)
Amity Institute of Biotechnology, Amity University, Noida, India
e-mail: dpaul@amity.edu

S. Mukherjee
University of Holy Cross, New Orleans, LA, USA

M. Reddy
Department of Microbiology, Palamuru University, Mahabubnagar, Telangana, India



P. K. Arora (ed.), *Microbial Products for Health, Environment and Agriculture*, Microorganisms for Sustainability 31, https://doi.org/10.1007/978-981-16-1947-2_11

## 11.1 Introduction

It is believed that the biological means for treatment of toxic waste effluents are more effective and economic as compared to physico-chemical means (Dua et al. 2002; Paul et al. 2005a). Sometimes, however, due to the minimal availability of contaminants for uptake by microbial cells, clean up is incomplete or inefficient (Lewis et al. 2004; Pieper and Reineke 2000). This problem may be sufficiently ameliorated by using biofilms where bacteria act as a group to enhance bioremediation as compared to free planktonic cells. Bioremediation mediated by biofilm formation is a proficient and safer alternative to use of planktonic microorganisms, because cells in a biofilm adapt and survive better owing to the protection provided by the matrix.

Stabilization of biofilms is one of the major challenges encountered in the process of increasing the bioavailability of contaminants to bacteria. Advances in nanomaterials and hybridized nanomaterials can significantly influence the stabilization of biofilms for bioremediation purposes by increasing the ratio of exposed surface area to volume. However, the challenges due to toxicity and cost have deterred their use for abatement of pollution where only inexpensive and natural measures are preferable. Biocomposites formed by integrating microbial strains and nanoparticles or use of "combined" abiotic and biotic forces have been applied for treating persistent pollutants in a cost-effective but rapid and dependable process. Bioremediation and nano-biotechnology can potentially decrease the challenges faced in using nanomaterials for bioremediation purposes in the near future. Over the years, the levels and nature of pollution have changed and so have the means and measures for abatement (Fig. 11.1). Here we highlight the issues faced by various existing methods and compare with the upcoming methods of bioremediation.

## 11.2 Bioremediation: Application and Problems

Since the beginning of the industrial era in the nineteenth century, pollution and contamination have become major threats to our ecosystem and human survival alike (Guieysse et al. 2004; Lewis et al. 2004; Paul et al. 2005a). Use of chemical-based pesticides and fertilizers above the permissible limits has contaminated several major water bodies including the ground water for which physico-chemical treatment has proved to be a failure; however, bioremediation has now proven to be an effective and sustainable technique for environmental decontamination (Ritter 1990).

The major obstacles affecting the rate of bioremediation is the activity of bacteria, availability of contaminants, nutrient availability, adaptation of the bacteria to the complex environment, etc. Every biodegradation reaction depends on majorly three factors: (1) properties and growth of the microorganism(s) in question, i.e. concentration, diversity, activities, (2) properties of the pollutant (physical state, structure, chemistry, concentration), and (3) environmental conditions (Ghosal et al. 2016; Paul et al. 2005a, b). Bioavailability is one of the major factors that

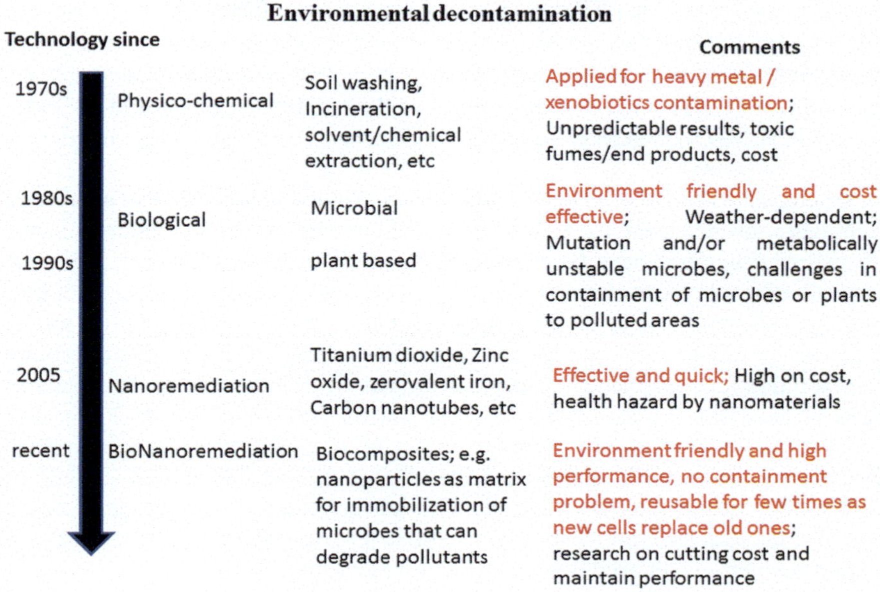


**Fig. 11.1** Technological progress in the area of treatment of pollutants over the past few decades

control the efficiency of bioremediation, mostly in case of hydrophobic hydrocarbons (Singh et al. 2006). The rate at which microbes convert contaminants during bioremediation depends on mass transfer, i.e. the rate of uptake of the contaminants followed by their transfer to degrading cells. Therefore, effective remediation is a consequence of appropriate bioavailability of contaminants to the degrading microorganisms (Paul et al. 2006).

### 11.2.1 *Importance of Bioavailability and Biofilm Stabilization for Microbial Bioremediation*

Hydrophobic organic hydrocarbons like PCBs and PAHs are the major contaminants in bioremediation sites and their decontamination is significantly affected by bioavailability. Severely polluted soils exhibit a distinct non-aqueous-phase liquid (NAPL) (Paul et al. 2006) that is usually present as droplets or films on soil particles. Bioavailability of hydrocarbons has been improved by using surfactants, e.g. Triton-X 100, Tergitol NPX, Brij 35, Tween 20, etc. (Lawniczak et al. 2013; Ward 2010). These are amphiphilic molecules that reduce the surface tension of the system and reduce free energy causing easy dispersion of the molecules in the matrix, thereby making them suitable solvents for dissolving and flushing down relatively insoluble contaminants from soil (Nguyen-Ngoc and Tran-Minh 2007). While some research groups have reported that the use of surfactants enhance biodegradation (Aronstein and Alexander 1993; Tiehm 1994; Volkering et al. 1995a, b; Churchill et al. 1995; Lantz et al. 1995), others report that these surfactants inhibit biodegradation (Laha and Luthy 1992; Tiehm 1994; Grimberg and Aitken 1995; Churchill et al. 1995; Wilson et al. 1995); therefore, the use of surfactants to increase the bioavailability is quite contradictory.

Discrepancy about the use of biosurfactants makes it imperative to find alternative techniques for increasing the bioavailability of contaminants. Uses and importance of microbial biofilms have been assessed as an alternate strategy to improve the bioavailability of contaminants to microbes. Apart from naturally occurring biofilms, microbial biofilms may be developed by suitably immobilizing the cells on specifically designed/fabricated surfaces, in "artificial systems." The potential of biofilm communities for bioremediation processes apart from industrial applications has been realized (Davey and O'Toole 2000; Decho 2000; Paul et al. 2006) as a prescient, effective, and safer alternative to using free cells. "Naturally occurring" biofilms comprising of a consortia made of bacteria and microalgae, capable of hydrocarbon degradation were exhibited on gravels found in the intertidal coastal region of the Arabian Gulf (Radwan et al. 2002). The occurrence and success of such natural biofilms encouraged their application in industries to facilitate the biodegradation of pollutants (Lazarova and Manem 2000).

The advantages of biofilms for improving bioavailability thus motivated further research to develop enhanced carrier material to create "artificial" biofilms using suitable surfaces that can form a thin film (Lazarova and Manem 2000). This article describes the immobilization of microbes as biofilms on various surfaces including synthetic polymers, fibers, resins, and nanomaterials which can be used as surfaces for bacterial adhesions.

## 11.3 Immobilization of Cells

### *11.3.1 Matrices*

Japan, USA, and few other countries successfully utilized microbial cells in immobilized state for adsorbing heavy metals from solutions (Gupta and Mohapatra 2003; Kamaludeen et al. 2003), for purification of sewage (Hunt et al. 2008) and for strengthening bio-based technologies to produce antibiotics, fermented products, organic acids, etc. (Akin 1987). Natural (activated C) and synthetic polymeric matrices have been screened and used for immobilizing microbial cells. Takata et al. (1977) tried natural polymers, e.g. κ-carrageenan, furcellaran, sodium and calcium alginate, ethyl succinylated cellulose, etc., to immobilize the cells of *Streptomyces phaeochromogenes;* κ-carrageenan was reported most suitable amongst all. Alginate from *Sargassum sinicolawas* (macroalgae) was for co-immobilization of *Chlorella sorokiniana* (microalgae) and *Azospirillum brasilense* (bacterium) for waste water treatment (Yabur et al. 2007)

### 11.3.2 Methods of Immobilization

New techniques for immobilization include the "Sol-gel entrapment method" (early 1990s) showed advantages over conventional immobilization methods. The Sol–gel entrapment method was tested for immobilization of atrazine-degrading *Pseudomonas* ADP in aqueous environments. The negative aspect of encapsulation was growth limitations due to nutrient and $O_2$ unavailability causing rapid aging of cells leading to impaired activity. Degradation rates could be restored by supply and amendment of nutrients to the trapped bacterium (Rietti-Shati et al. 1996). Innovative studies on utilizing various immobilization matrices were developed by researchers to increase degradation rates and enhance efficiency. Drizit (a matrix designed for immobilization) was successfully tested for bioremediation of oil contaminated saltwater using immobilized cells over free cells. Radwan et al. (2002) discovered immobilized oil-utilizing bacterial biofilms coated on microalgae found in the Gulf coast. This naturally induced method of bacterial immobilization protected the cells from getting washed-away or diluted and also provided them with a niche containing sufficient oxygen, nitrogenous/phosphorus, and vitamins. Kim et al. discovered that calcium alginate immobilized *Pseudomonas putida*MK1 (KCTC 12283) cells were better protected from co-contaminants (phenol) during the biodegradation of pyridine as compared to free cells.

### 11.3.3 Mechanisms of Attachment

Designing matrices for bioremediation applications requires an understanding of mechanisms involved in microbial attachments to various surfaces. Normally, during the formation of a biofilm, cells follow a particular pattern while making contact with the surface and with each other (Fig. 11.2). Van Loosdrecht et al. (1987) indicated that attachment increases as both negative charge (electrophoretic mobility) and hydrophobicity (contact angle) increases. The surface hydrophobicity of the matrix surface can be altered by acetylation and phosphorylation as suggested by Olivia et al. They also suggested that increasing hydrophilicity subsequently decreases the number of bacterial adhesions. Surface roughness is another parameter that governs the adhesion of cells to various matrices, especially in the cases where size of irregularities on the surface is similar to that of bacteria. The irregularities/ roughness much smaller than a bacterial cell may not impede initial attachment. The number of bacteria adhering to the etched surface was observed to increase by a factor of three according to Dineva et al. (2008).

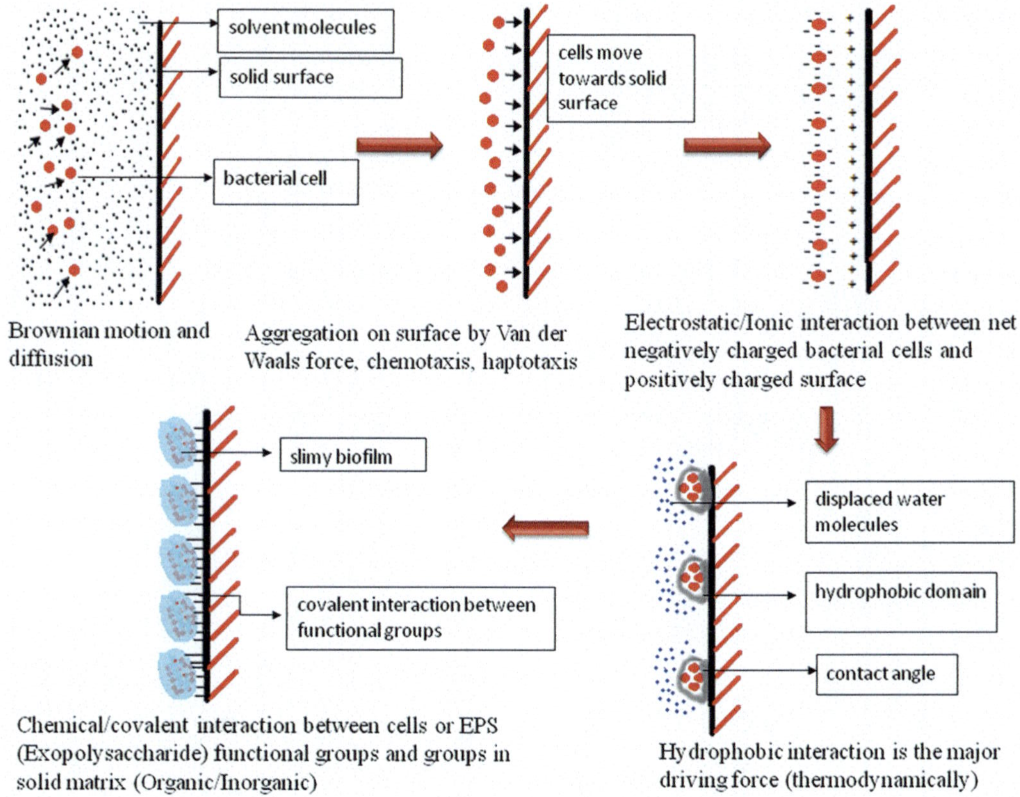


**Fig. 11.2** General mechanism of interaction of bacterial cells to a solid surface during formation of biofilms

## 11.4 Nanobioremediation: Nanomaterials for Microbial Immobilization to Aid Bioremediation

Nanomaterials (measuring 1–100 nm in dimension) have gained attention due to their unusual properties that are more advantageous as compared to their bulk counterparts (Daniel and Astruc 2004; Kato 2011). Nanomaterials are useful for electronic, optoelectronic, biomedical, pharmaceutical, and cosmetic applications, and they demonstrate significant prospects for environmental applications (Table 11.1). It is well-known that nanoparticles interact with bacterial cells or the community as a whole and enhance bioremediation by modifying the pathway(s) or substrate specificity of the bacterial cells. Microbial cells can biosynthesize nanoparticles under suitable conditions by enzymatically converting target metal ions from their environment into elemental metal. Synthesis of extracellular gold nanoparticles was possible using fungal cells of *Fusarium oxysporum* and actinomycete *Thermomonospora* sp., (Ahmad et al. 2003; Mukherjee et al. 2001), whereas *Verticillium* sp. produced intracellular gold nanoparticles (Mukherjee et al. 2002).

Nanomaterials not only aid in parental remediation by microbes, but some of them exhibit primary remediation properties, i.e. "nanoremediation." Recently zeolites, metal oxides, carbon nanotubes and fibers, noble metals [especially in the form of bimetallic nanoparticles (BNPs)], and titanium dioxide have been explored for remediation; of which, nanoscale zerovalent iron (nZVI) has been widely accepted (Lowry and Johnson 2004).

The search for suitable matrices led to "nanomaterials" that promised distinct advantages over other polymeric matrices for immobilization of cells for bioremediation. Magnetite ($Fe_3O_4$) nanoparticles mixed with magnetic gellan gum beads using traditional entrapment method for a carbazole-degrading strain *Sphingomonas* sp. XLDN2-5 resulted in poor substrate availability to the bacteria due to stearic hindrance. However, use of the same $Fe_3O_4$ magnetite nanoparticles for the immobilization of *Pseudomonas delafieldii* for the degradation of dibenzothiophene showed promising results (Shan et al. 2005).

Microbial cells are immobilized by entrapment or adsorption on nanomatrices, but the limitation of entrapment is limited diffusion and problems associated with mass transfer as discussed before. Table 11.1 shows various matrices and nanomaterials used for the immobilization of microbial cells for diverse applications.

Research has shown that etched iron nanomaterial, surface acetylated and coated with polyethyleneimine (cationic polymer) increases positive surface charge which is very strong to increase the adhesion of bacterial cells with net surface negative charge (D'Souza and Kamath 1988; Hsu et al. 2004; Chu et al. 2009). D'Souza and Kamath (1988) reported that immobilized yeast cells on a cotton cloth coated with PEI showed increased stability and higher enzyme activity. In contrast, Hsu et al. (2004) compared PEI-coated cotton and untreated cotton for its immobilization efficiency and xanthan production and reported that untreated cotton worked better. Later on Chu et al. (2009), reported stable immobilization of bioluminescent *E. coli* cells on PEI (0.667%) treated-viscose fiber including an increase in sensing time,

**Table 11.1** Different types of matrices and nanomaterials used for microbial immobilization for various applications

| S. No | Attachment matrix | Material | Strain/organism | Mode of attachment | Application | References |
|---|---|---|---|---|---|---|
| 1. | Alginate microbeads and silica membrane | Copolymer | *Saccharomyces cerevisiae*, *Oenococcus oeni* | Diffusion and entrapment | L-malic fermentation | Callone et al. (2008) |
| 2. | Lentikats® | Hydrogel based PVA | *Oenococcus oeni* | Entrapment | Malolactic fermentation | Rodríguez-Nogales et al. (2013) |
| 3. | Alginate beads | Biopolymer | *Acinetobacter junii*, *E. coli* and *Bacillus subtilis* | Adsorption | Adsorptive removal of Cr(VI) | |
| 4. | Sol–gel matrix | Gel matrix | *Aquincola tertiaricarbonis* L108 | Freeze-gelation | Degradation of recalcitrants (fuel oxygenates—Methyl *tert*-butyl ether (MTBE) and ethyl *tert*-butyl ether (ETBE)) | Pannier et al. (2010) |
| 5. | Tetraethylorthosilicate (TEOS+ heteropolysaccharide (AHPS) from the red microalga *Dixonella grisea*) | Hybrid sol–gel nanomaterial | Fungal (*Humicola lutea*) and bacterial (*Bacillus* sp.) cells | Entrapment | α-Galactosidase and Nitrilase activity | Djambaski et al. (2009) |
| 6. | Tetraethylorthosilicate (TEOS) and tetramethylorthosylicate (TMOS) | Sol-gel hybrid nanomaterial | Thermophilic bacterial strain UG-5B | Entrapment (100%) | Benzonitrile degradation | Kaibanova et al. (2006) |
| 7. | $Fe_3O_4$ nanoparticles (magnetite) | Supermagnetic nanoparticle | *Pseudomonas delafieldii* | Adsorption | Dibenzothiophene Degradation | Shan et al. (2005) |
| 8. | SWCNT (Applied Science Innovations Pvt. ltd., India) | Carbon nanotube (single walled) | *Escherichia coli* | Adsorption | Biosensor | Choudhury et al. (2012) |
| 9. | Iron nanoparticles on active carbon | Hybrid nanomaterial | *Shewanella putrefaciens* | Adsorption | Uranium-contaminated effluent (EF-03) | Baiget et al. (2013) |
| 10. | Pristine multi-wall carbon nanotubes | Hybrid nanomaterial | *Pseudomonas fluorescens* | Non-covalent bonding | Enantioselective transesterification in the industrial applications | Boncel et al. (2013) |

quicker response, and improved reproducibility of signals (Chu et al. 2009). These studies suggest that there is further need to explore and exploit various matrices for immobilization and also select for a proper procedure that would allow high productivity combined with extended operational stability.

Li et al. constructed an effective alternative, a "biocomposite" by accumulating nanoparticles of Iron (III) oxide on *Sphingomonas* sp.XLDN2-5 cells with 45.5 emu $g^{-1}$ saturation magnetization. The biocomposite comprising of microbial cell and Iron (III) oxide retained its ability for biodegradation and also bestowed the power of reusability as compared to free cells. Moreover, the microbial cell/$Fe_3O_4$ biocomposite could be easily separated and recycled by an external magnetic field due to the super-paramagnetic properties of $Fe_3O_4$ nanoparticle coating. The efficiency and adsorption characteristics of nanostructured carbonized material for microbial cells were analyzed and it was concluded that successful immobilization of microbial cells was a result of electrostatic and hydrophobic properties of the surface. The attachment of microbial cells to carbonized surfaces is mediated by a range of interactions and the immobilization process of biocomposites is a fine interplay of such versatile interactions leading to adaptation of biocomposite for various.

Another interesting study on carbon nanotubes (CNT) showed that they possess rare properties, e.g. miniscule size to high aspect ratio ($>$1000), and a wide range of electrical properties (Baughman et al. 2002). Adsorption of *E. coli* cells on different carbon nanoforms, including single walled carbon nanotubes (SWCNT), multi-walled carbon nanotubes (MWCNT), graphite, and mixed fullerene was studied. Diffusion of *E. coli* cells in SWCNT was determined to be the highest, that is about 3 times of MWCNT, double than for graphite and also approximately 1.5 times more than fullerene aggregates. These experiments proved CNT to be the best candidate for stable microbial adsorption (Choudhury et al. 2012). Several other studies on remediation of pollutants using nanomaterials in combination with biological agents have been tabulated in Table 11.2.

Utilizing nanoparticles in combination with microbes has the potential to produce suitable alternatives for wastewater treatment (He et al. 2017). Due to their small size the nano carbon particles are not susceptible to steric hindrance between the substrate (pollutant) and cells and might also allow stronger immobilization and decrease chances of cell wash out. Carbon nanoparticles coated with PEI would provide stronger binding by increasing the positive charges (Fig. 11.3). Singh et al. (2013) demonstrated degradation of γ-HCH in soil (99% cleanup in 6 days) using an integrated use of nano CMC-Pd/Fe and *Sphingomonas* sp. strain NM05. The nanoparticles promoted the growth of the cells and also increased the degradation rate of γ-HCH by approx. 1.7–2.1 times with respect to the controls, i.e. system containing either cells or the nanomaterial alone. $Fe_3O_4$/biochar nano-composites loaded with photosynthetic bacteria for decreasing the COD of waste water (He et al. 2017) and PVDF/$TiO_2$ nanocomposite membranes for immobilizing algal cells in A-MBRs (algal-membrane bioreactors) for preventing fouling of membranes while treating wastewater (Anjum et al. 2016), and bacterial cells magnetically immobilized on iron oxide nanoparticles for environmental remediation and other

**Table 11.2** The nanomaterials used for bioremediation purpose

| S. no. | Nanomaterial | Chemical/microbial process involved | Pollutant | References |
|---|---|---|---|---|
| 1 | Pd/Fe(0) bimetallic nanoparticle (CMC-Pd/nFe(0)) | *Sphingomonas sp. strain* NM05 | γ-HCH degradation | Singh et al. (2013) |
| 2 | Nano-sized Fe oxides | Anaerobic microorganisms | Several pollutants including heavy metals | Braunschweig et al. (2013) |
| 3 | Nano-porous silica beads | Microbial laccase enzyme | 2,4-dinitrophenol | Dehghanifard et al. (2013) |
| 4 | Nano-silica (NSi) | *Fusarium verticillioides* | Water treatment | Mahmoud et al. (2013) |
| 5 | Nano-sized magnetite particles | Chemical process | Removal of Cr (VI) | Nethaji et al. (2013) |
| 6 | Magnetite nanoparticles | *Rhodococcus erythropolis FMF and R. erythropolis IGTS8* | Desulfurization activity | Baughman et al. (2002) |
| 7 | Nano-copper (II) oxide and nanomagnesium oxide | Biochemical process | Activated sludge treatment | Liu and Wang (2012) |

applications (Ranmadugala et al. 2018) are promising upcoming technologies demonstrating the use of nanomaterials for decontamination and treatment of pollutants.

## 11.5 Conclusion and Future Perspectives

In the field of bioremediation, the application spectrum for immobilized cells is very large extending from the detection of toxic compounds, to waste water treatment, adsorption of heavy metals, etc. Cells living in biofilms exhibit better adaptation and survivability (during stress periods) due to the protection offered by matrices. Once formed, biofilms influence the degradation of various compounds due to their innate tendency to absorb water and inorganic/organic compounds (Paul et al. 2006), and this property may be exploited for developing suitable bioremediation technologies. The effective use of various types of matrices for artificially established biofilms is currently under research where, nanomaterials have proved their worth. The special properties of nanoparticles including particle aggregation, photoemission, electrical/heat conductivities along with desirable chemical properties (catalytic activity) have promoted their utility in drug delivery, diagnostics, whole cell biosensors, biochemical engineering apart from environmental clean up. Research has shown that addition of cationic polymers such as PEI or activated charcoal to nanomaterials

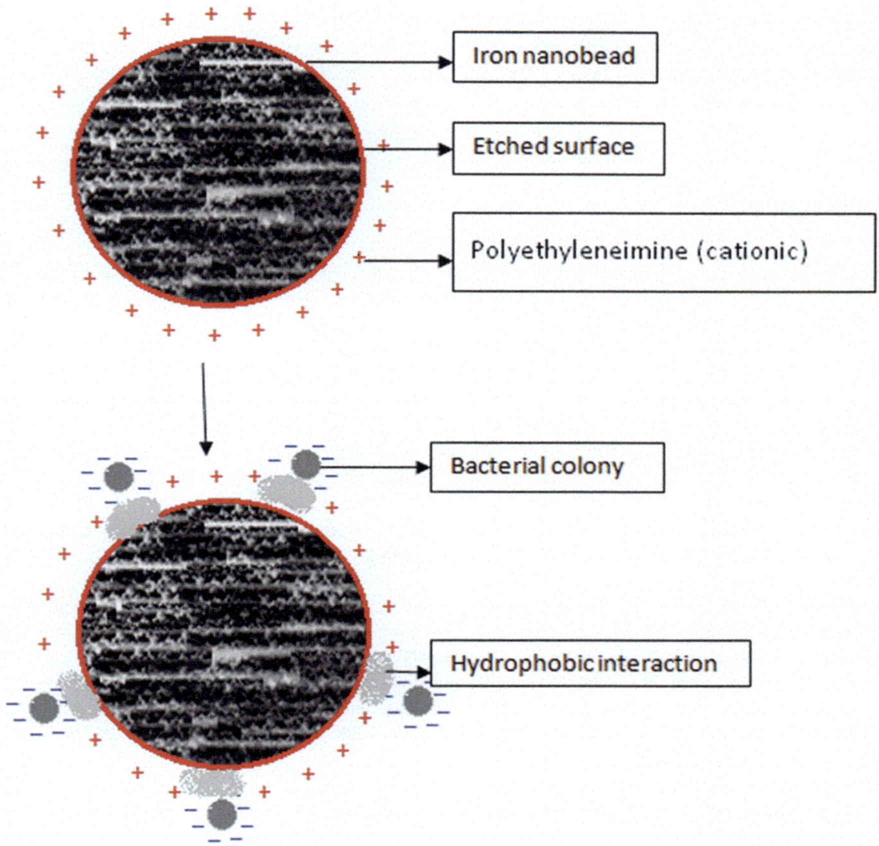


**Fig. 11.3** Concept for improved bioremediation using PEI in association with iron nanomaterials for effective microbial attachment

efficiently increases cellular attachment and colonization on the matrix. There are potential challenges to use nanomaterials alone for bioremediation owing to their tendency to form aggregates when used in-situ, causing soil particles to deposit and impeding transport across soil pores to reach the contaminated zones (Cecchin et al. 2017). Cost of the nanomaterials and toxicity also question the use of such constituents for bioremediation. However, the sustainability, reproducibility, reusability, and efficiency of the system are tremendously enhanced in cases where nanomaterials have been used as matrices for microbial attachment during bioremediation (He et al. 2017; Ranmadugula et al. 2018; Anjum et al. 2016; Singh et al. 2013). Few studies on nanobioremediation state the application of nanoparticles (e.g., nZVI) for removal of the pollutant and further treat with a biological agent (bacteria) and the combined effort (abiotic and biotic treatment) yielded enhanced degradation and the nanoparticles used were not toxic to the microbes in question

(Cecchin et al. 2017). Further studies are required to develop novel techniques for the efficient removal of pollutants using a combination of microbial potential along with nanoparticles.

## References

Ahmad A, Senapati S, Khan MI, Kumar R, Sastry M (2003) Extracellular biosynthesis of monodisperse gold nanoparticles by a novel extremophilic actinomycete, *Thermomonospora* sp. Langmuir 19(8):3550–3553

Akin C (1987) Biocatalysis with immobilized cells. Biotechnol Genet Eng Rev 5:319–367

Anjum M, Miandad R, Waqas M, Gehany F, Barakat MA (2016) Remediation of wastewater using various nano-materials. Arab J Chem 12:4897–4919. https://doi.org/10.1016/j.arabjc.2016.10.004

Aronstein BN, Alexander M (1993) Effect of a non-ionic surfactant added to the soil surface on the biodegradation of aromatic hydrocarbons within the soil. Appl Microbiol Biotechnol 39:386–390

Baiget M, Constantí M, López MT, Medina F (2013) Uranium removal from a contaminated effluent using a combined microbial and nanoparticle system. New Biotechnol 30:788–792. https://doi.org/10.1016/j.nbt.2013.05.003

Baughman RH, Zakhidov AA, Heer WA (2002) Carbon nanotubes- the route toward applications. Science 297(5582):789–792

Boncel S, Zniszczoł A, Szymańska K, Mrowiec-Białoń J, Jarzębski A, Walczak KZ (2013) Alkaline lipase from *Pseudomonas fluorescens* non-covalently immobilised on pristine versus oxidised multi-wall carbon nanotubes as efficient and recyclable catalytic systems in the synthesis of Solketal esters. Enzyme Microb Technol 53(4):263–270. https://doi.org/10.1016/j.enzmictec.2013.05.003

Braunschweig J, Bosch J, Meckenstock RU (2013) Iron oxide nanoparticles in geomicrobiology: from biogeochemistry to bioremediation. Nat Biotechnol 30(6):793–802. https://doi.org/10.1016/j.nbt.2013.03.008

Callone E, Campostrini R, Carturan G, Cavazza A, Guzzon R (2008) Immobilization of yeast and bacteria cells in alginate microbeads coated with silica membranes: procedures, physico-chemical features and bioactivity. J Mater Chem 18:4839–4848

Cecchin I, Reddy KR, Thome A, Tessaro EF, Schnaid F (2017) Nanobioremediation: Integration of nanoparticles and bioremediation for sustainable remediation of chlorinated organic contaminants in soils. Int Biodeterior Biodegrad 119:419–428

Choudhury MSH, Udding MA, Hasan MZ, Alam MS, Rashid MM (2012) Adsorption kinetics of *Escherichia Coli* on different carbon nanoforms. Carbon Sci Tech 4(1):161–166

Chu YF, Hsu CH, Pavan KS, Lo YM (2009) Immobilization of bioluminescent *Escherichia coli* cells using natural and artificial fibers treated with polyethylenimine. Bioresour Technol 100:3167–3174

Churchill PF, Dudley RJ, Churchill SA (1995) Surfactant-enhanced bioremediation. Waste Manag 15:371–377

D'Souza SF, Kamath N (1988) Cloth bioreactor containing yeast cells immobilized on cotton cloth using polyethylenimine. Appl Microb Biotechnol 29:136–140

Daniel MC, Astruc D (2004) Gold nanoparticles: assembly, supramolecular chemistry, quantum-size-related properties, and applications toward biology, catalysis, and nanotechnology. Chem Rev 104(1):293–346

Davey ME, O'Toole GA (2000) Microbial biofilms: from ecology to molecular genetics. Microbiol Mol Biol Rev 64:847–867

Decho AW (2000) Microbial biofilms in intertidal systems: an overview. Cont Shelf Res 20:1257–1273
Dehghanifard E, Jonidi Jafari A, Rezaei Kalantary R, Mahvi AH, Faramarzi MA, Esrafili A (2013) Biodegradation of 2,4-dinitrophenol with laccase immobilized on nano-porous silica beads. Iran J Environ Health Sci Eng 10(1):25. https://doi.org/10.1186/1735-2746-10-25
Dineva NM, Wang J, Mocanasu RC, Stoddart PR, Crawford RJ, Ivanova EP (2008) Impact of nano-topography on bacterial attachment. Biotechnol J 3:536–544
Djambaski P, Aleksieva P, Emanuilova E, Chernev G, Spasova D, Nacheva L, Kabaivanova L, Salvado IM, Samuneva B (2009) Sol-gel nanomaterials with algal heteropolysaccharide for immobilization of microbial cells, producing α-galactosidase and nitrilase. Biotechnol Biotech Eq 23(2):1270–1274
Dua M, Singh A, Sethunathan N, Johri AK (2002) Biotechnology and bioremediation: successes and limitations. Appl Microbiol Biotechnol 59:143–152
Ghosal D, Ghosh S, Dutta TK, Ahn Y (2016) Current state of knowledge in microbial degradation of polycyclic aromatic hydrocarbons (PAHs): a review. Front Microbiol 7:1369. https://doi.org/10.3389/fmicb.2016.01369
Grimberg SJ, Aitken MD (1995) Biodegradation kinetics of phenanthrene solubilized in surfactant micelles. In: Hinchee RE, Brockman FJ, Vogel CM (eds) Microbial processes for bioremediation. Battelle Press, Columbus, OH, pp 59–66
Guieysse B, Viklund G, Toes AC, Mattiasson B (2004) Combined UV biological degradation of PAHs. Chemosphere 55:1493–1499
Gupta R, Mohapatra H (2003) Microbial biomass: an economical alternative for removal of heavy metals from waste water. Indian J Exp Biol 41(9):945–966
He S, Zhong L, Duan J, Feng Y, Yang B, Yang L (2017) Front Microbiol 8:823
Hsu CH, Chu YF, Argin-Soysal S, Hahm TS, Lo YM (2004) Effect of surface characteristics and xanthan polymers on the immobilization of *Xanthomonas campestris* to fibrous matrices. J Food Sci 69:E441–E448
Hunt PG, Matheny TA, Ro KS, Stone KC, Vanotti MB (2008) Denitrification of agricultural drainage line water via immobilized denitrification sludge. *J Environ Sci Health A Tox Hazard Subst Environ Eng* 43(9):1077–1084
Kaibanova L, Dobreva E, Emanuilova E, Chernev G, Samuneva B, Salvado I (2006) Synthesis, structure and application of new hybrid nanomaterials for cell immobilization. Minerva Biotecnol 18(1):23–29
Kamaludeen SP, Arunkumar KR, Avudainayagam S, Ramasamy K (2003) Bioremediation of chromium contaminated environments. Indian J Exp Biol 41(9):972–985
Kato H (2011) In vitro assays: tracking nanoparticles inside cells. Nat Nanotechnol 6(3):139–140
Laha S, Luthy RG (1992) Effects of nonionic surfactants on the solubilization and mineralization of phenanthrene in soil-water systems. Biotechnol Bioeng 40:1367–1380
Lantz S, Lin JE, Mueller JG, Pritchard PH (1995) Effects of surfactants on fluoranthene mineralization by *Sphingomonas paucimobilis* strain EPA 505. In: Hinchee RE, Brockman FJ, Vogel CM (eds) Microbial processes for bioremediation. Battelle Press, Columbus, OH, pp 7–14
Lawniczak L, Marecik R, Chrzanowski L (2013) Contributions of biosurfactants to natural or induced bioremediation. Appl Microbiol Biotechnol 97:2327–2339
Lazarova V, Manem J (2000) Innovative biofilm treatment technologies for water and wastewater treatment. In: Bryers JD (ed) Biofilms II-process analysis and applications. Wiley, New York, pp 159–206
Lewis TA, Newcombe DA, Crawford RL (2004) Bioremediation of soils with explosives. J Environ Manag 70:291–307
Liu G, Wang J (2012) Effects of nano-copper(II) oxide and nanomagnesium oxide particles on activated sludge. Water Environ Res 84(7):569–576
Lowry GV, Johnson M (2004) Congener-specific dechlorination of dissolved PCBs by microscale and nanoscale zerovalent iron in a water/methanol solution. Environ Sci Technol 38 (19):5208–5216

Mahmoud ME, Yakout AA, Abdel-Aal H, Osman MM (2013) Immobilization of *Fusarium verticillioides* fungus on nano-silica (NSi-Fus): a novel and efficient biosorbent for water treatment and solid phase extraction of mg(II) and ca(II). Bioresour Technol 134:324–330. https://doi.org/10.1016/j.biortech.2013.01.171

Mukherjee P, Ahmad A, Mandal D, Senapati S, Sainkar SR, Khan MI et al (2001) Bioreduction of $AuCl_{(4)}(-)$ ions by the fungus *Verticillium* sp and surface trapping of the gold nanoparticles formed. Angew Chem Int Ed Engl 40:3585–3588

Mukherjee P, Senapati S, Mandal D (2002) Extracellular synthesis of gold nanoparticles by the fungus *Fusarium oxysporum*. Chembiochem 3(5):461–463

Nethaji S, Sivasamy A, Mandal AB (2013) Preparation and characterization of corn cob activated carbon coated with nano-sized magnetite particles for the removal of Cr(VI). Bioresour Technol 134:94–100. https://doi.org/10.1016/j.biortech.2013.02.012

Nguyen-Ngoc H, Tran-Minh C (2007) Sol-gel process for vegetal-cells encapsulation. Mat Sci Eng C 27:607–611

Pannier A, Oehm C, Fischer AR, Werner P, Soltmann U, Böttcher H (2010) Biodegradation of fuel oxygenates by sol–gel immobilized bacteria *Aquincola tertiaricarbonis* L108. Enzyme Microb Tech 47(6):291–296

Paul D, Pandey G, Jain RK (2005b) Suicidal genetically engineered microorganisms for bioremediation: need and prospective. BioEssays 27:563–573

Paul D, Pandey G, Pandey J, Jain RK (2005a) Accessing microbial diversity for bioremediation and environmental restoration. Trends Biotechnol 23:135–142

Paul D, Singh R, Jain RK (2006) Chemotaxis of *Ralstonia* sp. SJ98 towards *p*-nitrophenol in soil. Environ Microbiol 8:1797–1804

Pieper DH, Reineke W (2000) Engineering bacteria for bioremediation. Curr Opin Biotechnol 11:262–270

Radwan SS, Al-Hasan RH, Salamah S et al (2002) Bioremediation of oily sea water by bacteria immobilized in biofilms coating microalgae. Int Biodet Biodeg 50:55–59

Ranmadugala D, Ebrahiminezhad A, Manley-Harris M, Ghasemi Y, Berenjian A (2018) Magnetic immobilization of bacteria using iron oxide nanoparticles. Biotechnol Lett 40:237–248. https://doi.org/10.1007/s10529-017-2477-0

Rietti-Shati M, Ronen D, Mandelbaum RT (1996) Atrazine degradation by *Pseudomonas* strain ADP entrapped in sol-gel glass. J Sol-Gel Sci Technol 7(1-2):77–79

Ritter WF (1990) Pesticide contamination of ground water in the United States—a review. J Environ Sci Health B 25(1):1–29

Rodríguez-Nogales JM, Crespo JV, Fernández-Fernández E (2013) Immobilization of *Oenococcus oeni* in lentikats® to develop malolactic fermentation in wines. Biotechnol Progress 29:60–65

Shan G, Xing J, Zhang H, Liu H (2005) Biodesulfurization of dibenzothiophene by microbial cells coated with magnetite nanoparticles. Appl Environ Microbiol 71:4497–4502

Singh R, Manickam N, Mudiam MK, Murthy RC, Misra V (2013) An integrated (nano-bio) technique for degradation of $\gamma$-HCH contaminated soil. J Hazard Mater 258-259:35–41. https://doi.org/10.1016/j.jhazmat.2013.04.016

Singh R, Paul D, Jain RK (2006) Biofilms: implications in bioremediation. Trends Microbiol 14:389–397

Takata I, Tosa T, Chibata I (1977) Screening of matrix suitable for immobilization of microbial cells. J Solid Phase Biochem 2:225–236

Tiehm A (1994) Degradation of polycyclic aromatic hydrocarbons in the presence of synthetic surfactants. Appl Environ Microbiol 60:258–263

Van Loosdrecht MCM, Lyklema J, Norde W, Schraa G, Zehnder AJB (1987) Electrophoretic mobility and hydrophobicity as a measure to predict the initial steps of bacterial adhesion. Appl Environ Microbiol 53:1898–1901

Volkering F, Breure AM, Van Andel JG, Rulkens WH (1995b) Influence of nonionic surfactants on bioavailability and biodegradation of polycyclic aromatic hydrocarbons. Appl Environ Microbiol 61:1699–1705

Volkering F, Van de Wiel R, Breure AM, Van Andel JG, Rulkens WH (1995a) Biodegradation of polycyclic aromatic hydrocarbons in the presence of nonionic surfactants. In: Hinchee RE, Brockman FJ, Vogel CM (eds) Microbial processes for bioremediation. Battelle Press, Columbus, OH, pp 145–151

Ward OP (2010) Microbial biosurfactants and biodegradation. Adv Exp Med Biol 672:65–74

Wilson BH, Hutchins SR, West CC (1995) Surfactant use with nitrate-based bioremediation. In: Hinchee RE, Brockman FJ, Vogel CM (eds) Microbial processes for bioremediation. Battelle Press, Columbus, OH, pp 25–31

Yabur R, Gustavo YB, Carmona H (2007) Alginate from the macroalgae *Sargassum sinicola* as a novel source for microbial immobilization material in waste water treatment and plant growth promotion. J Appl Phycol 19:43–53

# Chapter 12
# Recent Advances in Microbial Remediation Techniques for Xenobiotics-Polluted Soil

**Naga Raju Maddela, Marcos Raúl Heredia Pinos, Chizoba Ignatius Ezugwu, Kondakindi Venkateswar Reddy, and Pabbati Ranjit**

**Abstract** Rapid industrialization, growth in the human population combinedly lead to bulk release of xenobiotic substances (e.g. drugs, food additives, hydrocarbons, pesticides and personal care products) into the environment. Global market projection rates strongly indicate that environmental burden posed by xenobiotics is in increasing trend. Thus, there is a high level of global concern over these environmental pollutants because of their human toxicity, ecotoxicity and long-term persistence in the environment. One of the immediately available and the most viable solutions for the removal of xenobiotics from environmental media (soil and water) is microbiologically mediated approach, called 'bioremediation'. Bioremediation offers undisputable benefits in the restoration of contaminated sites, in terms of cost, technology, operation. Keeping in view of the advantages of bioremediation, this chapter has been designed to address the emerging techniques in the area of bioremediation, and we are in opinion that such insights will decrease the gap between laboratory and field-level execution of bioremediation. Besides these, information related to the occurrence, impact and fate of xenobiotics in the environment, role of microorganisms in the degradation of xenobiotics has also been included in this chapter. Towards the end, future directions of research in microbial

N. R. Maddela (✉)
Departamento de Ciencias Biológicas, Facultad la Ciencias de la Salud, Universidad Técnica de Manabí, Portoviejo, Ecuador

Instituto de Investigación, Universidad Técnica de Manabí, Portoviejo, Ecuador
e-mail: raju.maddela@utm.edu.ec

M. R. H. Pinos
Facultad de Ingeniería Agropecuaria, Universidad Laica Eloy Alfaro de Manabí, Chone, Ecuador

C. I. Ezugwu
Department of Chemical Engineering, University of Alcala, Madrid, Spain

K. V. Reddy · P. Ranjit
Centre for Biotechnology, Institute of Science & Technology, Jawaharlal Nehru Technological University, Hyderabad, Telangana, India

 
P. K. Arora (ed.), *Microbial Products for Health, Environment and Agriculture*, Microorganisms for Sustainability 31, https://doi.org/10.1007/978-981-16-1947-2_12

removal of xenobiotics from the environment has been discussed. Overall, this chapter can be a single source of information to understand the future threat posed by xenobiotics and how to combat this problem through microbiological treatments.

**Keywords** Bioremediation · Xenobiotics · Petroleum hydrocarbons · Emerging approaches · Microorganisms

## Abbreviations

| | |
|---|---|
| BES | Bioelectrochemical system |
| CAGR | Compound annual growth rate |
| CAT | Catalase |
| DDD | Defined daily dose |
| DDT | Dichlorodiphenyltrichloroethane |
| DIMS | Direct injection mass spectrometry |
| EPSPS | 5-enolpyruvylshikimate-3-phosphate synthase |
| FAO-STATS | Food and Agriculture Organization of the United Nations Statistical Database |
| FISH | Fluorescence in situ hybridization |
| FT-IR | Fourier-transform infrared spectroscopy |
| GC | Gas chromatography |
| GPx | Glutathione peroxidase |
| GST | Glutathione S-transferase |
| HPLC | High performance liquid chromatography |
| HRP | Horseradish peroxidase |
| IUWS | Integrated Urban Wastewater System |
| Lac | Laccase |
| MFC | Microbial fuel cell |
| MP | Micro pollutants |
| MS | Mass spectroscopy |
| mT | million tons |
| NMR | Nuclear magnetic resonance |
| N-P-K | Nitrogen-phosphorous-potassium |
| OCI | Organochlorine insecticides |
| PHs | Petroleum hydrocarbons |
| PPCPs | Pharmaceutical and personal care products |
| qPCR | quantitative polymerase chain reaction |
| SIP | Stable isotope probing |
| SOD | Superoxide dismutase |
| TPHs | Total petroleum hydrocarbons |
| UMBBD | The University of Minnesota Biocatalysis/Biodegradation database |
| WW | Wastewater |

## 12.1 Introduction

Any chemical substance which is foreign to the body or to an ecosystem is called a '*xenobiotic*'. Xenobiotic is not a naturally produced chemical, however, a natural compound can become a xenobiotic if it is taken up by another organism. This can happen (Mansuy 2013) when natural human hormones are entered into the fish body at the downstream of sewage treatment plant outfalls, or natural chemicals produced by some organisms as protection from the predators. However, the term 'xenobiotic' is very often used in the contest of environmental pollutants (e.g. dioxins, polychlorinated biphenyls) which are not basically found in nature before they are synthesized by humans. In Greek language, 'xenos' means foreigner or stranger, and 'bios' means life. There are several compounds that come under the category of xenobiotics, e.g. heavy metals, hydrocarbons, dyes, pharmaceuticals, petroleum hydrocarbons, pesticides, etc. (Dhiman et al. 2020).

There is a vast release of xenobiotics into the environment. For example, regarding antibiotics, a recent investigation has revealed that the antibiotic defined daily dose (DDD) and antibiotic consumption rates (DDDs per 1000 inhabitants per day) have been increased by 65% and 39%, respectively for the period of 2000 and 2015 (Klein et al. 2018). More importantly, this hike is mainly observed in the low- and middle-income countries. Key xenobiotic compounds that have been identified in this investigation were glycyclines, oxazolidinones, carbapenems, and polymyxins. If there are no strict regulations on the antibiotic consumption, then consumption rates in 2030 could be 200% higher than that in 2015 (Klein et al. 2018). Similarly, global pesticide consumption (https://ourworldindata.org/pesticides) has also been increased by 34% between 2000 (3.06 million tons) and 2017 (4.11 million tons). According to the market projection rates (https://www.thebusinessresearchcompany.com/report/pesticides-market), global pesticide market size is growing rapidly. Global pesticide market size value in 2019 was US $ 84.5 billion, and the compound annual growth rate (CAGR) for 2015 and 2019 was 4.2%, and CAGR is expected to reach 11.5% with a market size value $ 130.7 billion by 2023. Global market size of beauty and personal care products (Source: Fior Market, 24 January 2020) for the year 2018 was US $ 493.34 billion, this value is expected to increase to $ 756.63 billion by 2026 with a CAGR of 5.81% for the period of 2019–2026. Also, world oil consumption has been increased by 25% from 2000 (28.2 billion barrels per year) to 2016 (35.4 billion barrels per year) (https://www.worldometers.info/oil/). Though these values indicate the consumption and market project rates of different xenobiotic compounds in the world, ultimately they are reached and accumulated in different environmental media (such as soil, water and air) and damage not only the ecosystem but also human health.

One of the ecologically viable techniques to clean up the polluted sites is 'bioremediation', which utilizes the natural biological (mostly by microorganisms) activity for this purpose. Bioremediation approach is relatively cost effective, low-technology technique, can be carried out on site, has been successfully used to remove the xenobiotics for the polluted media (soil and water), and feasibly possible to achieve the limits set by global health and regulatory agencies (Gupta and Pathak 2020). While removing the pollutant by microorganisms (mainly bacteria and fungi) different forms reactions are possible, such as partial degradation, mineralization, co-metabolic reaction etc. (Gouma et al. 2014). But it should be remembered that, not in all cases, bioremediation is a preferable approach, as it is a relatively slow process. For example, in certain incidences, pollutants need to be removed quickly (e.g. petroleum hydrocarbons) from the polluted site as they rapidly damage the health of human and environment, therefore, under these circumstances, bioremediation is not an immediate preferable option. Nevertheless, bioremediation is still widely used approach in the remediation of xenobiotic polluted sites (Gouma et al. 2014), and this approach is being updated on regular basis with the emergence of new insights, such as discovery of new pollutant degraders, identifying the optimum reaction conditions, application of novel 'omics' tools and genetic engineering. One strong evidence for this, scientific production (number of articles published as tracked by using a keyword 'bioremediation' in ScienceDirect on 9th August 2020) in the 2010 and 2019 were 949 and 2524, respectively. These values strongly suggest that bioremediation is a highly preferable approach for the restoration of xenobiotic polluted sites.

In view of the environmental pollution caused by xenobiotics and importance of bioremediation in the cleanup of polluted sites, the present chapter is intended to provide following key issues such as distribution and fate of xenobiotics in the environment, microbial degradation of xenobiotics with a special emphasis on petroleum hydrocarbons, advances in the bioremediation technologies in different areas such as bioelectrochemical system, microbial treatments, rhizoremediation, algal based technologies, microbial enzymes and 'omics' tools. Finally, future directions of research have been suggested along with conclusions.

## 12.2 Distribution and Fate of Xenobiotics

### *12.2.1 Occurrence*

Growth in population, industrialization, industrial processes, agricultural activities collectively lead to ever increasing xenobiotics in the environment (refer review (Mudhoo et al. 2020)) (Fig. 12.1). Large industries such as pharmaceuticals, fossil fuels, pulp and paper bleaching and agriculture-based industries are the principal sources of xenobiotic contributors to the environment (Díaz 2004). Environmental concentrations of different xenobiotics can be at high (μg $L^{-1}$ to mg $L^{-1}$) or low (ng $L^{-1}$ to μg $L^{-1}$) (Meckenstock et al. 2015). Gradual accumulation and recalcitrance nature of xenobiotics make them a major environmental concern. For example, in agriculture, there is an extensive use of pesticides, and the global pesticide consumption per annum is $3 \times 10^9$ kg, which is worth of US$ 40 b (Hussain et al. 2009). According to FAO-STATS, the major pesticide consumed counties in 2016 were found to be China, USA, Brazil, and Argentina (Khalid et al. 2020). Similarly,

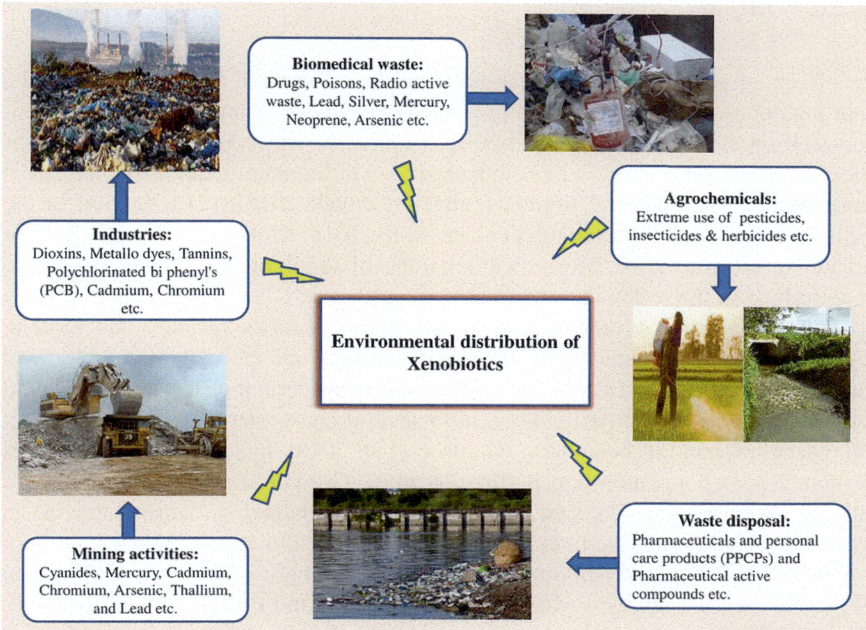


**Fig. 12.1** Occurrence of Xenobiotics in the environment (Embrandiri et al. 2016)

accumulation of microplastics in the ocean found to have significant threat to the biological systems in the aquatic system (Parthasarathy et al. 2019).

Similarly, in the terrestrial environments also, there is vast accumulation of plastic materials, attributed to the emergence of large usage of plastic bags and waste generated from the plastic packaging material (Jambeck et al. 2015). It is not surprising that the plastic film pollution in Chinese soil has been increased fourfolds in the last two decades (Liu et al. 2014), and the residues of plastic films accounts for the 10% of total area (Ramos et al. 2015). According to the estimations of trade association PlasticsEurope, the plastic production in the world has been increasing steadily, for example, plastic production (million tonnes, mT) in 1950, 2010 and 2018 were 1.5, 275 and 359, respectively (https://www.britannica.com/science/plastic-pollution). It is important to note that nearly 4.8–12.7 mT of plastics are thrown into the oceans. There were nearly, two million microplastics pieces per $m^2$ area in Tyrrhenian sea (Charles 2020). These insights clearly indicate the severity of the global plastic pollution, subsequently this situation demands to stiffen the regulations over the indiscriminate usage and disposal of plastic materials. Plastic materials are highly persistent, once they are released into the environment, they never go away easily, therefore, great threat pose to human health, groundwater, wildlife, food chain. Huan health disorders that are linked with the plastic pollution include birth defects, cancers, endocrine disruption, impaired immunity etc.

### 12.2.2 Impact

The effects of xenobiotics on human health and environment are very serious. The main intention of this section is to provide environmental damage posed by xenobiotics (Kuppusamy et al. 2020d, 2020f). The adverse effects of long-term accumulation of xenobiotics in the environment are – (1) biomagnification (in terrestrial plants/aquatic animals) and trophic transfer (Connell 2018), (2) reduction in the fertility of the soil (Maddela and Venkateswarlu 2018; Mohiddin et al. 2015; Raju and Venkateswarlu 2014). More specific effects of xenobiotics on the biotic system are described in the following para.

There are several adverse effects posed by xenobiotics on the plant system. Legume-rhizobium chemical signals are negatively affected by herbicides and fungicides (Eberbach 2018), this has a significant impact on the crop yield. It should also be remembered that pesticides exhibit several non-target effects, in this direction, pesticides can kill honeybees (Goulson et al. 2018), this subsequently reduces the honey bees population and pollination rates. Growth and developmental, and multiple toxicities were induced by Atrazine, desethylatrazine, hydroxyatrazine in *Arabidopsis thaliana* (Alberto et al. 2017). Glyphosate did adversely affect the root growth and 5-enolpyruvylshikimate-3-phosphate synthase (EPSPS) and shikimate pathway in *Fagopyrum esculentum* (Silva et al. 2019) and *Pouteria torta* (Rezende-Silva et al. 2019), respectively. A mixture of 17 PPCPs (pharmaceutical and personal care products) were known to negatively affect the root activity in cucumber plants (Sun et al. 2018). The other recent investigations (Alkimin et al. 2019; Bartucca et al. 2019; de Lima et al. 2017) revealed that xenobiotics (e.g. diclofenac, metribuzin, diuron) showed negative effects on physiological, biochemical parameters, chlorophyll content, concentration of phytosiderophores, wax production in different plant systems (e.g. *Lema minor*, *Zea mays* L, *Bauhinia variegata). These insights clearly suggest that the plant system is highly susceptible to xenobiotic pollution. Similarly, experiments in animals revealed that different xenobiotic compounds* (e.g. *doxycycline, dimethoate, chlorpyrifos, glyphosate-based herbicides) were known to adversely affected the juvenile's total number in earthworms* (Litskas et al. 2019), activities of enzymes (e.g. SOD, GST, GPx, CAT in freshwater amphipod i.e. *Gammarus pulex* (Serdar 2019), antioxidant enzymes and DNA in freshwater mussel i.e. *Viviparus bengalensis* (Al-Fanharawi et al. 2019), neurological and antioxidant system in freshwater fish i.e. *Rhamdia quelen* (Sobjak et al. 2017), neurons and other biomarkers in crustacean i.e. *Gammarus kischineffensis* (Demirci et al. 2018), growth and development in marbled crayfish i.e. *Procambarus fallax* f. *virginalis* (Velisek et al. 2017), etc. Also, there is several evidence for the adverse effects of xenobiotics in humans. For example, reduced mean red blood cells upon chronic exposure to airborn mercury (Faber et al. 2019), ROS-mediated neurotoxicity by paraquat, dieldrin, organochlorines and organophosphates (Yan et al. 2016). Exposure to pesticides also leads to childhood leukemia (Kumar et al. 2014), high risk of miscarriages, low birth weight, hypospadias, cryptorchidism and micropenis etc. (García et al. 2017). Furthermore, human exposure to pesticides is thought to be linked with Hodgkin's and non-Hodgkin lymphoma (Luo et al. 2016), Parkinson's disease (Brouwer et al. 2017), endocrine disruption (Mazur et al. 2015) etc. With these insights, it is clearly understood that xenobiotics have strong adverse effects on plants, animals and human beings, therefore, removal of xenobiotic-based pollutants from the environment and subsequent remediation of polluted sites is obligatory to protect the human health as well as to restore the naturality of the ecosystem.

### *12.2.3 Fate*

Terrestrial fate of xenobiotics is interesting (Fig. 12.2). Upon the release of xenobiotic molecules into the soil, they are mixed with the several thousands of natural organic molecules, some of them are biodegradable and some are recalcitrant to microbial degradation in the soil. According to short-term closed aerated laboratory soil systems with $^{14}$C-compounds, xenobiotics (e.g. chlorinated benzenes, non-chlorinated compounds, and pesticides) were subjected to different fates, such as biomineralization, biotransformation, and formation of bound residues (Scheunert 1991). In the recent time, several models have been proposed to understand the fate of xenobiotics. To describe their physicochemical properties and biodegradation, several parameters are being used such as, sorption, volatilization, first-order kinetics

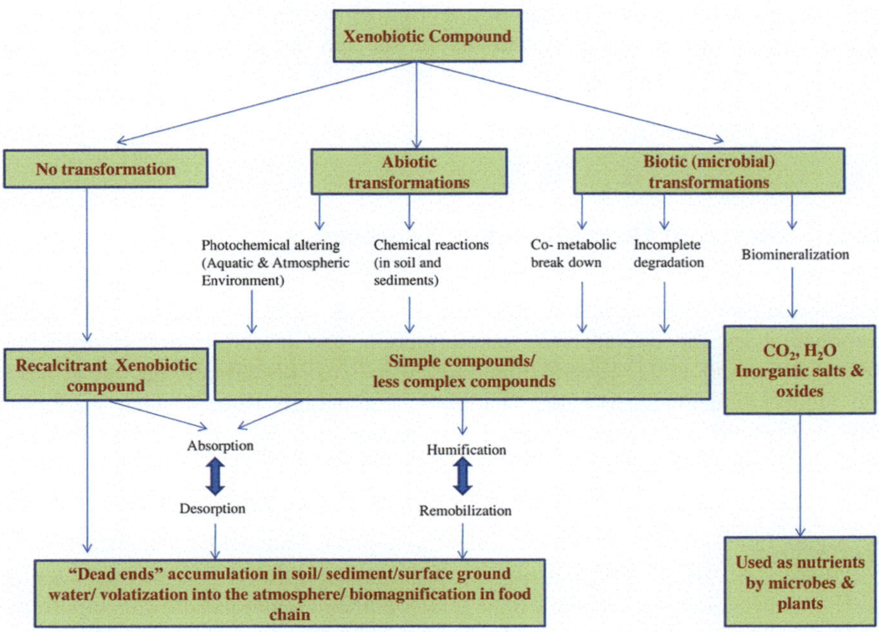


**Fig. 12.2** Possible environmental fate of a xenobiotic compound (modified after (Malla et al. 2018))

rates. For example, IUWS_MP model library system (IUWS - Integrated Urban Wastewater System; MP - Micro pollutants) is a dynamic model library system (Vezzaro et al. 2014), which helps in the simulation of MPs fluxes across the integrated WW systems, easy development of monitoring campaigns, evaluates the risk-benefits of other strategies. There are multiple factors that influence the fate of xenobiotics at the polluted site. For example, fate and distribution of a group of insecticides namely organochlorine insecticides (OCIs, e.g. dichlorodiphenyltrichloroethane (DDT), aldrin, dieldrin, endrin, chlordane, nonachlor, heptachlor, heptachlor-epoxide) are influenced by multiple factors such as climate, geochemistry of the site, nature of the chemical, magnitude of application and exposure period (Gopalan and Chenicherry 2018). It is important to note that understanding the fate of a pollutant is crucial in designing an effective remediation strategy. According to earlier reports, soil pH (Fredrickson and Shea 1986) changing humid climate (Ten Hulscher and Cornelissen 1996) have significant impact on the fate of OCIs. Changes in the soil pH greatly affects the lipophilicity of xenobiotics and subsequently affect their retention in the soil system.

On the other side, soil amendments are the additional factors that do affect the fate of xenobiotics in the soil (Khalid et al. 2020). For example, the impact of wastewater (WW) on pesticides fate and distribution has been reviewed in detail very recently (Peña et al. 2020). According to this review, it is clearly known that WW can alter the fate of pesticide in agriculture soils by impairing pesticide behavior and distribution. Therefore, experiments with pure xenobiotics may not reveal the exact picture of their fate since their fate is highly influenced by the coresident chemicals (Kuppusamy et al. 2020e). This warrants the site-specific studies for understanding the fate of xenobiotics, also such studies will aid the development of appropriate remediation strategy.

## 12.3 Microbial Degradation of Xenobiotics

Microorganisms have a capacity to produce a wide range of enzymes, which are responsible for the degradation, transformation, and mineralization of different xenobiotic compounds. Several microorganisms like bacteria, fungi, algae have the ability to degrade complex xenobiotic compounds through their enzymes. Different groups of microorganisms do interact with the xenobiotics (Dhiman et al. 2020) are as follows:

- Bacteria: Alcaligenes, Bacillus, Arthrobacter, Brevibacterium.
- Algae: Enteromorpha prolifera, Chlorella vulgaris, Nannochloris aculata.
- Fungi: Agaricus brasiliensis, Coriolus versicolor, Ganodermalucidum, Hymenoscyphus ericae, Marasmius quercophilus, Phanerochaete chrysoporium, Phanerochaete chrysosporium, Pichia pastoris, Pleurotus ostreatus, Pleurotus ostreatus, Rhizopogon vinicolor, Rigidoporus sp., Sckerogaster pacificus,

Trametes polyzona, Trametes versicolor, Trametes versicolor, and Trametes versicolor,

Details of different microorganisms and their interactions with different xenobiotic compounds have been described in Table 12.1, and mechanism of microbial cell-based bioremediation has been presented in Fig. 12.3.

## 12.3.1 Degradation of Petroleum Hydrocarbons

Xenobiotics degradation potential of microorganisms is confirmed in a series of in vitro and in vivo experiments. In our recent investigations, we observed the potential of bacteria and fungi to degrade petroleum hydrocarbons in the medium and soil (lab and land conditions). Two bacteria (*Bacillus cereus* and *B. thuringiensis*) and two fungi (*Geomyces pannorum* and *Geomyces* sp.) were isolated from the crude oil-polluted soils in Ecuadorian Amazon rainforest (Maddela et al. 2015a). When these cultures were tested in the laboratory conditions using using microbiological medium containing 1% diesel oil, removal efficiencies of 4 cultures were continuously 77.34% (*G. pannorum*), 68.55% (*Geomyces* sp.), 62.62% (*B. cereus*), and 49.71% (*B. thuringiensis*), and such results were might be due to the rapid utilization of petroleum hydrocarbons by fungi than bacteria. When two fungal strains were tested for the removal of crude oil from the medium (Maddela et al. 2015c), the per cent of removal was in the range of 24.0 to 43.4 after 30 days. However, a mixed culture of two fungal strains could remove 79.9% of TPHs from the soil in 30 days of post-treatment (Maddela et al. 2015c). We also found the optimum conditions for the sporulation of fungi in the medium containing PHs, for example, pH 5.0, 25 °C, 1-1.5% substrate (crude oil) and 4-6 g $L^{-1}$ N-P-K were found to stimulate optimum sporulation by fungi used in this study. It implies that cultural conditions in terms of pH, temperature and nutrients have the greatest impact on the degradation of pollutants by microbial cultures. When a mixed culture of two bacterial strains (*B. cereus* and *B. thuringiensis*) were tested for 30 days, they found to remove 84% and 28% of TPHs from crude oil- and spent lubricating oil-polluted soils, respectively (Raju et al. 2017), this could be attributed to the complex chemical nature of spent lubricant oil. When we mix all the above 4 strains, a higher per cent of crude oil (from soil in 30 days under laboratory conditions) was removed in the slurry phase (87.77%) than in the solid phase (79.47%) (Maddela et al. 2016). In general, slurry phase treatments provide optimum conditions for the interaction between pollutant and microbial cells. When soil slurry is mixed with nutrients and oxygen and subjected to stirring, microbial cells come in contact with the soil components very easily and there is a high-level possibility for the degradation of pollutants by microorganisms. The efficiency of these strains for the degradation of PHs at field-level has also been evaluated, results indicated that a mixed culture of 4 strains (*B. cereus, B. thuringiensis, G. pannorum, Geomyces* sp.) could remove 87.45% of TPHs after 90 days from crude oil polluted soil in an open

**Table 12.1** Role of microorganisms in the degradation of xenobiotic compounds

| Sl. No. | Microorganism | Pollutant | Significant outcomes | Reference |
|---|---|---|---|---|
| *Bacteria* | | | | |
| 1 | *Alkaliphilus metalliredigens* strain QYMF | Heavy metals | A metal-reducing bacterium capable of thriving in alkaline environments, a feature that is not commonly found in metal respiring microbes | Hwang et al. (2016) |
| 2 | *Arthrobacter* sp. strain SJCon | 2-Chloro-4-Nitrophenol | *Arthrobacter sp.* strains are useful in drafting the genetic pathways that are involved in the bioremediation of the aromatic compounds | Vikram et al. (2013) |
| 3 | *Arthrobacter* sp. strain SPG23 | Hydrocarbon degradation | *Arthrobacter* is a hydrocarbonoclastic gram-positive bacterium and is a potent bacterium for used for the remediation of the diesel fuels | Gkorezis et al. (2015) |
| 4 | *Bacillus* sp. CDB2, *Lysinibacillus sphaericus* | Arsenic | *Bacillus sp.* CDB2, *Lysinibacillus sphaericus* ahows high resistance to arsenic contamination and aids in treating arsenic poisoning | Rahman et al. (2016), Yang et al. (2013) |
| 5 | *Bacillus subtilis* SR1 | Polyaromatic hydrocarbon | *B. subtilis* SR1 is a bacterium besides showing resistance to heavy metals is also capable of degrading polyaromatic hydrocarbons | Kotoky et al. (2017) |
| 6 | *Brachybacterium* sp., *Cytophaga* sp., *Sphingomonas* sp., *Pseudomonas* sp. | Oil spills | The bacteria are capable of remediating the oil-contaminated sites via the processes of biostimulation and bioaugmentation | Angelim et al. (2013) |
| 7 | *Brevibacterium epidermidis* EZ-K02 | Industrial wastes | *B. epidermidis*, a bacterium that is capable of degrading waste waters contaminated with large scale | Ziganshina et al. (2018) |

| | | | | |
|---|---|---|---|---|
| | | | dissolution of chemical compounds and nitrocellulose particles | |
| 8 | *Caulobacter* sp. strain OR37 | Heavy metals | Possesses tolerance to elevated concentrations of heavy metals viz., cadmium, cobalt, uranium, nickel | Utturkar et al. (2013) |
| 9 | *Hyphomicrobium* sp. strain GJ21 | Dichloromethane | *Hyphomicrobium* degrades halogenated contaminants by making the use of dichloromethane as a source of both energy and carbon | Bringel et al. (2017) |
| 10 | *Microbacterium oleivorans* strain A9 | Radionuclides | *M. oleivorans* strain A9, a radionuclide-resistant actinobacterium capable of degrading uranium | Ortet et al. (2017) |
| 11 | *Microbacterium* sp. | Heavy metals | *Microbacterium sp.* plays an important role in phytoextraction and mobilization of heavy metals | Corretto et al. (2015) |
| 12 | *Moraxella saccharolytica, Alteromonas putrefaciens Klebsiella pneumonia, Pseudomonas fragi* | Diesel hydrocarbon | *M. saccharolytica, A. putrefaciens, K. pneumonia, P. fragi,* a complete bacterial consortium is proved to be one of the better and reliable choices for the rapid and complete remediation of the diesel hydrocarbon contaminated environments | Sharma and Rehman (2009) |
| 13 | *Mycobacterium dioxanotrophicus* | Heterocyclic organic compounds (Dioxane) | *M. dioxanotrophicus* is capable of remediating various heterocyclic organic compound contaminated environments by making the use of 1, 4-dioxane as a single source of energy and carbon and energy | He et al. (2017) |
| 14 | *Mycobacterium* sp. PYR-1 | Degradation of PAHs and pyrene | The strains of the *Mycobacterium sp.* although not producing the biosurfactants possess a strong ability | Chauhan et al. (2008) |

(continued)

**Table 12.1** (continued)

| Sl. No. | Microorganism | Pollutant | Significant outcomes | Reference |
|---|---|---|---|---|
| | | | to degrade even the low concentrations of aqueous-phase anthracene | |
| 15 | *Neisseria elongate, Acinetobacter faecalis, Staphylococcus* sp. | Crude petroleum oil | *N. elongate, A. faecalis, Staphylococcus. sp.*, are the potent bacterial isolates and carry up to 93, 94, and 95% of hydrocarbon degradation, respectively | Mukred et al. (2008) |
| 16 | *Ochrobactrum pseudogrignonense* | Arsenic pollutants | *O. pseudogrignonense*, a highly potent and efficient arsenate-resistant bacterium. This bacterium is involved in the degradation of the arsenic from arsenate-contaminated soils | Yang et al. (2013) |
| 17 | *Plantibacter flavus* strain 251 | Hydrocarbon-contaminated environments | *P. flavus* isolate 251 is known to possess novel biodegradation enzymes. This bacterium is anticipated to provide some of the novel insights into exploiting the hydrocarbon degrading pathways | Lumactud et al. (2017) |
| 18 | *Pseudomonas aeruginosa* | Organic and inorganic mercury | *P. aeruginosa* bacterium is one of the potent agent that carries out the bioremediation of both organic and inorganic mercury in highly contaminated mercury sites | Dash and Das (2012) |
| 19 | *Pseudomonas plecoglossicida* TND35 | Nicotine | *P. plecoglossicida* TND35 besides being an effective nicotine-degrading bacterium also has genes responsible for the degradation of heavy metals, | Raman et al. (2015) |

| | | | | |
|---|---|---|---|---|
| | | | aromatic compounds, and biosynthesis of butanol | |
| 20 | *Pseudomonas putida* strains, *P. putida* G7, *P.aeruginosa* PaK1, *P. putida* BS202, *Pseudomonas* sp. strain U2, *Rhodococcus* sp. NCIMB12038, *P. putida* OUS82, *Alcaligenes faecalis* AKF2, *Nocardiodes* sp. KP7 | Degradation of PAHs naphthalene and Phenanthrene | Isolation of these bacterial strains via the development of the modern genetic technologies provided a major breakthrough in the PAH remediation. The bacteria are capable of degrading substrates like phenanthrene, pyrene and benzo-pyrene, fluoranthene, naphthalene | Chauhan et al. (2008) |
| 21 | *Pseudomonas taeanensis* | Hydrocarbons (petroleum compounds) | *P. taeanensis* a bacterium that is able to degrade petroleum compounds like diesel, kerosene and gasoline | Lee et al. (2014) |
| 22 | *Pseudomonas veronii* strain 1YdBTEX2 | Aromatic solvents viz., benzene, toluene, ethyl benzene, xylene (BTEX) | *P. veronii* bacterium contains genes that carry out the degradation of aromatic solvents via a catabolic pathway | Junca and Pieper (2004), Morales et al. (2016), Moreno-Forero et al. (2016) |
| 23 | *Raoultella ornithinolytica*-TNT | Trinitrotoluene | *R. ornithinolytica*-TNT is a gram-negative bacterium. Strains of TNT make use of nitrate released from trinitrotoluene thereby making it less toxic. Hence is considered as a potent microbe in terms of bioremediation applications | Thijs et al. (2014) |
| 24 | *Sphingomonadaceae/Sphingomonas* | Degrades Hexachlorocyclohexae | *Sphingomonads* offer biostimulation of HCH polluted sites via addition of nutrients and aeration | Dadhwal et al. (2009) |
| *Fungi* | | | | |
| 25 | *Penicillium chrysogenum* | Monocyclic aromatic hydro carbons, benzene, toluene, ethyl benzene and xylene, phenol compounds | *P. chrysogenum* oxidize aromatic hydrocarbons using mono-oxygenases, forming a trans-diol | Abdulsalam et al. (2012), Pereira et al. (2014) |

(continued)

**Table 12.1** (continued)

| Sl. No. | Microorganism | Pollutant | Significant outcomes | Reference |
|---|---|---|---|---|
| 26 | *Phanerochaete chrysosporium* | Biphenyl and triphenylmethane | *P. chrysosporium* is able to degrade a broad spectrum of structurally diverse organo pollutants due to the presence of lignocellulolytic system | Wolski et al. (2012) |
| 27 | *Aspergillus niger, A. fumigatus, Fusarium solani* and *Penicillium funiculosum* | Hydrocarbon | *A. niger, A. fumigates* were play a major role in the detoxification of hydro carbons and the rate was 52.01% | Ai-Jawhari (2014) |
| 28 | *Coprinellus radians* | PAHs, methylnaphthalenes, and dibenzofurans | *C. radians* secretes the peroxygenases to reduce the environmental pollution by PAHs, methylnaphthalenes, and dibenzofurans | Aranda et al. (2010) |
| 29 | *Gleophyllum striatum* | Striatum pyrene, anthracene, 9-metil anthracene, Dibenzothiophene lignin peroxidasse | Lignin peroxidase oxidizes the compounds into nontoxic form | Yadav et al. (2011) |
| 30 | *Candida viswanathii* | Phenanthrene, benzopyrene | Detoxification of Phenanthrene, benzopyrene | Hesham et al. (2012) |
| 31 | *Candida glabrata, Candida krusei* | Crude oil | *C. glabrata, C. krusei* recorded the highest bio degradation of hydro carbons 60 and 61%, respectively | Burghal et al. (2016) |
| 32 | *Phanerochaete chrysosporium* | Industrial dyes | *P. chrysosporium* decolorizes the dye up to 98% | Yan et al. (2014) |
| 33 | *Penicillium ochrochloron* | Industrial dyes | *P. ochrochloron* decolorizes the dye up to 100% | Shedbalkar and Jadhav (2011) |
| 34 | *Aspergillus terreus* JAS1 | Chlorpyrifos (herbicide) | *A. terreus* degrade the 100% of herbicide after 48 hrs incubation period | Silambarasan and Abraham (2013) |

| | | | | |
|---|---|---|---|---|
| *Actinomycetes* | | | | |
| 35 | *Nocardiopsis* sp. MORSY1948 | Ni + Zn, Ni + Cr | The removal efficiency of *Nocardiopsis* sp. for Ni + Zn, Ni + Cr is 72.48% and 79.5% respectively | Said Mohamed et al. (2014) |
| 36 | *Nocardia* sp. MORSY2014 | Ni + Zn, Ni + Cr | The removal efficiency of *Nocardia* sp. for Ni + Zn, Ni + Cr is 69.12% and 60.65%, respectively | Simeonova et al. (2008) |

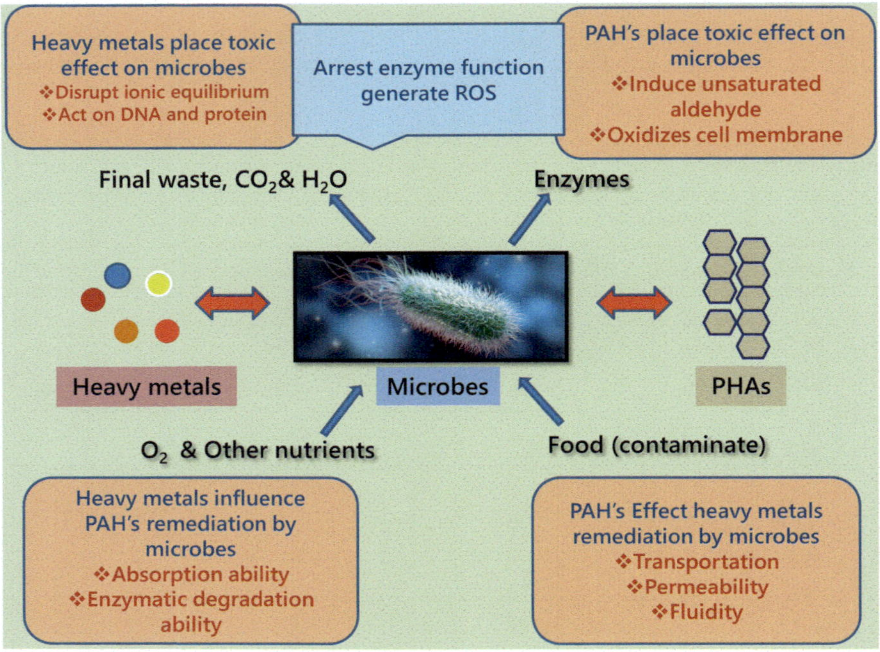


**Fig. 12.3** Mechanism of microbial cell based bioremediation (Modified after (Malla et al. 2018))

low-land area of Amazon rainforest (Maddela et al. 2017). Importantly, during this 90-day field-scale experiment, there was 631.6 mm of rainfall and the lowest and highest air temperatures recorded were 6 and 29 °C, respectively. Furthermore, two fungal strains (*G. pannorum* and *Geomyces* sp.) have shown 97-100% of biosorption of Cu from the medium (contained 5 mg $L^{-1}$ $CuSO_4.5H_2O$) in 7 days (Maddela et al. 2015b). Above results firmly confirm that microorganisms have potential to degrade PHs, thus the biodegradation is a key point in any bioremediation strategy (Raju and Scalvenzi 2017), however, environmental conditions cannot be ignored in the biodegradation studies. Site specific studies are obligatory before designing an effective remediation strategy. Also, bioavailability of the fractions of PHs is also an important factor which influences the biodegradation (Kuppusamy et al. 2020b), the rate of biodegradation is usually high when a pollutant has a high bioavailable fraction, and vice versa.

### *12.3.2 Degradation of Other Xenobiotics*

Likewise, microorganisms can interact with several types of xenobiotic compounds. For example, in a very recent investigation, microbial degradation of antiepileptic drug (e.g. carbamazepine) and antimicrobials (e.g. triclocarban and triclosan) was observed in 4 different agricultural soils (Thelusmond et al. 2019), where biodegradation of carbamazepine and triclocarban were degraded slowly (≤50% degradation occurred in 80 d) and triclosan was degraded rapidly (~80% degradation occurred in 25 d). Principal bacterial species that have identified in the degradation of above three xenobiotic compounds in this study (Thelusmond et al. 2019) were *Methylobacillus, Pseudomonas, Rhodococcus, Sphingomonas, Stenotrophomonas,* and *Streptomyces*. In fact, all these bacterial species have been reported for the degradation of carbamazepine, triclocarban and triclosan previously, where Pseudomonas has been identified as a potential bacterium due to its degradation potential of all three compounds. There are several genes and enzymes responsible for the degradation of xenobiotic compounds, as determined by emerging methods such as metatranscriptome analysis (Singh et al. 2018) and other OMIC tools (Kucharzyk et al. 2019). Various xenobiotic compounds are degraded by microbial enzymes such as microbial oxidoreductases, microbial oxygenases, monooxygenases, microbial dioxygenases, microbial laccases, microbial peroxidases, microbial lignin peroxidases, microbial manganese peroxidases, microbial lipases, esterase etc. (Gangola et al. 2019). These emerging techniques are also available to quantify either pollutant-degrading microorganisms or their enzymes. For example, methyl tertiary-butyl ether-degrading bacterium (e.g. *Methylibium petroleiphilum* PM1) or ETBE degradation gene (*ethB*) can be quantified by qPCR. Though this approach gives about the abundance information, does not tell about the degradation activity, therefore, advanced metagenomic analysis coupled with proteomics provide much more useful information about the abundance and activity of biodegradation (Kucharzyk et al. 2019). Thus, now-a-days, culture-independent methods have a key role in the determination of either abundance or activity of microorganisms involved in the degradation of xenobiotic compounds in various environments.

It should be remembered that the xenobiotic compounds are degraded by microorganisms by both aerobic and anaerobic mechanisms (Gangola et al. 2019). However, there is vast information available for aerobic pathways comparatively over anaerobic pathways, which is attributed to the effectiveness of aerobic pathways in the biodegradation of xenobiotics by microorganisms. This para provides key points of aerobic and anaerobic approaches in the degradation of xenobiotics. Key elements of aerobic degradation are – (1) incorporation of oxygen atom into pollutant by oxygenases and peroxidases; (2) formation of intermediates from the pollutant step-by-step, and subsequent entry of intermediates into tricarboxylic acid cycle; (3) Central precursor metabolites (e.g. acetyl-CoA, succinate, pyruvate) help in the formation of cell biomass. On the other side, key aspects of anaerobic degradation are methylation, hydroxylation, direct carboxylation, aromatic succinate production, β-oxidation (ring saturation) and fumerate addition.

## 12.4 Advances in the Bioremediation Technologies

There are several conventional approaches available for the remediation of xenobiotic polluted sites, but the important thing is a method which is effective at one site may not be effective at another site. Also, several things need to be remembered for the best selection of bioremediation approaches. For example, site background, receptors, contaminant's biotransformation, geochemistry, fate and transport, hydrogeology etc. are some of the important factors to consider while choosing the bioremediation approach. Bioremediation approaches offer certain undisputable benefits – (1) they are natural processes and eco-friendly with least side-effects, (2) is suitable for in situ method, therefore, there is excavation of pollutants to other sites, (3) there is less restoration time for the reuse of polluted sites, (4) there is high acceptability the public, regulatory and health agencies, (5) cost effective, and (6) these approaches consume less energy. However, in most occasions, bioremediation approaches are used in association with one or two physicochemical remediation approaches in order to get the even better results in a shortest time (Kuppusamy et al. 2020c). Traditional bioremediation approaches include bioventing, bioaugmentation, biostimulation, biopiling, compositing, slurry bioreactors, phytoremediation, etc. (Kuppusamy et al. 2020a; Raju and Scalvenzi 2017). Potential of these methods in the removal of xenobiotic compounds from the polluted sites have been furnished in the Table 12.2. Details of emerging bioremediation techniques have been provided below.

### *12.4.1 Bioelectrochemical System*

Bioelectrochemical systems (BES) has significant importance in the field of bioremediation, especially in the removal of pollutants from the wastewater and sediments (Li et al. 2017). There is a stimulation of pollutant-degrading bacteria through electrochemical reactions, which results in removal of contaminants by an enhanced metabolic activity of bacteria. It should be remembered that, in BES, pollutant's degradation efficiency of a bacterium is directly linked with the amount of electrons, there is a high rate of pollutant removal by bacterium in the presence of a high amount of electrons and vice versa (Li et al. 2020). Interestingly, pollutants will act as a electron donors (e.g. petroleum hydrocarbons) and carbon electrodes will function as electron acceptors, and such system has degraded 46% of PHs in three rounds of operations, the removal efficiency was 40% higher over the treatment done without electrodes (Mohanakrishna et al. 2019). In other words, this system can also be called as a microbial fuel cell (MFC). BES is also suitable for the removal of low biodegradability xenobiotics (such as dyes, polymers) (Zhang et al. 2015), metals (e.g. Cu (II) (Wang et al. 2016)), besides having in situ applications. However, future research is necessary in order to improve the overall process efficiency, and to

**Table 12.2** Potential of different bioremediation approaches in the removal of xenobiotics from the contaminated media

| S. No | Xenobiotics | Bioremediation technique | Microorganism involved | Results | Reference |
|---|---|---|---|---|---|
| 1 | Molinate (H) | Natural attenuation | Endogenous flora | 39% mineralized after 42 days | Lopes (2013) |
| | | Bioaugmentation | Microbial consortium: *G. molinativorax* ON4T, *Pseudomonas* (two strains), *Stenotrophomonas* and *Achromobacter* | 63% mineralized after 42 days | |
| 2 | Myclobutanil, tetraconazole and flusilazole (F) | Bioaugmentation | *Bacillus* strains namely DR-39, CS-126, TL-171, and TS-204 | 85% biodegraded after 20 days | Salunkhe et al. (2015) |
| 3 | Fenpropathrin (I) | Bioaugmentation | *Bacillus* sp. DG-02 | 93.3% biodegraded after 72 h | Chen et al. (2014) |
| 4 | 2,4-D (H) | Bioaugmentation | *Novosphingobium* strain DY4 | 50 and 95% biodegraded after 3 and 7 days, respectively | Dai et al. (2015) |
| 5 | Chlorpyrifos (H) | Bioaugmentation | *Bacillus cereus* Ct3, *Aspergillus terreus* JAS1 | 88% biodegraded after 7 days and 100% biodegraded after 48 h | Farhan et al. (2014), Silambarasan and Abraham (2013) |
| 6 | Bensulphuron-methyl (H) | Bioaugmentation | *Penicillium pinophilum* strain BP-H-02 | 87% biodegraded after 60 h | Peng et al. (2012) |
| 7 | Organochlorine pesticides | Biostimulation (soil macronutrients C:N:P 100:10:1) | Endogenous flora | – | Islas-García et al. (2015) |
| 8 | DDTs (I) | Natural attenuation | Endogenous flora | 23% biodegraded after 7 weeks | Ortíz et al. (2013) |
| | | Biostimulation (phenol, hexane and toluene) | | 23% biodegraded after 7 weeks | |
| 9 | Pentachlorophenol (H) | Biostimulation (lactate and anthraquinone-2,6-disulfonate) | Endogenous flora | Up to 97% biodegraded after 6 days | Chen et al. (2012) |
| 10 | Chlorpyrifos (H) | Bioaugmentation, | CS2 strain | 55% biodegraded after 6 days | Singh et al. (2016) |
| | | Bioaugmentation bioavailability enhancer | CS2 strain, bio-surfactant rhamnolipid | 82.3% biodegraded after 6 days | |

(continued)

**Table 12.2** (continued)

| S. No | Xenobiotics | Bioremediation technique | Microorganism involved | Results | Reference |
|---|---|---|---|---|---|
| 11 | Atrazine (H) | Bioaugmentation | Strain A6 (*Acinetobacter*) | 30% biodegraded after 6 days | Singh and Cameotra (2014) |
| | | Bioaugmentation bio-availability enhancer | Strain A6 (*Acinetobacter*) rhamnolipids and triton X-100 | 80% biodegraded after 6 days | |
| 12 | Diuron (H) | Bioaugmentation, | Bacterial consortium: *Arthrobacter* sp. N2, *Variovorax* sp. SRS16 | 45% mineralized after 120 days | Villaverde et al. (2012) |
| | | Bioaugmentation bio-availability enhancer | Bacterial consortium | 98% mineralized after 120 days | |
| 13 | Organochlorine pesticides (OCPs) | Biostimulation, bioavail-ability enhancer | Nitrate ($KNO_3$), methyl-β-cyclodextrin | 74.3% biodegraded after 180 days | Ye et al. (2014) |
| 14 | Diuron (H) | Biostimulation, composting, bioavailabil-ity enhancer | Micronutrients, sewage sludge mixed with pruning wastes, urban solid resi-dues, hydroxypropyl-β cyclodextrin | 46.5% mineralized after 140 days | Rubio-Bellido et al. (2015) |
| 15 | Linuron, diazinon and myclobutanil (H) | Composting | Sewage sludge, grape marc, spent mushroom substrate | Positive or negative effects on mineralization depending on organic amendment | Marín-Benito et al. (2014) |
| 16 | Lindane (H) | Aerobic soil slurry biore-actors biostimulation, bioaugmentation | Lindane-acclimated inoculum, final electron acceptor ($O_2$, $CO_2$ and $SO^{-2}$), CO-substrate (sucrose) | 55–70% biodegraded after 7 days | Varo-Arguello et al. (2012) |
| 17 | Pendimethalin (H) | Slurry bioreactor (composting) | Sewage sludge | 91% biodegraded after 10 days | Ramakrishnan et al. (2011) |
| 18 | Methoxychlor (H) | Slurry bioreactor (bioaugmentation) | *Actinobacteria* | N 60% biodegraded after 12 h | Fuentes et al. (2014) |
| 19 | Bentazone, boscalid, and pyrimethanil (H) | Biobed (biomixture) | Biofilter materials (mixtures of soil with digestate and/or biochar) | Desorption was hysteretic for all pesticides on these materials | Mukherjee et al. (2016) |
| 20 | Carbofuran (I) | Biobed (biomixture) | Lignocellulosic materials mixed with compost | 98.5% mineralized after 16 days | Chin-Pampillo et al. (2015) |

| | | | | | |
|---|---|---|---|---|---|
| 21 | Chlorothalonil (F) | Biobed (biomixture) | Spent mushroom substrate | DT50 7–9 days, biodegradation | Gao et al. (2015) |
| 22 | Atrazine (H) | Biobed (biomixture) | Soil, peat and straw, lignocellulosic residues. | 90% biodegraded after 90 days | Diez et al. (2013a), Diez et al. (2013b) |
| 23 | Oxyfluorfen (H) | Biobed (biomixture) | Vermicompost | 70% biodegraded after 30 days | Diaz et al. (2016) |
| 24 | Carbofuran (I) | Biobed (biomixture) | Ligninolytic fungus *Trametes versicolor*, compost | 100% biodegraded after 48 days | Madrigal-Zúñiga et al. (2016) |

*H* herbicide, *I* insecticide, *F* fungicide

resolve technological and economical barriers (Chandrasekhar et al. 2020) for the wide implications of BES in the field of bioremediation of xenobiotic polluted sites.

### *12.4.2 Novel Microorganisms*

It is necessary to search for the new microorganisms having capacity to degrade xenobiotic compounds for the successful bioremediation in diversified environments. In recent investigations, several novel microflora have been identified, for example, extremophiles (Giovanella et al. 2020), *Penicillium* spp. DC-F11 (Chang et al. 2020), *Gordonia* sp. QH-11 (Kong et al. 2019), *Marinobacter* sp. (Lopes et al. 2020), algal-bacterial consortium (*Chlorella* sp. MM3 and *Rhodococcus wratislaviensis* strain 9) (Subashchandrabose et al. 2019), Sargassum sp. (Saldarriaga-Hernandez et al. 2020) etc.

### *12.4.3 Rhizoremediation*

Rhizoremediation, also called rhizodegradation is a process where pollutants are degraded in the root zone (rhizosphere) of plants. Rhizoremediation has got wide acceptance as pollutants are mineralized in this process (Correa-García et al. 2018; Kiamarsi et al. 2020). Usually, certain pollutants (e.g. PHs) are less bioavailable due to their high hydrophobic nature. However, root exudates will change these equations, and make the pollutant more accessible to the microorganisms through their chemical nature. It is widely known that root exudates are rich in several substances (e.g. organic acids, enzymes, protons, sugars) which do support the activity and luxuriant growth of the rhizosphere microorganisms, and subsequently activated rhizosphere microorganisms can easily mineralize the pollutants (Hoang et al. 2021). Principally, there are 4 mechanisms by which root exudates favors the removal of pollutants (Martin et al. 2014) – (i) direct degradation of pollutants, (ii) enhanced bioavailability of pollutants due to presence of surfactants, (iii) stimulation of co-metabolic activities, and (iv) microbial activation via nutrient supply. Net result of rhizoremediation is, there is a plant resilience to pollutants, and there is reduced oxidative stress in the plants (Gkorezis et al. 2016). Thus, rhizoremediation of polluted sites is feasibly possible through a high-level synergistic relation between plant and rhizosphere microflora. With respect to rhizoremediation, research has been carried out in limited directions, such as worked with a specific pollutant (Turkovskaya and Muratova 2019), plant (Martin et al. 2014) and microbial group (Correa-García et al. 2018), or studying bioremediation in plant-independent condition (Hussain et al. 2018). It should not be ignored that there is application of several microbial inoculants to the soil as a part of sustainable agriculture, and least is known about the role of these inoculants in the rhizoremediation, therefore, future studies should investigate into the understanding the role of externally introduced microbial inoculants in the rhizoremoval of pollutants from the soil in the vicinity of plant system.

### 12.4.4 Algal Based Technologies

In the present days, microalgae (*Chlorella vulgaris*) and macroalgae (seaweeds) are the promising tools in the bioremediation of several microenvironments (e.g. sediments, effluents and sea surfaces) in marine environments (Chen et al. 2019; Sepehri et al. 2020). Seaweeds are the potential agents in the removal of several pollutants either by uptake (e.g. metals such as Fe, Zn, Ni, Cu, Mn and Co) or by positive attachment to its surfaces (e.g. oils such as Hg, Pb, Cd, Zn etc.) (Bilal et al. 2018). Surprisingly, algal members can concentrate metals 1000 times more than their biomass, which implies that algal cells are the good biosorbents, and subsequently metal recovery from the algal biomass is highly possible (Naja and Volesky 2009). Important factors those do govern the biosorption capacity of algae are pH, alkalinity, concentration of ions, contact period and temperature (Saldarriaga-Hernandez et al. 2020). The above insights clearly suggest that algal agents are the suitable tools for the sorption, desorption and recovery of several toxic metals, therefore, algal remediation could be an emerging approach if further integrated algal agents with industrial and land processes.

### 12.4.5 Microbial Enzymes

Now-a-days, there is much importance to enzyme-based bioremediation (Bilal et al. 2017; Sharma et al. 2018), this might be attributed to the effectiveness and specificity of enzyme treatment in the removal of pollutants from the contaminated media. Up ahead, enzymes are considered to be better tools than whole microbial cells, as enzymes show much better activity than the cells (Ye et al. 2019). Two most widely used enzymes in the bioremediation process are horseradish peroxidase (HRP) and laccase (Lac) (Zdarta et al. 2018), this might be due to their broad range of reactivity with different pollutants. Enzyme-based treatment has already been implicated in various environments, e.g. treatment of municipal wastewaters (Melo et al. 2016), however this treatment serious problems such as low stability of enzymes, economically unprofitable, less reuse and difficulty in the separation of enzymes from the reaction system (Liu et al. 2018). To overcome these problems, the trend has been shifted towards treatment with immobilized enzymes, where HRP and Lac enzymes are immobilized by different mechanisms such as adsorption (noncovalent), covalent, cross-linking, encapsulation, entrapment etc. For example, immobilized Lac has been tested for the removal of several xenobiotic compounds such as triclosan (Xu et al. 2014), diclofenac (Xu et al. 2015), carbamazepine (Ji et al. 2016), tetracycline (Yang et al. 2017), bisphenol A (Brugnari et al. 2018), malachite green coexistence with Cd (Wen et al. 2019) etc. Currently, investigations are focusing on the development of new substrate materials and methods for the effective immobilization of enzymes (Shakerian et al. 2020). Such developments are expected to improve the efficiency of enzyme (immobilized)-based bioremediation by improving the efficiency of degradation and immobilized enzyme properties (e.g. stability, reusability, easy separation).

### 12.4.6 Recombinant Technology

Microbial degradation of xenobiotic compounds is made easy with the emergence of several biotechnological approaches such as natural gene transfer or genetic engineering methods. These approaches allow the microorganisms to promote the production of specific enzymes to degrade the toxic xenobiotics (Pandotra et al. 2018). Genetic engineering is not new in the area of bioremediation (Chauhan et al. 2008; Magazine 1975), however this approach is still in practice (Ezezika and Singer 2010), this might be due to undisputed advantages offered by recombinant technology in the bioremediation.

For in situ monitoring of microbial community structure and function, there are several molecular techniques now available (Desai et al. 2010), for instance, fluorescence in situ hybridization (FISH), nucleic acids-based stable isotope probing (SIP), molecular biosensors/bioreporters etc. Dual benefits are offered by FISH technique, as it allows simultaneous phylogenetic classification plus gives abundance of active microbial population in the environmental sample. SIP permits the analysis of microbial community structure and function in a culture-independent manner. Whereas biosensors or bioreporters are the integrated devices, which do use different biomolecules (e.g. enzymes, antibodies, organs, tissues etc.) for measuring the interactions between pollutant and biological systems.

On the other side, "omics" techniques (e.g. genomics, transcriptomics, metatranscriptomics, proteomics, metaproteomics, metabolomics and fluxomics) playing a vital role in the characterization and monitoring of pollutant degraders, and identification of new pathways of biodegradation. Genomics relies on metagenomic libraries, pyrosequencing and DNA microarrays for the analysis of complete genetic information of a microbial cell (so called pollutant degraders). Transcriptomics and metatranscriptomics reveal the information about the mRNA transcriptional profiles of microbial communities which help us to understand more about the activities of microbial cells in the polluted environments. Whereas, protein composition and abundance in the microbial community can be revealed by proteomics and metaproteomics. Beyond all these, a new approach has emerged in the analysis of the complete repertoire of metabolites of a single microbial cell for the quantification of functional roles, the so-called approach is 'metabolomics'. Metabolomics relies on several analytical techniques (e.g. NMR, DIMS, FT-IR, HPLC, GC, capillary electrophoresis based MS (Mapelli et al. 2008)). It should also be remembered that several bioinformatics tools are available for the interpretation or correlation of -omics data, e.g. in silico software, pipelines, web resources, algorithms etc. There is one database, called The University of Minnesota Biocatalysis/Biodegradation database (UMBBD), is an important and freely available (http://umbbd.msi.umn.edu/) web resource of following microbial bioremediation information: Pathways = 187; Reactions = 1,287; Compounds = 1,195; Enzymes = 833; Microorganisms entries = 491; Biotransformation rules = 259 (Ellis et al. 2006). Nevertheless, emergence of molecular and "omics' techniques in the area of microbial remediation of polluted sites is a strong positive sign for the rapid restoration of contaminated sites and development of cleaner environment.

## 12.5 Future Research Directions

It is a well-known fact that microorganisms have a wider range of capabilities either to degrade or detoxify different types of xenobiotic compounds. However, most of these studies have been conducted at lab and small-scale level, therefore, bringing these technologies to field-level is always challenging. Following are some of the future research directions in the areas that have been discussed in Sect. 4.

1. BES-based bioremediation is known to be influenced by several factors such as type of contaminant, microbial community structure, distribution of electron donors and acceptors, nutrients, and hydraulic flow dynamics. As most of these insights are known through the lab-based experiments, still it is not known how to optimize the electrochemical simulation at in situ bioremediation. Suggested areas in this field (Li and Yu 2015) are, monitoring of microbial community using molecular approaches (e.g. Catalyzed Reporter Deposition-FISH, metagenomics, proteomics analysis). Modeling of contaminant flow and reaction under specific electrochemical stimulation scenarios. Modeling studies greatly help in the design of effective and optimal remediation strategies in the future. There should be proper emphasis on remediation efficiency, ecotoxicity, cost benefits of BES-based bioremediation. Field-level electro-bioremediation studies are obligatory in order to have reliable results.
2. In rhizoremediation, existing limiting factors (soil conditions, temperature, pH, soil organic matter, plant species and microbes involved, pollutants availability to microbes) can be overcome through biostimulation and bioaugmentation processes (Saravanan et al. 2020). Soil treatment with fertilizers, minerals and biosurfactants (so called biostimulation) will enhance the efficiency of rhizoremediation by modifying the soil physicochemical properties. Bioaugmentation is also a suitable process to enhance the rate of degradation of contaminants in the rhizosphere region, and possible routes for the introduction of microbial inoculants into soil are seed coating, soil drenching and root dipping. Another promising approach for the enhancement of efficiency of rhizoremediation in the removal of pollutants is use of transgenic plants, which have ability to produce more diverse root exudates for the establishment of better

plant-microbial interactions, which could have significant impact on the pollutant degradation subsequent removal.

3. As microalgae provide multiple advantages in the area of heavy metal bioremediation, more emphasis is needed to bring algal-based bioremediation closure to the field-level. Efforts should be made in the area of genetic, metabolic and molecular fields to increase the adaptability, specificity and robustness of microalgae in heavy metal removal (Leong and Chang 2020). Equilibrium constant and kinetics of heavy metal biosorption and bioaccumulation by microalgae are poorly understood. To make the algal-based bioremediation approach cost effective, innovation of harvesting technologies is greatly warranted. It is important to note that microalgae are the important sources of producing valuable products such as lipids, exopolymers, pigments, phytochelatin, and phytohormones, however, heavy metal bioaccumulation has significant interference at downstream purification level. Therefore, additional research is still needed in adapting effective downstream purification processes and for further production of value-added products.
4. Owing to the advantages (such as good and cost-effective alternatives) of enzyme-based bioremediation, future studies should emphasize on different microorganisms and their enzymes potential in the degradation of xenobiotic compounds. There are certain enzymes (e.g. nanozymes) which can perform the biodegradation even at remote conditions (Sharma et al. 2018), therefore, proper emphasis should be made on this type of enzymes. Metagenomics, metatranscriptomics and metaproteomics techniques should be implemented in the screening of novel microorganisms and enzymes in a culture-independent manner.
5. With respect to the molecular and 'omics' techniques, appropriate statistical algorithms and bioinformatics tools should be explored for the proper interpretation of massive data generated by molecular and 'omics' tools (Desai et al. 2010). There are certain gene editing tools (e.g. TALEN, ZFNs, CRISPR Cas 9) which allow us to identify and isolate desired microorganisms with special functions and genes encoding desirable functions (so called xenobiotics degradation) (Jaiswal et al. 2019). Possible implications of omics tools must be searched in depth as these tools contribute to the logical identification of pollutant degrading microorganisms. Multi-omic tools also provide an undisputable advantage in the selection of suitable hosts for the gene expression. Tailoring of gene expression, decrease in metabolic burden, optimization of degradation pathways are feasible by modeling studies (e.g. genome-scale and kinetic models) (Dvořák et al. 2017).

## 12.6 Conclusions

The main conclusions that have been drawn from the above insights are as follows:

- Global consumption and market projection rates are clearly in increasing trend. As a result, environmental concentrations of xenobiotics are in the range of ng to mg.
- Bioremediation is yet a promising approach for the clearance of xenobiotic substances from the contaminated environmental media as indicated by scientific production in the last decade.
- Biomagnification, trophic transfer, reduction of soil fertility, serious human health risks are the principal consequences of xenobiotics in the environment.
- Optimum reaction conditions are obligatory for the removal of pollutants from laboratory medium or soil (in vitro and in vivo) as discussed in the case of biodegradation of petroleum hydrocarbons.
- There should be more emphasis on the emerging bioremediation techniques such as Bioelectrochemical system, rhizoremediation, algal based technologies, implications of enzymes and omics tools. Such studies may take the bioremediation approach much nearer to the real and large-scale level.

## References

Abdulsalam S, Adefila SS, Bugaje IM, Ibrahim S (2012) Bioremediation of soil contaminated with used motor oil in a closed system. J Bioremed Biodegr 3(12):172–179

Ai-Jawhari IFH (2014) Ability of some soil fungi in biodegradation of petroleum hydrocarbon. J Appl Environ Microbiol 2(2):46–52

Alberto D, Couée I, Sulmon C, Gouesbet G (2017) Root-level exposure reveals multiple physiological toxicity of triazine xenobiotics in Arabidopsis thaliana. J Plant Physiol 212:105–114

Al-Fanharawi AA, Rabee AM, Al-Mamoori AMJ (2019) Multi-biomarker responses after exposure to organophosphates chlorpyrifos in the freshwater mussels Unio tigridis and snails viviparous benglensis. Hum Ecol Risk Assess Int J 25(5):1137–1156

Alkimin GD, Daniel D, Dionísio R, Soares AMVM, Barata C, Nunes B (2019) Effects of diclofenac and salicylic acid exposure on Lemna minor: is time a factor? Environ Res 177:108609

Angelim AL, Costa SP, Farias BCS, Aquino LF, Melo VMM (2013) An innovative bioremediation strategy using a bacterial consortium entrapped in chitosan beads. J Environ Manag 127:10–17

Aranda E, Ullrich R, Hofrichter M (2010) Conversion of polycyclic aromatic hydrocarbons, methyl naphthalenes and dibenzofuran by two fungal peroxygenases. Biodegradation 21(2):267–281

Bartucca ML, Mimmo T, Cesco S, Panfili I, Del Buono D (2019) Effect of metribuzin on nitrogen metabolism and iron acquisition in Zea mays. Chem Ecol 35(8):720–731

Bilal M, Asgher M, Iqbal HMN, Hu H, Zhang X (2017) Bio-based degradation of emerging endocrine-disrupting and dye-based pollutants using cross-linked enzyme aggregates. Environ Sci Pollut Res 24(8):7035–7041

Bilal M, Rasheed T, Sosa-Hernández JE, Raza A, Nabeel F, Iqbal H (2018) Biosorption: an interplay between marine algae and potentially toxic elements—a review. Mar Drugs 16(2):65

Bringel F, Postema CP, Mangenot S, Bibi-Triki S, Chaignaud P, Haque MFU, Gruffaz C, Hermon L, Louhichi Y, Maucourt B (2017) Genome sequence of the dichloromethane-degrading bacterium Hyphomicrobium sp. strain GJ21. Genome Announc 5(30)

Brouwer M, Huss A, van der Mark M, Nijssen PCG, Mulleners WM, Sas AMG, Van Laar T, de Snoo GR, Kromhout H, Vermeulen RCH (2017) Environmental exposure to pesticides and the risk of Parkinson's disease in the Netherlands. Environ Int 107:100–110

Brugnari T, Pereira MG, Bubna GA, de Freitas EN, Contato AG, Corrêa RCG, Castoldi R, de Souza CGM, de Moraes, M.d.L.T. and Bracht, A. (2018) A highly reusable MANAE-agarose-immobilized Pleurotus ostreatus laccase for degradation of bisphenol A. Sci Total Environ 634:1346–1351

Burghal AA, Abu-Mejdad N, Al-Tamimi WH (2016) Mycodegradation of crude oil by fungal species isolated from petroleum contaminated soil. Int J Innovat Res Sci Eng Technol 5 (2):1517–1524

Chandrasekhar K, Kumar G, Venkata Mohan S, Pandey A, Jeon B-H, Jang M, Kim SH (2020) Microbial electro-remediation (MER) of hazardous waste in aid of sustainable energy generation and resource recovery. Environ Technol Innovat 19:100997

Chang J, Shi Y, Si G, Yang Q, Dong J, Chen J (2020) The bioremediation potentials and mercury (II)-resistant mechanisms of a novel fungus Penicillium spp. DC-F11 isolated from contaminated soil. J Hazard Mater 396:122638

Charles M (2020) Plastic pollution. https://www.britannica.com/science/plastic-pollution/Plastic-pollution-in-oceans-and-on-land. Accessed 3 Aug 2020

Chauhan A, Oakeshott JG, Jain RK (2008) Bacterial metabolism of polycyclic aromatic hydrocarbons: strategies for bioremediation. Indian J Microbiol 48(1):95–113

Chen M, Shih K, Hu M, Li F, Liu C, Wu W, Tong H (2012) Biostimulation of indigenous microbial communities for anaerobic transformation of pentachlorophenol in paddy soils of Southern China. J Agric Food Chem 60(12):2967–2975

Chen S, Chang C, Deng Y, An S, Dong YH, Zhou J, Hu M, Zhong G, Zhang L-H (2014) Fenpropathrin biodegradation pathway in Bacillus sp. DG-02 and its potential for bioremediation of pyrethroid-contaminated soils. J Agric Food Chem 62(10):2147–2157

Chen B, Ye X, Zhang B, Jing L, Lee K (2019) World seas: an environmental evaluation. Elsevier, Amsterdam, pp 407–426

Chin-Pampillo JS, Ruiz-Hidalgo K, Masís-Mora M, Carazo-Rojas E, Rodríguez-Rodríguez CE (2015) Adaptation of biomixtures for carbofuran degradation in on-farm biopurification systems in tropical regions. Environ Sci Pollut Res 22(13):9839–9848

Connell DW (2018) Bioaccumulation of xenobiotic compounds: 0. CRC Press

Correa-García S, Pande P, Séguin A, St-Arnaud M, Yergeau E (2018) Rhizoremediation of petroleum hydrocarbons: a model system for plant microbiome manipulation. Microb Biotechnol 11(5):819–832

Corretto E, Antonielli L, Sessitsch A, Kidd P, Weyens N, Brader G (2015) Draft genome sequences of 10 Microbacterium spp., with emphasis on heavy metal-contaminated environments. Genome Announc 3(3)

Dadhwal M, Singh A, Prakash O, Gupta SK, Kumari K, Sharma P, Jit S, Verma M, Holliger C, Lal R (2009) Proposal of biostimulation for hexachlorocyclohexane (HCH)-decontamination and characterization of culturable bacterial community from high-dose point HCH-contaminated soils. J Appl Microbiol 106(2):381–392

Dai Y, Li N, Zhao Q, Xie S (2015) Bioremediation using Novosphingobium strain DY4 for 2,4-dichlorophenoxyacetic acid-contaminated soil and impact on microbial community structure. Biodegradation 26(2):161–170

Dash HR, Das S (2012) Bioremediation of mercury and the importance of bacterial mer genes. Int Biodeterior Biodegradation 75:207–213

de Lima DA, Müller C, Costa AC, Batista PF, Dalvi VC, Domingos M (2017) Morphoanatomical and physiological changes in Bauhinia variegata L. as indicators of herbicide diuron action. Ecotoxicol Environ Saf 141:242–250

Demirci Ö, Güven K, Asma D, Öğüt S, Uğurlu P (2018) Effects of endosulfan, thiamethoxam, and indoxacarb in combination with atrazine on multi-biomarkers in Gammarus kischineffensis. Ecotoxicol Environ Saf 147:749–758

Desai C, Pathak H, Madamwar D (2010) Advances in molecular and "-omics" technologies to gauge microbial communities and bioremediation at xenobiotic/anthropogen contaminated sites. Bioresour Technol 101(6):1558–1569

Dhiman N, Jasrotia T, Sharma P, Negi S, Chaudhary S, Kumar R, Mahnashi MH, Umar A, Kumar R (2020) Immobilization interaction between xenobiotic and Bjerkandera adusta for the biodegradation of atrazine. Chemosphere 257:127060

Díaz E (2004) Bacterial degradation of aromatic pollutants: a paradigm of metabolic versatility. Int Microbiol 7(3):173–180

Diaz JMC, Delgado-Moreno L, Núñez R, Nogales R, Romero E (2016) Enhancing pesticide degradation using indigenous microorganisms isolated under high pesticide load in bioremediation systems with vermicomposts. Bioresour Technol 214:234–241

Diez MC, Levio M, Briceño G, Rubilar O, Tortella G, Gallardo F (2013a) Biochar as a partial replacement of peat in pesticide-degrading biomixtures formulated with different soil types. J Biobaased Mater Bioenergy 7(6):741–747

Diez MC, Tortella GR, Briceño G, Castillo MDP, Díaz J, Palma G, Altamirano C, Calderón C, Rubilar O (2013b) Influence of novel lignocellulosic residues in a biobed biopurification system on the degradation of pesticides applied in repeatedly high doses. Electron J Biotechnol 16 (6):11–11

Dvořák P, Nikel PI, Damborský J, de Lorenzo V (2017) Bioremediation 3.0: engineering pollutant-removing bacteria in the times of systemic biology. Biotechnol Adv 35(7):845–866

Eberbach PL (2018) Pesticide interactions in crop production: beneficial and deleterious effects. CRC Press, pp 183–212

Ellis LBM, Roe D, Wackett LP (2006) The University of Minnesota biocatalysis/biodegradation database: the first decade. Nucl Acids Res 34(suppl_1):D517–D521

Embrandiri A, Kiyasudeen SK, Rupani PF, Ibrahim MH (2016) Plant responses to xenobiotics. Springer, Berlin, pp 1–18

Ezezika OC, Singer PA (2010) Genetically engineered oil-eating microbes for bioremediation: prospects and regulatory challenges. Technol Soc 32(4):331–335

Faber S, Fahrenholz T, Wolle MM, Kern JC, Pamuku M, Miller L, Jamrom J, Skip Kingston HM (2019) Chronic exposure to xenobiotic pollution leads to significantly higher total glutathione and lower reduced to oxidized glutathione ratio in red blood cells of children with autism. Free Radic Biol Med 134:666–677

Farhan M, Butt ZA, Khan AU, Wahid A, Ahmad M, Ahmad F, Kanwal A (2014) Enhanced biodegradation of chlorpyrifos by agricultural soil isolate. Asian J Chem 26(10):3013

Fredrickson DR, Shea PJ (1986) Effect of soil pH on degradation, movement, and plant uptake of chlorsulfuron. Weed Sci:328–332

Fuentes MS, Alvarez A, Saez JM, Benimeli CS, Amoroso MJ (2014) Bioremediation in Latin America. Springer, Berlin, pp 267–277

Gangola S, Joshi S, Kumar S, Pandey SC (2019) In: Bhatt P (ed) Smart bioremediation technologies. Academic Press, pp 169–189

Gao H, Gao X, Cao Y, Xu L, Jia L (2015) Influence of hydroxypropyl-β-cyclodextrin on the extraction and biodegradation of p, p′-DDT, o, p′-DDT, p, p′-DDD, and p, p′-DDE in soils. Water Air Soil Pollut 226(7):208

García J, Ventura MI, Requena M, Hernández AF, Parrón T, Alarcón R (2017) Association of reproductive disorders and male congenital anomalies with environmental exposure to endocrine active pesticides. Reprod Toxicol 71:95–100

Giovanella P, Vieira GAL, Ramos Otero IV, Pais Pellizzer E, de Jesus Fontes B, Sette LD (2020) Metal and organic pollutants bioremediation by extremophile microorganisms. J Hazard Mater 382:121024

Gkorezis P, Bottos EM, van Hamme JD, Thijs S, Rineau F, Franzetti A, Balseiro-Romero M, Weyens N, Vangronsveld J (2015) Draft genome sequence of Arthrobacter sp. strain SPG23, a hydrocarbon-degrading and plant growth-promoting soil bacterium. Genome Announc 3(6)

Gkorezis P, Daghio M, Franzetti A, Van Hamme JD, Sillen W, Vangronsveld J (2016) The interaction between plants and bacteria in the remediation of petroleum hydrocarbons: an environmental perspective. Front Microbiol 7:1836

Gopalan NK, Chenicherry S (2018) Fate and distribution of organochlorine insecticides (OCIs) in Palakkad soil, India. Sustain Environ Res 28(4):179–185
Goulson D, Thompson J, Croombs A (2018) Rapid rise in toxic load for bees revealed by analysis of pesticide use in Great Britain. PeerJ 6:e5255
Gouma S, Fragoeiro S, Bastos AC, Magan N (2014) In: Das S (ed) Microbial biodegradation and bioremediation. Elsevier, Oxford, pp 301–323
Gupta S, Pathak B (2020) In: Singh P, Kumar A, Borthakur A (eds) Abatement of environmental pollutants. Elsevier, Oxford, pp 127–149
He Y, Wei K, Si K, Mathieu J, Li M, Alvarez PJJ (2017) Whole-genome sequence of the 1, 4-dioxane-degrading bacterium Mycobacterium dioxanotrophicus PH-06. Genome Announc 5(35)
Hesham AE-L, Khan S, Tao Y, Li D, Zhang Y, Yang M (2012) Biodegradation of high molecular weight PAHs using isolated yeast mixtures: application of meta-genomic methods for community structure analyses. Environ Sci Pollut Res 19(8):3568–3578
Hoang SA, Lamb D, Seshadri B, Sarkar B, Choppala G, Kirkham MB, Bolan NS (2021) Rhizoremediation as a green technology for the remediation of petroleum hydrocarbon-contaminated soils. J Hazard Mater 401:123282
Hussain S, Siddique T, Arshad M, Saleem M (2009) Bioremediation and phytoremediation of pesticides: recent advances. Crit Rev Environ Sci Technol 39(10):843–907
Hussain I, Puschenreiter M, Gerhard S, Schöftner P, Yousaf S, Wang A, Syed JH, Reichenauer TG (2018) Rhizoremediation of petroleum hydrocarbon-contaminated soils: improvement opportunities and field applications. Environ Exp Bot 147:202–219
Hwang C, Copeland A, Lucas S, Lapidus A, Barry K, Detter JC, del Rio TG, Hammon N, Israni S, Dalin E (2016) Complete genome sequence of Alkaliphilus metalliredigens strain QYMF, an alkaliphilic and metal-reducing bacterium isolated from borax-contaminated leachate ponds. Genome Announc 4(6)
Islas-García A, Vega-Loyo L, Aguilar-López R, Xoconostle-Cázares B, Rodríguez-Vázquez R (2015) Evaluation of hydrocarbons and organochlorine pesticides and their tolerant microorganisms from an agricultural soil to define its bioremediation feasibility. J Environ Sci Health B 50(2):99–108
Jaiswal S, Singh DK, Shukla P (2019) Gene editing and systems biology tools for pesticide bioremediation: a review. Front Microbiol 10:87
Jambeck JR, Geyer R, Wilcox C, Siegler TR, Perryman M, Andrady A, Narayan R, Law KL (2015) Plastic waste inputs from land into the ocean. Science 347(6223):768–771
Ji C, Hou J, Wang K, Zhang Y, Chen V (2016) Biocatalytic degradation of carbamazepine with immobilized laccase-mediator membrane hybrid reactor. J Membr Sci 502:11–20
Junca H, Pieper DH (2004) Functional gene diversity analysis in BTEX contaminated soils by means of PCR-SSCP DNA fingerprinting: comparative diversity assessment against bacterial isolates and PCR-DNA clone libraries. Environ Microbiol 6(2):95–110
Khalid S, Shahid M, Murtaza B, Bibi I, Natasha AN, M. and Niazi, N.K. (2020) A critical review of different factors governing the fate of pesticides in soil under biochar application. Sci Total Environ 711:134645
Kiamarsi Z, Kafi M, Soleimani M, Nezami A, Lutts S (2020) Conjunction of Vetiveria zizanioides L. and oil-degrading bacteria as a promising technique for remediation of crude oil-contaminated soils. J Clean Product 253:119719
Klein EY, van Boeckel TP, Martinez EM, Pant S, Gandra S, Levin SA, Goossens H, Laxminarayan R (2018) Global increase and geographic convergence in antibiotic consumption between 2000 and 2015. Proc Natl Acad Sci 115(15):E3463
Kong X, Jin D, Tai X, Yu H, Duan G, Yan X, Pan J, Song J, Deng Y (2019) Bioremediation of dibutyl phthalate in a simulated agricultural ecosystem by Gordonia sp. strain QH-11 and the microbial ecological effects in soil. Sci Total Environ 667:691–700

Kotoky R, Singha LP, Pandey P (2017) Draft genome sequence of polyaromatic hydrocarbon-degrading bacterium Bacillus subtilis sr1, which has plant growth-promoting attributes. Genome Announc 5(49)

Kucharzyk KH, Rectanus HV, Bartling CM, Rosansky S, Minard-Smith A, Mullins LA, Neil K (2019) Use of omic tools to assess methyl tert-butyl ether (MTBE) degradation in groundwater. J Hazard Mater 378:120618

Kumar A, Vashist M, Rathee R (2014) Maternal factors and risk of childhood leukemia. Asian Pac J Cancer Prev 15(2):781–784

Kuppusamy S, Maddela NR, Megharaj M, Venkateswarlu K (2020a) Total petroleum hydrocarbons. Springer, Berlin, pp 167–205

Kuppusamy S, Maddela NR, Megharaj M, Venkateswarlu K (2020b) Total petroleum hydrocarbons. Springer, Berlin, pp 79–94

Kuppusamy S, Maddela NR, Megharaj M, Venkateswarlu K (2020c) Total petroleum hydrocarbons. Springer, Berlin, pp 225–256

Kuppusamy S, Maddela NR, Megharaj M, Venkateswarlu K (2020d) Total petroleum hydrocarbons. Springer, Berlin, pp 95–138

Kuppusamy S, Maddela NR, Megharaj M, Venkateswarlu K (2020e) Total petroleum hydrocarbons. Springer, Berlin, pp 57–77

Kuppusamy S, Maddela NR, Megharaj M, Venkateswarlu K (2020f) Total petroleum hydrocarbons. Springer, Berlin, pp 139–165

Lee S-Y, Kim SH, Lee D-G, Shin S, Yun SH, Choi C-W, Chung Y-H, Choi JS, Kahng H-Y, Kim SI (2014) Draft genome sequence of petroleum oil-degrading marine bacterium Pseudomonas taeanensis strain MS-3, isolated from a crude oil-contaminated seashore. Genome Announc 2(1)

Leong YK, Chang J-S (2020) Bioremediation of heavy metals using microalgae: recent advances and mechanisms. Bioresour Technol 303:122886

Li W-W, Yu H-Q (2015) Electro-assisted groundwater bioremediation: fundamentals, challenges and future perspectives. Bioresour Technol 196:677–684

Li T, Zhou L, Qian Y, Wan L, Du Q, Li N, Wang X (2017) Gravity settling of planktonic bacteria to anodes enhances current production of microbial fuel cells. Appl Energy 198:261–266

Li T, Chen F, Zhou Q, Wang X, Liao C, Zhou L, Wan L, An J, Wan Y, Li N (2020) Unignorable toxicity of formaldehyde on electroactive bacteria in bioelectrochemical systems. Environ Res 183:109143

Litskas VD, Karamanlis XN, Prousali SP, Koveos DS (2019) The xenobiotic doxycycline affects nitrogen transformations in soil and impacts earthworms and cultivated plants. J Environ Sci Health A 54(14):1441–1447

Liu EK, He WQ, Yan CR (2014) 'White revolution'to 'white pollution'—agricultural plastic film mulch in China. Environ Res Lett 9(9):091001

Liu D-M, Chen J, Shi Y-P (2018) Advances on methods and easy separated support materials for enzymes immobilization. TrAC Trends Anal Chem 102:332–342

Lopes AR (2013) Molinate biodegradation in soils: natural attenuation versus bioaugmentation. Appl Microbiol Biotechnol 97(6):2691–2700. 2013 v.2697 no.2696

Lopes EM, Fernandes CC, de Macedo Lemos EG, Kishi LT (2020) Reconstruction and in silico analysis of new Marinobacter adhaerens t76_800 with potential for long-chain hydrocarbon bioremediation associated with marine environmental lipases. Mar Genomics 49:100685

Lumactud R, Fulthorpe R, Sentchilo V, van der Meer JR (2017) Draft genome sequence of Plantibacter flavus strain 251 isolated from a plant growing in a chronically hydrocarbon-contaminated site. Genome Announc 5(17)

Luo D, Zhou T, Tao Y, Feng Y, Shen X, Mei S (2016) Exposure to organochlorine pesticides and non-Hodgkin lymphoma: a meta-analysis of observational studies. Sci Rep 6:25768

Maddela NR, Venkateswarlu K (2018) Insecticides– soil microbiota interactions. Springer, Berlin

Maddela NR, Masabanda M, Leiva-Mora M (2015a) Novel diesel-oil-degrading bacteria and fungi from the Ecuadorian Amazon rainforest. Water Sci Technol 71(10):1554–1561

Maddela NR, Reyes JJM, Viafara D, Gooty JM (2015b) Biosorption of Copper (II) by the microorganisms isolated from the crude-oil-contaminated soil. Soil Sediment Contamin 24 (8):898–908
Maddela NR, Scalvenzi L, Pérez M, Montero C, Gooty JM (2015c) Efficiency of indigenous filamentous fungi for biodegradation of petroleum hydrocarbons in medium and soil: laboratory study from Ecuador. Bull Environ Contam Toxicol 95(3):385–394
Maddela NR, Burgos R, Kadiyala V, Carrion AR, Bangeppagari M (2016) Removal of petroleum hydrocarbons from crude oil in solid and slurry phase by mixed soil microorganisms isolated from Ecuadorian oil fields. Int Biodeterior Biodegradation 108:85–90
Maddela NR, Scalvenzi L, Venkateswarlu K (2017) Microbial degradation of total petroleum hydrocarbons in crude oil: a field-scale study at the low-land rainforest of Ecuador. Environ Technol 38(20):2543–2550
Madrigal-Zúñiga K, Ruiz-Hidalgo K, Chin-Pampillo JS, Masís-Mora M, Castro-Gutiérrez V, Rodríguez-Rodríguez CE (2016) Fungal bioaugmentation of two rice husk-based biomixtures for the removal of carbofuran in on-farm biopurification systems. Biol Fertil Soils 52 (2):243–250
Magazine T (1975) Environment: oil-eating bug. Time
Malla MA, Dubey A, Yadav S, Kumar A, Hashem A, Abd-Allah EF (2018) Understanding and designing the strategies for the microbe-mediated remediation of environmental contaminants using omics approaches. Front Microbiol 9:1132
Mansuy D (2013) Metabolism of xenobiotics: beneficial and adverse effects. Biol Aujourdhui 207 (1):33–37
Mapelli V, Olsson L, Nielsen J (2008) Metabolic footprinting in microbiology: methods and applications in functional genomics and biotechnology. Trends Biotechnol 26(9):490–497
Marín-Benito JM, Herrero-Hernández E, Andrades MS, Sánchez-Martín MJ, Rodríguez-Cruz MS (2014) Effect of different organic amendments on the dissipation of linuron, diazinon and myclobutanil in an agricultural soil incubated for different time periods. Sci Total Environ 476:611–621
Martin BC, George SJ, Price CA, Ryan MH, Tibbett M (2014) The role of root exuded low molecular weight organic anions in facilitating petroleum hydrocarbon degradation: current knowledge and future directions. Sci Total Environ 472:642–653
Mazur CS, Marchitti SA, Zastre J (2015) P-glycoprotein inhibition by the agricultural pesticide propiconazole and its hydroxylated metabolites: implications for pesticide–drug interactions. Toxicol Lett 232(1):37–45
Meckenstock RU, Elsner M, Griebler C, Lueders T, Stumpp C, Aamand J, Agathos SN, Albrechtsen H-J, Bastiaens L, Bjerg PL (2015) Biodegradation: updating the concepts of control for microbial cleanup in contaminated aquifers. ACS Publications
Melo CF, Dezotti M, Marques MRC (2016) A comparison between the oxidation with laccase and horseradish peroxidase for triclosan conversion. Environ Technol 37(3):335–343
Mohanakrishna G, Al-Raoush RI, Abu-Reesh IM, Pant D (2019) A microbial fuel cell configured for the remediation of recalcitrant pollutants in soil environment. RSC Adv 9(71):41409–41418
Mohiddin GJ, Srinivasulu M, Maddela NR, Manjunatha B, Rangaswamy V, Kaiser ARK, Asqui JCM, Rueda OD (2015) Influence of the insecticides acetamiprid and carbofuran on arylamidase and myrosinase activities in the tropical black and red clay soils. Environ Monit Assess 187 (6):388
Morales M, Sentchilo V, Bertelli C, Komljenovic A, Kryuchkova-Mostacci N, Bourdilloud A, Linke B, Goesmann A, Harshman K, Segers F (2016) The genome of the toluene-degrading Pseudomonas veronii strain 1YdBTEX2 and its differential gene expression in contaminated sand. PLoS One 11(11):e0165850
Moreno-Forero SK, Rojas E, Beggah S, van der Meer JR (2016) Comparison of differential gene expression to water stress among bacteria with relevant pollutant-degradation properties. Environ Microbiol Rep 8(1):91–102

Mudhoo A, Ramasamy DL, Bhatnagar A, Usman M, Sillanpää M (2020) An analysis of the versatility and effectiveness of composts for sequestering heavy metal ions, dyes and xenobiotics from soils and aqueous milieus. Ecotoxicol Environ Saf 197:110587
Mukherjee S, Weihermüller L, Tappe W, Hofmann D, Köppchen S, Laabs V, Vereecken H, Burauel P (2016) Sorption–desorption behaviour of bentazone, boscalid and pyrimethanil in biochar and digestate based soil mixtures for biopurification systems. Sci Total Environ 559:63–73
Mukred AM, Hamid AA, Hamzah A, Yusoff WMW (2008) Development of three bacteria consortium for the bioremediation of crude petroleum-oil in contaminated water. Online J Biol Sci 8(4):73–79
Naja GM, Volesky B (2009) Treatment of metal-bearing effluents: removal and recovery. Heavy Metals Environ:247–292
Ortet P, Gallois N, Piette L, Long J, Berthomieu C, Armengaud J, Barakat M, Chapon V (2017) Draft genome sequence of microbacterium oleivorans strain A9, a bacterium isolated from Chernobyl radionuclide-contaminated soil. Genome Announc 5(14)
Ortíz I, Velasco A, Le Borgne S, Revah S (2013) Biodegradation of DDT by stimulation of indigenous microbial populations in soil with cosubstrates. Biodegradation 24(2):215–225
Pandotra P, Raina M, Salgotra RK, Ali S, Mir ZA, Bhat JA, Tyagi A, Upadhahy D (2018) Modern age environmental problems and their remediation. Springer, Berlin, pp 169–200
Parthasarathy A, Tyler AC, Hoffman MJ, Savka MA, Hudson AO (2019) Is plastic pollution in aquatic and terrestrial environments a driver for the transmission of pathogens and the evolution of antibiotic resistance? ACS Publications
Peña A, Delgado-Moreno L, Rodríguez-Liébana JA (2020) A review of the impact of wastewater on the fate of pesticides in soils: effect of some soil and solution properties. Sci Total Environ 718:134468
Peng X, Huang J, Liu C, Xiang Z, Zhou J, Zhong G (2012) Biodegradation of bensulphuron-methyl by a novel Penicillium pinophilum strain, BP-H-02. J Hazard Mater 213:216–221
Pereira P, Enguita FJ, Ferreira J, Leitão AL (2014) DNA damage induced by hydroquinone can be prevented by fungal detoxification. Toxicol Rep 1:1096–1105
Rahman A, Nahar N, Jass J, Olsson B, Mandal A (2016) Complete genome sequence of Lysinibacillus sphaericus B1-CDA, a bacterium that accumulates arsenic. Genome Announc 4 (1)
Raju MN, Scalvenzi L (2017) Recent insights in petroleum science and engineering. IntechOpen
Raju MN, Venkateswarlu K (2014) Effect of repeated applications of buprofezin and acephate on soil cellulases, amylase, and invertase. Environ Monit Assess 186(10):6319–6325
Raju MN, Leo R, Herminia SS, Moran RE, Venkateswarlu K, Laura S (2017) Biodegradation of diesel, crude oil and spent lubricating oil by soil isolates of Bacillus spp. Bull Environ Contam Toxicol 98(5):698–705
Ramakrishnan B, Megharaj M, Venkateswarlu K, Sethunathan N, Naidu R (2011) Reviews of environmental contamination and toxicology, vol 211. Springer, Berlin, pp 63–120
Raman G, Sakthivel N, Park S (2015) Draft genome sequence of a novel nicotine-degrading bacterium, Pseudomonas plecoglossicida TND35. Genome Announc 3(4)
Ramos L, Berenstein G, Hughes EA, Zalts A, Montserrat JM (2015) Polyethylene film incorporation into the horticultural soil of small periurban production units in Argentina. Sci Total Environ 523:74–81
Rezende-Silva SL, Costa AC, Dyszy FH, Batista PF, Crispim-Filho AJ, Nascimento KJT, da Silva AA (2019) Pouteria torta is a remarkable native plant for biomonitoring the glyphosate effects on Cerrado vegetation. Ecol Indic 102:497–506
Rubio-Bellido M, Madrid F, Morillo E, Villaverde J (2015) Assisted attenuation of a soil contaminated by diuron using hydroxypropyl-β-cyclodextrin and organic amendments. Sci Total Environ 502:699–705
Said Mohamed D, Amany Mohamed H, Neven Abd Elfatah E, Slah Ibrahem H (2014) Heavy metal adsorption of Streptomyces chromofuscus K101. J Coastal Life Med 2(6):431–437

Saldarriaga-Hernandez S, Hernandez-Vargas G, Iqbal HMN, Barceló D, Parra-Saldívar R (2020) Bioremediation potential of Sargassum sp biomass to tackle pollution in coastal ecosystems: circular economy approach. Sci Total Environ 715:136978

Salunkhe VP, Sawant IS, Banerjee K, Wadkar PN, Sawant SD (2015) Enhanced dissipation of Triazole and multiclass pesticide residues on grapes after foliar application of grapevine-associated Bacillus species. J Agric Food Chem 63(50):10736–10746

Saravanan A, Jeevanantham S, Narayanan VA, Kumar PS, Yaashikaa PR, Muthu CMM (2020) Rhizoremediation – A promising tool for the removal of soil contaminants: a review. J Environ Chem Eng 8(2):103543

Scheunert I (1991) In: Berthelin J (ed) Developments in geochemistry. Elsevier, pp 97–104

Sepehri A, Sarrafzadeh M-H, Avateffazeli M (2020) Interaction between Chlorella vulgaris and nitrifying-enriched activated sludge in the treatment of wastewater with low C/N ratio. J Clean Prod 247:119164

Serdar O (2019) The effect of dimethoate pesticide on some biochemical biomarkers in Gammarus pulex. Environ Sci Pollut Res 26(21):21905–21914

Shakerian F, Zhao J, Li S-P (2020) Recent development in the application of immobilized oxidative enzymes for bioremediation of hazardous micropollutants – a review. Chemosphere 239:124716

Sharma, A. and Rehman, M.B. (2009) Laboratory scale bioremediation of diesel hydrocarbon in soil by indigenous bacterial consortium

Sharma B, Dangi AK, Shukla P (2018) Contemporary enzyme based technologies for bioremediation: a review. J Environ Manag 210:10–22

Shedbalkar U, Jadhav JP (2011) Detoxification of malachite green and textile industrial effluent by Penicillium ochrochloron. Biotechnol Bioprocess Eng 16(1):196

Silambarasan S, Abraham J (2013) Ecofriendly method for bioremediation of Chlorpyrifos from agricultural soil by novel fungus Aspergillus terreus JAS1. Water Air Soil Pollut 224:1369

Silva V, Mol HGJ, Zomer P, Tienstra M, Ritsema CJ, Geissen V (2019) Pesticide residues in European agricultural soils–A hidden reality unfolded. Sci Total Environ 653:1532–1545

Simeonova A, Godjevargova T, Ivanova D (2008) Biosorption of heavy metals by dead Streptomyces fradiae. Environ Eng Sci 25(5):627–634

Singh AK, Cameotra SS (2014) Influence of microbial and synthetic surfactant on the biodegradation of atrazine. Environ Sci Pollut Res 21(3):2088–2097

Singh P, Saini HS, Raj M (2016) Rhamnolipid mediated enhanced degradation of chlorpyrifos by bacterial consortium in soil-water system. Ecotoxicol Environ Saf 134:156–162

Singh DP, Prabha R, Gupta VK, Verma MK (2018) Metatranscriptome analysis deciphers multifunctional genes and enzymes linked with the degradation of aromatic compounds and pesticides in the wheat rhizosphere. Front Microbiol 9:1331

Sobjak TM, Romão S, do Nascimento CZ, dos Santos AFP, Vogel L, Guimarães ATB (2017) Assessment of the oxidative and neurotoxic effects of glyphosate pesticide on the larvae of Rhamdia quelen fish. Chemosphere 182:267–275

Subashchandrabose SR, Venkateswarlu K, Venkidusamy K, Palanisami T, Naidu R, Megharaj M (2019) Bioremediation of soil long-term contaminated with PAHs by algal–bacterial synergy of Chlorella sp. MM3 and Rhodococcus wratislaviensis strain 9 in slurry phase. Sci Total Environ 659:724–731

Sun C, Dudley S, Trumble J, Gan J (2018) Pharmaceutical and personal care products-induced stress symptoms and detoxification mechanisms in cucumber plants. Environ Pollut 234:39–47

Ten Hulscher TEM, Cornelissen G (1996) Effect of temperature on sorption equilibrium and sorption kinetics of organic micropollutants-a review. Chemosphere 32(4):609–626

Thelusmond J-R, Strathmann TJ, Cupples AM (2019) Carbamazepine, triclocarban and triclosan biodegradation and the phylotypes and functional genes associated with xenobiotic degradation in four agricultural soils. Sci Total Environ 657:1138–1149

Thijs S, Van Hamme J, Gkorezis P, Rineau F, Weyens N, Vangronsveld J (2014) Draft genome sequence of Raoultella ornithinolytica TNT, a trinitrotoluene-denitrating and plant growth-promoting strain isolated from explosive-contaminated soil. Genome Announc 2(3)

Turkovskaya O, Muratova A (2019) Plant–bacterial degradation of polyaromatic hydrocarbons in the rhizosphere. Trends Biotechnol 37(9):926–930

Utturkar SM, Bollmann A, Brzoska RM, Klingeman DM, Epstein SE, Palumbo AV, Brown SD (2013) Draft genome sequence for Ralstonia sp. strain OR214, a bacterium with potential for bioremediation. Genome Announc 1(3)

Varo-Arguello WE, Camacho-Pérez B, Ríos-Leal E, Vazquez-Landaverde PA, Ponce-Noyola MT, Barrera-Cortés J, Sastre-Conde I, Rindernknecht-Seijas NF, Poggi-Varaldo HM (2012) Triphasic slurry bioreactors for the bioremediation of lindane-impacted soil under aerobic and anaerobic conditions. Environ Eng Manag J 11(10):1811–1823

Velisek J, Stara A, Zuskova E, Kouba A (2017) Effects of three triazine metabolites and their mixture at environmentally relevant concentrations on early life stages of marbled crayfish (Procambarus fallax f. virginalis). Chemosphere 175:440–445

Vezzaro L, Benedetti L, Gevaert V, De Keyser W, Verdonck F, De Baets B, Nopens I, Cloutier F, Vanrolleghem PA, Mikkelsen PS (2014) A model library for dynamic transport and fate of micropollutants in integrated urban wastewater and stormwater systems. Environ Model Softw 53:98–111

Vikram S, Kumar S, Vaidya B, Pinnaka AK, Raghava GPS (2013) Draft genome sequence of the 2-chloro-4-nitrophenol-degrading bacterium Arthrobacter sp. strain SJCon. Genome Announc 1 (2)

Villaverde J, Posada-Baquero R, Rubio-Bellido M, Laiz L, Saiz-Jimenez C, Sanchez-Trujillo, M.a.A. and Morillo, E. (2012) Enhanced mineralization of diuron using a cyclodextrin-based bioremediation technology. J Agric Food Chem 60(40):9941–9947

Wang Q, Huang L, Pan Y, Zhou P, Quan X, Logan BE, Chen H (2016) Cooperative cathode electrode and in situ deposited copper for subsequent enhanced Cd(II) removal and hydrogen evolution in bioelectrochemical systems. Bioresour Technol 200:565–571

Wen X, Du C, Wan J, Zeng G, Huang D, Yin L, Deng R, Tan S, Zhang J (2019) Immobilizing laccase on kaolinite and its application in treatment of malachite green effluent with the coexistence of Cd(II). Chemosphere 217:843–850

Wolski EA, Barrera V, Castellari C, González JF (2012) Biodegradation of phenol in static cultures by Penicillium chrysogenum ERK1: catalytic abilities and residual phytotoxicity. Revista Argentina de Microbiol 44(2):113–121

Xu R, Si Y, Wu X, Li F, Zhang B (2014) Triclosan removal by laccase immobilized on mesoporous nanofibers: strong adsorption and efficient degradation. Chem Eng J 255:63–70

Xu R, Tang R, Zhou Q, Li F, Zhang B (2015) Enhancement of catalytic activity of immobilized laccase for diclofenac biodegradation by carbon nanotubes. Chem Eng J 262:88–95

Yadav M, Singh SK, Sharma JK, Yadav KDS (2011) Oxidation of polyaromatic hydrocarbons in systems containing water miscible organic solvents by the lignin peroxidase of Gleophyllum striatum MTCC-1117. Environ Technol 32(11):1287–1294

Yan J, Niu J, Chen D, Chen Y, Irbis C (2014) Screening of Trametes strains for efficient decolorization of malachite green at high temperatures and ionic concentrations. Int Biodeterior Biodegradation 87:109–115

Yan D, Zhang Y, Liu L, Yan H (2016) Pesticide exposure and risk of Alzheimer's disease: a systematic review and meta-analysis. Sci Rep 6(1):1–9

Yang Y, Yu X, Zhang R (2013) Draft genome sequence of Ochrobactrum pseudogrignonense strain CDB2, a highly efficient arsenate-resistant soil bacterium from arsenic-contaminated cattle dip sites. Genome Announc 1(2)

Yang J, Lin Y, Yang X, Ng TB, Ye X, Lin J (2017) Degradation of tetracycline by immobilized laccase and the proposed transformation pathway. J Hazard Mater 322:525–531

Ye M, Sun M, Hu F, Kengara FO, Jiang X, Luo Y, Yang X (2014) Remediation of organochlorine pesticides (OCPs) contaminated site by successive methyl-β-cyclodextrin (MCD) and sunflower oil enhanced soil washing – Portulaca oleracea L. cultivation. Chemosphere 105:119–125

Ye X, Peng T, Feng J, Yang Q, Pratush A, Xiong G, Huang T, Hu Z (2019) A novel dehydrogenase 17β-HSDx from Rhodococcus sp. P14 with potential application in bioremediation of steroids contaminated environment. J Hazard Mater 362:170–177

Zdarta J, Meyer AS, Jesionowski T, Pinelo M (2018) Developments in support materials for immobilization of oxidoreductases: a comprehensive review. Adv Colloid Interf Sci 258:1–20

Zhang B, Tian C, Liu Y, Hao L, Liu Y, Feng C, Liu Y, Wang Z (2015) Simultaneous microbial and electrochemical reductions of vanadium (V) with bioelectricity generation in microbial fuel cells. Bioresour Technol 179:91–97

Ziganshina EE, Mohammed WS, Doijad SP, Shagimardanova EI, Gogoleva NE, Ziganshin AM (2018) Draft genome sequence of Brevibacterium epidermidis EZ-K02 isolated from nitrocellulose-contaminated wastewater environments. Data Brief 17:119–123

# Chapter 13
# Microbial Enzymes as Thrombolytics

**Prakash Kumar Sinha and Eshu Singhal Sinha**

**Abstract** An imbalance between blood coagulation and thrombolysis is known to disturb haemostasis in body resulting in thrombosis or blood clot formation which is a major cause of myocardial infarction and stroke. As a therapeutic measure, the enzymes that can dissolve blood clot are administered as thrombolytic agents in the patients. Fibrinolytic enzymes derived from different microbes including bacteria and fungi have gained considerable attention than physiological thrombolytic agents such as urokinase and tissue plasminogen activator because of their cost effectivity. This chapter focuses on different microbial proteases including Nattokinase, Streptokinase, Staphylokinase, *Streptococcus uberis* plasminogen activator and several fungi-derived proteases that are reported to exhibit potent thrombolytic activities.

**Keywords** Plasmin · Plasminogen activators · Nattokinase · Streptokinase · Staphylokinase

## 13.1 Introduction

Microorganisms are used for hundreds of years for the isolation of commercially useful organic products of industrial value. Besides being used for the production of herbicides, insecticides, plant hormones, enzymes, and vitamins; these organisms are also used for the production of therapeutics. For instance, antibiotics are isolated from fungi such as *Penicllium* and *Cephalosporium* and from some species of bacterium *Streptomyces*. A wide number of microbial species infect human beings by exploiting the ligand receptor interaction. These microbes either secrete specific molecules or express those molecules on their surfaces which have similarity or

---

Prakash Kumar Sinha and Eshu Singhal Sinha contributed equally.

P. K. Sinha · E. S. Sinha (✉)
Department of Biotechnology, Panjab University, Chandigarh, India



P. K. Arora (ed.), *Microbial Products for Health, Environment and Agriculture*, Microorganisms for Sustainability 31, https://doi.org/10.1007/978-981-16-1947-2_13

affinity for receptor expressed on the host cell surface. These molecules are exploited for many years as therapeutic molecules as anticoagulants, antidepressants, vasodilators etc. In USA alone, microorganisms are used for the production of billion dollars of drugs annually. Many of these drugs are natural, but several are modified form of these molecules. Established knowledge of microbial genomes, simple culture conditions and cost effectiveness make microbes the first choice of organisms to produce molecules of commercial and therapeutic importance.

To invade the host, the microbes need to invade the tissue barrier by degrading extracellular matrix and basement membrane. It requires the lysis of collagen fibrins, elastin, fibronectin, and proteoglycans, which occurs through proteinases. This property makes some specific microbial proteins efficient thrombolytics. Thrombolytic molecules are required for the treatment of coronary thrombosis for the dissolution of blood clots through degradation of fibrin present in blood clots.

When blood vessels are injured, platelets aggregate at the site of injury and stimulate coagulation pathway which eventually ends in the formation of a fibrin clot that can stop minor bleeding. Once the bleeding stops and the injured vessel is repaired; the growth of the thrombus is arrested, physiologically. Finally, the formed clot is dissolved by the action of intrinsic fibrinolytic system of the body (Riddel Jr. et al. 2007). Thus, a balance between thrombus formation and thrombolysis regulates hemostasis. When this physiological balance is disturbed due to altered physiology or pathological condition, blood clot/thrombus formation leads to blockage of blood flow that results in unavailability of oxygen and nutrient supply to respective tissue leading to ischemia. If thrombus formation occurs in blood vessels of heart muscles, restriction of blood flow to cardiac muscle results in myocardial infarction which may even lead to death (Peng et al. 2005). According to WHO report, 31% of all global deaths (17.9 million people) occur due to circulatory disorders (Sinha et al. 2019). Out of these 85% of deaths take place in low and middle income countries, and occur equally in men and women. An effective therapy is rapid clot dissolution to minimize tissue damage.

There are four therapy options for thrombosis: surgical removal of the clot, usage of anticoagulants, antiplatelet therapy, or administration of fibrinolytic enzymes. Generally, thrombolytic treatment through administration of fibrinolytic enzymes is preferred clinically to retrieve the function of clot affected tissue. Of the several known fibrinolytic enzymes, microbial fibrinolytic enzymes have attracted much more attention because of their cost-effectiveness and comparatively lesser side effects than other thrombolytic agents. Several fibrinolytic enzymes have been discovered from different microorganisms, especially from the genus *Bacillus*.

Before going in details about the microbes and their products, clot formation and clot lysis need to be understood first.

### 13.1.1 Clot Formation

The aggregation of platelets at the site of injury stimulates the coagulation pathway, which results in formation of a fibrin clot. Fibrinogen, a glycoprotein of Mr ~340 kDa is the precursor of fibrin and is present in plasma as well as platelet granules (Weisel 2005). Platelet aggregation provides the surface for the assembly and activation of coagulation complexes and generation of thrombin. Thrombin converts fibrinogen to fibrin and the fibrin strands thus generated, bind aggregated platelets (Riddel Jr. et al. 2007). Thrombin acts on the central domain of fibrinogen and liberates fibrinopeptides (Mosesson 2000). This is followed by non-covalent fibrin assembly involving progressive longitudinal and lateral elongation of polymer chains to form fibrin strands (Hermans and McDonagh 1982). At the end, cross-linking of factor XIIIa with fibrin further stabilizes fibrin (Weisel and Litvinov 2013). The fibrin mesh further recruits more platelets through receptor glycoproteins, fibronectin and platelet fibrinogen (Zucker et al. 1979). All these events contribute to formation of thrombus/blood clot. Additionally, factor XIIIa binds $\alpha_2$-plasmin inhibitor to make it mechanically more robust and more resistant to chemical and enzymatic degradation (Shen et al. 1975, 1977).

### 13.1.2 Thrombolysis

The thrombolytic pathway is a key component of haemostasis that serves to restrict the clot formation. It involves degradation of fibrin present inside blood clot. Therefore, it is alternatively referred to as fibrinolysis. This cascade involves zymogen-enzyme activation (LIJNEN 2006). Physiologically, fibrinolysis is done by the non-specific serine protease plasmin (PN). This requires activation of plasminogen (PG) to plasmin (PN) (Parry et al. 2000). PG is synthesized in liver and is converted to its active form PN with the help of PG activators (Rakoczi et al. 1978).

Fibrinolysis is regulated by two regulatory components; activators and the inhibitors. The activators include serine proteases: tissue type plasminogen activator (tPA) and urokinase type plasminogen activator (uPA) and the inhibitors include $\alpha_2$-antiplasmin and plasminogen activator inhibitor-1 (PAI-1). The activators are synthesized by endothelial cells while the inhibitors are mainly localized on the fibrin strands. These directly influence the degree of clot dissolution. A cardinal feature of both physiological fibrinolysis and therapeutic administration of PG activators is targeted generation of PN activity at the surface of the clot (fibrin selectivity) (Marder and Sherry 1988; Kiernan and Gersh 2007). PN degrades insoluble fibrin to small soluble fibrin degradation products (Fig. 13.1) that are swept into the circulation (Parry et al. 2000). PN activity is tightly regulated by $\alpha_2$-antiplasmin inhibitor which prevents widespread fibrinolysis.

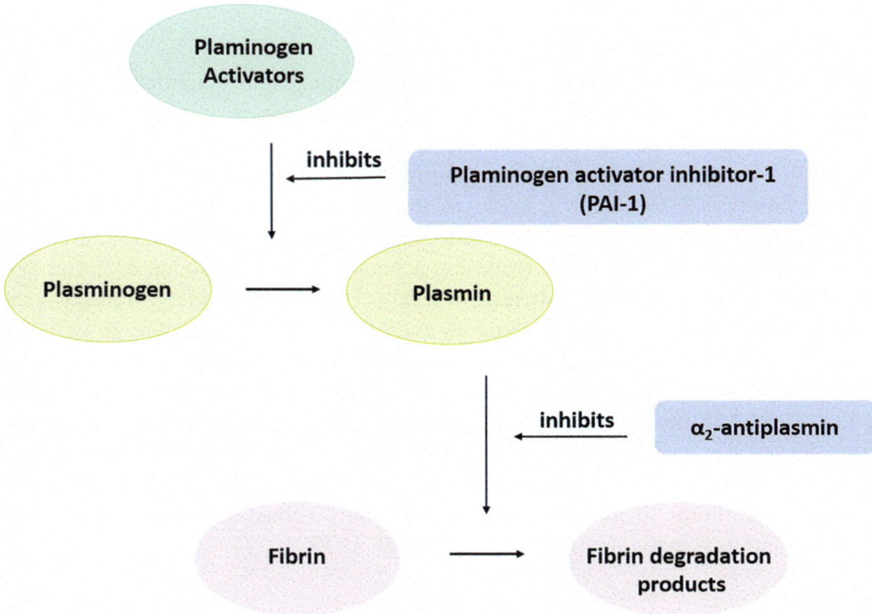


**Fig. 13.1** Schematic representation of fibrinolysis. Upon injury, vascular endothelial cells release plasminogen activators (tissue plasminogen activator and urokinase) which convert plasminogen to plasmin. Plasmin thus generated, degrades fibrin present in clots to fibrin degradation products. Fibrinolysis is negatively regulated by plasminogen activator inhibitor-1 (PAl-1) and plasmin inhibitor, $\alpha_2$-antiplasmin

## *13.1.3 PG Activators*

The activation of PG to PN by PG activators is the central event in the fibrinolytic cascade. PG activators can be broadly divided into two categories:

### 13.1.3.1 Direct PG Activators

These are generally intrinsic PG activators and show high degree of substrate specificity. They activate PG by direct cleavage of the Arg 561- Val 562 scissile peptide bond. These activators include tPA and uPA and their mutated or protein engineered derivatives.

### 13.1.3.2 Indirect PG Activators

These activators do not possess any enzymatic activity of their own. They form complexes with PG or PN and these complexes then act as PG activators. Indirect

PG activators include staphylokinase, *Streptococcus uberis* PG activator and streptokinase. Being bacterial in origin, the non-physiological PG activators are immunogenic in nature.

PN itself is a non-specific serine protease which is involved in tissue remodelling during development and can degrade various extracellular matrices and membranes (Lahteenmaki et al. 2001a). Some pathogenic microbes (*Streptococcus* and *Staphylococcus*) secrete PG activators which bind to PG/PN and inhibit its inactivation by $\alpha_2$-antiplasmin inhibitor (Lahteenmaki et al. 2001b). The bound PN can degrade metallo-protein and collagens which enables microbes to invade host (Boyle and Lottenberg 1997). This property of microbe's secretory protein is exploited by clinicians as thrombolytic drugs (Banerjee et al. 2004).

## 13.2 Bacterial Proteins as Thrombolytic Agents

Bacteria are a very important source of thrombolytic agents. Nattokinase (NK) from *Bacillus natto*, Streptokinase from *Streptococcus hemolyticus* and Staphylokinase from *Staphylococcus aureus* have proved to be very effective thrombolytic agents. Over the years, fibrinolytic enzymes from other bacteria have also been discovered such as subtilisin DFE and substilisin DJ-4 from *Bacillus amyloliquefacien*. However, the clinical implications of the latter enzymes is not fully established yet.

### *13.2.1 Nattokinase*

Nattokinase (NK) is produced by bacterium *Bacillus subtilis* during the fermentation of soyabeans to produce natto which is a traditional fermented food of Japan (Meruvu and Vangalapati 2011). It was the outcome of search for natural agent that could successfully dissolve thrombus. As a result, a Japanese researcher Hiroyuki Sumi discovered that natto has the capacity to degrade artificial fibrin. In 1987, he isolated an enzyme from natto that acted as a fibrinolytic enzyme which was then named "Nattokinase" (Sumi et al. 1987). Like most of the thrombolytic enzymes, NK is also serine protease in nature and belongs to subtilisin family of proteases. Mature NK protein has a molecular weight of 27.7 kDa, comprises of 275 amino acid residues and is devoid of cysteine residues (Chen and Chao 2006). Like other serine proteases, NK has a catalytic triad (D32, H64, and S221) and oxyanion hole (N155).

Unlike other thrombolytic enzymes such as tPA and uPA which are associated with various side effects including bleeding/haemorrhage, NK exhibits little to no side effects. Interestingly, NK can be taken orally as it is absorbed by intestine and exhibits strong fibrinolytic activity after its oral administration. It is a very potent thrombolytic enzyme that (1) acts directly on clots and degrades fibrin inside blood clot, (2) converts PG to PN (Fujita et al. 1993), (3) enhances fibrinolysis by

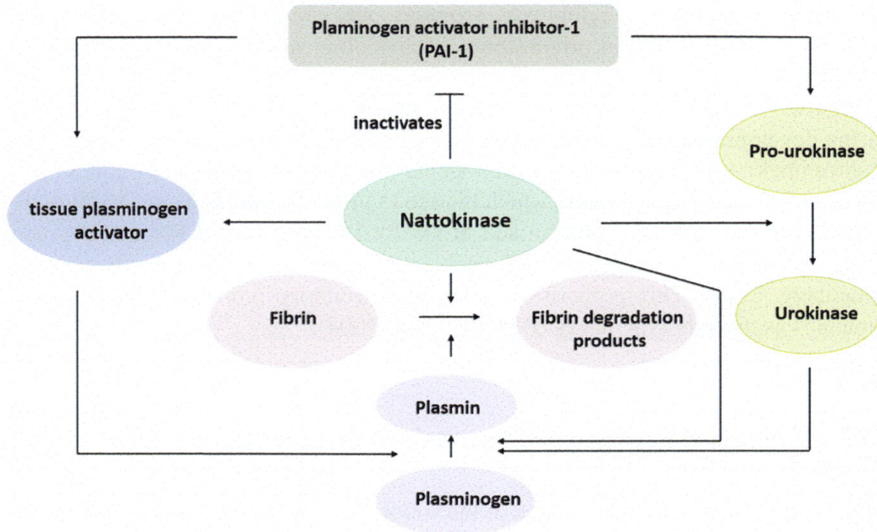


**Fig. 13.2** Mechanism of action of Nattokinase. Nattokinase dissolves blood clots by directly hydrolyzing fibrin to fibrin degradation products. It also converts plasminogen to plasmin which further degrades fibrin. Nattokinase increases level of tissue plasminogen activator and also converts endogenous prourokinase to urokinase. Being plasminogen activators, tissue plasminogen activator and urokinase convert plasminogen to plasmin. Nattokinase further supports fibrinolysis by inhibiting plasminogen activator inhibitor.

increasing the production of PG activators, uPA and tPA (Sumi et al. 1987; Fujita et al. 1993; Weng et al. 2017), and (4) enhances the half-life of tPA and uPA via inactivating PAI-1 (Fujita et al. 1993; Urano et al. 2001) (Fig. 13.2). Besides these NK leads to decrease in blood viscosity and decrease in RBC aggregation (Pais et al. 2006). It also reduces blood pressure via reducing systolic and diastolic pressure (Kim et al. 2008; Jensen et al. 2016).

### *13.2.2 Streptokinase*

Streptokinase (SK) is a single chain multi-domain protein cofactor secreted by various haemolytic *Streptococci* (Banerjee et al. 2004). It is composed of 414 amino-acids and has a molecular weight of ~47 kDa (Malke and Ferretti 1984). SK isolated from different microbes, differ in their sequences, structure and in their target host (Malke 1993; Huang et al. 1989). For therapeutic purpose, SK is isolated from *Streptococcus hemolyticus*.

SK is a fibrin-independent activator of PG/ PN (Reed et al. 1999) and therefore, can lead to activation of PG in blood (outside clot) where haemorrhage is regulated by α2-antiplasmin. SK is devoid of cystines and cysteines (Morgan and Henschen

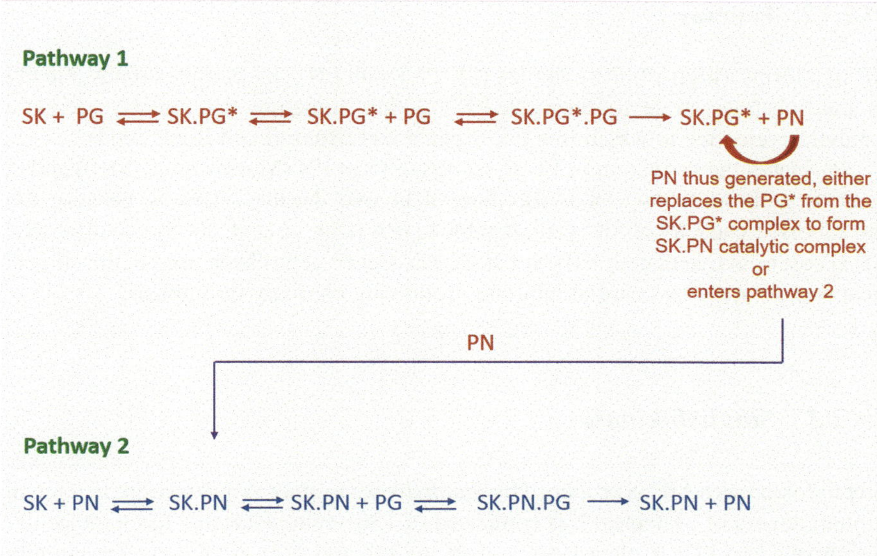


**Fig. 13.3** Mechanisms of plasminogen activation by Streptokinase. In pathway 1, SK forms an equimolar complex with PG (SK.PG*) which binds free PG molecule as a substrate forming SK.PG*.PG complex. PN is generated from the ternary complex (SK.PG*.PG) due to proteolytic cleavage of PG. PN thus generated, either binds free SK (in pathway 2) or replaces the PG* from the SK.PG* binary complex to form SK.PN catalytic complex. In pathway 2, SK forms a high affinity complex with PN (SK.PN) that binds free PG substrate to form SK.PN.PG, from which free PN is generated and liberated. *SK* Streptokinase, *PG* Plasminogen, *PN* Plasmin

1969) and does not have any proteolytic or esterase activity of its own (Dahiya et al. 2005). It acts as a protein cofactor to PG or PN. PN is a wide-spectrum non-specific protease. However, when SK binds with PG/PN, it modulates PN's specificity to using only PG as a substrate (Davidson et al. 1990). SK can act as a PG activator by two pathways (Fig. 13.3):

#### 13.2.2.1 Pathway 1

SK first forms a 1:1 molar complex with PG (SK.PG) to form a transition complex, SK.PG* that bears an active site in the zymogen (Summaria et al. 1982). In this step SK.PG gets converted to SK.PG* by undergoing a conformational transition without any proteolytic cleavage. This SK.PG* forms a ternary complex (SK.PG*.PG) upon binding to free PG as a substrate. The substrate PG gets cut at $Arg^{561}$-$Val^{562}$ scissile peptide bond, and it gets converted to active form of PG i.e. PN. The PN thus generated by SK.PG* either binds free SK or replaces the PG* from the SK.PG* complex to form SK.PN catalytic complex, as PN has ~11,000 fold higher affinity for SK in comparison to PG (Covarsi et al. 1978).

#### 13.2.2.2 Pathway 2

SK first forms a high affinity complex (SK.PN) with PN. This complex binds free PG in the substrate mode to form SK.PN.PG thus initiating the direct proteolytic catalytic cycle, due to which free PN is generated (Boxrud and Bock 2004).

SK causes the activation of PG to its active form PN (Murray et al. 2010) and is used as a thrombolytic agent for more than five decades. This is because the thrombolytic capacity of SK is comparable to tPA (Dahiya et al. 2005). Additionally, SK is cost effective than tPA (Baker Jr. 2002). Hence, it has been used in the clinical treatment of acute myocardial infarction following coronary thrombosis.

### *13.2.3 Staphylokinase*

Staphylokinase (SAK) is produced by *Staphylococcus aureus* and is a single domain protein devoid of cysteines. It is composed of 136 amino acids and has a molecular weight of ~15.5 kDa (Collen 1998; Collen and Lijnen 1994). Although it structurally matches with serine proteases, SAK is devoid of any enzymatic activity (Bokarewa et al. 2006). SAK acts as co-factor similar to SK (Collen et al. 1992). It is a thrombolytic molecule and forms 1:1 stoichiometry complex with PN. After forming SAK.PN complex, SK modulates the specificity of PN by generating PG activator complex which now acts on substrate PG. This leads to activation of substrate PG to PN which causes lysis of fibrin. N-terminal residues of SAK help in formation of SAK-PN complex. In contrast to SK-PG complex, the SAK-PG complex remains inactive until it gets converted to SAK-PN complex, which has specificity for PG activation (Grella and Castellino 1997). SAK has a great similarity with α and β domains of SK (Parry et al. 2000). Individual domains of SK are highly attenuated PG activators, whereas SAK is a single domain protein which is fully functional PG activator and has fibrin specificity (Dahiya et al. 2011). Inside circulatory system, $\alpha_2$-antiplasmin inhibits SAK-PN complex, but once SAK-PN complex gets attached to fibrin, it becomes resistant to $\alpha_2$-antiplasmin (Sakharov et al. 1996). Thus, unlike SK, SAK primarily activates fibrin-bound PG (Parry et al. 2000).

### *13.2.4* Streptococcus uberis *PG Activator*

As the name suggests, *Streptococcus uberis* PG Activator (SUPA) is isolated from *Streptococcus uberis,* a causative agent for bovine mastitis. SUPA has a two domain structure with molecular weight ~29 kDa which is an intermediate between single domain SAK ($M_r$ ~15.5 kDa) and three domain structure of SK ($M_r$ ~47 kDa). It cannot activate human PG but forms an activator complex with human PN (Zhang et al. 2012). SUPA shows activation kinetics similar to SK (Johnsen et al. 2000). In presence of fibrin the activity of SUPA-PN complex gets enhanced. Unlike SAK-PN complex, SUPA-PN complex is resistant to inhibition by $\alpha_2$-antiplasmin even in the absence of fibrin (Zhang et al. 2012). Similar to SK, SUPA generates non-proteolytic active site in host PG but at a slower pace as compared to SK. Catalytic efficiency of SUPA-PN is approximately 33% of SK-PN (Sazonova et al. 2001).

## 13.3 Fungal Proteins as Fibrinolytic Agents

Apart from bacterial sources, fungal proteases possessing fibrinolytic activity have also been reported. However, these are not used clinically as thrombolytics yet.

### *13.3.1 Proteases from* Fusarium sp

*Fusarium*, a filamentous fungi causes various diseases in plant and animals. Some novel enzymes with fibrinolytic activities have been isolated from different species of *Fusarium*. For instance, a novel protease, FP is produced *Fusarium sp*. BLB. The FP protease was firstly isolated in 2007 from tempeh which is a traditional Indonesian food produced by fermentation of soyabean by filamentous fungi, *Rhizopus sp*. and *Fusarium sp*. (Sugimoto et al. 2007). FP is a serine protease and has a molecular weight of 27 kDa, with maximal fibrinolytic activity at 50 °C and optimum pH of 9.5. This protease is a direct activator of PG, and thus forms PN from PG and dissolves the blood clot. The fibrinolytic activity of FP towards synthetic peptide is higher than that of Nattokinase and commercially available PN (Sugimoto et al. 2007).

*Fusarium oxysporum* N.R.C.1 also secretes a novel thrombolytic which acts optimally at 37 °C and at a pH of 7. It is a metallo-protease which requires $Co^{+2}$ as a cofactor for its activity. This protease can be inhibited by divalent chelator, ethylene diamine tetra acetic acid (EDTA) (Abdel-Fattah et al. 1993). Additionally, a fibrinolytic enzyme Fu-P was isolated from *Fusarium sp*. CPCC 480097. Fu-P has a molecular weight of 28 kDa, with optimum activity at 45 °C and at a pH of 8 (Wu et al. 2009). It is a direct thrombolytic enzyme and does not require PG for the dissolution of blood clot.

### *13.3.2 Proteases from* Rhizopus sp.

*Rhizopus sp*. are usually found in the organic rich soil. A serine metalloprotease enzyme was isolated in 2005 during brewing of rice wine which is fermented by *Rhizopus chinenesis* 12 in China. This protease has a molecular weight of 16–18 kDa and hydrolyzes fibrin clot directly without activation of PG activators. This novel fibrinolytic enzyme from *Rhizopus chinenesis* 12 degrades α, ß and γ chains of fibrin, simultaneously (Xiao-Lan et al. 2005). The fibrinolytic activity of this enzyme is comparable to Nattokinase. This makes it a promising therapeutic drug for thrombosis in future.

### *13.3.3 Proteases from* Pleurotus sp.

*Pleurotus sp.* are generally common edible oyster mushrooms. Some species of *Pleurotus* produce enzymes that can degrade fibrin. A fibrinolytic protease, PoFE (Mr 32 kDa) composed of 288 amino acid residues was isolated from *Pleurotus ostreatus* mycelia. The activity of this metalloprotease increases in the presence of $Ca^{+2}$, $Mg^{+2}$ and $Zn^{+2}$. PoFE is a direct fibrinolytic enzyme and does not need any PG activator. However, it hydrolyzes γ chain much slower than α and β chains of fibrin (Shen et al. 2007). Similar hydrolysis preference for α and β chains over γ chain of fibrin was also observed in case of a 14 kDa protease isolated from *Pleurotus eryngii*, which is also a direct fibrinolytic protease (Cha et al. 2010).

Potent thrombolytic enzymes have also been isolated from some other fungal species such as *Aspergillus species*, *Armillaria mellea*, *Cordyceps miltaris*, etc. Some novel fibrinolytic enzymes are isolated from *Fomitella fraxinea* mycelia (Lee et al. 2006). However, most of these enzymes are not completely studied to be established as clinically applicable thrombolytic drug candidates.

## 13.4 Comparison of Intrinsic PG Activators with Clot Busters of Microbial Origin

The choice of thrombolytic therapy depends on several factors including the stability/half-life of the agent, specificity for fibrin dissolution, immunogenicity, side effects associated with the therapy and cost of the thrombolytic agent. Thus, till today streptokinase, remains the preferred agent for thrombolytic therapy due to its cost-effectiveness especially in the developing countries.

The intrinsic PG activators, tPA and uPA are non-immunogenic but costly molecules. tPA is often accompanied with side effect of haemorrhage (Miller et al. 2011). Hence, alternative and safer thrombolytic molecules which are cost effective are sought. Despite being immunogenic, fibrinolytic proteins of microbial origin are preferred as these are easy to express and are cost-effective in nature. Thrombolytics of bacterial origin which are clinically used as therapeutic agents include nattokinase, staphylokinase and streptokinase. The administration of these bacterial proteins requires partial neutralisation of circulatory inhibitors (Roschlau 1972).

## 13.5 Advantages of Direct-Acting Thrombolytic Molecules over PG Activators

PG directed fibrin lysis is done by active form of PG i.e. PN. This property is exploited by PG activators. However, PG activators work efficiently on relatively small clots, *i.e* clot inside coronary artery in acute myocardial infarction (Novokhatny 2008). But in case of big, retracted clots present inside peripheral arterial occlusion, deep-vein thrombosis, and patients with deficiency in PG, direct fibrinolytic proteases or external supplementation of PN is required (Robbie et al. 1996; Potter Van Loon et al. 1992). In vitro study of tPA, PN and saline control on retracted clot suggests that tPA exhibits similar response as of saline, whereas PN shows dose dependent clot lysis (Novokhatny et al. 2003). The probable explanation of this experiment could be the limited or no availability of PG inside retracted clot (Robbie et al. 1996; Potter Van Loon et al. 1992). Therefore, direct fibrin lysis should be preferred, at least in cases where size of clot is large and unavailability of PG is there. One such example of direct fibrin lytic enzyme which is expressed in *E. coli.* is a truncated derivative TAL6003, developed by Talecris Biotherapeutics Inc., Research Triangle Park, NC. This derivative contains just one domain of PG covalently attached to serine protease domain of PN. It retains the fibrinolytic function, inhibition by $\alpha_2$-antiplasmin, and fibrin affinity as of full-length PN (Hunt et al. 2008). As this derivative has no glycosylation site and is composed of only two domains (as compared to seven domains in PG/PN), it is relatively easier to express in *E. coli.*

## References

Abdel-Fattah AF, Ismail AM, Saleh SA (1993) Production and some properties of fibrinolytic enzymes from Fusarium oxysporum N.R.C.1. Zentralbl Mikrobiol 148(2):117–122

Baker WF Jr (2002) Thrombolytic therapy. Clin Appl Thromb Hemost 8(4):291–314

Banerjee A, Chisti Y, Banerjee UC (2004) Streptokinase--a clinically useful thrombolytic agent. Biotechnol Adv 22(4):287–307

Bokarewa MI, Jin T, Tarkowski A (2006) Staphylococcus aureus: staphylokinase. Int J Biochem Cell Biol 38(4):504–509

Boxrud PD, Bock PE (2004) Coupling of conformational and proteolytic activation in the kinetic mechanism of plasminogen activation by streptokinase. J Biol Chem 279(35):36642–36649

Boyle MD, Lottenberg R (1997) Plasminogen activation by invasive human pathogens. Thromb Haemost 77(1):1–10

Cha WS, Park SS, Kim SJ, Choi DB (2010) Biochemical and enzymatic properties of a fibrinolytic enzyme from Pleurotus eryngii cultivated under solid-state conditions using corn cob. Bioresour Technol 101(16):6475–6481

Chen PT, Chao YP (2006) Enhanced production of recombinant nattokinase in Bacillus subtilis by the elimination of limiting factors. Biotechnol Lett 28(19):1595–1600

Collen D (1998) Staphylokinase: a potent, uniquely fibrin-selective thrombolytic agent. Nat Med 4 (3):279–284

Collen D, Lijnen HR (1994) Staphylokinase, a fibrin-specific plasminogen activator with therapeutic potential? Blood 84(3):680–686
Collen D, Silence K, Demarsin E, De Mol M, Lijnen HR (1992) Isolation and characterisation of natural and recombinant staphylokinase. Fibrinol Proteol 6(4):203–213
Covarsi A, Gonzalo A, Sarrias J, Carrera M (1978) Berger's disease: study of 12 cases. Rev Clin Esp 148(1):69–72
Dahiya M, Rajamohan G, Dikshit KL (2005) Enhanced plasminogen activation by staphylokinase in the presence of streptokinase beta/betagamma domains: plasminogen kringles play a role. FEBS Lett 579(7):1565–1572
Dahiya M, Singh S, Rajamohan G, Sethi D, Ashish, Dikshit KL (2011) Intermolecular interactions in staphylokinase-plasmin(ogen) bimolecular complex: function of His43 and Tyr44. FEBS Lett 585(12):1814–1820
Davidson DJ, Higgins DL, Castellino FJ (1990) Plasminogen activator activities of equimolar complexes of streptokinase with variant recombinant plasminogens. Biochemistry 29 (14):3585–3590
Fujita M, Nomura K, Hong K, Ito Y, Asada A, Nishimuro S (1993) Purification and characterization of a strong fibrinolytic enzyme (nattokinase) in the vegetable cheese natto, a popular soybean fermented food in Japan. Biochem Biophys Res Commun 197(3):1340–1347
Grella DK, Castellino FJ (1997) Activation of human plasminogen by staphylokinase. Direct evidence that preformed plasmin is necessary for activation to occur. Blood 89(5):1585–1589
Hermans J, McDonagh J (1982) Fibrin: structure and interactions. Semin Thromb Hemost 8 (1):11–24
Huang TT, Malke H, Ferretti JJ (1989) Heterogeneity of the streptokinase gene in group A streptococci. Infect Immun 57(2):502–506
Hunt JA, Petteway SR, Scuderi P, Novokhatny V (2008) Simplified recombinant plasmin: production and functional comparison of a novel thrombolytic molecule with plasma-derived plasmin. Thromb Haemost 100(3):413–419
Jensen GS, Lenninger M, Ero MP, Benson KF (2016) Consumption of nattokinase is associated with reduced blood pressure and von willebrand factor, a cardiovascular risk marker: results from a randomized, double-blind, placebo-controlled, multicenter north american clinical trial. Integr Blood Press Control 9:95–104
Johnsen LB, Rasmussen LK, Petersen TE, Etzerodt M, Fedosov SN (2000) Kinetic and structural characterization of a two-domain streptokinase: dissection of domain functionality. Biochemistry 39(21):6440–6448
Kiernan TJ, Gersh BJ (2007) Thrombolysis in acute myocardial infarction: current status. Med Clin N Am 91(4):617–637
Kim JY, Gum SN, Paik JK, Lim HH, Kim KC, Ogasawara K et al (2008) Effects of nattokinase on blood pressure: a randomized, controlled trial. Hypertens Res 31(8):1583–1588
Lahteenmaki K, Kukkonen M, Korhonen TK (2001a) The Pla surface protease/adhesin of Yersinia pestis mediates bacterial invasion into human endothelial cells. FEBS Lett 504(1–2):69–72
Lahteenmaki K, Kuusela P, Korhonen TK (2001b) Bacterial plasminogen activators and receptors. FEMS Microbiol Rev 25(5):531–552
Lee JS, Baik HS, Park SS (2006) Purification and characterization of two novel fibrinolytic proteases from mushroom, Fomitella fraxinea. J Microbiol Biotechnol 16(2):264–271
Lijnen HR (2006) Elements of the fibrinolytic system. Ann N Y Acad Sci
Malke H (1993) Polymorphism of the streptokinase gene: implications for the pathogenesis of post-streptococcal glomerulonephritis. Zentralbl Bakteriol 278(2–3):246–257
Malke H, Ferretti JJ (1984) Streptokinase: cloning, expression, and excretion by Escherichia coli. Proc Natl Acad Sci U S A 81(11):3557–3561
Marder VJ, Sherry S (1988) Thrombolytic therapy: current status. N Engl J Med 318 (23):1512–1520
Meruvu H, Vangalapati M (2011) Nattokinase: a review on fibrinolytic enzyme. Int J Chem Environ Pharm Res 2(1):61–66

Miller DJ, Simpson JR, Silver B, Silver B (2011) Safety of thrombolysis in acute ischemic stroke: a review of complications, risk factors, and newer technologies. Neurohospitalist 1(3):138–147

Morgan FJ, Henschen A (1969) The structure of streptokinase I. Cyanogen bromide fragmentation, amino acid composition and partial amino acid sequences. BBA Protein Struct 181(1):93–104

Mosesson MW (2000) Fibrinogen functions and fibrin assembly. Fibrinol Proteol 14(2–3):182–186

Murray V, Norrving B, Sandercock PA, Terent A, Wardlaw JM, Wester P (2010) The molecular basis of thrombolysis and its clinical application in stroke. J Intern Med 267(2):191–208

Novokhatny V (2008) Structure and activity of plasmin and other direct thrombolytic agents. Thromb Res 122(3):S3–S8

Novokhatny V, Taylor K, Zimmerman TP (2003) Thrombolytic potency of acid-stabilized plasmin: superiority over tissue-type plasminogen activator in an in vitro model of catheter-assisted thrombolysis. J Thromb Haemost 1(5):1034–1041

Pais E, Alexy T, Holsworth RE, Meiselman HJ (2006) Effects of nattokinase, a pro-fibrinolytic enzyme, on red blood cell aggregation and whole blood viscosity. Clin Hemorheol Microcircul 35(1–2):139–142

Parry MA, Zhang XC, Bode I (2000) Molecular mechanisms of plasminogen activation: bacterial cofactors provide clues. Trends Biochem Sci 25(2):53–59

Peng Y, Yang X, Zhang Y (2005) Microbial fibrinolytic enzymes: an overview of source, production, properties, and thrombolytic activity in vivo. Appl Microbiol Biotechnol 69(2):126–132

Potter Van Loon BJ, Rijken DC, Brommer EJP, Van Der Maas APC (1992) The amount of plasminogen, tissue-type plasminogen activator and plasminogen activator inhibitor type 1 in human thrombi and the relation to ex-vivo lysibility. Thromb Haemost 67(1):101–105

Rakoczi I, Wiman B, Collen D (1978) On the biological significance of the specific interaction between fibrin, plasminogen and antiplasmin. Biochim Biophys Acta 540(2):295–300

Reed GL, Houng AK, Liu L, Parhami-Seren B, Matsueda LH, Wang S et al (1999) A catalytic switch and the conversion of streptokinase to a fibrin-targeted plasminogen activator. Proc Natl Acad Sci U S A 96(16):8879–8883

Riddel JP Jr, Aouizerat BE, Miaskowski C, Lillicrap DP (2007) Theories of blood coagulation. J Pediatr Oncol Nurs 24(3):123–131

Robbie LA, Bennett B, Croll AM, Brown PAJ, Booth NA (1996) Proteins of the fibrinolytic system in human thrombi. Thromb Haemost 75(1):127–133

Roschlau WH (1972) Thrombolytic properties and side effects of brinase (fibrinolytic enzyme from Aspergillus oryzae) in the dog. J Clin Pathol 25(7):635–636

Sakharov DV, Lijnen HR, Rijken DC (1996) Interactions between staphylokinase, plasmin(ogen), and fibrin. Staphylokinase discriminates between free plasminogen and plasminogen bound to partially degraded fibrin. J Biol Chem 271(44):27912–27918

Sazonova IY, Houng AK, Chowdhry SA, Robinson BR, Hedstrom L, Reed GL (2001) The mechanism of a bacterial plasminogen activator intermediate between streptokinase and staphylokinase. J Biol Chem 276(16):12609–12613

Shen LL, Hermans J, McDonagh J, McDonagh RP, Carr M (1975) Effects of calcium ion and covalent crosslinking on formation and elasticity of fibrin cells. Thromb Res 6(3):255–265

Shen LL, McDonagh RP, McDonagh J, Hermans J (1977) Early events in the plasmin digestion of fibrinogen and fibrin. Effects of plasmin on fibrin polymerization. J Biol Chem 252 (17):6184–6189

Shen MH, Kim JS, Sapkota K, Park SE, Choi BS, Kim S et al (2007) Purification, characterization, and cloning of fibrinolytic metalloprotease from Pleurotus ostreatus mycelia. J Microbiol Biotechnol 17(8):1271–1283

Sinha PK, Sinha ES, Kaur J (2019) Biomarkers of cardiac health and disease. Crit Rev Biomed Eng 47(5):395–407

Sugimoto S, Fujii T, Morimiya T, Johdo O, Nakamura T (2007) The fibrinolytic activity of a novel protease derived from a tempeh producing fungus, Fusarium sp. BLB. Biosci Biotechnol Biochem 71(9):2184–2189

Sumi H, Hamada H, Tsushima H, Mihara H, Muraki H. A novel fibrinolytic enzyme (nattokinase) in the vegetable cheese Natto; a typical and popular soybean food in the Japanese diet. Experientia 1987;43(10):1110–1

Summaria L, Wohl RC, Boreisha IG, Robbins KC (1982) A virgin enzyme derived from human plasminogen. Specific cleavage of the arginyl-560-valyl peptide bond in the diisopropoxyphosphinyl virgin enzyme by plasminogen activators. Biochemistry 21 (9):2056–2059

Urano T, Ihara H, Umemura K, Suzuki Y, Oike M, Akita S et al (2001) The profibrinolytic enzyme subtilisin NAT purified from Bacillus subtilis cleaves and inactivates plasminogen activator inhibitor type 1. J Biol Chem 276(27):24690–24696

Weisel JW (2005) Fibrinogen and fibrin. Adv Protein Chem 70:247–299

Weisel JW, Litvinov RI (2013) Mechanisms of fibrin polymerization and clinical implications. Blood 121(10):1712–1719

Weng Y, Yao J, Sparks S, Wang KY (2017) Nattokinase: an oral antithrombotic agent for the prevention of cardiovascular disease. Int J Mol Sci 18(3):523

Wu B, Wu L, Chen D, Yang Z, Luo M (2009) Purification and characterization of a novel fibrinolytic protease from Fusarium sp. CPCC 480097. J Ind Microbiol Biotechnol 36 (3):451–459

Xiao-Lan L, Lian-Xiang D, Fu-Ping L, Xi-Qun Z, Jing X (2005) Purification and characterization of a novel fibrinolytic enzyme from Rhizopus chinensis 12. Appl Microbiol Biotechnol 67 (2):209–214

Zhang Y, Gladysheva IP, Houng AK, Reed GL (2012) Streptococcus uberis plasminogen activator (SUPA) activates human plasminogen through novel species-specific and fibrin-targeted mechanisms. J Biol Chem 287(23):19171–19176

Zucker MB, Mosesson MW, Broekman MJ, Kaplan KL (1979) Release of platelet fibronectin (cold-insoluble globulin) from alpha granules induced by thrombin or collagen; lack of requirement for plasma fibronectin in ADP-induced platelet aggregation. Blood 54(1):8–12

# Chapter 14
# Plant Growth Promoting Microbes and their Potential Application in Biotechnology

**Hafida Baoune, Mabrouka Bouafiane, and Thinhinane Fecih**

**Abstract** In the last decades, quite a number of plant associated microbes have been intensively studied. Those microbes can benefit their host plant directly or indirectly. Direct mechanisms could be by helping plants with nutrient acquisition and uptake, alleviating environmental stress and regulating related phytohormones. Indirectly, they are able to suppress the growth of plant pathogens with the production of antibiotics, anti-fungus, hydrolytic enzymes, nutrients competition, and stimulation of plant defense system. Plant growth promoting microbes and their metabolites have been used to improve world food production, soil fertilization, environment clean-up, etc. Consequently, these emerged microbes can serve the same purpose or even do better than the chemical approaches. In this chapter, we introduce plant growth promoting microbes, their metabolites and the potential to be used in different ecosystems. Plant associated microbes are a most promising microbial resource to be exploited as a sustainable alternative technology.

**Keywords** Plant growth promoting microbes · PGP metabolites · Phytohormones · Endophytes

## Abbreviations

| | |
|---|---|
| ACC | 1-aminocyclopropane-1-decarboxylate |
| DFP | Deoxyfusapyrone |
| FP | Fusapyrone |
| HCN | Hydrogen cyanide |
| HPLC | High Performance Liquid Chromatography |
| IAA | Indole-3-acetic acid |
| $NH_3^+$ | Ammonia |

H. Baoune (✉) · M. Bouafiane · T. Fecih
Université Kasdi Merbah, Ouargla, Algeria
e-mail: baounehafida@hotmail.fr



P. K. Arora (ed.), *Microbial Products for Health, Environment and Agriculture*, Microorganisms for Sustainability 31, https://doi.org/10.1007/978-981-16-1947-2_14

| | |
|---|---|
| $NH_4^+$ | Ammonium |
| PGPB | Plant growth promoting bacteria |
| PGPM | Plant growth promoting microbes |
| PGPR | Plant growth promoting rhizobacteria |
| SAM | S-adenosyl methionine |

## 14.1 Introduction

Some of the microorganisms present in the soil can develop beneficial associations with members of their ecosystems such as plant. Such microorganisms are a class of bacteria or fungi that provide several benefits to their host plant and vice versa. These microbes can live in the extern or the intern ecosystems of their host plant, those living externally are either epiphytic, or rhizospheric, whereas, those living inside plants are called endophytes (Afzal et al. 2019). Enormous variety of those microorganisms play a significant role in the growth, tolerance, and disease protection of plants (Rilling et al. 2019). These microbes can enhance plant growth directly by either facilitating resource acquisition (fixing nitrogen, iron uptake, phosphate solubilize) or modulating phytohormone levels, or indirectly by reducing the inhibitory effects of phytopathogens on plant growth and development, acting as biocontrol agents (Glick 2012).

The increase of both environmental damage and human population pressure has negative effect on the ecosystem, therefore, many studies have been published on the use of plant growth promoting microbes as bioinoculants or their products not only to increase plant growth but also to improve nutrients uptake and to reduce the environmental damages (Barac et al. 2004; Afzal et al. (2014a, b); Basu et al. 2018; Baoune et al. 2019). The positive results obtained of plant growth promoting microbes (PGPM) studies at laboratory scale have encouraged their use in diverse aspects of biotechnological axes (Rilling et al. 2019). Furthermore, the use of PGPM to mitigate the environmental stress is eco-friendly and cost effective strategy comparing to physico-chemical approaches (Singh et al. 2018). PGPM serve as bio-fertilizers, biocontrol, phyto-stimulator, which enhance the fertility of the soil, protect plants from phytopathogens, and enhance the growth of plants (Barac et al. 2004). This chapter is divided into the following sections: First, we briefly introduce plant associated microbes, in the second section, we summarize some metabolites produced by those microbes. Finally, we focus on their potential use in biotechnology.

### 14.1.1 Rhizospheric Microbes

Rhizo-microbes are classified according to their proximity to the roots to four types; (1) microbes living near the roots (rhizosphere), (2) microbes colonizing the root surface (rhizoplane), (3) microbes residing in the root tissue (endophytes), and (4) microbes residing inside cells in specialized root structures or nodules; the latter category is further divided into two groups: *Rhizobia* associated with leguminous and *Frankia* associated with woody plants. Microbes belonging to any of these groups are able to enhance plant growth either directly by developing volatile stimulating compounds and phytohormones, reducing plant ethylene levels, enhancing plant nutrient status (solubilizing phosphates from insoluble sources; fixing nitrogen), and activating pathways for disease resistance (induced systemic resistance) (Korpi et al. 2009). On the other hand, other mechanisms are applied by some rhizobacteria such as the induction of host plant resistance to phytopathogens and abiotic stress alleviation, the activation of other beneficial symbioses, or the protection of the plant through inhibitory degradation of xenobiotics, those traits are referred as plant growth promoting rhizobacteria (PGPR). Several rhizobacteria have been reported to have those features as, the genera *Bacillus*, *Pseudomonas*, *Erwinia*, *Caulobacter*, *Serratia*, *Arthrobacterium*, *Flavobacterium*, *Chromobacterium*, *Agrobacterium*, *Rhizobium*, *Streptomyces*, and *Rhodococcus* (Egamberdieva et al. 2015; Ponmurugan et al. 2016).

### 14.1.2 Endophyte Microbes

Endophytes are defined as microorganisms that live asymptomatically within a plant at least for a part of their life cycle. Those microbes grow inter- or intracellularly without causing visible manifestations of disease (Afzal et al. 2014a, b). It was demonstrated that every single plant is able to harbor endophytes and their diversity depends on several factors, such as plant species, plant density, plant age, nutrient availability, abiotic conditions, and interactions with soil microbiota (De Silva et al. 2019).

Endophytic bacteria have been isolated and characterized from different type of host plants, including agronomic crops, prairie plants, plants growing in extreme environments, and wild and perennial plants (Afzal et al. 2019). Further, endophytes can be considered as a subset of rhizospheric bacteria with the ability to invade plant roots after establishing in the rhizosphere. This colonization is determined by different bacterial traits which resumed in a complex communication process between the two partners, involving the recognition of some specific compounds in the root exudates by the endophytic bacteria (Compant et al. 2010). Apart from bacterial colonization, a variety of factors could determine the diversity of a particular plant species, including host plant age, genotype, geographical location as well as the changes in climate (Peng et al. 2013). Endophytic bacteria diversity has been reported in several studies. Generally, Proteobacteria is the most predominant phylum including α, β, and gamma proteobacteria. Members of Actinobacteria Bacteroidetes, and Firmicutes are also among the most commonly found as endophytes (De Silva et al. 2019).

### *14.1.3 Plant Growth Promoting Metabolites*

Plant associated microbes have demonstrated positive effects on plant growth and development when applied as bioinoculants for the seeds. Endophytes are able to thrive the plant interior tissue, and benefiting their host by providing nutrients. Both rhizomicrobes and endophytes may enhance plant development directly or indirectly. Direct mechanisms include assisting resource acquisition and modulating plant hormone levels. While, the indirect mechanisms involved the inhibition of the growth of various plant pathogens (Ramakrishna et al. 2019). The metabolites produced by PGPM are the following:

#### 14.1.3.1 Siderophores

In the environment, iron is an insoluble element (ferric oxide/hydroxide complexes), it is required as a cofactor of many enzymes. Some bacteria produce small molecules called siderophores with high affinity for ferric form ($Fe^{3+}$) of iron (Scavino and Pedraza 2013). Insoluble iron binds with siderophores outside the cell, the complex transported into the cell and converted into soluble ferrous iron form ($Fe^{2+}$). As well as, siderophores are involved in the acquisition of ferric citrate (Ghosh et al. 2020). Siderophores producing bacteria can suppress pathogen proliferation by reducing the amount of iron available in the environment and therefore enhancing plant growth and development (Dimkpa et al. 2009).

#### 14.1.3.2 Phosphate Solubilization

One of the essential elements for the plant growth is phosphorus. Although many strategies have been applied to fertilize soil, soil phosphorus still not available for the plant (Jog et al. 2014). Some plant associated bacteria are able to solubilize and mineralize insoluble soil phosphate and release soluble phosphorus and increasing its availability to plants as well as improving the fertility and the agriculture of soils (Han and lee 2005; Alori et al. 2017).

#### 14.1.3.3 Phytohormones

Ethylene is a plant hormone regulates growth, senescence plant at low concentration, its production more than its threshold level causes stress which affect negatively the plant growth (Glick 2005). This hormone is produced from 1-aminocyclopropane-1-decarboxylate (ACC). PGP bacteria produce ACC deaminase which is a key hormone to reduce plant stress by uptake of the ACC and metabolize it to α-ketobutyrate and $NH_3$ (Naik et al. 2019). Additionally, the indole-3-acetic acid is an auxin hormone produced by PGPB may improve the growth of roots, or activate the transcription of the enzyme ACC synthase which increases the level of ethylene. However, the presence of ACC deaminase producers decreases the plant ethylene level and the IAA continues as plant growth promoting hormones (Orozco-Mosqueda et al. 2020). IAA is synthesized from the amino acid tryptophan present in plant root exudates at a different concentration based on the genotype of the plant. Furthermore, it functions in root initiation (lateral and adventitious), cell division, stem and root elongation (Olanrewaju et al. 2017). The mechanisms of ACC deaminase and IAA are illustrated in Fig. 14.1.

#### 14.1.3.4 Nitrogen Fixation

Some of PGPM found in the rhizosphere can contribute in the fixation of atmospheric nitrogen by symbiotic or non-symbiotic processes. Rhizospheric microbes

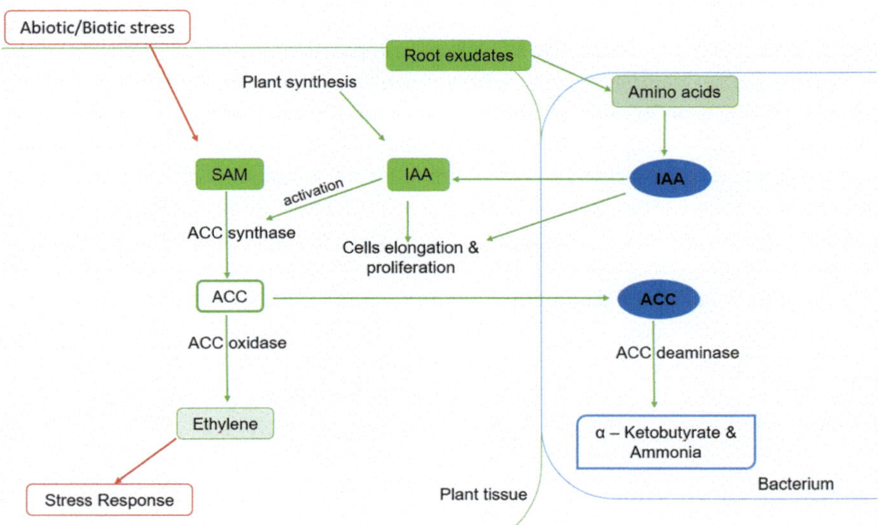


**Fig. 14.1** Schematic representation of the role of IAA and ACC deaminase to enhance the plant growth. SAM (S-adenosyl methionine), IAA (indole-3-acetic acid), ACC (1-aminocyclopropane-1-decarboxylate) (Baoune 2021)

have been well known for their ability of symbiotic nitrogen fixation such as *Azotobacter* spp., *Bacillus* spp., *Beijerinckia* spp., whereas free-living diazotrophs like *Azospirillum*, *Pseudomonas*, and *Burkholderia* are known for non-symbiotic nitrogen fixation (Choudhary et al. 2016). Free-living diazotrophs can improve the agricultural crops with their ability to fix small amount of atmospheric nitrogen. Symbiotic nitrogen-fixing microorganisms are those diazotrophs that live in close proximity to plant roots (i.e., in the rhizosphere or in plants) and can gain plant energy materials. It was demonstrated that the genus *Rhizobia* associated to leguminous can provide large quantities of nitrogen, the total amount of fixed N from this association in terrestrial ecosystem is 70 million tons per year (Werner and Newton 2005). This kind of symbiosis produces specific organs called nodules which appear on the root or on the stem (Nieder and Benbi 2008).

Nitrogen-fixing microorganisms have specific enzymes which are called nitrogenases that transform the atmospheric nitrogen into bioavailable ammonium ($NH_4^+$). Nitrogenase biosynthesis is encoded by the set of *nif* genes, which are arranged in a single cluster of approximately 20–24 kb in many diazotrophic bacteria with seven separated operons which encode for around 20 different proteins. The gene *nif*HDK encodes for the structural components of the complex enzyme molybdenum nitrogenase. The NifH protein is synthesized by the *nif*H gene, it is also reported that the *nif*DK gene and the Fe Protein of the nitrogenase enzyme codes for the FeMo protein (Antonio Castellano-Hinojosa and Bedmar 2017).

#### 14.1.3.5 Other Metabolites

The indirect mechanisms used by PGPM to promote plant growth could be resumed in the production of antibiotics, nutrient competitions. Some plant growth promoting microbe produces antibiotics to inhibit the growth of phytopathogens and therefore, preventing plant damages. Additionally, the competition between PGPM and phytopathogenes for nutrients or for biding sites on the plant root is considered as an indirect mechanism to promote plant growth (Olanrewaju et al. 2017).

Other PGPM produce hydrogen cyanide (HCN) which suppresses the proliferation of phytopathogens, inhibiting cytochrome c oxidase and other metallo-enzymes. Other produce cell wall degrading enzymes such as chitinase, peroxidases, and β-1,3-glucanase (Olanrewaju et al. 2017). Under different conditions in greenhouse studies, PGPM were found to be able to apply different effects on the plants, such as increasing the root and shoot length, absorption of mineral elements, fresh and dry biomass (Singh et al. 2018).

## 14.2 Biotechnological Uses of Plant Promoting Products

Microbial secondary metabolites are low-molecular-mass organic compounds produced by the most groups of microbes during the idiophase (stationary phase), including antibiotics, pigments, toxins, and others. The role of the secondary metabolites is poorly recognized in comparison with the primary metabolites which are known as essential for growth of microorganisms. Some studies were able to determine the role of the secondary metabolites as antibiotics agents, herbicides, insecticides, and plant growth promoters (Singh et al. 2019). Different environmental stress conditions can be found, such as the presence of phytopathogens, salinity, drought, presence of contaminants (Singh et al. 2018). Lately, great attention has been focused on the use of PGPM to reduce the stress produced from external factors.

### *14.2.1 Plant Growth Promoters*

In agriculture, plant growth promoting microbes have become a safe alternative technology comparing with chemical pesticides (Shiva et al. 2018). Microorganisms can promote plant growth by several direct and/or indirect mechanisms as mentioned below. The direct mechanisms include nutrient acquisition enhancement or stimulation of plant defense mechanisms against pathogens, which are known as biofertilization and phytostimulation (Hossain et al. 2017), whereas the indirect mechanisms comprise the inhibition of phytopathogenic agents growth (Vacheron et al. 2013). Some studies have reported that the secondary metabolites produced by cyanobacteria are low molecular weight organic molecules, the polyketides and nonribosomal peptides are the important classes of cyanobacterial secondary metabolites, which have a significant activity for plant growth promoting (Davies and Ryan 2011; Kumar et al. 2019). Also, plant associated actinobacteria isolated from different environment produce around 45% of microbial secondary metabolites, which play interesting role in plant growth promotion. Many studies have been conducted on the isolation and the characterization of secondary metabolites from actinobacteria. El-tarabily et al. (1997) have isolated seven actinobacteria from carrot rhizosphere, with the ability to produce non-volatile antifungal metabolites, those strains belonged to different genera; *Streptomyces, Streptoverticillium, Actinomadura*, *Actinoplanes*, *Micromonospora,* and *Streptosporangium*. In the study of Gopalakrishnan et al. (2014), six actinobacteria isolated from herbal vermicomposting were able to produce hydrogen cyanide (HCN), IAA indole-3-acetic acid (IAA), siderophore, and ß-1,3-glucanase. Several studies have reported that 80% of rhizospheric microorganisms promote plant growth by producing phytohormones such as auxins, cytokinins, and gibberellins. Bacteria from the genus of *Azospirillum*, *Paenibacillus*, *Pseudomonas*, *Streptomyces* have been reported to be an efficient candidate to improve plant growth (Patten and Glick

2002; Fuentes-Ramirez and Caballero-Mellado 2005). Furthermore, the indole-3-acetic acid (IAA) and the cytokinins produced by *Pseudomonas fluorescens* enhanced root elongation of canola (*Brassica napus L.*), in pure culture and in the rhizosphere of canola under gnotobiotic conditions (Pallai et al. 2012). Salisbury (1994) demonstrated that the indole-3-acetic acid is involved in root initiation, cell division, and cell elongation. Besides that, studies conducted in saline stress conditions reported that PGPM stimulate plant antioxidant defense mechanisms, regulating the activity of superoxide dismutase (SOD), catalase, and peroxidase. While, in drought conditions, those microbes produce some phytohormones, proteins, and polysaccharides to adjust the physiological and biochemical levels in plants such as changes in phytohormones concentrations and antioxidant defense systems (Singh et al. 2018).

### *14.2.2 Biocontrol*

Plant disease could be caused by various organisms like viruses, bacteria, fungi, insects, nematodes, etc., which become serious issue worldwide due to the major threat to food security (Agrillo et al. 2019). One of the tools to control plant diseases with minimal impact on the environment is the use of beneficial microbes, their genes, and/or products (De Silva et al. 2019). The use of endophytes as biocontrol agent has gained a strong attention as a friendly approach (Hardoim et al. 2008). As well as, rhizobacteriaare known by their capability to produce a variety of secondary metabolites such as lytic enzymes, toxins, gases, volatile organic compounds, and other metabolites which play a role in controlling nematodes (Marin-Bruzos and Grayston 2019). It the study of Oliveira et al. (2014), a metabolites called Uracil, 9H-purine which is a dichloromethane soluble metabolite identified by high performance liquid chromatography (HPLC) coupled to mass spectrometry and produced by *Bacillus cereus* and *Bacillus subtilis*, was reported as an active compound against nematodes. In other study, it was reported that *Pseudomonas fluorescens* is able to produce some compounds such as HCN, siderophores, and mainly pyoverdine and pyochelin which contribute to suppress several plant diseases such as black root rot of tobacco caused by the fungus *Thielaviopsis basicola* (Nandhini et al. 2012).

In the family of actinobacteria, the genus *Streptomyces* has been well known as active producers of antibiotics and volatile organic compounds with great potential for controlling various fungal and bacterial phytopathogens (Shiva et al. 2018). Besides that, Fungi have been well known by their enormous production of a variety of metabolites which can be used in different area (Keller et al. 2005; Nawar, 2016). This attribute might be due to their versatility and their huge enzymatic language to produce a wide range of bioactive natural compounds (Sidorova and Voronina 2019). It has been reviewed that about 17% of natural bioactive compounds are produced by fungi (Bérdy 2012). α-Pyrones, viz. fusapyrone (FP) and deoxyfusapyrone (DFP) are two secondary metabolites produced by the rhizofungi *Fusarium semitectum*, inhibit the growth of many pathogen filamentous fungi such

**Table 14.1** Examples of microorganisms and their secondary metabolites

| Microorganisms | | Groups of secondary metabolite | Secondary metabolites names | References |
|---|---|---|---|---|
| Fungus | *Flammulina velutipes (Curtis) singer (macrofungi)* | Terpenoids | Enokipodins A–D | Ishikawa et al. (2001) |
| | *Fusarium semitectum* | Heterocyclic compounds | Deoxyfusapyrone | Evidente et al. (1999) |
| | *Trichoderma viride* | Lactone | 6-pentyl-α-pyrone | Prapulla et al. (1992) |
| | *Trichoderma* sp. | Volatile compounds | 1β-vinylcyclopentane-1α,3α-diol | Yang et al. (2012) |
| | *Verticillium biguttatum* | Phenol | Bigutol | Morris et al. (1995) |
| Bacteria | *Pseudomonas aurantiaca* | Aromatic carboxylic acid | Phenazine-1-carboxylic acid | Shahid et al. (2017) |
| | *Pseudomonas fluorescens* | Polyketide | 2,4-diacetylphloroglucinol | Siddiqui and Shaukat (2003) |
| | *Pseudomonas putida* | Nonribosomal peptide | Pyoluteorin | Hassan et al. (2011) |
| | *Nostoc* sp. | Depsipeptide | Cryptophycin | Biondi et al. (2004) |

as *Cladosporium cucumerinum, Alternaria alternata, Ascochyta rabiei* (Evidente et al. 1999; Altomare et al. 2000). The endophyte *Fusarium* is well known as a producer of some toxins such as nivalenol, T-2, neosolaniol, HT-2, and diacetoxyscirpenol that inhibit the growth of plant parasites of *Orobanche ramosa,* weed plant (Zonno and Vurro 2002). Plant growth promoting *Bacillus amyloliquefaciens* produces some secondary metabolites belonging to the chemical family of aldehydes, ketones, and benzenes which are active against *Fusarium oxysporum* (Yuan et al. 2012). Moreover, some members of the family Clavicipitaceae produce a mycotoxin called ergot alkaloids which improve host plant resistance to herbivores and cause toxicity to consuming livestock (Schardl et al. 2013a, b). Bacteria isolated from different type of environment might be a promising source for the production of metabolites involved in biocontrol of plant disease (Table 14.1) (Agrillo et al. 2019). Taken together all the data, we conclude that several plant growth promoting microbial secondary metabolites are involved in the promotion of plant growth indirectly by the suppression of plant pathogens growth.

### 14.2.3 *Phytoremediation*

The presence of both organic and inorganic xenobiotics in the environment presents a serious issue which has a negative effect on the ecosystem. Their degradation might occur within the plant or in the rhizosphere (Barac et al. 2004). Plant associated microbes can enhance the phytoremediation of contaminants. The capability of endophytes to thrive the plant tissue makes those microbes protected from the stress caused by the presence of xenobiotics. Plant growth promoting bacteria improve the nutrient acquisition during phytoremediation which increase the bioaccumulation of contaminants (Agnello et al. 2016; Cristaldi et al. 2017). It was demonstrated that endophytes enhance the phytoextraction of metals by using siderophores which can bind to some metals such as copper, zinc, and cadmium, increasing their solubility (Li et al. 2012). As well as, they can reduce ethylene levels by the production of ACC deaminase. Afzal et al. (2011) showed that the inoculated plants with bacteria having ACC deaminase exhibit a massive gene expression *acdS* responsible of ACC cleavage, thus reducing stress caused by the presence of contaminants. Many studies have been outlining the beneficial effect of endophytes in the enhancement of plant growth and improvement of phytoremediation due to their plant growth promoting or/and xenobiotic mineralization or degradation ability (Polti et al. 2011; Afzal et al. 2014a, b; Baoune et al. 2019). Successful laboratory experiments have emphasized the importance of plant associated bacteria in phytoremediation, however, we still lack an integrated understanding of in situ studies level.

## 14.3 Conclusion

The use of plant growth promoting microbes is an integral component of ecosystems as it is a technology whose time has come to be applied. Those microbes are already known a success in different field as it expected to grow. Thus, it is logic to expect their increase use in various strategies. However, their vast use in the worldwide and the need to study different factors that might be taken in consideration while going from laboratory scale to field trails, but the future of this approach looks extremely bright.

## References

Afzal M, Yousaf S, Reichenauer TG, Kuffner M, Sessitsch A (2011) Soil type affects plant colonization, activity and catabolic gene expression of inoculated bacterial strains during phytoremediation of diesel. J Hazard Mater 186:1568–1575. https://doi.org/10.1016/j.jhazmat.2010.12.040

Afzal M, Khan QM, Sessitsch A (2014a) Endophytic bacteria: prospects and applications for the phytoremediation of organic pollutants. Chemosphere 117(décembre):232–242. https://doi.org/10.1016/j.chemosphere.2014.06.078

Afzal M, Shabir G, Tahseen R, Islam E-u, Iqbal S, Khan QM, Khalid ZM (2014b) Endophytic *Burkholderia* Sp. strain PsJN improves plant growth and phytoremediation of soil irrigated with textile effluent: bacterial inoculation enhanced plant biomass production and soil remediation. Clean (Weinh) 42(9):1304–1310. https://doi.org/10.1002/clen.201300006

Afzal I, Shinwari ZK, Sikandar S, Shahzad S (2019) Plant beneficial endophytic bacteria: mechanisms, diversity, host range and genetic determinants. Microbiol Res 221(avril):36–49. https://doi.org/10.1016/j.micres.2019.02.001

Agnello AC, Bagard M, van Hullebusch ED, Esposito G, Huguenot D (2016) Comparative bioremediation of heavy metals and petroleum hydrocarbons co-contaminated soil by natural attenuation, phytoremediation, bioaugmentation and bioaugmentation-assisted phytoremediation. Sci Total Environ 563–564(septembre):693–703. https://doi.org/10.1016/j.scitotenv.2015.10.061

Agrillo B, Mirino S, Tatè R, Gratino L, Gogliettino M (2019) An alternative biocontrol agent of soil-borne phytopathogens: a new antifungal compound produced by a plant growth promoting bacterium isolated from North Algeria. Microbiol Res 221(January):60–69. https://doi.org/10.1016/j.micres.2019.02.004

Alori ET, Glick BR, Babalola OO (2017) Microbial phosphorus solubilization and its potential for use in sustainable agriculture. Front Microbiol 8(juin):971. https://doi.org/10.3389/fmicb.2017.00971

Altomare C, Perrone G, Zonno MC, Evidente A, Pingue R, Fanti F, Polonelli L (2000) Biological characterization of fusapyrone and deoxyfusapyrone, two bioactive secondary metabolites of Fusarium semitectum. J Nat Prod 63:1131–1135

Antonio Castellano-Hinojosa A, Bedmar EJ (2017) Methods for evaluating plant growth-promoting rhizobacteria traits. In: Singh HB, Sarma BK, Keswani C (eds) Advances in PGPR Research. CABI international, Oxford, pp 255–274

Baoune H, Aparicio JD, Acuña A, El Hadj-khelil AO, Sanchez L, Polti MA, Alvarez A (2019) Effectiveness of the zea mays-streptomyces association for the phytoremediation of petroleum hydrocarbons impacted soils. Ecotoxicol Environ Saf 184(novembre):109591. https://doi.org/10.1016/j.ecoenv.2019.109591

Baoune H (2021). Potential of endophytic actinobacteria for biological use in remediation of soil contaminated with petroleum hydrocarbons. Doctoral dissertation. Université Kasdi Merbah Ouargla.Algeria

Barac T, Taghavi S, Borremans B, Provoost A, Oeyen L, Colpaert JV, Vangronsveld J, van der Lelie D (2004) Engineered endophytic bacteria improve phytoremediation of water-soluble, volatile, organic pollutants. Nat Biotechnol 22(5):583–588. https://doi.org/10.1038/nbt960

Basu S, Rabara RC, Negi S, Shukla P (2018) Engineering PGPMOs through gene editing and systems biology: a solution for phytoremediation? Trends Biotechnol 36(5):499–510. https://doi.org/10.1016/j.tibtech.2018.01.011

Bérdy J (2012) Thoughts and facts about antibiotics: where we are now and where we are heading. J Antibiot 65:385–395. https://doi.org/10.1038/ja.2012.27

Biondi N, Piccardi R, Margheri MC, Rodolfi L, Smith GD, Tredici MR (2004) Evaluation of Nostocstrain ATCC 53789 as a potential source of natural pesticides. Appl Environ Microbiol 70:3313–3320

Choudhary DK, Varma A, Tuteja N, éd. (2016) Plant-microbe interaction: an approach to sustainable agriculture. Springer Singapore, Singapore. https://doi.org/10.1007/978-981-10-2854-0

Compant S, Clément C, Sessitsch A (2010) Plant growth-promoting bacteria in the rhizo- and endosphere of plants: their role, colonization, mechanisms involved and prospects for utilization. Soil Biol Biochem 42(5):669–678. https://doi.org/10.1016/j.soilbio.2009.11.024

Cristaldi A, Conti GO, Jho EH, Zuccarello P, Grasso A, Copat C, Ferrante M (2017) Phytoremediation of contaminated soils by heavy metals and PAHs. A brief review. Environ Technol Innov 8(novembre):309–326. https://doi.org/10.1016/j.eti.2017.08.002

Davies J, Ryan KS (2011) Introducing the parvome: bioactive compounds in the microbial world. ACS Chem Biol 7(2):252–259

De Silva NI, Brooks S, Lumyong S, Hyde KD (2019) Use of endophytes as biocontrol agents. Fungal Biol Rev 33(2):133–148. https://doi.org/10.1016/j.fbr.2018.10.001

Dimkpa CO, Merten D, Svatoš A, Büchel G, Kothe E (2009) Siderophores mediate reduced and increased uptake of cadmium by Streptomyces tendae F4 and sunflower (Helianthus Annuus), Respectively. J Appl Microbiol 107:1687-1696. https://doi.org/10.1111/j.1365-2672.2009.04355.x

Egamberdieva D, Shrivastava S, Varma A (eds) (2015) Plant-growth-promoting rhizobacteria (PGPR) and medicinal plants. Soil biology, vol vol 42. Springer, Cham. https://doi.org/10.1007/978-3-319-13401-7

El-Tarabily KA, Hardy GESJ, Sivasithamparam K, Hussein AM, Kurtböke ID (1997) The potential for the biological control of cavity spot disease of carrots caused by Pythium coloratumbystreptomycete and non-streptomycete actinomycetes in Western Australia. New Phytol 137:495–507

Evidente A, Amalfitano C, Pengue R, Altomare C (1999) High performance liquid chromatography for the analysis of Fusapyrone and Deoxyfusapyrone, two antifungal a-Pyrones from Fusarium semitectum. Nat Toxins 7:133–137

Fuentes-Ramirez LE, Caballero-Mellado J (2005) Bacterial biofertilizers. In: Siddiqui ZA (ed) PGPR: biocontrol and biofertilization. Springer, Dordrecht

Ghosh SK, Bera T, Chakrabarty AM (2020) Microbial Siderophore – a boon to agricultural sciences. Biol Control 144(mai):104214. https://doi.org/10.1016/j.biocontrol.2020.104214

Glick BR (2005) Modulation of plant ethylene levels by the bacterial enzyme ACC deaminase. FEMS Microbiol Lett 251:1–7. https://doi.org/10.1016/j.femsle.2005.07.030

Glick BR (2012) Plant growth-promoting bacteria: mechanisms and applications. Scientifica 2012:1–15. https://doi.org/10.6064/2012/963401

Gopalakrishnan S, Vadlamudi S, Bandikinda P, Sathya A, Vijayabharathi R, Rupela O, Kudapa H, Katta K, Varshney RK (2014) Evaluation of Streptomyces strains isolated from herbal vermicompost for their plant growth-promotion traits in rice. Microbiol Res 169(1):40–48. https://doi.org/10.1016/j.micres.2013.09.008

Han HS, Lee KD (2005) Phosphate and potassium solubilizing bacteria effect on mineral uptake, soil availability, and growth of egg plant. Res J Agric Biol Sci 1:176–180

Hardoim PR, van Overbeek LS, van Elsas JD (2008) Properties of bacterial endophytes and their proposed role in plant growth. Trends Microbiol 16(10):463–471. https://doi.org/10.1016/j.tim.2008.07.008

Hassan MN, Afghan S, Hafeez FY (2011) Biological control of red rot in sugarcane by native pyoluteorin-producing Pseudomonas putidastrain NH-50 under field conditions and its potential modes of action. Pest Manag Sci 67:1147–1154

Hossain MM, Sultana F, Islam S (2017) Plant growth-promoting fungi (PGPF): phytostimulation and induced systemic resistance. In: Singh D, Singh H, Prabha R (eds) Plant-microbe interactions in agro-ecological perspectives. Springer, Singapore

Ishikawa NK, Fukushi Y, Yamaji K, Tahara S, Takahashi K (2001) Antimicrobial cuparene-type sesquiterpenes, enokipodins C and D, from a mycelial culture of Flammulina velutipes. J Nat Prod 64:932–934

Jog R, Pandya M, Nareshkumar G, Rajkumar S (2014) Mechanism of phosphate solubilization and antifungal activity of Streptomyces spp. isolated from wheat roots and rhizosphere and their application in improving plant growth. Microbiol 160:778–788. https://doi.org/10.1099/mic.0.074146-0

Keller N, Turner G, Bennett J (2005) Fungal secondary metabolism — from biochemistry to genomics. Nat Rev Microbiol 3:937–947. https://doi.org/10.1038/nrmicro1286

Korpi A, Jarnberg J, Pasanen AL (2009) Microbial volatile organic compounds. Crit Rev Toxicol 39:139–193. https://doi.org/10.1080/10408440802291497

Kumar G, Teli B, Mukherjee A, Bajpai R, Sarma BK (2019) Secondary metabolites from cyanobacteria: a potential source for plant growth promotion and disease management. In: Singh H, Keswani C, Reddy M, Sansinenea E, García-Estrada C (eds) Secondary metabolites of plant growth promoting rhizomicroorganisms. Springer, Singapore

Li H-Y, Wei D-Q, Shen M, Zhou Z-P (2012) Endophytes and their role in phytoremediation. Fungal Divers 54(1):11–18. https://doi.org/10.1007/s13225-012-0165-x

Marin-Bruzos M, Grayston S (2019) Biological control of nematodes by plant growth promoting rhizobacteria: secondary metabolites involved and potential applications. In: Harikesh BS, Keswani C, Reddy MS, Sansinenea E, García-Estrada C (eds) Secondary metabolites of plant growth promoting rhizomicroorganisms, Springer, Singapore. 253-264. https://doi.org/10.1007/978-981-13-5862-3_13

Morris RA, Ewing DF, Whipps JM, Coley-Smith JR (1995) Antifungal hydroxymethyl-phenols from the mycoparasite Verticillium biguttatum. Phytochemistry 39(5):1043–1048

Naik K, Mishra S, Srichandan H, Singh PK, Sarangi PK (2019) Plant growth promoting microbes: potential link to sustainable agriculture and environment. Biocatal Agric Biotechnol 21 (septembre):101326. https://doi.org/10.1016/j.bcab.2019.101326

Nandhini S, Sendhilvel V, Babu S (2012) Endophytic bacteria from tomato and their efficacy against Fusarium oxysporumf. sp. lycopersici, the wilt pathogen. J Biopest 5:178–185

Nawar LS (2016) Phytochemical and SDS-dissociated proteins of pathogenic and nonpathogenic Fusarium oxysporumisolates. Int J Chem Tech Res 9(6):165–172

Nieder R, Benbi DK (2008) Carbon and nitrogen in the terrestrial environment. Springer, Dordrecht

Olanrewaju OS, Glick BR, Babalola OO (2017) Mechanisms of action of plant growth promoting bacteria. World J Microbiol Biotechnol 33(11):197. https://doi.org/10.1007/s11274-017-2364-9

Oliveira DF, Santos Junior HM, Dos Nunes AS et al (2014) Purification and identification of metabolites produced by Bacillus cereusand B. subtilisactive against Meloidogyne exigua, and their in silico interaction with a putative phosphoribosyl transferase from M. incognita. An Acad Bras Cienc 86:525–538

Orozco-Mosqueda MDC, Glick BR, Santoyo G (2020) ACC deaminase in plant growth-promoting bacteria (PGPB): an efficient mechanism to counter salt stress in crops. Microbiol Res 235 (mai):126439. https://doi.org/10.1016/j.micres.2020.126439

Pallai R, Hynes RK, Verma B, Nelson LM (2012) Phytohormone production and colonization of canola (Brassica napus L.) roots by pseudomonas fluorescens 6-8 under gnotobiotic conditions. Can J Microb 58(2):170–178

Patten CL, Glick BR (2002) Role of pseudomonas putida indoleacetic acid in development of the host plant root system. Appl Environ Microbiol 68(8):3795–3801. https://doi.org/10.1128/AEM.68.8.3795-3801.2002

Peng A, Liu J, Gao Y, Chen Z (2013) Distribution of endophytic bacteria in alopecurus aequalis sobol and oxalis corniculata L. from soils contaminated by polycyclic aromatic hydrocarbons. Édité par Raffaella Balestrini. PLoS One 8(12):e83054. https://doi.org/10.1371/journal.pone.0083054

Polti MA, Atjián MC, Amoroso MJ, Abate CM (2011) Soil chromium bioremediation: synergic activity of actinobacteria and plants. Int Biodeterior Biodegradation 65(8):1175–1181. https://doi.org/10.1016/j.ibiod.2011.09.008

Ponmurugan P, Elango V, Sathya A, Vijayabharathi R, Gopalakrishnan S (2016) Evaluation of plant growth-promoting actinomycetes on Vigna. In: Plant growth promoting actinobacteria: a new avenue for enhancing the productivity and soil fertility of grain legumes. Springer, Singapore, pp 275–286

Prapulla SG, Karanth NG, Engel KH, Tressl R (1992) Production of 6-pentyl-α-pyrone by *Trichoderma viride*. Flavour Fragr J 7:231–234. https://doi.org/10.1002/ffj.2730070412

Ramakrishna W, Yadav R, Li K (2019) Plant growth promoting bacteria in agriculture: two sides of a coin. Appl Soil Ecol 138:10–18. https://doi.org/10.1016/j.apsoil.2019.02.019

Rilling JI, Acuña JJ, Nannipieri P, Cassan F, Maruyama F, Jorquera MA (2019) Current opinion and perspectives on the methods for tracking and monitoring plant growth–promoting bacteria. Soil Biol Biochem 130(mars):205–219. https://doi.org/10.1016/j.soilbio.2018.12.012

Salisbury FB (1994) The role of plant hormones. In: Wilkinson RE (ed) Plant-environment interaction. Dekker, New York, pp 39–81

Scavino AF, Pedraza RO (2013) The role of siderophores in plant growth-promoting bacteria. In: Maheshwari D, Saraf M, Aeron A (eds) Bacteria in agrobiology: crop productivity. Springer, Berlin, Heidelberg

Schardl CL, Young CA, Pan J, Florea S, Takach JE, Panaccione DG, Farman ML, Webb JS, Jaromczyk J, Charlton ND, Nagabhyru P, Chen L, Shi C, Leuchtmann A (2013a) Currencies of mutualisms: sources of alkaloid genes in vertically transmitted epichloae. Toxins (Basel) 5:1064–1088. https://doi.org/10.3390/toxins5061064

Schardl CL, Young CA, Hesse U, Amyotte SG, Andreeva K et al (2013b) Plant-symbiotic fungi as chemical engineers: multi-genome analysis of the Clavicipitaceae reveals dynamics of alkaloid loci. PLoS Genet 9:e1003323. https://doi.org/10.1371/journal.pgen.1003323

Shahid I, Rizwan M, Baig DN, Saleem RS, Malik KA, Mehnaz S (2017) Secondary metabolites production and plant growth promotion by Pseudomonas chlororaphisand P. aurantiacastrains isolated from cactus, cotton, and para grass. J Microbiol Biotechnol 27(3):480–491

Shiva S, Prasad K, Giovanardi D, Stefani E (2018) Plant growth promoting and biocontrol activity of Streptomyces spp. as endophytes. Int J Mol Sci 19(4):952. https://doi.org/10.3390/ijms19040952

Siddiqui IA, Shaukat SS (2003) Non-pathogenic Fusarium solanirepresses the biosynthesis of nematicidal compounds in vitro and reduces the biocontrol of Meloidogyne javanicaby Pseudomonas fluorescensin tomato. Lett Appl Microbiol 37:109–114

Sidorova I, Voronina E (2019) Bioactive secondary metabolites of basidiomycetes and its potential for agricultural plant growth promotion. Springer, Cham

Singh VK, Singh AK, Singh PP, Kumar A (2018) Interaction of plant growth promoting bacteria with tomato under abiotic stress: A review. Agric Ecosyst Enviro 267:129–40. https://doi.org/10.1016/j.agee.2018.08.020

Singh HB, Keswani C, Reddy MS, Royano ES, Estrada CG (2019) Secondary metabolites of plant growth promoting rhizomicroorganisms: discovery and applications. Springer-Nature, Singapore, p 392

Vacheron J, Desbrosses G, Bouffaud ML et al (2013) Plant growth-promoting rhizobacteria and root system functioning. Front Plant Sci 4:365. https://doi.org/10.3389/fpls.2013.00356

Werner D, Newton WE (2005) Nitrogen fixation in agriculture, forestry, ecology, and the environment. Springer Netherlands, Dordrecht. https://doi.org/10.1007/1-4020-3544-6

Yang Z, Yu Z, Lei L, Xia Z, Shao L, Zhang K, Li G (2012) Nematicidal effect of volatiles produced by Trichodermasp. J Asia Pac Entomol 15:647–650

Yuan J, Raza W, Shen QR, Huang QW (2012) Antifungal activity of Bacillus amyloliquefaciens NJN-6 volatile compounds against Fusarium oxysporumf. sp cubense. Appl Environ Microbiol 78:5942–5944

Zonno MC, Vurro M (2002) Inhibition of germination of *Orobanche ramosa se*eds by *Fusarium to*xins. Phytoparasitica 30:519–524. https://doi.org/10.1007/BF02979757

# Chapter 15
# Advances in the Bioremediation of Pharmaceuticals and Personal Care Products (PPCPs): Polluted Water and Soil

**Mahendar Porika, Pabbati Ranjit, Radhika Tippani, and Kondakindi Venkateswar Reddy**

**Abstract** Pharmaceutical and personal care products (PPCPs) comprising a variety of organic categories, like antibiotics, hormones, antimicrobials, synthetic musks, etc., have promoted significant interest in recent times that their constant input of such substances has become a serious threat to the public and the environment, extending from surface water, sludge, sewage, aquatic bodies, treatment plants, sediments, soil, humans, and wildlife. In recent years, the existence of PPCPs has been attracting growing focus, leading to considerable concern about their frequency, development, fate, and danger in the ecosystems. However, in recent years the substantial use of such molecules has contributed to their accumulation in our ecosystem, and they are now observed in living organisms via their intervention in food chains and food webs. As bioactive in nature, these substances have significant toxicological and precarious impacts on ecological systems, environments, and human health. A number of methodologies have been widely studied for eliminating PPCPs from the environment, including physical, biological, and chemical methods. The categories, functions, and representatives of the commonly observed PPCPs within the environments were stated in this chapter. This chapter also structured to explicate the incidence, fate, and bioremediation (BR) of PPCPs from waste water and soil utilizing microorganisms (MOs) and plants, and also the experimental methods used to assess the PPCPs in nature.

**Keywords** Bioremediation · Gas chromatography · High-performance liquid chromatography (HPLC) · Microorganisms · PPCPs · Phytoremediation · Toxicity · Wastewater

Mahendar Porika and Pabbati Ranjit contributed equally with all other contributors.

M. Porika · R. Tippani
Department of Biotechnology, Kakatiya University, Warangal, Telangana, India

P. Ranjit · K. V. Reddy (✉)
Centre for Biotechnology, Institute of Science & Technology, Jawaharlal Nehru Technological University, Hyderabad, Telangana, India



P. K. Arora (ed.), *Microbial Products for Health, Environment and Agriculture*, Microorganisms for Sustainability 31, https://doi.org/10.1007/978-981-16-1947-2_15

## Abbreviation

| | |
|---|---|
| ACC | 1-Aminocyclopropane-1-carboxylate |
| AChE | Acetylcholinesterase |
| ATN | Artemisine |
| BTF | Biotransformation |
| CBZ | Carbamazepine |
| CDN | Codeine |
| CFN | Caffeine |
| CSCM | Circular supply chain management |
| CWs | Constructed wetlands |
| DFC | Diclofenac |
| DOS | Dioctyl sebacate |
| E2α | 17α-Ethinylestradiol |
| EROD | Ethoxyresorufin-*O*-deethylase |
| EST | Estradiol |
| FLU | Flumequine |
| GC | Gas chromatography |
| GFZ | Gemfibrozil |
| HPLC | High-performance liquid chromatography |
| HRAP | High rate algal pond |
| HRT | Hydraulic retention time |
| IBU | Ibuprofen |
| LLE | Liquid–liquid extraction |
| LLME | Liquid–liquid micro-extraction |
| MBR | Membrane bioreactors |
| MCs | Mixed cultures |
| NP | Nonylphenol |
| NPX | Naproxen |
| PGPH | Plant growth-promoting hormone |
| PPCPs | Pharmaceuticals and personal care products |
| ROS | Reactive oxygen species |
| SDG | Sustainable development goal |
| SMX | Sulfamethoxazole |
| SPE | Solid-phase extraction |
| SPME | Solid-phase micro-extraction |
| STP | Sewage treatment plants |
| TCS | Triclosan |
| UV | Ultraviolet |
| WRF | White rot fungi |
| WWTP | Wastewater treatment plants |

## 15.1 Introduction

Bioremediation (BR) is an evolving and ground breaking technology due to its economic viability, enhanced expertise, and friendliness with the legitimate habitat. The system uses numerous eco-friendly microbial techniques to cope with the ever-growing issue of environmental contamination (Singh et al. 2020). In these strategies, microbes adjust upon toxic wastes and naturally grow environmentally acclimated microbial strains that eventually turn an extensive range of noxious substances into non-harmful forms. Microbial xenobiotic degradation/decadence is dependent on the activity of enzymes (Singh et al. 2020). Nutrients are provided to the polluted zone during the BR process to accelerate the growth of the suitable microorganisms (MOs) which promote the BR of the contaminating noxious substances. For cases where no present MO is capable of breaking down a pollutant, scientists incorporate a MO proven to eliminate the noxious substances. BR has been effectively utilized to clean up pollutants such as, pesticides, crude oil, sewage, gasoline, and chlorinated solvents utilized in cleaning supplies etc.

## 15.2 Importance of Bioremediation

In recent decades, rapid industrialisation, urbanization, and indiscriminate use of resources by an ever-growing population of human has intensified contamination of the surface and ground waters, ambience, and land surfaces. The widespread decadence of natural resources poses a major threat to global public health. Pesticides, heavy metals, synthetic hydrocarbons, and vast quantities of harmful industrial effluents are the main pollutants affecting the soil and water system. Such anthropogenic xenobiotics are inherently recalcitrant. The regeneration of water, soil resources, degraded land is only feasible through safe and environmentally sensitive approaches in the current scenario. Among the numerous recent methods used, BR is known as an emerging technique for the regeneration of degraded habitats to reduce pollution. Nevertheless, its applicability at ground level is restricted due to various climatic factors. Different MOs kill obstinate contaminants under aerobic or anaerobic conditions by using contaminants as their carbon sources by full mineralization or co-metabolism. Bacteria and fungi were noticed as advantageous and worthy target for decadence of organic pollutants found in polluted areas both in situ and ex situ. In addition, the microbes may be genetically engineered to effectively degrade environmental contaminants. Moreover, the wide-ranging applicability of genetically engineered organisms is restricted by broad political and ethical concerns. New prospects in BR development are emerging biotechnological developments include the use of qualified indigenous microbes, application of unique enzymes, microbial consortia, biosurfactant, and rhizoremediation.

## 15.2.1 *Classes of Bioremediation*

BR lists in two classes.

#### 15.2.1.1 In Situ

In situ BR corresponds to the handling of hazardous waste at their source point. Soil may be contaminated, for instance. Instead of removing the soil from its place of origin, it is contained right as it is. The lead for in situ treatment is that during the removal and transfer of the infected material, it stops contaminants from spreading.

#### 15.2.1.2 Ex Situ

Ex situ BR corresponds to treatment that takes place after removal of the polluted waste to a treatment area. The soil can be collected and transferred to a region where the BR can be implemented to use the soil as an example again. The key benefit of this is that it helps to isolate and monitor the goods for BR and makes available the region that was polluted.

## 15.3 Pharmaceuticals and Personal Care Products (PPCPs)

In recent decades, unparalleled developments in the medicine, livestock, aquaculture, and cosmetics industries have modernized beautification services and healthcare, leading to a wide variety of PPCPs being developed (Petryna et al. 2006; Wang and Wang 2016). Because of their wide application and inadequate elimination via traditional biological wastewater treatment plants (WWTPs), they are pervasive in the ecosystems. Table 15.1 listed the specific classes, corresponding motive, and principal properties of PPCPs. Since the 2000s PPCPs have received increasing attention (Wang and Wang 2016). PPCPs were initially contemplated micro-pollutants (i.e., orders of ng/L up to several μg/L), although their persistent discharge into the habitat is of huge importance to ecologists and environmentalists (Nikolaou et al. 2007). Contemporary investigations have demonstrated enormous amount of PPCPs in the habitat and their possible implications on lifeforms (Kummerer 2009; Corcoran et al. 2010; Brausch and Rand 2011; Blair et al. 2013; Munze et al. 2017). The United States Environmental Protection Agency (USEPA) monitors all of the PPCPs in terms of the quality of potable water; furthermore, one antibiotic and eight hormones were proposed but are still uncontrolled, which may pose significant health issues in the future (Nguyen et al. 2019a, b). After rainfall incidents, the USEPA and the European Union (EU) have already identified a list of priorities of harmful substances found in wastewater and runoff water, which can

**Table 15.1** The classification, functions, and main properties of PPCPs. Reproduced from (Wang and Wang 2016). Copyright © 2016 with permission from Elsevier

| Classification of PPCPs | Functions | Representatives frequently detected | Molecular weight | LogKow |
|---|---|---|---|---|
| *Pharmaceuticals* | | | | |
| Antibiotics | Kill bacteria | Sulfamethoxazole | 253.3 | 0.89 |
| | | Trimethoprim | 290.3 | 0.91 |
| | | Amoxicillin | 365.4 | 0.87 |
| | | Erythromycin | 733.9 | 3.06 |
| | | Ofloxacin | 361.4 | −0.39 |
| | | Ciprofloxacin | 331.3 | 0.28 |
| | | Ampicillin | 349.4 | 1.45 |
| | | Doxycycline | 444.4 | −0.02 |
| | | Difloxacine | 399.4 | n.a. |
| | | Tylosin | 916.1 | 1.63 |
| | | Enoxacin | 320.1 | −0.2 |
| | | Sulfapyridine | 250.3 | 0.35 |
| | | Cefalexin | 347.4 | 0.65 |
| | | Cefaclor | 367.8 | 0.4 |
| | | Mecillinam | 325.4 | 1.3 |
| | | Tetracycline | 444.4 | −1.37 |
| Anticonvulsants | Treat mood disorders | Carbamazepine | 236.3 | 13.9 |
| | | Primidone | 218.3 | 1.12 |
| | | Dilantin | 252.3 | 2.47 |
| | | Phenobarbital | 232.2 | 1.47 |
| | | Cabapentin | 171.2 | −1.1 |
| Antidepressants | Improve the physical disorders | Diazepam | 284.7 | 3.08 |
| | | Doxepin | 279.4 | 3.84 |
| | | Imipramine | 280.4 | 4.28 |
| | | Amitriptyline | 277.4 | 4.81 |
| | | Fluxetine | 309.3 | 3.96 |
| | | Meprobamate | 218.3 | 0.93 |
| | | Oxazepam | 286.7 | 2.92 |
| | | Thioridazine | 370.6 | n.a. |
| Antineoplastics | Control or kill neoplastic cells | Epirubicin | 543.5 | 1.85 |
| | | Ifosfamide | 261.1 | 0.86 |
| | | Methotrexate | 454.4 | −1.28 |
| | | Tamoxifen | 371.5 | 6.3 |
| | | Cyclophosphamide | 261.1 | 0.73 |
| Beta-blockers | Inhibit the hormone adrenalin and the neurotransmitter noradrenalin | Atenolol | 266.3 | 0.16 |
| | | Metoprolol | 267.4 | 9.7 |
| | | Nadolol | 309.4 | −0.6 |
| | | Pindolol | 248.3 | 1.75 |
| | | Acebutolol | 336.4 | 1.71 |
| | | Propranolol | 259.3 | 3.48 |
| | | Sotalol | 272.4 | 0.24 |

(continued)

**Table 15.1** (continued)

| Classification of PPCPs | Functions | Representatives frequently detected | Molecular weight | LogKow |
|---|---|---|---|---|
| Diagnostic contrast media | Enhancement of vascular on magnetic resonance (MR) | Iopromide | 791.1 | −2.1 |
| | | Iomeprol | 777.1 | n.a. |
| | | Diatrizoate acid | 613.9 | 1.37 |
| Hormones | Regulation of metabolism; control of the sexual development; keep homeostasis | Estriol | 288.4 | 2.45 |
| | | Mestranol | 310.4 | 4.68 |
| | | Estrone | 270.4 | 3.13 |
| | | 17-b estradiol | 272.4 | 4.01 |
| | | Testosterone | 288.2 | 3.32 |
| | | Androstenedione | 286.2 | 2.75 |
| Lipid regulators | Regulation of triglycerides and cholesterol in blood | Clofibrate | 242.7 | 3.02 |
| | | Benzafibrate | 361.8 | 3.61 |
| | | Clorfibric acid | 214.6 | 2.88 |
| | | Gemfibrozil | 250.3 | 4.77 |
| | | Simvastatin | 418.6 | 4.68 |
| | | Furosemide | 330.7 | 1.51 |
| | | Bendroflumethiazide | 421.4 | 1.89 |
| Nonsteroidal anti-inflammatory drugs | Reduce pain and inflammation | Ibuprofen | 206.3 | 3.97 |
| | | Diclofenac | 296.1 | 4.5 |
| | | Acetaminophen | 151.2 | 0.46 |
| | | Aspirin | 180.2 | 1.19 |
| | | Indomethacin | 357.8 | 4.27 |
| | | Naproxen | 230.3 | 3.18 |
| | | Nimesulide | 308.3 | 2.6 |
| | | Phenazone | 188.2 | 0.38 |
| | | Salicylic acid | 138.1 | 2.26 |
| | | Paracetamol | 151.2 | 0.33 |
| | | Ketoprofen | 254.3 | 2.8 |
| *Personal care products* | | | | |
| Disinfectants | Destroy and kill unwanted germs and parasites | Triclosan | 289.5 | 4.76 |
| | | 2-Phenylphenol | 170.2 | 3.09 |
| | | 4-chlorocresol | 520.2 | n.a. |
| | | Chloroprene | 88.5 | 2.2 |
| | | Bromoprene | 132.9 | n.a. |
| | | 4-chloroxylenol | 156.6 | 3.27 |
| Fragrances | Create a pleasant odour | Musk xylene | 297.3 | 4.4 |
| | | Musk ketone | 294.3 | 4.3 |
| Preservatives | Prevent decomposition by microbial growth or by undesirable chemical changes | 2-phenoxyethanol | 138.2 | 1.16 |
| | | Ethyl 4-hydroxybenzoate | 166.2 | 2.47 |
| | | Propyl 4-hydroxybenzoate | 180.2 | 3.04 |
| | | Isopropyl 4-hydroxybenzoate | 180.2 | n.a. |
| | | Butyl 4-hydroxybenzoate | 194.2 | 3.57 |
| | | Isobutyl 4-hydroxybenzoate | 194.2 | n.a. |

(continued)

**Table 15.1** (continued)

| Classification of PPCPs | Functions | Representatives frequently detected | Molecular weight | LogKow |
|---|---|---|---|---|
| Sunscreen agents | Protect the skin from the sun's ultraviolet radiation, and reduces sunburn and other skin damage | Octocrylene | 361.5 | 6.9 |
| | | Ethylhexylmethoxycinnamate | 290.4 | 6.1 |
| | | Oxybenzone | 228.2 | 3.79 |

pose a dangerous hazard to water reception. The EU Water System Directive (WFD) primarily specified 33 substances in 2000 as a control/guide in the next 20 years. Many PPCPs, namely iopamidol, carbamazepine (CBZ), diclofenac (DFC), and musks, were later identified as potential targets for future surveillance in 2007. So it was proposed that triclosan (TCS), bisphenol A, ibuprofen (IBU), phthalates, and clofibric acid be included in the existing list of PPCPs (Ellis 2008; Nguyen et al. 2019a, b). Based on criteria of persistence, bioaccumulation, and toxicity (PBT), the English and Wales Environment Agency (EA) have also listed ten compounds as potentially hazardous substances: aminophylline, clotrimazole, dextropropoxyphene, lofepramine, paracetamol, procyclidine, thioridazine, tramadol, tamoxifen, and mebeverine (Ashton et al. 2004). Traditionally, safety risks of PPCPs date back to the USA in the 1970s, and England in the 1980s. Nevertheless, the subject remained controversial unless significant damage has been reported by estrogens in the fish inhabitants and by DFC in vultures (Hignite and Azarnoff 1977; Richardson and Bowron 1985; Aherne et al. 1990). An extensive range of PPCPs have since been identified in various ecological systems around the globe (Cizmas et al. 2015; Munze et al. 2017). PPCPs typically contain drugs (prescription and non-prescription), anticonvulsants, antidepressants, antihypertensives, fragrances (e.g., nitro- and polycyclic-musks), hormones, insect repellent, lipid regulators, moisturizers, nonsteroidal, opioids, soaps, UV blockers, veterinary medications, and their ensuing metabolites/conjugates (Daughton and Ternes 1999; Loraine and Pettigrove 2006; Cizmas et al. 2015). Annually thousands of PPCPs are generated globally and therefore their release into the atmosphere becomes an inevitable result of modern life (Nikolaou et al. 2007; Caldwell et al. 2014). As per a survey, in 2014, the USA solely distribute more than 76.9 million metformin prescription medications, likely to result in contamination of all surface water resources, inclusive of tap water at levels that exceed 50% of the permissible limits recommended by the Rhine River Basin Agency (Trautwein et al. 2014). The condition is mastering worse as the use of PPCPs is predictable to increase every day, with annual revenue projected to rise by 5% (Scudellari 2015). PPCPs have been detected worldwide continually since the 1990s in drinking water, groundwater, surface water, and wastewater (Cai et al. 2015). The PPCPs can introduce surface water by releasing it directly into the surface water by hospitals, households, industries, and WWTPs and by land runoff in the case of biosolids distributed over farm land that can enter the surface by leaching or filtration by banks (Mompelat et al. 2009). Sediment can accumulate PPCPs within the container of surface water, because it has an assortment of binding sites (Kastner et al. 2014). Soil also could be a sink for PPCPs. PPCPs may be transferred to the soil through irrigation with treated or untreated wastewater that contains PPCPs. PPCPs are accepted as pseudo insistent organic contaminants in the habitat, which are capable of posing the same hazard to the habitat as genuinely insistent organic contaminants due to their continued emergence into the surroundings by various ways like STPs. Even though concentration levels of PPCPs in the habitat are quite low as noted above, they can still influence the quality of water and the equilibrium of the ecosystems, and even affect the portable water supplies. It is important to elimination of the PPCPs from wastewater to reduce the possible danger of the PPCPs. As a result, numerous approaches have been developed to eliminate the PPCPs during WWTPs, such as physical, chemical, and biological methods.

### *15.3.1 PPCPs in the Environment/Ecosystems*

PPCPs may attain the habitat by various routes (Fig. 15.1). In the field the origins of PPCPs are of two forms (point and diffuse). PPCPs persist in the habitat as components of human and/or animal waste (e.g. discharge and also residential, industrial wastewater, and hospital) in point source pollution. While most PPCPs are intended to evoke a biological reaction in their original forms, large percentages of these molecules are discharged from human and animal bodies without decadence (Carballa et al. 2004). In addition, PPCPs can also reach the habitat as disseminated origins, like industrial, medical, agricultural, and/or household waste leftovers, thus providing another pervasive explanation for their presence in the habitat (Taylor and Senac 2014). Water and sludge are generally subjected as a source of nutrients to agricultural land after treatment. Thus, PPCPs can join groundwater explicitly through leaching or runoff from sludge-treated farming land (Nikolaou et al. 2007; La Farre et al. 2008). The PPCP metabolites are often commonly found in the atmosphere and also exist at concentration greater than those of their original compounds. Weigel et al. (2004), for example, analyzed the anti-inflammatory medication IBU and its hydroxyibuprofen and carboxyibuprofen metabolites in various samples of sewage. Carboxyibuprofen was determined to be more prevalent in the influent samples than hydroxyibuprofen and IBU. Among the PPCPs, the antibiotics, anticonvulsants, antidepressants, antineoplastics, Beta-blockers, diagnostic contrast media, disinfectants, fragrances, hormones, lipid regulators, nonsteroidal anti-inflammatory drugs, preservatives, and sunscreen agents are common groups excreted into the habitat (Wang et al. 2016). When in the atmosphere, PPCPs adopt various paths, such as dissolution, decadence, photolysis, and/or binding to the solid substrates, depending on their physicochemical properties and the environmental compartment characteristics through which these contaminants are released

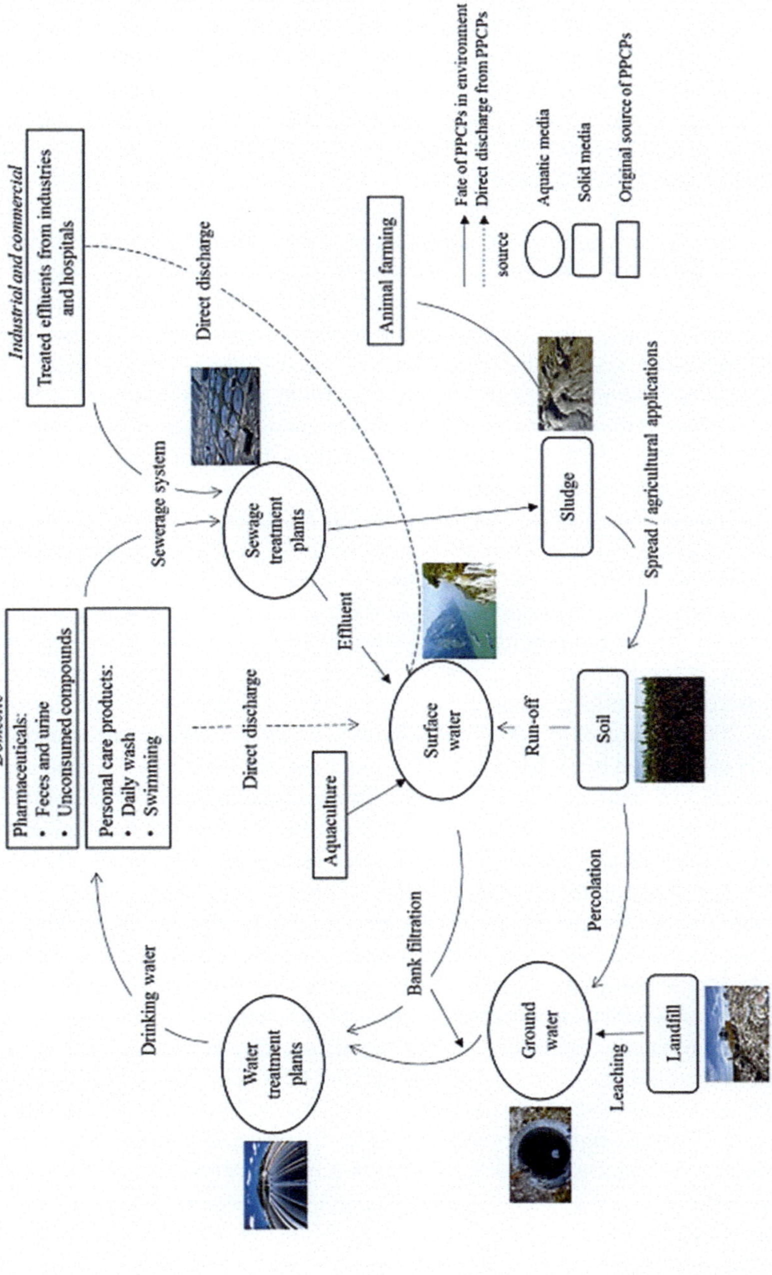


**Fig. 15.1** Sources and pathways of PPCPs (Yang et al. 2017). Copyright © 2017 with permission from Elsevier

(La Farre et al. 2008). As per figures, America alone generates 8–106 tons of dry sludge annually, approximately 50% added specifically to agronomic land (Kinney et al. 2008). Most of the medicines in biosolids worldwide were identified at elevated levels. Roughly 5000 mg/kg thiabendazole and equivalent quantities of other medications, like CBZ and caffeine (CFN), are tested in sewage sludge, as reported previously (Diaz-Cruz and Barcelo 2008; Lapworth et al. 2012). In agriculture, the use of biosolids and treated wastewater is a significant source of PPCPs that can pollute groundwater (Heberer 2002; Pedersen et al. 2005). Several experiments have shown that certain PPCPs in sediments, like TCS, sulfamethoxazole (SMX), ciprofloxacin, and CBZ, are more soluble when compared to water (Conkle et al. 2012).

Antibiotic transmission to the atmosphere will produce antibiotic resistance genes, resistance genes, and multiple antibiotic resistant super-integrates that present significant risks to the ecosystem (Pruden et al. 2006; Kemper 2008). The hormones are also known to induce detrimental consequences of endocrine disorders, particularly reproductive system disorders (Khanal et al. 2006). Investigations have indicated that, severe reproductive disorders in trout, minnow, medaka, carp, and turtle due to estrogen exposure in aquatic environments (Tabata et al. 2001; Irwin et al. 2001; Zha et al. 2008; Liu et al. 2011). Additionally, these pollutants are more likely to impact benthic ecosystems as they are continuously exposed to toxins in sediments and overlying water (Gilroy et al. 2012). Recent investigation revealed important effects on macro-invertebrate populations of the pollutants found in WWTP effluents (Munze et al. 2017). While research into the consequences of field-relevant pharmaceutical application to the aquatic environment is scarce, substantial impacts have been described on ecosystem services (Bundschuh et al. 2009; Painter et al. 2009; Wilson et al. 2004). *Daphnia magna* was reported to have the greater toxicological effects when exposed to CBZ, DFC, and IBU simultaneously (Cleuvers 2008). Munz et al. (2017) also identified adverse impact of diazinon and DFC in rivers affected by storm water. The toxicity of the PPCPs in the food supply chain has adverse consequences by trophic stage transmission. For example, the decline in population of the white-backed vulture (by 95 percent) was linked with renal collapse due to DFC, an anti-inflammatory medicine (Oaks et al. 2004). The studies on the decrease in population of the vulture across South Asia confirmed these results (Swan Gerry et al. 2006; Taggart et al. 2007). Because most PPCPs are found in an open space, they may endure chemical changes and interfere with each other and with the biotic and essential environmental elements. A host of processing products with new chemical properties can result in chemical and biological reactions. Few bacteria convert these compounds and generate new metabolites that are either biocompatible or stable on the atmosphere. Garcia-Galan et al. (2012) indicated that the compound of N4-acetylsulfapyridine, sulfapyridine metabolites, is higher toxic than the original molecule in algae. Likewise, the metabolites of CBZ (i.e., acridine) and naproxen photodegradation have been revealed to be not only higher hazardous than the original substance, and also possess cancer causing and mutation inducing effects (Chiron et al. 2006; Isidori et al. 2005). Such findings illustrate the significance of bioassay and check of the environment when detecting toxic chemicals.

### 15.3.2 Environmental and Health Risks

The prevalent incidence of PPCPs in water sources is a growing issue because of their effect on the atmosphere and public health. PPCPs are commonly found in drainage, rivers, reservoirs, and groundwater. These may harm health of human and animal as their residues will potentially reach and persist in the food chain by effluent release and the rehash of treated wastewater and sludge for agronomic applications (Rajapaksha et al. 2014; Vithanage et al. 2014). Even though a small amount of PPCPs are present in WTPs ranging from ng/L to μg/L, residues of PPCPs can have significant detrimental health effects and human acquaintance to these substances has unfamiliar long-term consequences (Boxall et al. 2012). Most PPCPs dissipate quickly in the atmosphere, however, their prolonged use leads to water pseudo-persistence and major impacts on water living things on the ecosystem (Kostich et al. 2014). Identification of high level PPCPs in STPs from untreated effluent and treated water (Chen et al. 2013) is induced by worldwide usage (Liu and Wong 2013), little human metabolic ability (Borova et al. 2014), inadequate disposal (Ternes et al. 2004), and bioactive structures (McClellan and Halden 2010). When recycled water and organic manures from waste sludge are used, other plant species take up different PPCPs. There have been earlier reports on detrimental impact of PPCPs on health and the environment (Tanoue et al. 2012; Jiang et al. 2013; Rajapaksha et al. 2015). PPCP residues were identified in plant eatable parts when biosolids or manure-amended soils were utilized or waste water was used for irrigation (Rajapaksha et al. 2014). In 28% of 27,000 PCPs, the United States Environmental Working Group (EWG 2008) erect 1,4-dioxane, a recognized cancer causing compound. They also supervised a study that involved 20 girls aged 14–19. The EWG decisive that 16 toxic chemicals were prevalent in the females' parts of the body because of the use of cosmetics, such as synthetic musk, 2-benzenedicarboxylic salt, and TCS. A report by the United States Environmental Protection Agency (USEPA) identified some types of drugs of interest in US water supplies, like antibiotics, antimicrobials, estrogenic steroids, and anti-epileptic drugs (EWG 2009).

Some PPCPs may result in bioaccumulation in fish and other aquatic creatures, triggering several unforeseen interferences with them. Individual and mixed PPCP substances have therefore been reported to induce adverse reproductive effects and histological changes in zebrafish (Galus et al. 2013a, b; Overturf et al. 2015). PPCPs also exert harmful aggregate impact on terrestrial and aquatic ecosystems (Hernando et al. 2004, 2006). The adverse effects of PPCPs on habitats are important for human health, as traces of PPCPs were found in our food chain (fruits, vegetables and potable water) (Hernando et al. 2006; Carmona et al. 2014; Awad et al. 2016).

### 15.3.3 Analytical Methods of PPCPs

Analytical techniques are essential for explore the destiny of PPCPs in the habitat. The investigation of focused mixes generally involves three stages. The initial step is to choose the appropriate analytical instrument. The more commonly used instruments are gas chromatography (GC) and high-performance liquid chromatography (HPLC). For HPLC, the sample preparation requirement is simpler than for GC. Water samples may be directly injected into the HPLC after filtration (usually 0.45 mm filter). The collection of instruments for the specific compound depends on the physicochemical properties of the intended compounds or molecules. HPLC can usually be used to determine compounds vulnerable to heat and non-volatile, whereas GC can evaluate volatile compounds. Certain organic compounds like dioctyl sebacate (DOS) can be tested by both HPLC and GC, however, introduced various sensitivities. Earlier reviews mentioned the mechanism and application of the three analytic methods (Carr and Purcell 1954; Stahl 1967; Efremov et al. 2008). Upon filtration the next step for the particulate fractions is to isolate and purify the samples upon deciding which instruments to choose. Solid-phase extraction (SPE), liquid–liquid extraction (LLE), liquid–liquid micro-extraction (LLME), and solid-phase micro-extraction (SPME) are the most extensively utilized procedures until now. The detailed knowledge on the use of every extraction process was added in previous studies, respectively (Huddleston and Rogers 1998; Thurman and Mills 1998; Lord and Pawliszyn 2000; Mohammadhosseini et al. 2006). The final step is to refine the measuring parameters to get the best execution. This move a tedious procedure, however, it must be finished. Recently, advancements in combination of chromatography with mass spectrometry, for example, liquid chromatography-mass spectrometry/mass spectrometry (LC-MS/MS), and gas chromatography-mass spectrometry/mass spectrometry (GC-MS/MS) extend the spectrum of confirmation and detection of PPCPs. LC-MS/MS and GC-MS/MS will achieve the detection limit of ng/L. The investigative strategies of PPCPs reported in references are outlined in Table 15.2. With improved awareness of people's health, the analytical techniques ought to be additionally enhanced to expand the detection limit of PPCPs, which first needs to remove the influence of complicated matrix on the environment. The sampling protocols could also be improved, in addition to analytical techniques. Sampling procedures are very important in assessing the true environmental variability of PPCPs. The measurement of PPCPs in nature requires refined scientific strategies as well as sufficient testing conventions.

### 15.3.4 Microbial Remediation

Although MOs may be an essential biological system of environment and provide eco-friendly solutions against pollution of PPCPs, little work has been done on the depletion of PPCPs by MOs in water and soils. The most extensively utilized method

**Table 15.2** The analytical methods of PPCPs presented in the literatures. Reproduced from (Wang and Wang 2016). Copyright © 2016 with permission from Elsevier

| | |
|---|---|
| HPLC | Estriol; Mestranol; Sulfamethoxazole, Thioridazine; 4-chlorocresol |
| LC-MS/MS | Androstenedione; Difloxacin; Clorfibric acid; Doxepin; Imipramine; Amitriptyline; Fluxetine; Oxazepam; Dilantin; Epirubicin; Ifosfamide; Tamoxifen; Iomeprol; Diatrizoate acid; Gabapentin |
| HPLC-MS/MS | Testosterone |
| GC-MS/MS | Musk xylene; Musk ketone |
| HPLC, LC-MS | Trimethoprim; Amoxicillin; Metoprolol |
| HPLC, LC-MS/MS | Estrone; 17-b Estradiol; Enoxacin; Sulfapyridine; Cephalexin; Cefaclor; Gemfibrozil; Simvastatin; Diclofenac; Acetaminophen; Aspirin; Indomethacin; Nimesulide; Phenazone; Paracetamol; Atenolol; Nadolol; Pindolol; Acebutolol; Sotalol; Iopromide |
| LC-MS, LC-MS/MS | Ofloxacin; Methotrexate; Cyclophosphamide |
| HPLC, GC-MS | Clofibrate; 2-phenoxyethanol; Salicylic acid |
| GC, GC-MS | Chloroprene |
| GC-MS, LC-MS/MS | Methylparaben; Ethyl 4-hydroxybenzoate; Propyl 4-hydroxybenzoate; Isopropyl 4-hydroxybenzoate; Butyl 4-hydroxybenzoate; Isobutyl 4-hydroxybenzoate |
| GC-MS/MS, LC-MS/MS | Ethylhexylmethoxycinnamate; Meprobamate |
| HPLC, LC-MS, LC-MS/MS | Erythromycin; Ciprofloxacin; Ampicillin; Tylosin; Mecillinam; Benzafibrate; Furosemide; Bendroflumethiazide |
| HPLC, GC-MS, LC-MS/MS | Doxycycline; Chloramphenicol; Naproxen; Propanolol; Diazepam; Carbamazepine; Primidone; Phenobarbital; Triclosan; Oxybenzone |
| HPLC, GC-MS/MS, LC-MS/MS | Octocrylene |
| HPLC, LC-MS, GC-MS, LC-MS/MS | Ibuprofen |

for the elimination of such substances/pollutants is the application of physicochemical techniques (Molina et al. 2020). BR, however, has the sign of efficient and affordable ways to eliminate PPCPs, such that a well-directed and systemic method is erect to research and incorporate these diverse processes in existing or new WWTPs and aquatic and in situ water and earthly systems. The comfort of utilizing consolidated physical, chemical, and biological procedures ought to be assessed.

#### 15.3.4.1 Remediation by Bacteria

The prokaryotes have been utilized in advanced systems of remediation, like active sludge, bioreactors or built wetlands. Majority of the medications are noxious to strains of bacteria (especially antibiotics), but other naturally happening bacteria can biodegrade these pollutants. The commonly identified PPCPs were removed using bacteria isolates from water, soil, and sediment (usually polluted) (Wang and Wang 2016). The pollutants can be removed by aerobic and anaerobic oxidative decadence, intracellular and extracellular decadence, and biosorption (absorption and adsorption) (Fig. 15.2). Prokaryote adaptation is a fundamental consideration to increase the efficiency of method. Several enzymes interfered in extracting PPCPs are inducible so that prior interaction with PPCPs with the MOs is necessary. For instance, the incidence of TCS induces ammonia monooxygenase production, that can degrade this substance (Roh et al. 2009). However, fundamental mechanisms, like oxidation through cytochrome P450 (Barth et al. 2004; Shrestha et al. 2008) or bacterial aerobic lignocellulolytic enzymes (Popa et al. 2014; Woo et al. 2014), are also implicated in bacterial decadence methods. Many pathways have been proposed for the reduction of PPCPs, for example, desorption and mobilization by biosurfactants (Guo et al. 2018), photodegradation coupled to biodegradation (Norvill et al. 2016; Bai and Acharya 2019; Huang et al. 2016), and a degradative pathway by biogenic manganese oxides (Furgal et al. 2014; Tran et al. 2018). A large portion of the PPCPs-degrading bacteria belong to the Proteobacteria or Actinobacteria group even less frequently utilized are Planococcus (little used) and Bacillus genera both Firmicutes (Molina et al. 2020). Overall, once the decadence ability of a MO is observed under in vitro conditions, it is assessed as being incredibly valuable in biotechnology. Thus, bacterial taxa were first investigated under in vitro condition and then applied in situ, with the exception of Actinobacteria which were often identified directly in active sludge or bioreactors (Fig. 15.3a). Among the most utilized, *Pseudomonas spp.* can be emphasized subsequently they are capable to degrade, among other molecules, CBZ, TCS, cephalexin, CFN, SMX, and diagnostic contrast media (Quandt et al. 2015; Thelusmond et al. 2019; Devatha and Pavithra 2019; Kumari and Ghosh Sachan 2019; Mahmoud et al. 2020; Gonzalez-Benitez et al. 2020). *Pseudomonas spp.* reported a 47% decadence of CBZ in 20 days (Li et al. 2013). *P. putida* activates effective decadence of 17-α-ethinylestradiol (E2α) through redox reactions intervened by biogenic manganese oxide (Furgal et al. 2014; Tran et al. 2018) (Fig. 15.2). The synergistic activity of these techniques with genetically modified *Escherichia coli* cells transformed with a *P. syringae* guiding motif allows bisphenol A and NP to be totally mineralized (Zhang et al. 2019a, b). Recently, 17β-hydroxysteroid dehydrogenase and its regulators were characterized (Wang et al. 2019). *Bacillus thuringiensis* is adequate for degrading naproxen (NPX), IBU, sulfonamides, trimethoprim, NP, TCS, and GFZ, among others (Grenni et al. 2014; Bai and Acharya 2019; Kjeldal et al. 2016; Marchlewicz et al. 2016; Liu et al. 2018; Zheng et al. 2018; Wang et al. 2018a, b). A new route for NPX utilization by *B. thuringiensis* has been depicted recently (Gorny et al. 2019). Numerous *Sphingomonas* are linked with effective biodegradation of PPCPs (Murdoch and Hay 2013; Thelusmond et al. 2016; Zhou et al. 2014; Kim et al. 2011; Bai et al. 2018). They were utilized as bioinoculants with successful results in recovery processes (Cirja et al. 2009; Zhou et al. 2014). The utilization of nanoparticles (NPs) in *Sphingomonas spp.* may be a possible approach for enhancing decadence (Murugesan et al. 2011). It has recently been immobilized on polydopamine-coated $Fe_3O_4$ iron NPs, demonstrating great effectiveness in the removal of NP polyethoxylates and through a more prominent amount of cycles. Moreover, separation and recycling for immobilized cells were more promptly accomplished compared with free cells. Other bacteria like *Stenotrophomonas*

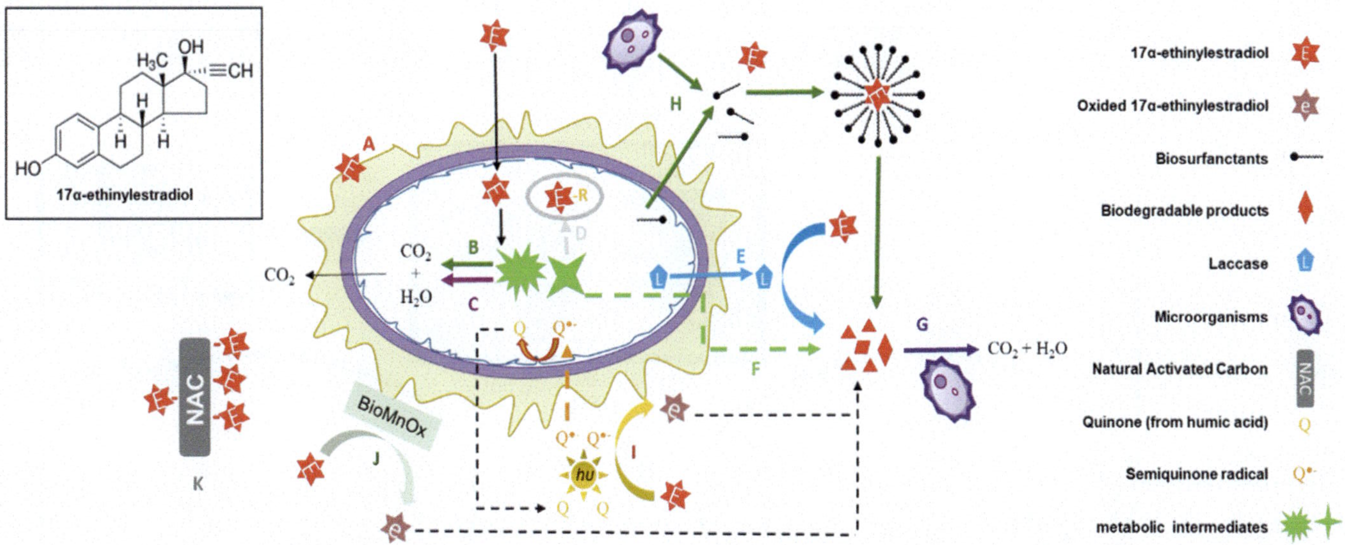


**Fig. 15.2** Various mechanisms for bio-removal of PPCPs, instance of 17α-ethinylestradiol from wastewater (Molina et al. 2020). (A) hydrophobic bioadsorption (Huang et al. 2019a, b); (B) Absorption and bacterial decadence (intracellular) (Sarmah and Halling-Sørensen 2007; Combalbert and Hernandez-Raquet 2010); (C) Absorption and microalgae decadence (intracellular) (Shi et al. 2010); (D) bioaccumulation (Al-Ansari et al. 2010); (E) Extracellular degradation by laccase, certain lignolytic enzymes may not tend to require (Eldridge et al. 2017; Otto et al. 2015; Matamoros et al. 2016); (F) Intermediate release of metabolites; (G) Synergistic decadence regulated by additional MOs (Wang and Wang 2016; Khunjar et al. 2011; Mikeskova et al. 2012); (H), Desorption and mobilization by biosurfactants (Guo et al. 2018); (I) photodegradation coupled to biodegradation (Norvill et al. 2016; Huang et al. 2019a, b; Bai and Acharya 2019); (J) Degradative way by biogenic manganese oxides (Furgal et al. 2014; Tran et al. 2018); and, (K) adsorption to natural activated carbon (Rovani et al. 2014). Discontinuous lines are possibly degrading, but not confirmed routes

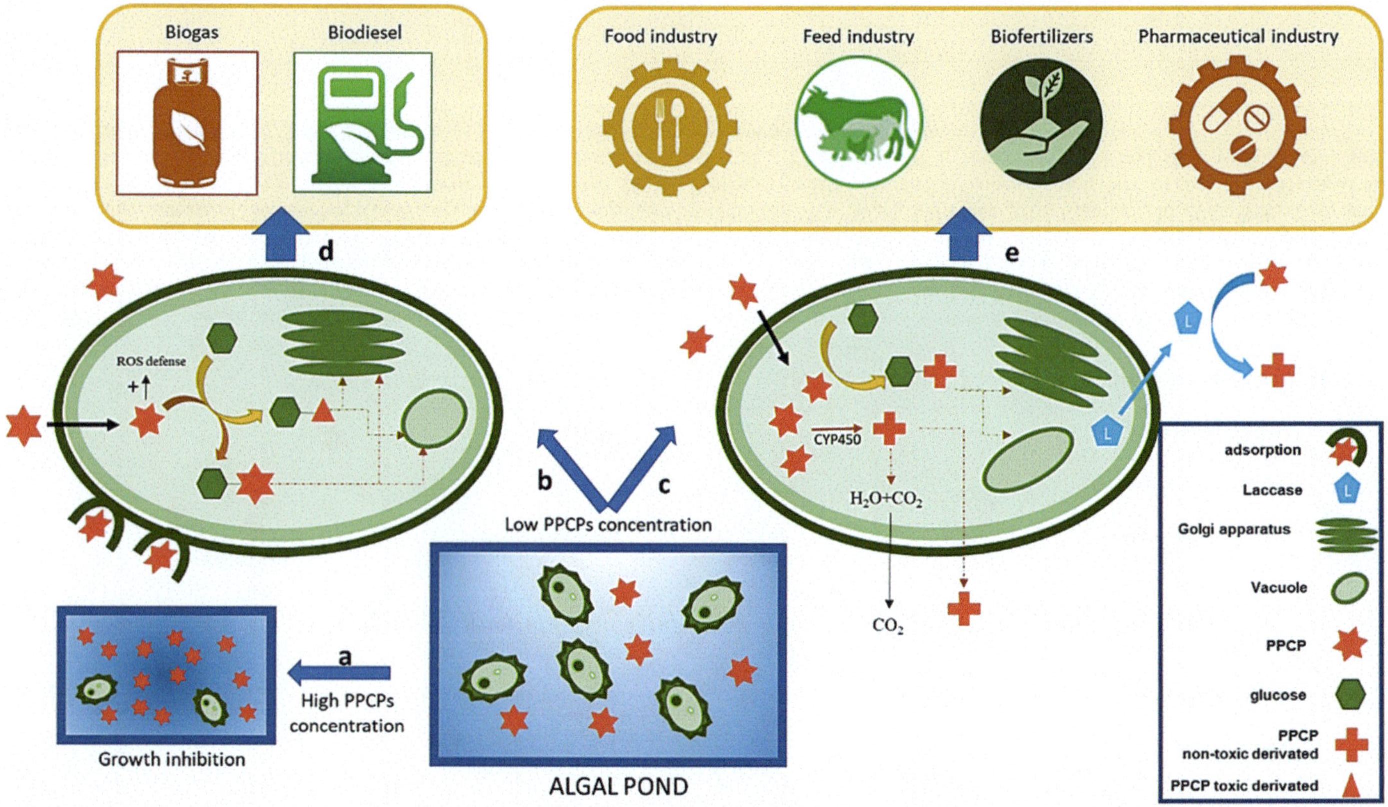


**Fig. 15.3** Towards the reactivation of the circular supply chain management (CSCM) (Molina et al. 2020). (a) Large concentrations of PPCPs inhibit growing algae (Nguyen et al. 2016; Olicon-Hernandez et al. 2019) However, small amounts may allow bioaccumulation (b) or Metabolism (c) and thus, biomass increase (Moon and Song 2012). (b) PPCP bioadsorption and bioabsorption associated to glucosil-conjugation bioaccumulation of PPCP or PPCP toxic derivatives (Wang et al. 2018a, b). Increase of ROS defense reactions (Gao et al. 2017). (d) Significantly rise in algal biomass production, beneficial by acclimatization in biodiesel and biogas production (Golveia et al. 2018). (c) Bioabsorption and decadence intracellular by cytochrome P450 and the related enzymes (Xiong et al. 2016) and extracellular decadence mediated by laccases (Otto et al. 2015; Matamoros et al. 2016). (e) Appropriate biomass in food and feed sectors (Lozano-Garcia et al. 2019), biofertilizants (de Siqueira Castro et al. 2020), and pharmaceutical industry (Abd El-Hakim et al. 2018)

*maltophilia* can degrade CBZ, triclocarban, TCS (Thelusmond et al. 2019), and NP (Wang et al. 2015), are less commonly used but with greater ability. *S. Maltophilia* can degrade NPX by enzymatic induction and co-metabolism (Wojcieszynska et al. 2014). *Acinetobacter* sp. may degrade sulfadiazine, sulfamethazine (SMT), and SMX but with varying mineralization efficiencies (Wang et al. 2018a, b) and degrade E2α in cometabolization with other hormones (Pauwels et al. 2008). *Arthrobacter denitrificans* BTF of sulfonamides yielded high decadence values (Reis et al. 2014; Reis et al. 2018).

In a metagenomic strategy, the draft genome of two sulfadiazine *Arthrobacter* bacteria was compared with other bacterial genomes which could help identify the functional genes involved in the decadence of these PPCPs (Deng et al. 2016). Bioaugmented *Achromobacter denitrificans* into bioreactors of membranes on a laboratory scale increases the rate of decadence of SMX (Nguyen et al. 2019b). *Nitrosomonas europaea* is an ammonia-oxidizing bacterium often linked to hormone decadence (Khunjar et al. 2011). Research on the emergence of microbial species with 17β-estradiol has identified a link between the involvement of certain bacteria (*Nitrosomonas*, *Bacillus*, *Pseudomonas*, *Sphingomonas*, *Novosphingobium*, *Alcaligenes*, *Rhodanobacter,* and *Mycobacterium*) and pollutant biomineralization (Navrozidou et al. 2019; Zhang et al. 2019a, b). Total sequencing of various species of *Sphingobium* is being performed to classify and juxtapose the expression behaviours of NP-degrading genes (Ootsuka et al. 2018). Recently, the full genome of *Cupriavidus sp.* have been sequenced, a caffeine-using bacterium (Watahiki and Kimura 2017) and other PPCPs such as CBZ (Gonzalez-Benitez et al. 2020). Purple phototrophic bacteria were identified as predicament pollutant degraders (Fig. 15.2) and have a greater value added as they generate ingredients of profitable attention from pollutants (de las Heras et al. 2017; Puyol et al. 2017). *Rhodobacter sphaeroides* have been shown to be efficacious in ameliorating toxic contaminants found in medicinal wastewater (Madukasi et al. 2010). *Rhodopseudomonas palustris* and *Rhodobacter capsulatus* are also worthy of attenuating evolving chemical molecules in domestic wastewater, creating efficiencies or interactions of rivalry based on organic oxides (Wang et al. 2014; Merugu et al. 2014). A recent study reveals that *Rhodopseudomonas* contained genes linked to xenobiotic degrading pathways (Thelusmond et al. 2019).

#### 15.3.4.2 Remediation by Fungi

The fungi, or their elements, are a BR tool and have been used in complex processes in 40 percent of them. The most frequently, white rot fungi are used. They belong to Basidiomycota and are known for their tolerance to great pollutant ranges (Tortella et al. 2015) and for their capability to metabolize and degrade a vast diversity of obstinate organic molecules. This outstanding capability has been associated to extracellular enzymes (ligninolytic enzymes) with low substrate precision (Tortella et al. 2015), and they can be amplified an 80% using redox mediators (Yang et al. 2018; Vasiliadou et al. 2019). The extracellular matrix of fungi consists of three enzymes: laccase, manganese peroxidase, and lignin peroxidase, the relative importance of which is species-specific in the decadence processes (Vasiliadou et al. 2016; Yang et al. 2013). It also described bioabsorption and intracellular decadence, as well as bioadsorption (Asif et al. 2017; Olicon-Hernandez et al. 2017; Lucas et al. 2018) (Fig. 15.2). Some of them have been effectively used in the decadence of PPCPs (Camacho-Morales et al. 2017). Key reactions are intricate in these white rot fungi medicinal alterations include formylation, hydroxylation, dehalogenation, deamination, conjugation, and polymerization (Olicon-Hernandez et al. 2017; Cruz-Morato et al. 2013). Many applications have been documented in the literature about white rot fungi (Tortella et al. 2015). The most widely studied basidiomycete is *Trametes versicolor*, which displays a high oxidative capacity to fully degrade PPCPs (Yang et al. 2013). More than 20 PPCPs susceptible of decadence were described by *Trametes versicolor* (Kumari and Ghosh Sachan 2019; Marchlewicz et al. 2016; Liu et al. 2018). Because *Trametes* laccases are more effective than other fungi and bacteria (Margot et al. 2013), they have been used in many biotechnological and commercial applications (Rodarte-Morales et al. 2012). *Phanerochaete chrysosporium* also can degrade numerous PPCPs under varying conditions of aeration. Isolated laccases of *Trametes versicolor* and *Phanerochaete chrysosporium* is used to form hybrid NPs that can remove both inorganic and drug molecule from polluted water (Maryskova et al. 2016; Ardao et al. 2015). Other extensively described white rot fungi are *Phanerochaete sordina*, *Pleurotus ostreatus,* and *Bjerkandra* species. White rot fungi, like *Panus tigrinus* and *Dichomitus squales,* are being assessed as effectively degrading fungi in recent years. *Ganoderma lucidum* was evaluated as a degrader of various drugs, with good DFC and ifosfamide results (Castellet-Rovira et al. 2018). *Irpex lactatus* is an appropriate white rot fungi in the biotransformation (BTF) of PPCPs like FLU (Cvancarova et al. 2013), hormones (Prenosilova et al. 2013; Loffredo et al. 2016) NP, TCS (Cajthaml et al. 2009; Moon and Song 2012), CBZ, and DFC (Castellet-Rovira et al. 2018). Recently, other white rot fungi like *Stropharia rugosoannulata*, *Gymnopilus luteofolius*, *Agrocybe erebia* (Castellet-Rovira et al. 2018), and *Moniliophra roreri* (Bronikowski et al. 2017) have recently been evaluated for their capability as PPCPs-degraders. Strains of *Rhodococcus rhodochrous* and *Aspergillus niger* have described the elimination of CBZ up to 10% (Gauthier et al. 2010). *Aspergillus* species biotransform DFC, TCS, estradiol, CFN, NP (Yang et al. 2018; Asif et al. 2017; Zhou et al. 2018; Aracagok et al. 2018; Ertit Tastan and Donmez 2015; Pai et al. 2013; Hussain et al. 2011).

Genetically manipulated *Aspergillus* escalates production of laccase that can be used for biotechnological purposes (Rodriguez et al. 2008; Nguyen et al. 2016; Asif et al. 2017). Some *Penicillium* species used successfully in TCS, NP, and hormones decadence (Tian et al. 2018; Kuzikova et al. 2017; Zhang et al. 2016; Tastan et al. 2016; Shan et al. 2016). The yeast *Yarrowia lipolytica* has been investigated for the countenance of cytochrome P450 and capability of degrading DFC and NPX (Aracagok et al. 2017, 2018). *Trichoderma harzianum* is able to degrade 17 β-estradiol (Chatterjee and Abraham 2019) and CBZ with a similar performance to that of *Pleurotus ostreatus* (Buchicchio et al. 2016). Estrogens (Eldridge et al. 2017; Lloret et al. 2012), DFC, NPX (Lloret et al. 2010), and antibiotics (Garcia-Delgado et al. 2018) are removed by Free laccase from *Myceliophthora thermophile* and *Lentinula edodes*.

Regarding Mucoromycota, *Cunninghamella elegans* can transform ATN (Parshikov et al. 2004), GFZ (Kang et al. 2009), NPX (Zhong et al. 2003). *Fusarium* species degrade estrogens and NP (Shi et al. 2002; Dubroca et al. 2005). *Umbellopsis isabellina* can degrade and decrease the hormones and NP toxicity (Janicki et al. 2018; Janicki et al. 2016). Several Ascomycota and Mucoromycota may be stronger degrading PPCPs than basidiomycetes, not just because of their strong resistance to severe conditions, but also because of the great prevalence of cytochrome P450 family (CYP) epoxidases and transferases intricate in xenobiotics metabolism in both classes (Olicon-Hernandez et al. 2017). However, the advancement of such filamentous fungi by commercial product development, biotechnological advances or industrial-scale applications is in its infancy.

#### 15.3.4.3 Remediation by Algae

Microalgae cultures have historically not been used to remove PPCPs, as these molecules are also poisonous to photosynthetic species (Czarny et al. 2019; Gosset et al. 2019; Gojkovic et al. 2019). In addition, due to their vulnerability to toxins some of them are used as species checking for ecotoxicity. *Scenedesmus obliquus* and *Chlamydomonas mexicana* are capable of eliminating CBZ at low levels, but higher levels supressed the growth of algae (Xiong et al. 2016). *Navicula* sp. has induced algal production by the IBU at lesser concentrations, but at high concentrations growth declined dramatically (Fig. 15.3). Further, *Navicula* sp. inoculation into water systems can inhibit IBU decadence, implying this anti-inflammatory drug could prolong the stress time (Ding et al. 2017).

Several experiments have demonstrated that algal treatment of wastewater dramatically decreases a large number of PPCPs, but the harmfulness of the water perseveres, due to the incidence of ammonium, a metabolism toxic derivative (Shi et al. 2010; Hamjinda et al. 2018; Escapa et al. 2017). While this toxic agent has been removed at a laboratory scale in certain experiments (Matamoros et al. 2016), the number of experimental studies will be expanded to validate these findings. Many of the research rely primarily on persistent open ponds (Lopez-Serna et al. 2019; de Godos et al. 2012). Extracellular pathways, like the fungi and bacteria, seem to be mediated by laccases. But, in the presence of a redox mediator, *Tetracystis* laccase transforms bisphenol A, 17α-ethinylestradiol, NP, and TCS (Otto et al. 2015). The intracellular degrading metabolic pathway will be regulated by cytochrome P450 and coupled with the inference of numerous other enzymes (Xiong et al. 2016; Fig. 15.2). Laccases and other extracelluar enzymes may perform external digestion (Otto et al. 2015; Matamoros et al. 2016). Metabolic pathways for SMT and SMX phycodegradation were advocated elsewhere (Xiong et al. 2019a, b). Degradation of such antibiotics has been greater at higher levels, implying that biodegradation may be an effective mechanism for acclimatizing microalgae to antibiotics (Xiong et al.

2019a, b). *Desmodesmus subspicatus* takes over and biotransforms the E2α, forming a highly toxic brominated compound (Maes et al. 2014). *Desmodesmus* sp.: Green algae and *Scenedesmus obliquus* was utilized for TCS hydrolysis and reduction dechlorination, indicating that these algae may be mineralized (Wang et al. 2018a, b). Likewise, (Escapa et al. 2018) not just to describe the BR of DFC utilizing *Chlorella sorokiniana*, *Chlorella vulgaris,* and *Scenedesmus obliquus*, but also demonstrate a significant decrease in effluent toxicity, particularly with *S. obliquus*. Particularly non-living *S. obliquus* may be an efficient alternative in bioadsorption removal of DFC and other PPCPs (Coimbra et al. 2018; Ali et al. 2018), as compounds with cationic groups are strongly drawn to the cell wall by electrostatic interactions (Xiong et al. 2018). The accumulation of PPCPs in microalgae cells will induce the formation of reactive oxygen species (ROS) linked to mechanisms of defence and adaptation (Gao et al. 2017). In addition, natural-isolated Nordic microalgae can eliminate lipophilic active medicinal components equally or more effectively than the culture collection strains under investigation. While *Coelastrella* sp. and *Coelastrum astroideum* was maximum effective in the aggregation of such molecules in its biomass, two species of algae, *Chlorella vulgaris* and *Chlorella saccharophila*, were not only exceptionally productive in the elimination of all 19 medicinal compounds, but even low quantities of mentioned molecules stored in their biomass for further application (Gojkovic et al. 2019). *Nannochloris* sp. mediated photo and biodegradation were the primary pathways for the elimination of 17α-, 17β-estradiol and salicylic acid, but the inadequate elimination of all steroid hormones shows that the potential for endocrine disruption in the ecosystem remains (Bai and Acharya 2019).

#### 15.3.4.4 Remediation by Mixed Culture

Full mineralization of raising pollutants is accomplished by the use of mixed cultures (MCs), as the synergistic enzymatic actions of consortium members are always great successful than each single action. Though, the reaction of MCs depends upon many factors, so the results may not be good if synergistic interactions are not produced or if competitive connections arises (Ha et al. 2016). Our knowledge has been broadened by controlled co-cultures. *Trametes versicolor* and *Ganoderma lucidum* were therefore used to eliminate 13 various medicinal compounds and to produce biodiesel from the sludge produced. Joining of both strains escalate the efficacy of elimination because of the interactions established between them (Vasiliadou et al. 2016). The antibacterial property of antibiotics can be eradicated after treatments with pure and co-culture of *P. chrysosporium* and *Pycnoporus sanguineus* (Gao and Shi 2018). Co-culture of *Alcaligenes faecalis* and *P. sanguineus* degrade SMX well (Li et al. 2016). *Cunninghamella* sp. a MC of WRF may be implicated in the depletion of the endocrine damaging compounds (Cabana et al. 2007). The total mineralization of E2α by metabolization was accomplished by ammonia-oxidizing bacteria and heterotrophic bacteria (Khunjar et al. 2011). The interactions among *Arthrobacter* sp. and *Pimelobacter* sp. also permit for the whole mineralization of sulfadiazine. Synergistic results were noted in mixed bacterial and fungal cultures, which could efficiently and exclusively eliminate some PPCPs (Mikeskova et al. 2012; Hai et al. 2012). Compared with the individual MOs, the consortia of cyanobacteria/microalgae and bacteria can be effective in detoxifying organic contaminants from wastewaters. Seven pharmaceutical molecules were analyzed for decadence by chlorella-Aspergillus pellets. In this laboratory examination, only one was degraded, and in the final degradation, the incidence of algae did not presume an advantage (Bodin et al. 2016). One of the problems relating to the utilization of microalgae in WWTP is their successive assortment. The association of microalga-duckweeds can eradicate estrogens completely from wastewater procedures that mediate sorption and biodegradation (Shi et al. 2010). Hydrophobic bioabsorption (Huang et al. 2019a, b) and bioaccumulation in PPCPs can result in biomass stabilization, white bioabsorption and biodegradation up to mineralization ($CO_2$ + $H_2O$) being able to eliminate pollutants from the ecosystem completely. Essentially, bacteria (Sarmah and Halling-Sørensen 2007; Combalbert and Hernandez-Raquet 2010) and also microalgae (Shi et al. 2010) generally implement intracellular decadence, while fungi metabolize this molecule by extracellular decadence that is regulated by laccase. No description has yet been given of the interference of other lignolithic enzymes (Eldridge et al. 2017). A string of synergistic reactions mediated by other MOs may naturally produce intermediate metabolites into the medium and biodegrade them (Wang and Wang 2016; Khunjar et al. 2011). Natural adsorbents, like activated carbon, promote the decadence of hormone (Rovani et al. 2014) and assist with its mobilization and eventual removal by biosurfactants (Guo et al. 2018). A biogenic manganese oxide degrading pathway has been identified (Furgal et al. 2014; Tran et al. 2018).

#### 15.3.4.5 Remediation by Plants

Plants can accumulate PPCPs in their roots by adsorption, or by hydrophilicity-regulated transport by xylems translocate soluble pollutants into the water (Wu et al. 2013). Plant uptake of over 100 PPCPs has been documented in both soil and water systems (Al-Farsi et al. 2017). Vegetables are documented to effectively eliminate gaseous contaminants like benzene (Treesubsuntorn and Thiravetyan 2018) although systems must be improved (Treesubsuntorn et al. 2017). Many plants can absorb, stabilize, and retain contaminants (Fig. 15.4a), metabolize, mineralize, volatilize, secrete, or detoxify pollutants (Fig. 15.4b) via in situ treatments (Bhatnagar and Bhatnagar 2013). Because of such methods, certain plants may use these contaminants as nutrients (source of carbon, nitrogen, and phosphorus) which produce an improvement in their biomass and eliminate the noxious impact which have on other ecosystem species (Fig. 15.4b).

Though phytoremediation is an outstanding biotechnology technique in soil and water environments, it does not restore 100% of the polluted area and involves additional approaches to minimize the amount of toxins in the field prior to planting. Furthermore, when choosing a plant for phytoremediation, it is essential to

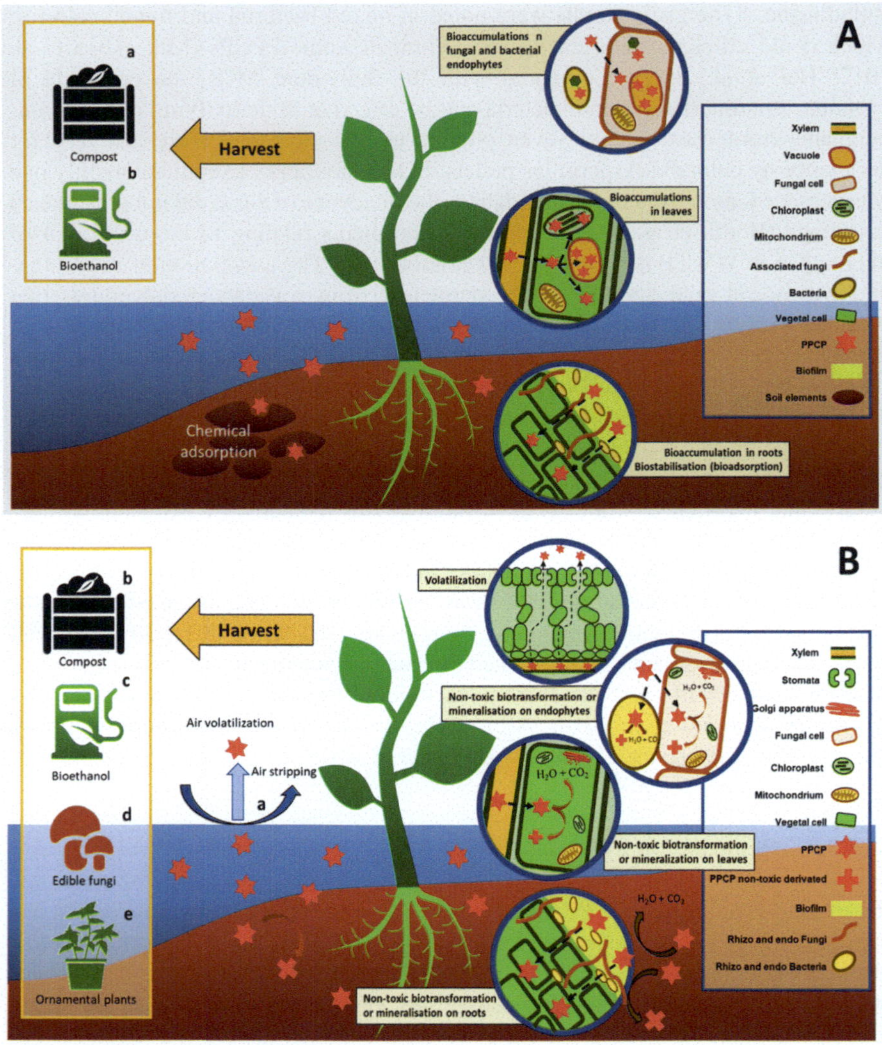


**Fig. 15.4** A. Towards the reactivation of circular supply chain management (CSCM) (Molina et al. 2020). (**a**). Taking and bioaccumulation of PPCPs in plants (roots, leaves, and endophytes) and chemical adsorption on substrates. Biomass harvest of macrophytes, use in secondary usages. (*a*) (Kouki et al. 2016); (*b*) (Sanchez-Galvan and Bolanos-Santiago 2018). (**b**). Road of PPCPs taking up and full mineralization (metabolic interactions between plants and related MOs), air and plant volatilization in built wetlands. B. Biomass harvest of macrophytes, including of secondary applications. (*a*) (Matamoros et al. 2016); (*b*) (Kouki et al. 2016); (*c*) (Sanchez-Galvan and Bolanos-Santiago 2018); (*d*) (da Luz et al. 2013; Hultberg et al. 2018); (*e*) (Calheiros and Bessa 2015; Macci et al. 2015)

contemplate its physiological needs for growth, its position as a native or introduced plant (Salamanca et al. 2015), and also its efficacy towards a distinct contaminant, and also the secondary added value of the plants (Fig. 15.4), (Kouki et al. 2016; Hultberg et al. 2018; Sopajarn and Sangwichien 2015). Microbiome (endosphere, rizhosphere, and phylosphere) is composed mostly of prokaryotes and fungi. Endophytes play a significant role mostly in detoxification process. Adequate studies concentrated on heavy metals, metalloids, and organic pollutants (Mallick et al. 2018; Barac et al. 2004; Afzal et al. 2014; Feng et al. 2017; Dolphen and Thiravetyan 2019; Ma et al. 2016; Molina et al. 2019). The *Burkholderia kururiensis* and *Agrobacterium rhizogenes* are phenolic resistant and decaying bacteria with respect to organic compounds. Some of the rizhophytic and endophytic MOs associated rhizosphere bacteria greatly improve the decadence of certain PPCPs (Toyama et al. 2011; Toyama et al. 2013). A further study explained a successful BTF of anti-inflammatory drugs by *Plantago lanceolata* isolated endophytic and epiphytic fungi (Gonda et al. 2016). A fascinating research done by Hurtado et al. (2016) suggested that the enatiometric ibuprofen phytodegradation by endophytic bacteria could implicate in its metabolism through lettuce. *Rhizobium radiobacter* and *Chriseobacterium nitroreducens*, endophytes in horseradish, improve CBZ elimination plant. Of the four metabolic pathways discussed for CBZ decadence in plants (Klampfl 2019), at least one (acridine pathway) is activated by endophyte presence (Sauvetre et al. 2018; Sauvetre and Schroder 2015). The need for more detailed research on the synergistic impact between plants and PPCPs-degrading endophytic bacteria needs to be highlighted (Klampfl 2019); but, as far as we know, only the results alluded to in this chapter are accessible.

## 15.4 Conclusion and Future Prospectives

According to its complex chemical compositions, steady release into the atmosphere and low environmental concentrations, the biodegradation of PPCPs poses many challenges. Nevertheless, substantial progress has been made in recognizing the role of microbial metabolism and plants in converting and eliminating PPCPs in WWTPs, natural aquatic environments, and wetlands. Microbes (aerobic and anaerobic bacteria, basidiomycete and ascomycete fungi, and certain resistant algae) may serve as eco-factories, able to restore environments by clean, low-cost technologies. Since PPCP's pathways of mineralization can be complex, more work is required to systematically develop mutual relationships between MOs. Such studies, on a meso- and macro-scale basis, may allow full metabolization while preventing intermediate metabolites that are also often ecotoxic substances with numerous target cells (Leng et al. 2020). In addition, the integration of these MOs with macroorganisms (plants and fungi) in sequence systems improves the mechanisms of decadence and exacerbates the potentials for development (Chen and Wong 2016). The engineering of the systems and also the development of the techniques in the field are lines of investigation that can be ameliorated in the coming years. In addition, in this kind of process, rhizo and endophytes not only specifically degrade PPCPs but also promote plant growth by developing growth-promoting enzymes and, thus, the capacity for remediation of CWs (Nguyen et al. 2019a, b), a study line in the initial phases of development. On the other side, some microalgae have been advocated as good eco-factories, not only because of their capability as pollutant molecule degraders, and also as a basis of bioenergy and natural pigments, particularly in circumstances of saline stress, i.e. for the treatment of contaminated seawater. These must be improved in their effectiveness and profitability (Vo et al. 2020). Finally, it should be mentioned that many reports merely focus on eliminating the parental molecules by nursing the levels of PPCPs in influent and effluent. The pathway to decadence and the PPCP intermediates are still not explicit. Further attempts should also be made to study the decadence mechanism and PPCP intermediates, which will lead to a deeper understanding of the environmental condition and vulnerability of PPCPs.

## References


Abd El-Hakim YM, Mohamed WA, El-Metwally AE (2018) Spirulina platensis attenuates furan reprotoxicity by regulating oxidative stress, inflammation, and apoptosis in testis of rats. Ecotoxicol Environ Saf 161:25–33

Afzal M, Khan QM, Sessitsch A (2014) Endophytic bacteria: prospects and applications for the phytoremediation of organic pollutants. Chemosphere 117:232–242

Aherne GW, Hardcastle A, Nield A (1990) Cytotoxic drugs and the aquatic environment: estimation of bleomycin in river and water samples. J Pharm Pharmacol 42:741–742

Al-Ansari AM, Saleem A, Kimpe LE et al (2010) Bioaccumulation of the pharmaceutical 17-α-ethinylestradiol in shorthead redhorse suckers (*Moxostoma macrolepidotum*) from the St. Clair River, Canada. Environ Pollut 158:2566–2571

Al-Farsi RS, Ahmed M, Al-Busaidi A, Choudri BS (2017) Translocation of pharmaceuticals and personal care products (PPCPs) into plant tissues: a review. Emerg Contam 3:132–137

Ali MEM, Abd El-Aty AM, Badawy MI, Ali RK (2018) Removal of pharmaceutical pollutants from synthetic wastewater using chemically modified biomass of green alga *Scenedesmus obliquus*. Ecotoxicol Environ Saf 151:144–152

Aracagok YD, Goker H, Cihangir N (2017) Biodegradation of micropollutant naproxen with a selected fungal strain and identification of metabolites. Z Naturforsch 72:173–179

Aracagök YD, Goker H, Cihangir N (2018) Biodegradation of diclofenac with fungal strains. Arch Environ Prot 44:55–62

Ardao I, Magnin D, Agathos SN (2015) Bioinspired production of magnetic laccase-biotitania particles for the removal of endocrine disrupting chemicals. Biotechnol Bioeng 112:1986–1996

Ashton D, Hilton M, Thomas KV (2004) Investigating the environmental transport of human pharmaceuticals to streams in the United Kingdom. Sci Total Environ 333:167–184

Asif MB, Hai FI, Singh L, Price WE et al (2017) Degradation of pharmaceuticals and personal care products by white-rot fungi-a critical review. Curr Pollut Rep 3:88–103

Awad YM, Ok YS, Igalavithana AD et al (2016) Sulphamethazine in poultry manure changes carbon and nitrogen mineralisation in soils. Chem Ecol 32:899–918

Bai X, Acharya K (2019) Removal of seven endocrine disrupting chemicals (EDCs) from municipal wastewater effluents by a freshwater green alga. Environ Pollut 247:534–540

Bai N, Wang S, Sun P et al (2018) Degradation of nonylphenol polyethoxylates by functionalized $Fe_3O_4$ nanoparticle-immobilized *Sphingomonas* sp. Y2. Sci Total Environ 615:462–468

Barac T, Taghavi S, Borremans B et al (2004) Engineered endophytic bacteria improve phytoremediation of water-soluble, volatile, organic pollutants. Nat Biotechnol 22:583–588
Barth S, Fischer M, Schmid RD, Pleiss J (2004) Sequence and structure of epoxide hydrolases: a systematic analysis. Proteins Struct Funct Genet 55:846–855
Bhatnagar N, Bhatnagar N (2013) Phytoremediation: a promising Technology of Environmental Pollution Control. Glob J Microb Biotechnol 4:39–42
Blair BD, Crago JP, Hedman CJ, Klaper RD (2013) Pharmaceuticals and personal care products found in the Great Lakes above concentrations of environmental concern. Chemosphere 93:2116–2123
Bodin H, Daneshvar A, Gros M, Hultberg M (2016) Effects of biopellets composed of microalgae and fungi on pharmaceuticals present at environmentally relevant levels in water. Ecol Eng 91:169–172
Borova VL, Maragou NC, Gago-Ferrero P et al (2014) Highly sensitive determination of 68 psychoactive pharmaceuticals, illicit drugs, and related human metabolites in wastewater by liquid chromatography–tandem mass spectrometry. Anal Bioanal Chem 406:4273–4285
Boxall A, Rudd MA, Brooks BW et al (2012) Pharmaceuticals and personal care products in the environment: what are the big questions? Environ Health Perspect 120:1221–1229
Brausch JM, Rand GM (2011) A review of personal care products in the aquatic environment: environmental concentrations and toxicity. Chemosphere 82:1518–1532
Bronikowski A, Hagedoorn PL, Koschorreck K, Urlacher VB (2017) Expression of a new laccase from *Moniliophthora roreri* at high levels in *Pichia pastoris* and its potential application in micropollutant degradation. AMB Express 7:73
Buchicchio A, Bianco G, Sofo A et al (2016) Biodegradation of carbamazepine and clarithromycin by *Trichoderma harzianum* and *Pleurotus ostreatus* investigated by liquid chromatography - high-resolution tandem mass spectrometry (FTICR MS-IRMPD). Sci Total Environ 557:733–739
Bundschuh M, Hahn T, Gessner MO, Schulz R (2009) Antibiotics as a chemical stressor affecting an aquatic decomposer–detritivore system. Environ Toxicol Chem 28:197–203
Cabana H, Jones JP, Agathos SN (2007) Elimination of endocrine disrupting chemicals using white rot fungi and their lignin modifying enzymes: a review. Eng Life Sci 7:429–456
Cai MQ, Wang R, Feng L, Zhang LQ (2015) Determination of selected pharmaceuticals in tap water and drinking water treatment plant by high-performance liquid chromatography-triple quadrupole mass spectrometer in Beijing, China. Environ Sci Pollut Res 22:1854–1867
Cajthaml T, Kresinova Z, Svobodova K, Moder M (2009) Biodegradation of endocrine-disrupting compounds and suppression of estrogenic activity by ligninolytic fungi. Chemosphere 75:745–750
Caldwell DJ, Mastrocco F, Margiotta-Casaluci L, Brooks BW (2014) An integrated approach for prioritizing pharmaceuticals found in the environment for risk assessment, monitoring and advanced research. Chemosphere 115:4–12
Calheiros CSC, Bessa VS, Mesquita RBR et al (2015) Constructed wetland with a polyculture of ornamental plants for wastewater treatment at a rural tourism facility. Ecol Eng 79:1–7
Camacho-Morales RL, Gerardo-Gerardo JL, Guillen Navarro K, Sanchez JE (2017) Ligninolytic enzyme production by white rot fungi during paraquat (herbicide) degradation. Rev Argent Microbiol 49:189–196
Carballa M, Omil F, Lema JM et al (2004) Behavior of pharmaceuticals, cosmetics and hormones in a sewage treatment plant. Water Res 38:2918–2926
Carmona E, Andreu V, Pico Y (2014) Occurrence of acidic pharmaceuticals and personal care products in Turia River basin: from waste to drinking water. Sci Total Environ 484:53–63
Carr HY, Purcell EM (1954) Effects of diffusion on free precession in nuclear magnetic resonance experiments. Phys Rev 94:630–638
Castellet-Rovira F, Lucas D, Villagrasa M et al (2018) Stropharia rugosoannulata and *Gymnopilus luteofolius*: promising fungal species for pharmaceutical biodegradation in contaminated water. J Environ Manag 207:396–404

Chatterjee A, Abraham J (2019) Mycoremediation of 17β-estradiol using *Trichoderma citrinoviride* strain AJAC3 along with enzyme studies. Environ Prog Sustain Energy 8:13142
Chen RZ, Wong MH (2016) Integrated wetlands for food production. Environ Res 426:13–31
Chen W, Lu S, Jiao W et al (2013) Reclaimed water: a safe irrigation water source? Environ Dev 8:74–83
Chiron S, Minero C, Vione D (2006) Photodegradation processes of the antiepileptic drug carbamazepine, relevant to estuarine waters. Environ Sci Technol 40:5977–5983
Cirja M, Hommes G, Ivashechkin P et al (2009) Impact of bio-augmentation with *Sphingomonas* sp. strain TTNP3 in membrane bioreactors degrading nonylphenol. Appl Microbiol Biotechnol 84:183–189
Cizmas L, Sharma VK, Gray CM, McDonald TJ (2015) Pharmaceuticals and personal care products in waters: occurrence, toxicity, and risk. Environ Chem Lett 13:381–394
Cleuvers M (2008) Chronic mixture toxicity of pharmaceuticals to daphnia—the example of nonsteroidal anti-inflammatory drugs. In: Kümmerer K (ed) Pharmaceuticals in the environment: sources, fate, effects and risks. Springer, Berlin Heidelberg, pp 277–284
Coimbra RN, Escapa C, Vazquez NC et al (2018) Utilization of non-living microalgae biomass from two different strains for the adsorptive removal of diclofenac from water. Water 10:1401
Combalbert S, Hernandez-Raquet G (2010) Occurrence, fate, and biodegradation of estrogens in sewage and manure. Appl Microbiol Biotechnol 86:1671–1692
Conkle JL, Gan J, Anderson MA (2012) Degradation and sorption of commonly detected PPCPs in wetland sediments under aerobic and anaerobic conditions. J Soils Sediments 12:1164–1173
Corcoran J, Winter MJ, Tyler CR (2010) Pharmaceuticals in the aquatic environment: a critical review of the evidence for health effects in fish. Crit Rev Toxicol 40:287–304
Cruz-Morato C, Ferrando-Climent L, Rodriguez-Mozaz S et al (2013) Degradation of pharmaceuticals in non-sterile urban wastewater by *Trametes versicolor* in a fluidized bed bioreactor. Water Res 47:5200–5210
Cvancarova M, Moeder M, Filipova A et al (2013) Biotransformation of the antibiotic agent flumequine by ligninolytic fungi and residual antibacterial activity of the transformation mixtures. Environ Sci Technol 47:14128–14136
Czarny K, Szczukocki D, Krawczyk B (2019) Toxic effects of single animal hormones and their mixtures on the growth of *Chlorella vulgaris* and Scenedesmus armatus. Chemosphere 224:93–102
da Luz JMR, Paes SA, Torres DP et al (2013) Production of edible mushroom and degradation of antinutritional factors in jatropha biodiesel residues. LWT Food Sci Technol 50:575–580
Daughton CG, Ternes TA (1999) Pharmaceuticals and personal care products in the environment: agents of subtle change? Environ Health Perspect 107:907–938
de Godos I, Munoz R, Guieysse B (2012) Tetracycline removal during wastewater treatment in high-rate algal ponds. J Hazard Mater 229:446–449
de las Heras I, Padrino B, Molina R et al (2017) Efficient treatment of synthetic wastewater contaminated with emerging pollutants by anaerobic purple phototrophic bacteria. Front Wastewater Treat Model 1:324–330
de Siqueira Castro J, Calijuri ML, Mattiello EM et al (2020) Algal biomass from wastewater: soil phosphorus bioavailability and plants productivity. Sci Total Environ 711:135088
Deng Y, Mao Y, Li B et al (2016) Aerobic degradation of sulfadiazine by *arthrobacter* spp.: kinetics, pathways, and genomic characterization. Environ Sci Technol 50:9566–9575
Devatha CP, Pavithra N (2019) Isolation and identification of *Pseudomonas* from wastewater, its immobilization in cellulose biopolymer and performance in degrading Triclosan. J Environ Manag 232:584–591
Diaz-Cruz MS, Barcelo D (2008) Trace organic chemicals contamination in ground water recharge. Chemosphere 72:333–342
Ding T, Yang M, Zhang J et al (2017) Toxicity, degradation and metabolic fate of ibuprofen on freshwater diatom *Navicula* sp. J Hazard Mater 330:127–134

Dolphen R, Thiravetyan P (2019) Reducing arsenic in rice grains by leonardite and arsenic–resistant endophytic bacteria. Chemosphere 223:448–454

Dubroca J, Brault A, Kollmann A et al (2005) Biotransformation of nonylphenol surfactants in soils amended with contaminated sewage sludges. In: Environmental chemistry: green chemistry and pollutants in ecosystems. Springer-Verlag, Berlin

Efremov EV, Ariese F, Gooijer C (2008) Achievements in resonance Raman spectroscopy: review of a technique with a distinct analytical chemistry potential. Anal Chim Acta 606:119–134

Eldridge HC, Milliken A, Farmer C et al (2017) Efficient remediation of 17 α-ethinylestradiol by *Lentinula edodes* (shiitake) laccase. Biocatal Agric Biotechnol 10:64–68

Ellis JB (2008) Assessing sources and impacts of priority PPCP compounds in urban receiving waters. In: 11th international conference on urban drainage, Edinburgh, Scotland, vol 1. pp 1–10

Environmental Working Group (EWG) (2008) Almost half of all 'natural' personal care products contain known carcinogen. http://www.ewg.org/news/testimony-official-correspondence/study-almost-half-all-%E2%80% 98natural%E2%80%99-personal-care-products. Accessed 23 Mar 2016

Environmental Working Group (EWG) (2009) Hearing on the testing by the Department of Environmental Protection for the presence of pharmaceuticals and personal care products in the NYC drinking water supply. http://www.ewg.org/news/testimony-officialcorrespondence/testing-pharmaceuticalsand-personal-care-products-new-york. Accessed 23 Mar 2016

Ertit Tastan B, Donmez G (2015) Biodegradation of pesticide triclosan by *A. versicolor* in simulated wastewater and semi-synthetic media. Pestic Biochem Physiol 118:33–37

Escapa C, Coimbra RN, Paniagua S et al (2017) Paracetamol and salicylic acid removal from contaminated water by microalgae. J Environ Manag 203:799–806

Escapa C, Torres T, Neuparth T et al (2018) Zebrafish embryo bioassays for a comprehensive evaluation of microalgae efficiency in the removal of diclofenac from water. Sci Total Environ 640–641:1024–1033

Feng NX, Yu J, Zhao HM et al (2017) Efficient phytoremediation of organic contaminants in soils using plant–endophyte partnerships. Sci Total Environ 583:352–368

Furgal KM, Meyer RL, Bester K (2014) Removing selected steroid hormones, biocides and pharmaceuticals from water by means of biogenic manganese oxide nanoparticles in situ at ppb levels. Chemosphere 136:321–326

Galus M, Jeyaranjaan J, Smith E (2013a) Chronic effects of exposure to a pharmaceutical mixture and municipal wastewater in zebrafish. Aquat Toxicol 132–133:212–222

Galus M, Kirischian N, Higgins S et al (2013b) Chronic, low concentration exposure to pharmaceuticals impacts multiple organ systems in zebrafish. Aquat Toxicol 132–133:200–211

Gao T, Shi X (2018) Preparation of a synthetic seed for the common reed harboring an endophytic bacterium promoting seedling growth under cadmium stress. Environ Sci Pollut Res 25:8871–8879

Gao QT, Wong YS, Tam NFY (2017) Antioxidant responses of different microalgal species to nonylphenol-induced oxidative stress. J Appl Phycol 29:1317–1329

Garcia-Delgado C, Eymar E, Camacho-Arevalo R et al (2018) Degradation of tetracyclines and sulfonamides by stevensite- and biochar-immobilized laccase systems and impact on residual antibiotic activity. J Chem Technol Biotechnol 93:3394–3409

Garcia-Galan MJ, Gonzalez Blanco S, Lopez Roldan R et al (2012) Ecotoxicity evaluation and removal of sulfonamides and their acetylated metabolites during conventional wastewater treatment. Sci Total Environ 437:403–412

Gauthier H, Yargeau V, Cooper DG (2010) Biodegradation of pharmaceuticals by *Rhodococcus rhodochrous* and *Aspergillus niger* by co-metabolism. Sci Total Environ 408:1701–1706

Gilroy EAM, Balakrishnan VK, Solomon KR et al (2012) Behaviour of pharmaceuticals in spiked lake sediments-effects and interactions with benthic invertebrates. Chemosphere 86:578–584

Gojkovic Z, Lindberg RH, Tysklind M, Funk C (2019) Northern green algae have the capacity to remove active pharmaceutical ingredients. Ecotoxicol Environ Saf 170:644–656

Golveia JCS, Santiago MF, Sales PTF et al (2018) Cupuaçu (*Theobroma grandiflorum*) residue and its potential application in the bioremediation of 17-A-ethinylestradiol as a *Pycnoporus sanguineus* laccase inducer. Prep Biochem Biotechnol 48:541–548

Gonda S, Kiss-Szikszai A, Szucs Z et al (2016) Efficient biotransformation of non-steroid anti-inflammatory drugs by endophytic and epiphytic fungi from dried leaves of a medicinal plant, Plantago lanceolata L. Int Biodeterior Biodegrad 108:115–121

Gonzalez-Benitez N, Molina MC, Arrayas M (2020) Empirical evidence and mathematical modelling of carbamazepine degradative kinetics by a wood-rotting microbial consortium. Waste Biomass Valori. https://doi.org/10.1007/s12649-020-01041-1

Gorny D, Guzik U, Hupert-Kocurek K, Wojcieszynska D (2019) A new pathway for naproxen utilisation by *Bacillus thuringiensis* B1 (2015b) and its decomposition in the presence of organic and inorganic contaminants. J Environ Manag 239:1–7

Gosset A, Durrieu C, Barbe P (2019) Microalgal whole-cell biomarkers as sensitive tools for fast toxicity and pollution monitoring of urban wet weather discharges. Chemosphere 217:522–533

Grenni P, Patrolecco L, Ademollo N (2014) Capability of the natural microbial community in a river water ecosystem to degrade the drug naproxen. Environ Sci Pollut Res 21:13470–13479

Guo YP, Hu YY, Lin H, Ou XL (2018) Sorption and desorption of 17α-ethinylestradiol onto sediments affected by rhamnolipidic biosurfactants. J Hazard Mater 344:707–715

Ha H, Mahanty B, Yoon S, Kim CG (2016) Degradation of the long-resistant pharmaceutical compounds carbamazepine and diatrizoate using mixed microbial culture. J Environ Sci Heal Part A Toxic/Hazardous Subst Environ Eng 51:467–471

Hai FI, Modin O, Yamamoto K et al (2012) Pesticide removal by a mixed culture of bacteria and white-rot fungi. J Taiwan Inst Chem Eng 43:459–462

Hamjinda NS, Chiemchaisri W, Watanabe T et al (2018) Toxicological assessment of hospital wastewater in different treatment processes. Environ Sci Pollut Res 25:7271–7279

Heberer T (2002) Occurrence, fate, and removal of pharmaceutical residues in the aquatic environment: a review of recent research data. Toxicol Lett 131:5–17

Hernando MD, Petrovic M, Fernandez-Alba AR, Barcelo D (2004) Analysis by liquid chromatography–electrospray ionization tandem mass spectrometry and acute toxicity evaluation for β-blockers and lipid-regulating agents in wastewater samples. J Chromatogr A 1046:133–140

Hernando MD, Mezcua M, Fernandez-Alba AR, Barcel D (2006) Environmental risk assessment of pharmaceutical residues in wastewater effluents, surface waters and sediments. Talanta 69:334–342

Hignite C, Azarnoff DL (1977) Drugs and drug metabolites as environmental contaminants: chlorophenoxyisobutyrate and salicylic acid in sewage water effluent. Life Sci 20:337–341

Huang W, Zhang Y, Zhang Y et al (2016) Optimization of the measurement of particle-bound reactive oxygen species with 2′,7′-dichlorofluorescin (DCFH). Water Air Soil Pollut 227:1–10

Huang B, Lai C, Dai H et al (2019a) Microbially reduced humic acid promotes the anaerobic photodegradation of 17α-ethinylestradiol. Ecotoxicol Environ Saf 171:313–320

Huang Y, Guo J, Yan P et al (2019b) Sorption-desorption behavior of sulfamethoxazole, carbamazepine, bisphenol A and 17A-ethinylestradiol in sewage sludge. J Hazard Mater 368:739–745

Huddleston J, Rogers R (1998) Room temperature ionic liquids as novel media for 'clean'liquideliquid extraction. Chem Commun 16:1765–1766

Hultberg M, Prade T, Bodin H et al (2018) Adding benefit to wetlands – valorization of harvested common reed through mushroom production. Sci Total Environ 637:1395–1399

Hurtado C, Dominguez C, Perez-Babace L et al (2016) Estimate of uptake and translocation of emerging organic contaminants from irrigation water concentration in lettuce grown under controlled conditions. J Hazard Mater 305:139–148

Hussain J, Muhammad Z, Ullah R et al (2011) Biotransformation of ß-estradiol isolated from Sonchus eruca. African J Biotechnol 10:5529–5533

Irwin LK, Gray S, Oberdorster E (2001) Vitellogenin induction in painted turtle, Chrysemys picta, as a biomarker of exposure to environmental levels of estradiol. Aquat Toxicol 55:49–60

Isidori M, Lavorgna M, Nardelli A et al (2005) Ecotoxicity of naproxen and its phototransformation products. Sci Total Environ 348:93–101

Janicki T, Krupinski M, Długonski J (2016) Degradation and toxicity reduction of the endocrine disruptors nonylphenol, 4-tert-octylphenol and 4-cumylphenol by the non-ligninolytic fungus *Umbelopsis isabellina*. Bioresour Technol 200:223–229

Janicki T, Długonski J, Krupinski M (2018) Detoxification and simultaneous removal of phenolic xenobiotics and heavy metals with endocrine-disrupting activity by the non-ligninolytic fungus *Umbelopsis isabellina*. J Hazard Mater 360:661–669

Jiang JQ, Zhou Z, Sharma VK (2013) Occurrence, transportation, monitoring and treatment of emerging micro-pollutants in waste water - a review from global views. Microchem J 110:292–300

Kang SI, Kang SY, Kanaly RA et al (2009) Rapid oxidation of ring methyl groups is the primary mechanism of biotransformation of gemfibrozil by the fungus *Cunninghamella elegans*. Arch Microbiol 191:509–517

Kastner M, Nowak KM, Miltner A (2014) Classification and modelling of nonextractable residue (NER) formation of xenobiotics in soilea synthesis. Crit Rev Environ Sci Technol 44:2107–2171

Kemper N (2008) Veterinary antibiotics in the aquatic and terrestrial environment. Ecol Indic 8:1–13

Khanal SK, Xie B, Thompson ML, Sung S, Ong S-K, van Leeuwen J (2006) Fate, transport, and biodegradation of natural estrogens in the environment and engineered systems. Environ Sci Technol 40:6537–6546

Khunjar WO, Mackintosh SA, Baik S et al (2011) Elucidating the relative roles of ammonia oxidizing and heterotrophic bacteria during the biotransformation of 17alpha-ethinylestradiol and trimethoprim. Environ Sci Technol 45:3605–3612

Kim YM, Murugesan K, Schmidt S et al (2011) Triclosan susceptibility and co-metabolism - a comparison for three aerobic pollutant-degrading bacteria. Bioresour Technol 102:2206–2212

Kinney CA, Furlong ET, Kolpin DW et al (2008) Bioaccumulation of pharmaceuticals and other anthropogenic waste indicators in earthworms from agricultural soil amended with biosolid or swine manure. Environ Sci Technol 42:1863–1870

Kjeldal H, Zhou NA, Wissenbach DK et al (2016) Genomic, proteomic, and metabolite characterization of gemfibrozil-degrading organism *Bacillus* sp. GeD10. Environ Sci Technol 50:744–755

Klampfl CW (2019) Metabolization of pharmaceuticals by plants after uptake from water and soil: a review. TrAC Trends Anal Chem 111:13–26

Kostich MS, Batt AL, Lazorchak JM (2014) Concentrations of prioritized pharmaceuticals in effluents from 50 large wastewater treatment plants in the US and implications for risk estimation. Environ Pollut 184:354–359

Kouki S, Saidi N, M'hiri F et al (2016) Co-composting of macrophyte biomass and sludge as an alternative for sustainable management of constructed wetland by-products. Clean Soil Air Water 44:694–702

Kumari R, Ghosh Sachan S (2019) Bioconversion of toxic micropollutant triclosan to 2,4-dichlorophenol using a wastewater isolate *Pseudomonas aeruginosa* KS2002. Int J Environ Sci Technol 16:7663–7672

Kummerer K (2009) The presence of pharmaceuticals in the environment due to human use–present knowledge and future challenges. J Environ Manag 90:2354–2366

Kuzikova I, Safronova V, Zaytseva T, Medvedeva N (2017) Fate and effects of nonylphenol in the filamentous fungus *Penicillium expansum* isolated from the bottom sediments of the Gulf of Finland. J Mar Syst 171:111–119

La Farre M, Perez S, Kantiani L, Barcelo D (2008) Fate and toxicity of emerging pollutants, their metabolites and transformation products in the aquatic environment. TrAC Trends Anal Chem 27:991–1007

Lapworth D, Baran N, Stuart M, Ward R (2012) Emerging organic contaminants in groundwater: a review of sources, fate and occurrence. Environ Pollut 163:287–303

Leng L, Wei L, Xiong Q et al (2020) Use of microalgae based technology for the removal of antibiotics from wastewater: a review. Chemosphere 238:124680

Li H, Sumarah MW, Topp E (2013) Persistence of the tricyclic antidepressant drugs amitriptyline and nortriptyline in agriculture soils. Environ Toxicol Chem 32:509–516

Li X, Xu QM, Cheng JS, Yuan YJ (2016) Improving the bioremoval of sulfamethoxazole and alleviating cytotoxicity of its biotransformation by laccase producing system under coculture of *Pycnoporus sanguineus* and *Alcaligenes faecalis*. Bioresour Technol 220:333–340

Liu J-L, Wong M-H (2013) Pharmaceuticals and personal care products (PPCPs): a review on environmental contamination in China. Environ Int 59:208–224

Liu S, Ying G-G, Zhao J-L et al (2011) Trace analysis of 28 steroids in surface water, wastewater and sludge samples by rapid resolution liquid chromatography–electrospray ionization tandem mass spectrometry. J Chromatogr A 1218:1367–1378

Liu Q, Li M, Liu X et al (2018) Removal of sulfamethoxazole and trimethoprim from reclaimed water and the biodegradation mechanism. Front Environ Sci Eng 12:6

Lloret L, Eibes G, Lu-Chau TA (2010) Laccase-catalyzed degradation of anti-inflammatories and estrogens. Biochem Eng J 51:124–131

Lloret L, Eibes G, Feijoo G et al (2012) Degradation of estrogens by laccase from *Myceliophthora thermophila* in fed-batch and enzymatic membrane reactors. J Hazard Mater 213:175–183

Loffredo E, Castellana G, Taskin E (2016) A two-step approach to eliminate pesticides and estrogens from a wastewater and reduce its phytotoxicity: adsorption onto plant-derived materials and fungal degradation. Water Air Soil Pollut 227:188

Lopez-Serna R, Posadas E, Garcia-Encina PA, Munoz R (2019) Removal of contaminants of emerging concern from urban wastewater in novel algal-bacterial photobioreactors. Sci Total Environ 662:32–40

Loraine GA, Pettigrove ME (2006) Seasonal variations in concentrations of pharmaceuticals and personal care products in drinking water and reclaimed wastewater in southern California. Environ Sci Technol 40:687–695

Lord H, Pawliszyn J (2000) Evolution of solid-phase microextraction technology. J Chromatogr A 885:153–193

Lozano-Garcia DF, Cuellar-Bermudez SP, del Rio-Hinojosa E et al (2019) Potential land microalgae cultivation in Mexico: from food production to biofuels. Algal Res 39:101459

Lucas D, Castellet-Rovira F, Villagrasa M et al (2018) The role of sorption processes in the removal of pharmaceuticals by fungal treatment of wastewater. Sci Total Environ 610–611:1147–1153

Ma Y, Oliveira RS, Freitas H, Zhang C (2016) Biochemical and molecular mechanisms of plant-microbe-metal interactions: relevance for phytoremediation. Front Plant Sci 7:753

Macci C, Peruzzi E, Doni S et al (2015) Ornamental plants for micropollutant removal in wetland systems. Environ Sci Pollut Res 22:2406–2415

Madukasi EI, Dai X, He C, Zhou J (2010) Potentials of phototrophic bacteria in treating pharmaceutical wastewater. Int J Environ Sci Technol 7:165–174

Maes HM, Maletz SX, Ratte HT et al (2014) Uptake, elimination, and biotransformation of 17 α-ethinylestradiol by the freshwater alga *Desmodesmus subspicatus*. Environ Sci Technol 48:12354–12361

Mahmoud IS, Altaif KI, Sini MKA et al (2020) Determination of antimicrobial drug resistance among bacterial isolates in two hospitals of Baghdad. Jordan J Pharm Sci 13:1–9

Mallick I, Bhattacharyya C, Mukherji S et al (2018) Effective rhizoinoculation and biofilm formation by arsenic immobilizing halophilic plant growth promoting bacteria (PGPB) isolated from mangrove rhizosphere: a step towards arsenic rhizoremediation. Sci Total Environ 610:1239–1250

Marchlewicz A, Domaradzka D, Guzik U, Wojcieszyńska D (2016) *Bacillus thuringiensis* B1 (2015b) is a gram-positive bacteria able to degrade naproxen and ibuprofen. Water Air Soil Pollut 227:1–8

Margot J, Bennati-Granier C, Maillard J et al (2013) Bacterial versus fungal laccase: potential for micropollutant degradation. AMB Express 3:63

Maryskova M, Ardao I, Garcia-Gonzalez CA et al (2016) Polyamide 6/chitosan nanofibers as support for the immobilization of *Trametes versicolor* laccase for the elimination of endocrine disrupting chemicals. Enzym Microb Technol 89:31–38

Matamoros V, Uggetti E, Garcia J, Bayona JM (2016) Assessment of the mechanisms involved in the removal of emerging contaminants by microalgae from wastewater: a laboratory scale study. J Hazard Mater 301:197–205

McClellan K, Halden RU (2010) Pharmaceuticals and personal care products in archived U.S. biosolids from the 2001 EPA national sewage sludge survey. Water Res 44:658–668

Merugu R, Prashanthi Y, Sarojini T, Badgu N (2014) Bioremediation of waste waters by the anoxygenic photosynthetic bacterium *Rhodobacter sphaeroides* SMR 009. Int J Res Environ Sci Technol 4:16–19

Mikeskova H, Novotny C, Svobodova K (2012) Interspecific interactions in mixed microbial cultures in a biodegradation perspective. Appl Microbiol Biotechnol 95:861–870

Mohammadhosseini M, Tehrani MS, Ganjali MR (2006) Preconcentration, determination and speciation of chromium (III) using solid phase extraction and flame atomic absorption spectrometry. J Chin Chem Soc 53:549–557

Molina MC, Bautista LF, Belda I et al (2019) Bioremediation of soil contaminated with arsenic. In: Ashok K, Swati S (eds) Microbes and enzymes in soil health and bioremediation. Springer, Singapore, pp 321–351

Molina MC, Bautista LF, Catala M et al (2020) From laboratory tests to the ecoremedial system: the importance of microorganisms in the recovery of PPCPs-disturbed ecosystems. Appl Sci 10:3391

Mompelat S, Le Bot B, Thomas O (2009) Occurrence and fate of pharmaceutical products and by-products, from resource to drinking water. Environ Int 35:803–814

Moon DS, Song HG (2012) Degradation of alkylphenols by white rot fungus Irpex lacteus and its manganese peroxidase. Appl Biochem Biotechnol 168:542–549

Munz NA, Burdon FJ, de Zwart D et al (2017) Pesticides drive risk of micropollutants in wastewater impacted streams during low flow conditions. Water Res 110:366–377

Munze R, Hannemann C, Orlinskiy P et al (2017) Pesticides from wastewater treatment plant effluents affect invertebrate communities. Sci Total Environ 599-600:387–399

Murdoch RW, Hay AG (2013) Genetic and chemical characterization of ibuprofen degradation by *Sphingomonas* Ibu-2. Microbiology 159:621–632

Murugesan K, Bokare V, Jeon JR et al (2011) Effect of Fe-Pd bimetallic nanoparticles on *Sphingomonas* sp. PH-07 and a nano-bio hybrid process for triclosan degradation. Bioresour Technol 102:6019–6025

Navrozidou E, Melidis P, Ntougias S (2019) Biodegradation aspects of ibuprofen and identification of ibuprofen-degrading microbiota in an immobilized cell bioreactor. Environ Sci Pollut Res 26:14238–14249

Nguyen LN, Van de Merwe JP, Hai FI et al (2016) Laccase-syringaldehyde-mediated degradation of trace organic contaminants in an enzymatic membrane reactor: removal efficiency and effluent toxicity. Bioresour Technol 200:477–484

Nguyen PM, Afzal M, Ullah I et al (2019a) Removal of pharmaceuticals and personal care products using constructed wetlands: effective plant-bacteria synergism may enhance degradation efficiency. Environ Sci Pollut Res 26:21109–21126

Nguyen PY, Silva AF, Reis AC et al (2019b) Bioaugmentation of membrane bioreactor with *Achromobacter denitrificans* strain PR1 for enhanced sulfamethoxazole removal in wastewater. Sci Total Environ 648:44–55

Nikolaou A, Meric S, Fatta D (2007) Occurrence patterns of pharmaceuticals in water and wastewater environments. Anal Bioanal Chem 387:1225–1234

Norvill ZN, Shilton A, Guieysse B (2016) Emerging contaminant degradation and removal in algal wastewater treatment ponds: identifying the research gaps. J Hazard Mater 313:291–309

Oaks JL, Gilbert M, Virani MZ et al (2004) Diclofenac residues as the cause of vulture population decline in Pakistan. Nature 427:630–633

Olicon-Hernandez DR, González-Lopez J, Aranda E (2017) Overview on the biochemical potential of filamentous fungi to degrade pharmaceutical compounds. Front Microbiol 8:1–7

Olicon-Hernandez DR, Camacho-Morales RL, Pozo C et al (2019) Evaluation of diclofenac biodegradation by the ascomycete fungus *Penicillium oxalicum* at flask and bench bioreactor scales. Sci Total Environ 662:607–614

Ootsuka M, Nishizawa T, Hasegawa M et al (2018) Comparative analysis of the genetic basis of branched nonylphenol degradation by *Sphingobium amiense* DSM 16289 T and *Sphingobium cloacae* JCM 10874 T. Microbes Environ 33:450–454

Otto B, Beuchel C, Liers C et al (2015) Laccase-like enzyme activities from chlorophycean green algae with potential for bioconversion of phenolic pollutants. FEMS Microbiol Lett 362:fnv072

Overturf MD, Anderson JC, Pandelides Z et al (2015) Pharmaceuticals and personal care products: a critical review of the impacts on fish reproduction. Crit Rev Toxicol 45:469–491

Pai PV, Pai A, Pai S et al (2013) Effect of glucose and nitrogen source on caffeine degradation by four filamentous fungi. Indian J Biotechnol 12:432–434

Painter MM, Buerkley MA, Julius ML et al (2009) Antidepressants at environmentally relevant concentrations affect predator avoidance behavior of larval fathead minnows (*Pimephales promelas*). Environ Toxicol Chem 28:2677–2684

Parshikov IA, Muraleedharan KM, Avery MA, Williamson JS (2004) Transformation of artemisinin by *Cunninghamella elegans*. Appl Microbiol Biotechnol 64:782–786

Pauwels B, Wille K, Noppe H et al (2008) 17 α-ethinylestradiol cometabolism by bacteria degrading estrone, 17β-estradiol and estriol. Biodegradation 19:683–693

Pedersen JA, Soliman M, Suffet IH (2005) Human pharmaceuticals, hormones, and personal care product ingredients in runoff from agricultural fields irrigated with treated wastewater. J Agric Food Chem 53:1625–1632

Petryna A, Kleinman A, Lakoff A (2006) The pharmaceutical nexus. In: Global pharmaceuticals: ethics, markets, practices. Duke University Press, Durham, pp 1–32

Popa C, Favier L, Dinica R et al (2014) Potential of newly isolated wild *Streptomyces* strains as agents for the biodegradation of a recalcitrant pharmaceutical, carbamazepine. Environ Technol (U K) 35:3082–3091

Prenosilova L, Kresinova Z, Amemori AS et al (2013) Transcriptional response of lignin-degrading enzymes to 17 α-ethinyloestradiol in two white rots. Microb Biotechnol 6:300–306

Pruden A, Pei R, Storteboom H, Carlson KH (2006) Antibiotic resistance genes as emerging contaminants: studies in northern Colorado. Environ Sci Technol 40:7445–7450

Puyol D, Batstone DJ, Hulsen T et al (2017) Resource recovery from wastewater by biological technologies: opportunities, challenges, and prospects. Front Microbiol 7:1–23

Quandt EM, Summers RM, Subramanian MV, Barrick JE (2015) Draft genome sequence of the bacterium *Pseudomonas putida* CBB5, which can utilize caffeine as a sole carbon and nitrogen source. Genome Announc 3:5–6

Rajapaksha AU, Vithanage M, Lim JE et al (2014) Invasive plant-derived biochar inhibits sulfamethazine uptake by lettuce in soil. Chemosphere 111:500–504

Rajapaksha AU, Vithanage M, Ahmad M et al (2015) Enhanced sulfamethazine removal by steam-activated invasive plant-derived biochar. J Hazard Mater 290:43–50

Reis PJM, Reis AC, Ricken B et al (2014) Biodegradation of sulfamethoxazole and other sulfonamides by *Achromobacter denitrificans* PR1. J Hazard Mater 280:741–749

Reis AC, Cvancarova M, Liu Y et al (2018) Biodegradation of sulfamethoxazole by a bacterial consortium of *Achromobacter denitrificans* PR1 and Leucobacter sp. GP. Appl Microbiol Biotechnol 102:10299–10314

Richardson ML, Bowron JM (1985) The fate of pharmaceutical chemicals in the aquatic environment. J Pharm Pharmacol 37:1–12

Rodarte-Morales AI, Feijoo G, Moreira MT, Lema JM (2012) Biotransformation of three pharmaceutical active compounds by the fungus *Phanerochaete chrysosporium* in a fed batch stirred reactor under air and oxygen supply. Biodegradation 23:145–156

Rodriguez E, Ruiz-Duenas FJ, Kooistra R et al (2008) Isolation of two laccase genes from the white-rot fungus *Pleurotus eryngii* and heterologous expression of the pel3 encoded protein. J Biotechnol 134:9–19

Roh H, Subramanya N, Zhao F et al (2009) Biodegradation potential of wastewater micropollutants by ammonia-oxidizing bacteria. Chemosphere 77:1084–1089

Rovani S, Censi MT, Pedrotti SL et al (2014) Development of a new adsorbent from agro-industrial waste and its potential use in endocrine disruptor compound removal. J Hazard Mater 344:350–359

Salamanca EJP, Madera-Parra CA, Avila-Williams CA et al (2015) Phytoremediation using terrestrial plants. In: Phytoremediation. Springer, Berlin, pp 305–319

Sanchez-Galvan G, Bolanos-Santiago Y (2018) Phytofiltration of anaerobically digested sugarcane ethanol stillage using a macrophyte with high potential for biofuel production. Int J Phytoremediation 20:805–812

Sarmah AK, Halling-Sørensen B (2007) Biodegradation of selected emerging organic contaminants in the environment-an overview. In: Pawley LE (ed) Leading-edge environmental biogradation research. Nova Science Publishers, Hauppauge, NY

Sauvetre A, Schroder P (2015) Uptake of carbamazepine by rhizomes and endophytic bacteria of *Phragmites australis*. Front Plant Sci 6:1–11

Sauvetre A, May R, Harpaintner R et al (2018) Metabolism of carbamazepine in plant roots and endophytic rhizobacteria isolated from *Phragmites australis*. J Hazard Mater 342:85–95

Scudellari M (2015) Humans have spiked ecosystems with a flood of active pharmaceuticals. The drugs are feminizing male fish, confusing birds, and worrying scientists. The Scientist (Magazine)

Shan L, Li Y, Chen Y et al (2016) Microbial hydroxylation of 17β-estradiol by *Penicillium brevicompactum*. Biocatal Biotransformation 34:137–143

Shi JH, Suzuki Y, Lee BD, Nakai S, Hosomi M (2002) Isolation and characterization of the ethynylestradiol-biodegrading microorganism *Fusarium proliferatum* strain HNS-1. Water Sci Technol 45:175–179

Shi W, Wang L, Rousseau DPL, Lens PNL (2010) Removal of estrone, 17α-ethinylestradiol, and 17ß-estradiol in algae and duckweed-based wastewater treatment systems. Environ Sci Pollut Res 17:824–833

Shrestha P, Oh TJ, Liou K, Sohng JK (2008) Cytochrome P450 (CYP105F2) from *Streptomyces peucetius* and its activity with oleandomycin. Appl Microbiol Biotechnol 79:555–562

Singh P, Singh VK, Singh R et al (2020) Bioremediation: a sustainable approach for management of environmental contaminants. In: Abatement of environmental pollutants. Elsevier, Amsterdam, pp 1–23

Sopajarn A, Sangwichien C (2015) Optimization of enzymatic saccharification of alkali pretreated *Typha angustifolia* for glucose production. Int J Chem Eng Appl 6:232–236

Stahl E (1967) Thin-layer chromatography. A laboratory handbook, 2nd edn. Springer, Berlin

Swan Gerry E, Cuthbert R, Quevedo M et al (2006) Toxicity of diclofenac to Gyps vultures. Biol Lett 2:279–282

Tabata A, Kashiwada S, Ohnishi Y et al (2001) Estrogenic influences of estradiol-17b, pnonylphenol and bis-phenol-A on Japanese Medaka (*Oryzias latipes*) at detected environmental concentrations. Water Sci Technol 43:109–116

Taggart MA, Cuthbert R, Das D (2007) Diclofenac disposition in Indian cow and goat with reference to Gyps vulture population declines. Environ Pollut 147:60–65

Tanoue R, Sato Y, Motoyama M et al (2012) Plant uptake of pharmaceutical chemicals detected in recycled organic manure and reclaimed wastewater. J Agric Food Chem 60:10203–10211

Tastan BE, Ozdemir C, Tekinay T (2016) Effects of different culture media on biodegradation of triclosan by *Rhodotorula mucilaginosa* and *Penicillium* sp. Water Sci Technol 74:473–481

Taylor D, Senac T (2014) Human pharmaceutical products in the environment–the "problem" in perspective. Chemosphere 115:95–99

Ternes TA, Herrmann N, Bonerz M et al (2004) A rapid method to measure the solid–water distribution coefficient (Kd) for pharmaceuticals and musk fragrances in sewage sludge. Water Res 38:4075–4084

Thelusmond JR, Strathmann TJ, Cupples AM (2016) The identification of carbamazepine biodegrading phylotypes and phylotypes sensitive to carbamazepine exposure in two soil microbial communities. Sci Total Environ 571:1241–1252

Thelusmond JR, Strathmann TJ, Cupples AM (2019) Carbamazepine, triclocarban and triclosan biodegradation and the phylotypes and functional genes associated with xenobiotic degradation in four agricultural soils. Sci Total Environ 657:1138–1149

Thurman EM, Mills MS (1998) Solid-phase extraction: principles and practice, vol 16. Wiley, New York

Tian H, Ma YJ, Li WY, Wang JW (2018) Efficient degradation of triclosan by an endophytic fungus *Penicillium oxalicum* B4. Environ Sci Pollut Res 25:8963–8975

Tortella G, Duran N, Rubilar O (2015) Are white-rot fungi a real biotechnological option for the improvement of environmental health? Crit Rev Biotechnol 35:165–172

Toyama T, Murashita M, Kobayashi K et al (2011) Acceleration of nonylphenol and 4-tert-octylphenol degradation in sediment by *phragmites australis* and associated rhizosphere bacteria. Environ Sci Technol 45:6524–6530

Toyama T, Ojima T, Tanaka Y et al (2013) Sustainable biodegradation of phenolic endocrine-disrupting chemicals by *Phragmites australis*-rhizosphere bacteria association. Water Sci Technol 68:522–529

Tran TN, Kim DG, Ko SO (2018) Synergistic effects of biogenic manganese oxide and Mn(II)-oxidizing bacterium *Pseudomonas putida* strain MnB1 on the degradation of 17 A-ethinylestradiol. J Hazard Mater 344:350–359

Trautwein C, Berset JD, Wolschke H, Kummerer K (2014) Occurrence of the antidiabetic drug metformin and its ultimate transformation product guanylurea in several compartments of the aquatic cycle. Environ Int 70:203–212

Treesubsuntorn C, Thiravetyan P (2018) Botanical biofilter for indoor toluene removal and reduction of carbon dioxide emission under low light intensity by using mixed C3 and CAM plants. J Clean Prod 194:94–100

Treesubsuntorn C, Dolphen R, Dhurakit P et al (2017) Green technology innovation in a developing country. AIP Conf Proc 1908:030004. https://doi.org/10.1063/1.5012704

Vasiliadou IA, Sanchez-Vazquez R, Molina R et al (2016) Biological removal of pharmaceutical compounds using white-rot fungi with concomitant FAME production of the residual biomass. J Environ Manag 180:228–237

Vasiliadou IA, Molina R, Pariente MI et al (2019) Understanding the role of mediators in the efficiency of advanced oxidation processes using white-rot fungi. Chem Eng J 359:1427–1435

Vithanage M, Rajapaksha AU, Tang X et al (2014) Sorption and transport of sulfamethazine in agricultural soils amended with invasive-plant-derived biochar. J Environ Manag 141:95–103

Vo HNP, Ngo HH, Guo W et al (2020) Microalgae for saline wastewater treatment: a critical review. Crit Rev Environ Sci Technol 50:1224–1265

Wang J, Wang S (2016) Removal of pharmaceuticals and personal care products (PPCPs) from wastewater: a review. J Environ Manag 182:620–640

Wang D, Sui Q, Lu SG et al (2014) Occurrence and removal of six pharmaceuticals and personal care products in a wastewater treatment plant employing anaerobic/anoxic/aerobic and UV processes in Shanghai, China. Environ Sci Pollut Res 21:4276–4285

Wang Z, Dai Y, Zhao Q et al (2015) Nonylphenol biodegradation, functional gene abundance and bacterial community in bioaugmented sediment: effect of external carbon source. Environ Sci Pollut Res 22:12083–12091

Wang L, Peng Y, Nie X et al (2016) Gene response of CYP360A, CYP314, and GST and whole-organism changes in Daphnia magna exposed to ibuprofen comparative biochemistry and physiology. Comp Biochem Physiol C Toxicol Pharmacol 179:49–56

Wang S, Hu Y, Wang J (2018a) Biodegradation of typical pharmaceutical compounds by a novel strain *Acinetobacter* sp. J Environ Manag 217:240–246

Wang S, Poon K, Cai Z (2018b) Removal and metabolism of triclosan by three different microalgal species in aquatic environment. J Hazard Mater 342:643–650

Wang P, Zheng D, Peng W (2019) Characterization of 17β-hydroxysteroid dehydrogenase and regulators involved in estrogen degradation in *Pseudomonas putida* SJTE-1. Appl Microbiol Biotechnol 103:2413–2425

Watahiki S, Kimura N (2017) Draft genome sequence of a caffeine-utilizing bacterium, *Cupriavidus* sp. strain D384. Genome Announc 5:e00370–e00317

Weigel S, Berger U, Jensen E et al (2004) Determination of selected pharmaceuticals and caffeine in sewage and seawater from Tromso/Norway with emphasis on ibuprofen and its metabolites. Chemosphere 56:583–592

Wilson CJ, Brain RA, Sanderson H et al (2004) Structural and functional responses of plankton to a mixture of four tetracyclines in aquatic microcosms. Environ Sci Technol 38:6430–6439

Wojcieszynska D, Domaradzka D, Hupert-Kocurek K, Guzik U (2014) Bacterial degradation of naproxen - undisclosed pollutant in the environment. J Environ Manag 145:157–161

Woo HL, Hazen TC, Simmons BA, DeAngelis KM (2014) Enzyme activities of aerobic lignocellulolytic bacteria isolated from wet tropical forest soils. Syst Appl Microbiol 37:60–67

Wu X, Ernst F, Conkle JL, Gan J (2013) Comparative uptake and translocation of pharmaceutical and personal care products (PPCPs) by common vegetables. Environ Int 60:15–22

Xiong JQ, Kurade MB, Abou-Shanab RAI et al (2016) Biodegradation of carbamazepine using freshwater microalgae *Chlamydomonas mexicana* and *Scenedesmus obliquus* and the determination of its metabolic fate. Bioresour Technol 205:183–190

Xiong JQ, Kurade MB, Jeon BH (2018) Can microalgae remove pharmaceutical contaminants from water? Trends Biotechnol 36:30–44

Xiong JQ, Govindwar S, Kurade MB et al (2019a) Toxicity of sulfamethazine and sulfamethoxazole and their removal by a green microalga, *Scenedesmus obliquus*. Chemosphere 218:551–558

Xiong JQ, Kim SJ, Kurade MB et al (2019b) Combined effects of sulfamethazine and sulfamethoxazole on a freshwater microalga, *Scenedesmus obliquus*: toxicity, biodegradation, and metabolic fate. J Hazard Mater 370:138–146

Yang S, Hai FI, Nghiem LD et al (2013) Understanding the factors controlling the removal of trace organic contaminants by white-rot fungi and their lignin modifying enzymes: a critical review. Bioresour Technol 359:203–212

Yang Y, Ok YS, Kim KH et al (2017) Occurrences and removal of pharmaceuticals and personal care products (PPCPs) in drinking water and water/sewage treatment plants: a review. Sci Total Environ 596:303–320

Yang Z, Shi Y, Zhang Y et al (2018) Different pathways for 4-n-nonylphenol biodegradation by two *Aspergillus* strains derived from estuary sediment: evidence from metabolites determination and key-gene identification. J Hazard Mater 359:203–212

Zha J, Sun L, Zhou Y et al (2008) Assessment of 17α-ethinylestradiol effects and underlying mechanisms in a continuous, multigeneration exposure of the Chinese rare minnow (*Gobiocypris rarus*). Toxicol Appl Pharmacol 226:298–308

Zhang D, Luo J, Lee ZMP et al (2016) Characterization of microbial communities in wetland mesocosms receiving caffeine-enriched wastewater. Environ Sci Pollut Res 23:14526–14539

Zhang H, Wang L, Li Y et al (2019a) Background nutrients and bacterial community evolution determine 13 C-17β-estradiol mineralization in lake sediment microcosms. Sci Total Environ 651:2304–2311

Zhang Z, Ruan Z, Liu J et al (2019b) Complete degradation of bisphenol A and nonylphenol by a composite of biogenic manganese oxides and *Escherichia coli* cells with surface-displayed multicopper oxidase CotA. Chem Eng J 362:897–908

Zheng G, Wang T, Niu M et al (2018) Biodegradation of nonylphenol during aerobic composting of sewage sludge under two intermittent aeration treatments in a full-scale plant. Environ Pollut 238:783–791

Zhong DF, Sun L, Liu L, Huang HH (2003) Microbial transformation of naproxen by *Cunninghamella* species. Acta Pharmacol Sin 24:442–447

Zhou NA, Lutovsky AC, Andaker GL et al (2014) Kinetics modeling predicts bioaugmentation with *Sphingomonad* cultures as a viable technology for enhanced pharmaceutical and personal care products removal during wastewater treatment. Bioresour Technol 166:158–167

Zhou B, Ma C, Wang H, Xia T (2018) Biodegradation of caffeine by whole cells of tea-derived fungi *Aspergillus sydowii*, *Aspergillus niger* and optimization for caffeine degradation. BMC Microbiol 18:1–10

# Chapter 16
# Screening of Microbial Enzymes and Their Potential Applications in the Bioremediation Process

**Raj Saini, Varsha Rani, Sneh Sharma, and Madan L. Verma**

**Abstract** Industrialization and the faster urbanization lead to the release of hazardous substances in our ecosystem which is affecting our health and the environment. Bioremediation technology utilizes the microorganisms for removing these hazardous chemicals from our ecosystem in an eco-friendly way. Enzymes of microbial origin have proved to be powerful tools for the process of bioremediation. Microbial enzymes showed higher specificity for a broader range of pollutants, higher activity in the presence of inhibitors and higher effectiveness at low pollutant concentration. Process of bioremediation uses recombinant as well as natural microorganisms to degrade the toxic and hazardous substances by aerobic and anaerobic means. Microorganisms availability, contaminants accessibility, and the conducive environment are the factors which governs the success of bioremediation process.

**Keywords** Bioremediation · Hazardous · Industrialization · Contaminants · Ecosystem

Raj Saini and Varsha Rani contributed equally to this work.

R. Saini
Department of Basic Sciences, Dr. Y.S. Parmar University of Horticulture and Forestry, Neri, Hamirpur, Himachal Pradesh, India

V. Rani
Department of Biotechnology, Shoolini University, Solan, Himachal Pradesh, India

S. Sharma
Department of Biotechnology, Dr. Y.S. Parmar University of Horticulture and Forestry, Neri, Hamirpur, Himachal Pradesh, India

M. L. Verma (✉)
Department of Biotechnology, School of Basic Sciences, Indian Institute of Information Technology, Una, Himachal Pradesh, India
e-mail: madanverma@iiitu.ac.in

 
P. K. Arora (ed.), *Microbial Products for Health, Environment and Agriculture*, Microorganisms for Sustainability 31, https://doi.org/10.1007/978-981-16-1947-2_16

## 16.1 Introduction

Pollution of our environment is increasing day by day with the development of industries. Efforts are being made to develop new, eco-friendly and cost-effective technology for the reduction as well as the elimination of pollutants from the soil, water and air. Microorganisms, particularly enzymes, have been found to detoxify and transform the pollutants effectively from the environment, and also recognized to transform the pollutants at detectable rates as well as suitable for restoring the polluted environment. Bioremediation is a microorganism mediated process of transforming or degrading the pollutants into non-hazardous substances (Fig. 16.1). Pollutants are enzymatically transformed by microorganisms, and thus converting these contaminants into harmless products via the process of bioremediation. Conversion of harmful pollutants into harmless substances via the process of bioremediation is a very slow process and is shown by certain strains of microorganisms. Bioremediation of pollutants can be performed in aerobic as well as anaerobic environmental settings (Karigar and Rao 2011).

Bioremediation process is using the diverse type of microorganisms and their specific enzymes, plants and living organisms. It has appeared as an appealing technology for treating the pollutants. Enzymes act as the main effectors for the biotransformation occurring in the biota. Enzymes can be applied to the larger range of different compounds because of their broader specificity. Extensive

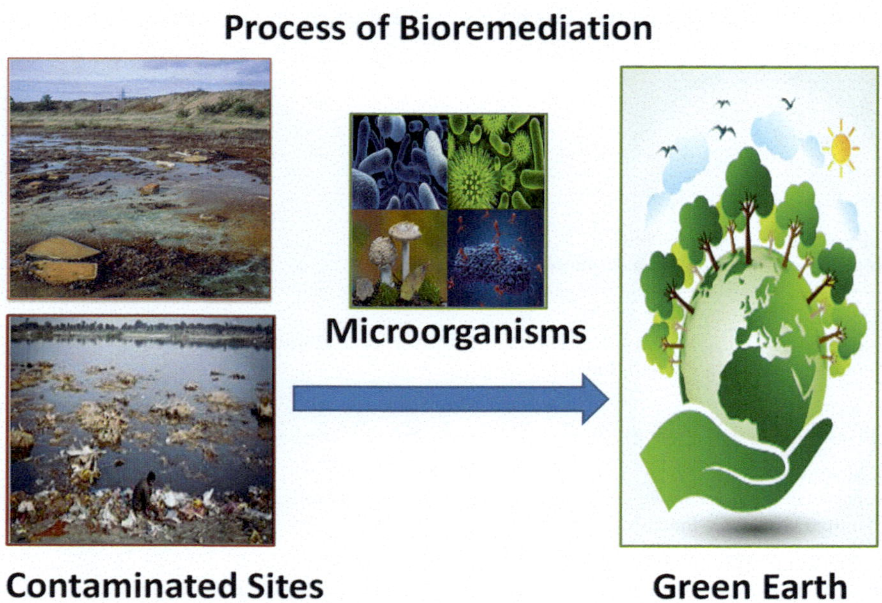


**Fig. 16.1** Conversion of contaminated sites into the green earth with the intervention of environmental microorganisms

transformations can be done by using enzymes such as toxicological as well as the structural properties of various contaminants can be transformed and harmless products can be formed by complete conversion of various contaminants into safer compounds. Enzymes have been found to show their effects at lower pollutants concentrations and are active even in the presence of microbial predators. Because of the presence of all these characteristics in enzymes, enzymatic techniques are eco-friendly and safe. The most representative classes of enzymes for the treatment of contaminants are dehalogenases, hydrolases, oxidoreductases and transferases. This chapter deals with the microbial enzymes and their potential applications in the bioremediation of the environmental pollutants.

## 16.2 Screening of Microbial Enzymes

Microbial enzymes are nature's biocatalysts catalysing different types of chemical reactions at higher rate, under mild temperature conditions and with greater specificity in environment friendly way. Applications of commercial enzymes are continuously increasing, despite the suboptimal performance of many enzymes occurring naturally under the industrial conditions (Leemhuis et al. 2009). A vast number of microbial enzymes are provided by nature. Biodiversity is widespread among the natural microbial populations. Isolation, recognition and screening of microbes in pure cultures are very important to conserve the gene pools of these microorganisms as well as contributing towards the biotechnological progress (Srinivasan 1994). Increased efforts in the screening of microbes and their biocatalysts (enzymes) provide a higher range of novel biocatalysts with different enzyme activities, suitable for use under the harsh industrial conditions like extreme pH, high temperature, and organic solvents resistance as well as more catalysts with improved characteristics would be discovered. Improved efforts and expertise in the process of screening would supplement the other emerging techniques such as protein engineering, and thus reducing the manpower and other resources needed to carry out the screening process (Cheetham 1987). The study of immense biodiversity among the biocatalysts depends upon the tools available for searching new enzymes. Meta-genomic approach-basedpowerful screening has been emerged as a high throughput microbe screening process by which a genome library has been prepared from environmental DNA. This genomic library will be screened for the open reading frames present in this library, encoding putative novel enzymes (Gilbert and Dupont 2011). Meta-genomic screening is based on either sequence or function approaches.

Function-based screening is a straightforward way for isolating genes which shows a desirable function by phenotypic detection, induced gene expression and heterologous complementation (Li et al. 2012). While sequence-based screening has been performed either by hybridization process or by polymerase chain reaction.

Success in the genome sequencing programs has resulted in various sequence databases which provides information about the newly discovered natural products including enzymes by database mining (Van Lanen and Shen 2006). Next generation sequencing programs are holding the promise to reduce the cost as well as the time of genome sequencing. Two approaches are being followed for the discovery of new enzymes, genome hunting and data mining. Open reading frames have been searched in a genome of microorganism in the genome hunting. Annotated sequences are subjected to cloning, over expression and screening for activity. Data mining is based on the homology among the sequences deposited in the databases. Different bioinformatics tools can be used to search the conserved regions between the sequences. Blast is, an example of such bioinformatics tool, used to find the conserved regions among the protein as well as nucleic acid sequences (Luo et al. 2012; Adrio and Demain 2014).

Extremophiles are the microorganisms which are living in the extreme conditions like temperature upto 110 °C, pH of lesser than 2 and greater than 9, higher salt conditions, higher pressure and radiations. So, extremophiles are great source of extremozymes (enzymes present in the extremophiles) which are stable at extreme conditions (Kumar et al. 2011; Pikuta et al. 2007). Thermophilic lipases, amylases, proteases and cellulases are being used in different industrial applications. *Taq DNA polymerase* is the example of such enzymes which is isolated from *Thermus aquaticus* used for amplifying the DNA in polymerase chain reaction (de Carvalho 2011; Verma and Kanwar 2012).

Applications of enzymes are continuously expanding sector and creating a demand of new and improved biocatalysts (de Carvalho 2011). Enzyme necessarily does not fulfil all the process requirements to achieve the industrial scale production. The problems faced while working with the enzymes are, stability, substrate/product inhibition and narrow substrate specificity (Marrs et al. 1999). To overcome these hurdles while working with enzymes, recombinant DNA technology and genetic modifications can be done. There are two different ways namely, rational redesign and combinatorial methods by which enzymes can be modified (Singhania et al. 2010). Rational design approach includes site directed mutagenesis by which target amino acid substitutions are performed. Three-dimensional structure of the protein as well as the reactions performed by the protein should be well known to perform the site directed mutagenesis. Sequence of the new biocatalyst can be compared with thousands of related proteins in the databases and structurally as well as functionally related proteins can be found (Cedrone et al. 2000). Combinatorial method such as directed evolution does not required extensive knowledge about the enzyme concerned. Directed evolution creates multiple variants based on catalytic efficiency, enantioselectivity, solubility, catalytic rate, enzyme stability and specificity. Directed evolution is an inexpensive and faster way tofind enzyme variants which work better than the naturally occurring enzymes under specific conditions (Kumar and Singh 2013; Schmidt et al. 2010; Dalby 2011). Genetic diversity can be achieved through a range of molecular biology techniques via directed evolution. Random mutagenesis of the protein encoding gene can be achieved by different techniques including errorprone polymerase chain reaction, chemical agents, and repeated oligonucleotide directed mutagenesis. Random point mutations can be accomplished via error prone polymerase chain reaction in a population of enzymes.

In vitro random homologous recombination is allowed by such molecular breeding techniques between the parent genes having more than 70% homology (Ness et al. 2000).

A large collection of enzyme variants is available after cloning and expression of the concerned enzymes and further subjected for screening and selection. Directed evolution techniques make use of smaller enzyme variant libraries designed by semi rational or rational methods to reduce the screening efforts without the likelihood of getting better variants (Dalby 2011).

Screening of large number of populations or cells and genetic engineering are crucial bottleneck in today's applied microbiology and system biology. Scientist are relying on high-throughputstrategies which are minimizing the experiments to nano- and pico-liter scales as well as single cell level instead of using the standard methods in flasks, bottles, or 96 well plates. Throughput of the genome editing has been increased significantly as more individuals and genes can be engineered simultaneously via multiplex automated genome evolution (MAGE) and CRISPR/Cas systems (Vervoort et al. 2017).Ultra-low volume nano reactors use increased the genotyping and phenotyping of single cells as well as populations. Thousands or millions of variants can be screened by using ultra-low volume nanoreactors. Generation of billions of mutants by repeated deletion, insertion or DNA mutations at multiple chromosomal sites has been made by multiplex automated genome evolution systems (Wang et al. 2009).

Testing and production of a wider range of mutations in specific genes or pathways without changing other genes is the major application of multiplex automated genome evolution. Multiplex automated genome evolution can this way create genetic diversity with greater possibilities to find more efficient microorganisms as well as improved synthetic pathways can be created. One of the pioneer studies which used multiplex automated genome evolution had been carried out in *Escherichia coli* showing increased production of lycopene via fine tuning of several genes (Wang et al. 2009). Lycopene or L-DOPA is an aromatic compound used to treat Parkinson's disease (Wei et al. 2016).

CRISPR/Cas revolution in high-throughput microbial genome engineering is well known. CRISPR/Cas technique causes alteration in genomes at specific sites by using RNA- guided nuclease activity and allows cheaper, faster and more efficient genome engineering as compared to the traditional techniques (Wright et al. 2016; Kim 2016). Genome of *Sacchromyces cerevisiae* was altered using CRISPR/Cas technique to increase mevalonate or (R-R)-2,3-butanediol production and also enhances xylose utilization (Shi et al. 2016).

## 16.3 Applications of Microbial Enzymes towards Bioremediation Process

As the quality of life is concerned on the planet earth, it is dependent on the quality of environment in which we are residing. Progress in technology, science and at the industrial level, a large amount of sewage to the nuclear waste is dumped in the ecosystem which is a serious threat to the survival of mankind on earth. Large number of microbial enzymes has been reported for the biodegradation of these toxic organic pollutants. Bioremediation is aneco- friendly and cost-effective biotechnology empowered by the enzymes of microbial origin. A number of research activities are being carried out for the development of such bioprocess technologies which can reduce the toxicity of pollutants (Karigar and Rao 2011).

Degradation of toxic substances as well as the management of waste has been reported via a number of microbial enzymes. Effluent from various industries and the domestic waste contains so many chemical commodities which are producing harmful effects on our health as well as our ecosystem. Enzymes of microbial origin alone or in combinations are being utilized for treating the effluents obtained from various industries containing aromatic amines, phenols, nitriles etc. (Raj et al. 2006; Rubilar et al. 2008; Pandey et al. 2011). Waste treatment has been reported via using enzymes like, amidases, cellulases, proteases, lipases, nitrile hydratases and amyloglucosidases (Karigar and Rao 2011). Industrial effluents containing chlorinated phenolic compounds has been reported to be removed via using enzymes such as tyrosinases, laccases, manganese peroxidases, and lignin peroxidases (Piontek et al. 2001; Le Roes-Hill and Prins 2016). Wei and Zimmermann (2018) reported that biocatalysts from microorganisms are involved in the process of degradation synthetic plastics polystyrene, polyurethane, polyethylene terephthalate and polyethylene. As depicted in Table 16.1, these microbial enzymes are playing a very important role in the development of green environment (Singh et al. 2016).

Oxidoreductases from the microorganisms detoxify the toxic organic compounds via the process of oxidative coupling (Gianfreda et al. 1999). Microorganisms extract energy by using their enzymes via the energy yielding reactions for cleaving the chemical bonds and assist the transfer of electrons from the organic substrate which is reduced to another chemical compound. Contaminants are oxidized via these oxidation reduction reactions into the harmless substances. Oxidoreductases also take part in the process of humification of different phenolic substances which are produced in the soil via decomposition of lignin. Xenobiotics such asanilinic and phenolic compounds can be detoxified by oxidoreductases via the process of coploymerization, polymerization with other substrates or by binding with the humic substances (Park et al. 2006). Enzymes of microbial origin are reported to degrade as well as in the decolorization ofazo dye (Williams 1977; Husain 2006). Effluents from the paper and pulp industry generates large amount of recalcitrant wastes such as chlorinated phenolic compounds. Partial degradation of lignins, produces thesechlorinated phenolic compounds during the bleaching of pulp in the paper and pulp industry. Oxidoreductases like, lignin peroxidase, manganese

**Table 16.1** Applications of microbial enzymes in the bioremediation process

| S. No. | Enzymes | Functions | Microorganisms |
|---|---|---|---|
| 1. | Amidase | Degradation of nitriles containing wastes | *Rhodococcuserythropolis* |
| 2. | Lipase | Degradation of crude oil hydrocarbons | *Aspergillus oryzae, Candida tropicalis* |
| 3. | Amylase | Bioremediation of vegetables wastes | *B. licheniformis, Aspergillus sp* |
| 4. | Nitrile hydratase | Degradation of nitriles containing wastes | *Rhodococcus sp.* |
| 5. | Amyloglucosidase | Starch hydrolysis for bioremediation | *Aspergillus niger* |
| 6. | Protease | Bioremediation of keratinic wastes | *Chrysosporiumkeratinophilum* |
| 7. | Cutinase | Degradation of plastics, Polycaprolactone | *Fusarium solani f. Pisi* |
| 8. | Laccase | Degradation of waste containing olefin unit, polyurethane and phenolic compounds | *Trametes versicolor* |
| 9. | Oxygenase | Degradation of halogenated contaminants | *Pseudomonas sp., Rhodococcus sp.* |
| 10. | Lignin peroxidise | Degradation of phenolic compounds | *Phanerochaetechrysosporium, Coprinus cinereus* |
| 11. | Manganese peroxidise | | |

peroxidases and laccases from fungal species are reported to be suitable for removing chlorinated phenolic compounds from the contaminated environments (Rubilar et al. 2008). Microbial oxygenases are found to play a key role in metabolizing the organic compounds via increasing their solubility in water, enhancing their reactivity as well as by the cleavage of their aromatic ring. Oxygenases belong to broader substrate range enzymes and are found active against a wider range of compounds which includes chlorinated aliphatic. The aromatic rings present in the organic pollutants are cleaved by introducing oxygen via oxygenases. Mono- and dioxygenases of microbial origin are among the most studied enzymes for bioremediation process (Arora et al. 2009; Fetzner and Lingens 1994; Fetzner 2003). Halogenated organic compounds are among the largest group of compounds which are polluting our environment and are produced by the widespread use of insecticides, herbicides, fungicides, plasticizers, heat transfer fluids, and hydraulic fluids. Degradation of theses halogenated organic compounds has been achieved by using oxygenases. Oxygenases in the association of other multifunctional enzymes, mediate the dehalogenation of halogenated ethylene, ethane and methane (Fetzner and Lingens 1994). Due to versatile nature of monooxygenases, theses can catalyse the oxidative reactions of simple alkanes to the complex molecules like fatty acids and steroids. Monooxygenases have high region selectivity, thus acting like biocatalysts for the process of bioremediation. Monooxygenases are the enzymes which require only molecular oxygen to catalyse the reaction and substrate is utilized as a reducing agent (Arora et al. 2010; Cirino and Arnold 2002). Monooxygenases catalyses biotransformation, biodegradation, denitrification, dehalogenation, desulfurization, hydroxylation and ammonification reactions (Arora et al. 2010). Oxidative dehalogenation reactions are carried out by monooxygenases under the oxygen rich environment while reductive dechlorination takes place under lower oxygen concentration (Fetzner 1998; Jones et al. 2001). Catechol dioxygenases participates in the nature's strategy in the degradation of aromatic compounds and are present in the bacteria present inside the soil (Que and Ho 1996).

Microbial laccases are reported to catalyse oxidation of wider range of aromatic and phenolic compounds (Mai et al. 2000). Intracellular and extracellular laccases produced from microorganisms are known to oxidise theaminophenol, ortho and diphenols, aryl diamines, lignins, polyamines, polyphenols and some inorganic ions (Ullah et al. 2000; Rodrıguez Couto and Toca Herrera 2006). Laccases perform demethylation and decarboxylation of methoxyphenolic and phenolic acids along with the oxidation of these compounds. Laccases represents an interesting group of biological agents offering great potential in bioremediation and biotechnological applications (Gianfreda et al. 1999; Kim et al. 2002). Reagents like cyanide, azide, halides and hydroxides are known to inhibit the activity of laccases (Xu 1996). Production of laccases is found sensitive to the nitrogen concentration. Higher amount of laccase production requires higher nitrogen concentrations (Gianfreda et al. 1999).

Oxidation of phenolic compounds and lignin has been reported by expanding hydrogen peroxide via peroxidases of microbial origin (Hiner et al. 2002). Peroxidases are of different types based on their activity and source of production. Manganese dependent peroxidise, lignin peroxidase and versatile peroxidase are among the most studied peroxidases and have higher potential of degrading the toxic substances present in the environment. White rot fungus is found to secrete lignin peroxidases containing heme as a secondary metabolite. This lignin peroxidise is reported to degrade lignin and other phenolic compounds in the presence of hydrogen peroxide as a co-substrate and veratryl alcohol LiP as a mediator. Hydrogen peroxide got reduced by the gain of electron from LiP into water and LiP return to its native state via gaining electron from veratryl alcohol; thus, forming veratryl aldehyde. This reaction leads to the oxidation of polycyclic aromatic compounds, halogenated phenolic compounds and some aromatic compounds (Yoshida 1998; Ten Have and Teunissen 2001). Manganese peroxidases are the lignin degrading enzymes produced via basidiomycetes fungus. Manganese peroxidase is an heme containing extracellular enzyme which oxidizes $Mn^{2+}$ into $Mn^{3+}$. Production of manganese peroxidase has been stimulated by $Mn^{2+}$and also act as a substrate for manganese peroxidise. During the process of oxidation of phenolic compounds, $Mn^{3+}$acts as a mediator (Ten Have and Teunissen 2001). Versatile peroxidases of microbial origin are the enzymes which directly oxidizethe phenolic compounds, $Mn^{2+}$, aromatic substrates and methoxybenzenes. Versatile peroxidases do not need the presence of manganese like other peroxidases for the oxidation of its substrates. Versatile peroxidases not only oxidise the phenolic compounds but are able to oxidize the non-phenolic lignin model dimmers (Ruiz-Dueñas et al. 2007). So, it can be concluded that among the peroxidases, versatile peroxidases are the enzymes having extraordinarily broader substrate specificity and desired for the process of bioremediation of contaminants in the environment (Tsukihara et al. 2006).

Industrial chemicals and hydrocarbons from petroleum are polluting our environment and is a serious problem. Using bioremediation process to remove these contaminants from our environment is proved as a best technology as compared to the physical chemical treatments which are commonly used for the removal of pollutants from our ecosystem. Microorganisms are used as the main agents for hydrolysing the organic pollutants (Table 16.2). Extracellular microbial biocatalysts play a key role in the degradation of the organic pollutants (Vasileva-Tonkova and Galabova 2003). The main chemical bonds of the toxic compounds have been disrupted by the hydrolytic enzymes which resulted in the reduction of the toxicity of these toxic compounds. This kind of the mechanism is found effective for degrading the oil spill, carbamate and organophosphate insecticides. Insecticides like heptachlor and DDT can be degraded in the absence of air because they are found stable under the aerobic conditions in the soil (Vasileva-Tonkova and Galabova 2003; Lal and Saxena 1982). Alcoholysis and condensation are the reactions catalysed by the hydrolases for removing the pollutants from our environment. Hydrolases are the biocatalysts having the properties like easy availability, tolerate the addition of water miscible solvents and lacking cofactor stereoselectivity (http://www.wiley-vch.de/publish/dt/). Glycosidases, cellulases, and hemicellulases are the enzymes which find applicability for degrading the biomass (Schmidt 2006).

Lipases are the enzymes which degrade lipids derived from plants, animals and microorganisms (Verma 2019; Verma et al. 2008, 2009; Verma and Kanwar 2008, 2010). These lipases are found to be closely related with the organic pollutants of the soil. Hydrocarbons present in the contaminated soil have been drastically reduced via the action of lipases (Margesin et al. 1999; Riffaldi et al. 2006). Lipases isolated from the microbial sources are found more versatile as compared to the lipases isolated from the other sources. Reactions performed by lipases are esterification, hydrolysis, interesterification, aminolysis and alcoholysis (Prasad and Manjunath 2011). Reaction catalysed by the lipases is the hydrolysis of triacylglycerol into glycerol and free fatty acids. Main component of the natural fat or oil is the triglyceride. Triglyceride has been hydrolysed via the action of lipases into glycerol, monoacylglycerol, diacylglycerol and the fatty acids (Hermansyah et al. 2007; Sharma et al. 2011). Activity of the lipases has been found to be most useful parameter which indicates the degradation of hydrocarbons in the contaminated soil (Margesin et al. 1999; Riffaldi et al. 2006). Besides, the various uses of lipases in chemical, detergent manufacturing, food, paper and cosmetics, lipases are found useful in the process of bioremediation of pollutants in the soil (Sharma et al. 2011; Joseph et al. 2006).

Cellulases from the microbial origin have the potential to convert the waste cellulosic materials into the food which is a subject of concern because of the increasing population (Bennet et al. 2002). Some bacteria and fungi are reported to show the expression of cellulases, pectinases and hemicellulases at very lower levels (Adriano-Anaya et al. 2005). Cellulases are the mixture of various enzymes which

**Table 16.2** Applications of microbial enzymes in the process of bioremediation of pollutants

| S. No. | Enzyme | Substrate | Reaction |
|---|---|---|---|
| 1. | Oxidoreductases family | | |
| 1.1 | Oxygenases | | |
| 1.1.1 | Monooxygenases | Aromatic compounds, steroids, alkane and fatty acids | Incorporating oxygen atom to the substrate and utilize the substrate as a reducing agent. Desulfurization, denitrification, dehalogenation, hydroxylation of substrate and ammonification |
| 1.1.2 | Dioxygenases | Aromatic compounds | Introducing two oxygen atoms to the substrate which results in extradiol cleaving and intradiol cleaving thereby forming aliphatic compounds. |
| 1.2 | Peroxidases | | |
| 1.2.1 | Mangnese peroxidase | Phenolic compounds and lignin | In the presence of $Mn^{2+}$ and hydrogen peroxidethe co-substrate catalyses the oxidation of $Mn^{2+}$into $Mn^{3+}$resulting in an $Mn^{3+}$chelateoxalate, which in turn perform the oxidation of the phenolic substrates. |
| 1.2.2. | Lignin peroxidase | Polycyclic aromatic compounds, halogenated phenolic compounds and other aromatic compounds | Oxidation of substrate in the presence of cosubstratehydrogen peroxide and veratrylalcohol as a mediator. |
| 1.2.3. | Versatile peroxidase | Phenolic, aromatic compounds and methoxybenzenes | This enzyme catalyses the transfer of electron from an oxidizable substrate, by forming and reducingthe compound I and compound II intermediates. |
| 1.3 | Laccases | Aminophenols,polyphenols, ortho and paradiphenols, lignins, polyamines and aryldiamines | Oxidation, demethylation and decarboxylation of the substrate. |
| 2. | Hydrolases family | | |
| 2.1. | Cellulase | Cellulosic substances | Hydrolyses the substrate into simple carbohydrates |
| 2.2. | Lipases | Organic pollutants like oilspill | The hydrolysis of triacylglycerols to glycerolsand free-fatty acids |
| 2.3. | Pectinases | Proteins | Enzymes which hydrolyse peptide bonds in aqueous environment |

act together to perform their functions. Hydrolysis process involves the use of three major groups of cellulases, endoglucanase, β glucosidase and exoglucanase. Regions of low crystallinity have been attacked by endoglucanase in the cellulose fiber thereby creating the free chain ends. Exoglucanase or cellobiohydrolase removes the cellobiose units from the ends of free chains thus degrading the cellulose molecule. β glucosidase is the third one which hydrolysis the cellobiose into the glucose units. Cellulasesdegrade the cellulose into the reducing sugars by enzymatic hydrolysis and these reducing sugars are further fermented by bacteria and yeast into ethanol (Sun and Cheng 2002).

George Robinson has reported the use of microorganisms in the process of bioremediation and uses microorganisms for the consumption of oil spill (USMicrobics 2003). The process of bioremediation is removing the soil contaminants by using microorganisms and the products of microorganisms (USEPA Mine Waste Technology Program 2002; Leung 2004). Native soil microorganisms are found to play a key role in the bioremediation of the soil for transforming the complex organic compounds into the simpler inorganic compounds. Technology of bioremediation utilizes the microorganisms for eliminating, reducing and transforming the benign pollutants present in the water, soil, air and sediments (Shanahan 2004). Detoxification of pollutants via the process of bioremediation targets the harmful chemicals by transformation, mineralization or alternation (Shannon and Unterman 1993). In the past, civilizations had used natural bioremediation for treating the waste water but the reduction of harmful wastes intentionally is a recent development. There is the production of energy via the redox reactions within the cells of microorganisms. One electron acceptor, a source of energy and nutrients are required by the delivery system in the process of bioremediation. Microbial electron acceptor classes of various types involved in the process of bioremediation are manganese, oxygen, iron (III), nitrate, carbon dioxide reducing and sulphate (Adams et al. 2015). Venosa et al. (2002) reported that the most important requirement for the bioremediation of oil spills is the appropriateness of the metabolic capabilities of the microorganisms used. Microbial communities exposed to the hydrocarbons become selectively enriched, genetically modified and adapted to the environment present. These adapted communities of microorganisms respond to the hydrocarbon presence within hours and possess higher rates of biodegradation as compared to the microbial communities which never face the hydrocarbon contamination (Leahy and Colwell 1990). The most active oil degrading microorganisms can be isolated from that particular environment and these microbes can be utilized for the bioremediation of petroleum polluted sites. Crude oil is made up of a mixture of compounds; thus, requires a mixture of bacterial consortia for the process of bioremediation of petroleum hydrocarbons because individual microorganisms can only metabolize a limited hydrocarbons range (Al-Saleh and Drobiovaand Obuekwe 2009; Bordenave et al. 2007). This process of bioremediation depends on the availability of nutrients and optimum conditions such as temperature, concentration of contaminants, bioavailability, redox potential and oxygen content and moisture content which supports the biological functions (Adams et al. 2015).

Joergensen et al. (1995) has reported the use of catalase enzyme of microbial origin for the process of bioremediation in which enzyme act as an indicator of hydrocarbon degradation of the soil contaminated by crude oil. This catalase enzymes have also been reported to remove the reactive oxygen species hydrogen peroxide from textile industries effluents and also provide oxygen for the process of aerobic bioremediation by breaking the hydrogen peroxide into carbon dioxide and water (Rila 2008; Achuba and Okoh 2014). Metabolism of hydrogen peroxide is being regulated by the catalase enzyme. Catalase enzyme is found to have the highest turnover among all enzymes as it can decompose one million molecules of hydrogen peroxide per molecule of the catalase enzyme (Kaushal et al. 2018). Kaushal et al. (2018) has reported that catalase enzyme of microbial origin act as an indicator of hydrocarbon degradation in the soil contaminated from the crude oil and provides oxygen for the occurrence of aerobic bioremediation process as well as for the removal of hydrogen peroxide from the effluent of bleaching industry.

Achuba and Okoh (2014) had reported that Cr (VI) is a pollutant of environment because of the use of chromium compounds in the tanning process and other industries. Glutathione reductase enzyme is found to reduce Cr(VI) and forms Cr (V), which is a highly unstable radical thus forming reactive oxygen species. Garcia-Arellano et al. reported the use of chromate reductase enzyme which reduces the highly toxic Cr(VI) to insoluble Cr(III) which is less toxic. Enzymatic treatments for the removal of pollutants from the contaminated sites have the minimal impact on our ecosystem as there is no risk of biological contamination. Furthermore, enzymes work on a wider range of temperature, pH and ionic strength and also found active in the presence of organic solvents of higher concentrations which has soluble pollutant molecules (Torres et al. 2003). Nowadays researchers are very interested to face the problem of contaminants in our environment by using the enzymes based on the bioremediation techniques (Bilal et al. 2017a, 2017b; Ashrafi et al. 2013; Sharma et al. 2018). Degradation of contaminants can be better performed via using enzymes instead of using the microorganisms (Ye et al. 2019). Reactions performed by the enzymes requires mild conditions in water and lower activation energy is needed (Sheldon and van Pelt 2013). Thus, enzymes are playing very important role in developing the complementary oralternative biotechnological processes which have applications in the polluting industries. Despite the potential applications of microbial enzymes in the process of bioremediation and the clean processes, the oxidative enzymes activity is limited by the environmental conditions. Strategies to enhance the catalytic activity of enzymes has been reported in case of peroxidases which includes the chemical modifications of the enzymes (Vandertol-Vanier et al. 2002; Tinoco and Vazquez-Duhalt 1998) and via the genetic tools (Harford-Cross et al. 2000). Desired enzymes can be produced via the cloning and expression in the suitable host to facilitate the enzymatic characteristics as well as use of enzymes for environmental applications (Arnold 2001).

Shakerian et al. (2020) has reported that laccases and horseradish peroxidase are the enzymes which are most widely used for degrading the chemical pollutants because of their low substrate specificity as well as these enzymes catalysis the oxidation of a wider range of compounds as compared to the specific oxidative enzymes. Horseradish peroxidases are the enzymes which work on a broader range of temperature and pHin the presence of other compounds present in municipal waste water (Melo et al. 2016). Laccases are the oxidoreductases which contains copper and are found in various plants and microorganisms including fungi (Zdarta et al. 2018). The major disadvantage of using these enzymes are lower stability, low possibility of reuse, higher price and their separation is also difficult from the reaction media (Liu et al. 2018). Enzyme immobilization is a technique used for improving these problems via the fixation of enzymes in solid supports which makes their separation easy and also improves the reusability and stability (DiCosimo et al. 2013; Verma and Kanwar 2012).

Property of stability should be higher in case of enzymes if we want to reuse them. Immobilized enzymes can be economic, stable and reusable as compared to the free enzymes (Liu et al. 2018). Horseradish peroxidases and laccases are the most popular and nonspecific biocatalysts used for degrading the chemicals (Becker et al. 2016; Varga et al. 2019). Covalent immobilization makes stronger binding between the support and enzymes as well as represents good efficiency (Maryskova et al. 2016; Bayramoglu et al. 2018). But most of the cases in the covalent binding need additional pre-treatment steps as compared to the adsorption which is very complicated. Binding between the enzyme and support is weaker in adsorption methods as compared to the covalent immobilization but usually, adsorption methods are simpler, faster, economic and in many cases adsorbed enzymes showed good performance and reusability when compared with the covalently immobilized enzymes. Methods based on new materials like nano-particles, magnetic particles, biopolymers and composite have been developed. García-Morales et al. (2018) immobilized laccase by using titania nanoparticles. It is reported that titania nanoparticles use for immobilization provides larger surface area, reduces mass transfer resistance for the substrates and also allows higher enzyme loading. Titania has found to have higher chemical stability. Use of bifunctional crosslinking reagents can make covalent binding in it. Biotransformation of acetaminophen and diclofenac has been achieved via using laccases immobilized with titania nanoparticles. Immobilization of horseradish peroxidise with PEGylated magnetic composite microsphere has been reported by Xie et al. (2019). Immobilization of horseradish peroxidase with PEGylated magnetic composite microsphere showed excellent improvement in the storage ability, stability as well as reusability of the enzyme. This immobilized horseradish peroxidase has been utilized to degrade phenol. Horseradish peroxidise had immobilized onto the self-fabricated polyvinyl alcohol-alginate beads by the use of sodium nitrate as a cross linker. Degradation of azo dye i.e. methyl orange has been achieved by using immobilized horseradish peroxidase onto the self-fabricated polyvinyl alcohol-alginate beads (Bilal et al. 2017a, 2017b). Olajuyigbe et al. (2019) immobilized laccases on copper and calcium alginate beads via entrapment method and thus used for the degradation of biphenol A. Calcium alginate beads exhibited better storage stability as compared to the copper alginate beads. Higher thermal stability with good kinetic parameters has been reported by this kind of immobilization. Immobilization of the laccase on the hollow mesoporous carbon nanospheres has been achieved by Shao et al. (2019);

and they used this immobilized laccase for the antibiotic degradation. Immobilized laccase with mesoporous carbon nanospheres showed higher activity in acidic as well as basic pH conditions when compared with the free enzyme laccase. This immobilization also enhanced the laccases storage stability. Shakerian et al. (2020) showed that higher amount of micropollutants can be faced by immobilized oxidative enzymes and these enzymes are broadly accepted as a green way to tackle the problem of micropollutants present in our environment. Manganese peroxide enzyme isolated from fungi, *Anthracophyllum* has been reported to be used for the waste water treatment. This manganese peroxidise has been immobilized on the surface of magnetic nanocomposite $Fe_3O_4$/chitosan. The prepared MnP/$Fe_3O_4$/chitosan nanocomposite has proved to show promising results for the textile wastewater bioremediation (Siddeeg et al. 2019).

## 16.4 Conclusion and Future Directions

Bioremediation process has proved as an excellent approach to remove the pollutants of our ecosystem. It is a natural and safer way for removing the contaminants which produces no side effects. A variety of microbial enzymes are being utilized for the removal of pollutants from our environment but still there are certain factors which are limiting the enzymatic activities. So, the current goal of research should be on enhancing the activity and the performance of these biocatalysts. Researchers should develop highly stable, efficient and less time-consuming technologies for treating the pollutants.

**Acknowledgement** One of the authors (Dr Madan Verma) is grateful to the Director of the Indian Institute of Information Technology Una Himachal Pradesh for providing the necessary facility to pursue the present work.

## References

Achuba FI, Okoh PN (2014) Effect of petroleum products on soil catalase and dehydrogenase activities. Open J Soil Sci 4:399–406. https://doi.org/10.4236/ojss.2014.412040

Adams GO, Fufeyin PT, Okoro SE, Ehinomen I (2015) Bioremediation, biostimulation and bioaugmention: a review. Int J Environ Bioremed Biodegradat 3:28–39. https://doi.org/10.12691/ijebb-3-1-5

Adriano-Anaya M, Salvador-Figueroa M, Ocampo JA, Garcıa-Romera I (2005) Plant cell-wall degrading hydrolytic enzymes of *Gluconacetobacterdiazotrophicus*. Symbiosis 40:151–156. http://www.bashanfoundation.org/contributions/Ocampo-J/ocampoplantcellwall.pdf

Adrio JL, Demain AL (2014) Microbial enzymes: tools for biotechnological processes. Biomol Ther 4:117–139. https://doi.org/10.3390/biom4010117

Al-Saleh H, Drobiovaand Obuekwe C (2009) Predominant culturable crude oil-degrading bacteria in the coast of Kuwait. Int Biodeterior Biodegrad 63:400–406. https://doi.org/10.1016/j.ibiod.2008.11.004

Arnold FH (2001) Combinatorial and computational challenges for biocatalyst design. Nature 409:253–257. https://doi.org/10.1038/35051731

Arora PK, Kumar M, Chauhan A, Raghava GP and Jain RK (2009) OxDBase: a database of oxygenases involved in biodegradation. BMC Res Notes 2. https://bmcresnotes.biomedcentral.com/articles/10.1186/1756-0500-2-67

Arora PK, Srivastava A, Singh VP (2010) Application of monooxygenases in dehalogenation, desulphurization, denitrification and hydroxylation of aromatic compounds. J Bioremed Biodegr 1:1–8. https://www.omicsonline.org/application-of-monooxygenases-in-dehalogenation-desulphurization-denitrification-and-hydroxylation-of-aromatic-compounds-2155-6199.1000112.php?aid=464

Ashrafi SD, Rezaei S, Forootanfar H, Mahvi AH, Faramarzi MA (2013) The enzymatic decolorization and detoxification of synthetic dyes by the laccase from a soil-isolated ascomycete, Paraconiothyriumvariabile. Int Biodeterior Biodegrad 85:173e181. https://doi.org/10.1016/j.ibiod.2013.07.006

Bayramoglu G, Karagoz B, Arica MY (2018) Cyclic-carbonate functionalized polymer brushes on polymeric microspheres: immobilized laccase for degradation of endocrine disturbing compounds. J Ind Eng Chem 60:407e417. https://doi.org/10.1016/j.jiec.2017.11.028

Becker D, Giustina SVD, Rodriguez-Mozaz S, Schoevaart R, Barcelo D, de Cazes M, Belleville MP, Sanchez-Marcano J, de Gunzburg J, Couillerot O, Volker J, Oehlmann J, Wagner M (2016) Removal of antibiotics in wastewater by enzymatic treatment with fungal laccase degradation of compounds does not always eliminate toxicity. Bioresour Technol 219:500e509

Bennet JW, Wunch KG, Faison BD (2002) Use of Fungi biodegradation. ASM Press, Washington

Bilal M, Asgher M, Iqbal HMN, Hu H, Zhang X (2017a) Bio-based degradation of emerging endocrine-disrupting and dye-based pollutants using cross-linked enzyme aggregates. Environ Sci Pol 24:7035e7041. https://doi.org/10.1007/s11356-017-8369-y

Bilal M, Rasheed T, Iqbal HMN, Hu H, Wang W, Zhang X (2017b) Novel characteristics of horseradish peroxidase immobilized onto the polyvinyl alcohol-alginate beads and its methyl orange degradation potential. Int J Biol Macromol 105:328e335. https://doi.org/10.1016/j.ijbiomac.2017.07.042

Bordenave MS, Goñi-Urriza P, Caumette R (2007) Effects of heavy fuel oil on the bacterial community structure of a pristine microbial mat. Appl Environ Microbiol 73:6089–6097. https://doi.org/10.1128/AEM.01352-07

Cedrone F, Menez A, Quemeneur E (2000) Tailoring new enzyme functions by rational redesign. CurrOpin Struct Biol 10:405–410. https://doi.org/10.1016/s0959-440x(00)00106-8

Cheetham PSJ (1987) Screening for novel biocatalysts. Enzymes Microbial Technol 9:194–312. https://doi.org/10.1016/0141-0229(87)90016-0

Cirino PC, Arnold FH (2002) Protein engineering of oxygenases for biocatalysis. Curr Opin Chem Biol 6:130–135. https://doi.org/10.1016/s1367-5931(02)00305-8

Dalby PA (2011) Strategy and success for the directed evolution of enzymes. Curr Opin Struct Biol 21:473–480. https://doi.org/10.1016/j.sbi.2011.05.003

de Carvalho CC (2011) Enzymatic and whole cell catalysis: finding new strategies for old processes. Biotechnol Adv 29:75–83. https://doi.org/10.1016/j.biotechadv.2010.09.001

DiCosimo R, McAuliffe J, Poulose AJ, Bohlmann G (2013) Industrial use of immobilized enzymes. Chem Soc Rev 42:6437e6474. https://doi.org/10.1039/c3cs35506c

Fetzner S (1998) Bacterial dehalogenation. Appl Microbiol Biotechnol 50:633–657. https://doi.org/10.1007/s002530051346

Fetzner S (2003) Oxygenases without requirement for cofactors or metal ions. Appl Microbiol Biotechnol 60:243–257. https://doi.org/10.1007/s00253-002-1123-4

Fetzner S, Lingens F (1994) Bacterial dehalogenases: biochemistry, genetics, and biotechnological applications. Microbiol Rev 58:641–685. https://doi.org/10.1007/s002530051346

García-Morales R, García-García A, Orona-Navar C, Osma JF, Nigam KDP, Ornelas-Soto N (2018) Biotransformation of emerging pollutants in groundwater by laccase from P. sanguineus CS43 immobilized onto titania nanoparticles. J Environ Chem Eng 6:710e717

Gianfreda L, Xu F, Bollag JM (1999) Laccases: a useful group of oxidoreductive enzymes. Biorem J 3:1–25. https://doi.org/10.1080/10889869991219163
Gilbert JA, Dupont CL (2011) Microbial metagenomics: beyond the genome. Annu Rev Mar Sci 3:347–371. https://doi.org/10.1146/annurev-marine-120709-142811
Harford-Cross CF, Carmichael AB, Allan FK, England PA, Rouch DA, Wong LL (2000) Protein engineering of cytochrome P450(cam) (CYP101) for the oxidation of polycyclic aromatic hydrocarbons. Protein Eng 13:121–128. https://doi.org/10.1093/protein/13.2.121
Hermansyah H, Wijanarko A, Gozan M (2007) Consecutive reaction model for triglyceride hydrolysis using lipase. J Teknol 2:151–157. https://www.researchgate.net/publication/268363296_Consecutive_Reaction_Model_For_Triglyceride_Hydrolysis_Using_Lipase
Hiner ANP, Ruiz JH, Rodri JN (2002) Reactions of the class II peroxidases, lignin peroxidase and *Arthromycesramosus*peroxidase, with hydrogen peroxide: catalase-like activity, compound III formation, and enzyme inactivation. J Biol Chem 277:26879–26885. https://doi.org/10.1074/jbc.m200002200
Husain Q (2006) Potential applications of the oxidoreductive enzymes in the decolorization and detoxification of textile and other synthetic dyes from polluted water: a review. Crit Rev Biotechnol 26:201–221. https://doi.org/10.1080/07388550600969936
Joergensen RG, Schmaedeke F, Windhorst K, Meyer B (1995) Biomass and activity of microorganisms in a fuel-oil contaminated soil. Soil Biol Biochem 27:1137–1143. https://doi.org/10.1016/0038-0717(95)00040-L
Jones JP, Hare EJO, Wong LL (2001) Oxidation of polychlorinated benzenes by genetically engineered CYP101 (cytochrome P450cam). Eur J Biochem 268:1460–1467. https://doi.org/10.1046/j.1432-1327.2001.02018.x
Joseph B, Ramteke PW, Kumar PA (2006) Studies on the enhanced production of extracellular lipase by *Staphylococcusepidermidis*. J Gen Appl Microbiol 52:315–320. https://doi.org/10.2323/jgam.52.315
Karigar CS, Rao SS (2011) Role of microbial enzymes in the bioremediation of pollutants: a review. Enzym Res. https://doi.org/10.4061/2011/805187
Kaushal J, Mehandia S, Singh G, Raina A, Arya SK (2018) Catalase enzyme: application in bioremediation and food industry. Biocatal Agric Biotechnol 16:192–199. https://doi.org/10.1016/j.bcab.2018.07.035
Kim JS (2016) Genome editing comes of age. Nat Protoc 11:1573–1578. https://doi.org/10.1038/nprot.2016.104
Kim JS, Park JW, Lee SE, Kim JE (2002) Formation of bound residues of 8-hydroxybentazon by oxidoreductive catalysts in soil. J Agric Food Chem 50:3507–3511. https://doi.org/10.1021/jf011504z
Kumar A, Singh S (2013) Directed evolution: tailoring biocatalysis for industrial application. Crit Rev Biotechnol 33:365–378. https://doi.org/10.3109/07388551.2012.716810
Kumar L, Awasthi G, Singh B (2011) Extremophiles: a novel source of industrially important enzymes. Biotechnology 10:1–15. https://doi.org/10.3923/biotech.2011.121.135
Lal R, Saxena DM (1982) Accumulation, metabolism and effects of organochlorine insecticides on microorganisms. Microbiol Rev 46:95–127. http://www.wiley-vch.de/publish/dt/
Le Roes-Hill M, Prins A (2016) Biotechnological potential of oxidative enzymes from Actinobacteria. https://doi.org/10.5772/61321
Leahy JG, Colwell RR (1990) Microbial degradation of hydrocarbons in the environment. Microbial Rev 53:305–315. https://www.ncbi.nlm.nih.gov/pmc/articles/PMC372779/
Leemhuis H, Kelly RM, Dijkhuizen L (2009) Directed evolution of enzymes: library screening strategies. IUBMB Life 61:222–228. https://doi.org/10.1002/iub.165
Leung M (2004) Bioremediation: techniques for cleaning up a mess. J Biotechnol 2:18–22. https://www.bioteach.ubc.ca/Journal/V02I01/bioremediation.pdf
Li S, Yang X, Yang S, Zhu M, Wang X (2012) Technology prospecting onenzymes: application, marketing and engineering. Comput Struct Biotechnol J 2:1–11. https://doi.org/10.5936/csbj.201209017

Liu DM, Chen J, Shi YP (2018) Advances on methods and easy separated support materials for enzymes immobilization. TrAC Trends Anal Chem 102:332e342. https://doi.org/10.1016/j.trac.2018.03.011
Luo XJ, Yu HL, Xu JH (2012) Genomic data mining: an efficient way to find new and better enzymes. Enzyme Eng 1:104–108. https://doi.org/10.4172/2329-6674.1000104
Mai C, Schormann W, Milstein O, Huttermann A (2000) Enhanced stability of laccase in the presence of phenolic compounds. Appl Microbiol Biotechnol 54:510–514. https://doi.org/10.1007/s002530000452
Margesin R, Zimmerbauer A, Schinner F (1999) Soil lipase activity a useful indicator of oil biodegradation. Biotechnol Techniques 13:859–863
Marrs B, Delagrave S, Murphy D (1999) Novel approaches for discovering industrial enzymes. Curr Opin Microbiol 2:241–245. https://doi.org/10.1016/S1369-5274(99)80042-3
Maryskova M, Ardao I, García-Gonzalez CA, Martinova L, Rotkova J, Sevcu A (2016) Polyamide 6/chitosan nanofibers as support for the immobilization of trametes versicolor laccase for the elimination of endocrine disrupting chemicals. Enzym Microb Technol 89:31e38. https://doi.org/10.1016/j.enzmictec.2016.03.001
Melo CF, Dezotti M, Marques MRC (2016) A comparison between the oxidationwith laccase and horseradish peroxidase for triclosan conversion. Environ Technol 37:335e343. https://doi.org/10.1080/09593330.2015.1069897
Ness JE, del Cardayre SB, Minshull J, Stemmer WP (2000) Molecular breeding: the natural approach to protein design. Adv Protein Chem 55:261–292. https://doi.org/10.1016/S0065-3233(01)55006-8
Olajuyigbe FM, Adetuyi OY, Fatokun CO (2019) Characterization of free and immobilized laccase from Cyberlindnerafabianii and application in degradation of bisphenol A. Int J Biol Macromol 125:856e864. https://doi.org/10.1016/j.ijbiomac.2018.12.106
Pandey D, Singh R, Chand D (2011) An improved bioprocess forsynthesis of acetohydroxamic acids using DTT (dithiothreitol)treated resting cells of Bacillus sp. APB-6. Bioresour Technol 102:6579–6586. https://doi.org/10.1016/j.biortech.2011.03.071
Park JW, Park BK, Kim JE (2006) Remediation of soil contaminated with 2,4-dichlorophenol by treatment of minced shepherd's purse roots. Arch Environ Contamin Toxicol 50:191–195. https://doi.org/10.1007/s00244-004-0119-8
Pikuta EV, Hoover RB, Tang J (2007) Microbial extremophiles at the limit of life. Crit Rev Microbiol 33:183–209. https://doi.org/10.1080/10408410701451948
Piontek K, Smith AT, Blodig W (2001) Lignin peroxidase structureand function. Biochem Soc Trans 29:111–116. https://doi.org/10.1042/0300-5127:0290111
Prasad MP, Manjunath K (2011) Comparative study onbiodegradation of lipid-rich wastewater using lipase producingbacterial species. Indian J Biotechnol 10:121–124. http://hdl.handle.net/123456789/10960
Que L, Ho RYN (1996) Dioxygen activation by enzymes with mononuclear non-heme iron active sites. Chem Rev 96:2607–2624. https://doi.org/10.1021/cr960039f
Raj J, Prasad S, Bhalla TC (2006) Rhodococcusrhodochrous PA-34:a potential biocatalyst for acrylamide synthesis. Process Biochem 41:1359–1363. https://link.springer.com/article/10.1007/s12088-009-0033-x
Riffaldi R, Levi-Minzi R, Cardelli R, Palumbo S, Saviozzi A (2006) Soil biological activities in monitoring the bioremediation of diesel oil-contaminated soil. Water Air Soil Pollut 170:3–15. https://doi.org/10.1007/s11270-006-6328-1
Rila JP (2008) Development and application of bioassays for a site-specific risk assessment of contaminated soil. Utrecht University
Rodrıguez Couto S, Toca Herrera JL (2006) Industrial and biotechnological applications of laccases: a review. Biotechnol Adv 24:500–513. https://doi.org/10.1016/j.biotechadv.2006.04.003

Rubilar O, Diez MC, Gianfreda L (2008) Transformation ofchlorinated phenolic compounds by white rot fungi. Crit Rev Environ Sci Technol 38:227–268. https://doi.org/10.1080/10643380701413351

Ruiz-Dueñas FJ, Morales M, Pérez-Boada M (2007) Manganese oxidation site in *Pleurotuseryngii*versatile peroxidase: a site-directedmutagenesis, kinetic, and crystallographic study. Biochemistry 46:66–77. https://doi.org/10.1021/bi061542h

Schmidt O (2006) Wood and tree Fungi. Springer, Berlin. https://www.springer.com/gp/book/9783540321385

Schmidt M, Boettcher D, Bornscheuer UT (2010) In: Soetaert W, Vandamme EJ (eds) Industrial biotechnology: sustainable growth and economic success. Wiley, Weinheim, pp 155–187. https://www.wiley.com/en-us/Industrial+Biotechnology%3A+Sustainable+Growth+and+Economic+Success-p-9783527314423

Shakerian F, Zhao J, Li SP (2020) Recent development in the application of immobilized oxidative enzymes for bioremediation of hazardous micropollutants: a review. Chemosphere 239:124716. https://doi.org/10.1016/j.chemosphere.2019.124716

Shanahan P (2004) Bioremediation. Waste containment and remediation technology. Massachusetts Institute of Technology, MIT OpenCourseWare

Shannon MJ, Unterman R (1993) Evaluating bioremediation: distinguishing fact from fiction. Annu Rev Microbiol 47:715. https://www.annualreviews.org/doi/abs/10.1146/annurev.mi.47.100193.003435

Shao B, Liu Z, Zeng G, Liu Y, Yang X, Zhou C, Chen M, Liu Y, Jiang Y, Yan M (2019) Immobilization of laccase on hollow mesoporous carbon nanospheres: noteworthy immobilization, excellent stability and efficacious for antibiotic contaminants removal. J Hazard Mater 362:318e326. https://doi.org/10.1016/j.jhazmat.2018.08.069

Sharma D, Sharma B, Shukla AK (2011) Biotechnologicalapproach of microbial lipase: a review. Biotechnology 10:23–40. https://doi.org/10.3923/biotech.2011.23.40

Sharma B, Dangi AK, Shukla P (2018) Contemporary enzyme based technologies for bioremediation: a review. J Environ Manag 210:10e22

Sheldon RA, van Pelt S (2013) Enzyme immobilisation in biocatalysis: why, what and how. Chem Soc Rev 42:6223e6235. https://doi.org/10.1039/c3cs60075k

Shi S, Liang Y, Zhang MM, Ang EL, Zhao H (2016) A highly efficient single-step, markerless strategy for multi-copy chromosomal integration of large biochemical pathways in *Saccharomyces cerevisiae*. Metab Eng 33:19–27. https://doi.org/10.1016/j.ymben.2015.10.011

Siddeeg SM, Tahoon MA, Mnif W, Rebah FB (2019) Iron oxide/chitosan magnetic nanocomposite immobilized manganese peroxidase for decolorization of TextileWastewater. PRO 8:2–12. https://doi.org/10.3390/pr8010005

Singh R, Kumar M, Mittal A, Mehta PK (2016) Microbial enzymes: industrial progress in 21st century. Biotech 6:174. https://doi.org/10.1007/s13205-016-0485-8

Singhania RR, Patel AK, Pandy A (2010) Industrial biotechnology: sustainable growth and economic success. In: Soetaert W, Vandamme EJ (eds) . Wiley, Weinheim, pp 207–226. http://dl.booktolearn.com/ebooks2/science/biology/9783527314423_industrial_biotechnology_1c93.pdf

Srinivasan MC (1994) Microbial biodiversity and its relevance to screening for novel industrially useful enzymes. Curr Sci 66:137–140. https://www.jstor.org/stable/24109241

Sun Y, Cheng J (2002) Hydrolysis of lignocellulosic materialsfor ethanol production: a review. Bioresour Technol 83:1–11. https://doi.org/10.1016/S0960-8524(01)00212-7

Ten Have R, Teunissen PJM (2001) Oxidative mechanisms involved in lignin degradation by white-rot fungi. Chem Rev 101:3397–3413. https://doi.org/10.1021/cr000115l

Tinoco R, Vazquez-Duhalt R (1998) Chemical modification of cytochrome *c* improves their catalytic properties in the oxidation of polycyclic aromatic hydrocarbons. Enzyme Microb Technol 22:8–12. https://doi.org/10.1016/S0141-0229(97)00073-2

Torres E, Bustos-Jaimes I, Le Borgne S (2003) Potential use of oxidative enzymes for the detoxification of organic pollutants. Appl Catal B Environ 46:1–15. https://doi.org/10.1016/S0926-3373(03)00228-5

Tsukihara T, Honda Y, Sakai R, Watanabe T, Watanabe T (2006) Exclusive overproduction of recombinant versatile peroxidase MnP2 by genetically modified white rot fungus, *Pleurotusostreatus*. J Biotechnol 126:431–439. https://doi.org/10.1016/j.jbiotec.2006.05.013

Ullah MA, Bedford CT, Evans CS (2000) Reactions of pentachlorophenol with laccase from *Coriolus versicolor*. Appl Microbiol Biotechnology 53:230–234. https://doi.org/10.1007/s002530050013

USEPA Mine Waste Technology Program (2002) Activity III, project 12: sulfatereducing bacteria reactive wall demonstration. Final report. http://www.epa.gov/ORD/NRMRL/std/mtb/mtbdocs/actiiiproj12.pdf

USMicrobics (2003) Annual report FY-2003. http://www.bugsatwork.com/USMX/BUGS%20Report%20PRINT%20(07-13 04)%20Hawaii%20(paginate%201-8).pdf

Van Lanen S, Shen B (2006) Microbial genomics for the improvement of natural product discovery. Curr Opin Microbiol 9:252–260. https://doi.org/10.1016/j.mib.2006.04.002

Vandertol-Vanier HA, Vazquez-Duhalt R, Tinoco R, Pickard MA (2002) Enhanced activity by poly (ethylene glycol) modification of *Coriolopsis gallica* laccase. J Ind Microbiol Biotechnol 29:214–220. https://doi.org/10.1038/sj.jim.7000308

Varga B, Somogyi V, Meiczinger M, Kovats N, Domokos E (2019) Enzymatictreatment and subsequent toxicity of organic micropollutants usingoxidoreductases-A review. J Clean Prod 221:306e322. https://agris.fao.org/agris-search/search.do?recordID=US201900215708

Vasileva-Tonkova E, Galabova D (2003) Hydrolytic enzymes and surfactants of bacterial isolates from lubricantcontaminated wastewater. Z Naturforsch 58:87–92. https://doi.org/10.1515/znc-2003-1-216

Venosa AD, Lee K, Suidan MT, Garcia-Blanco S, Cobanli S, Moteleb M, Haines JR, Tremblay G, Hazelwood M (2002) Bioremediation and biorestoration of a crude freshwater wetland on the St. Lawrence River. Bioremed J 6. https://doi.org/10.1080/10889860290777602

Verma ML (2019) Biotechnological applications of lipases in flavour and fragrance ester production. Microbial Technology for the Welfare of Society. Microorganism for sustainability, vol 17. Springer, Singapore, pp 1–22

Verma ML, Kanwar SS (2008) Properties and application of*Poly*(MAc-2011*co*-DMA-*cl*-MBAm) hydrogel immobilized *Bacillus cereus* MTCC 8372 lipase for synthesis of geranyl acetate. J Appl Polym Sci 110:837–846

Verma ML, Kanwar SS (2010) Purification and characterization of a low molecular mass alkaliphilic lipase of *Bacillus cereus* MTCC 8372. Acta Microbiol et Immunol Hungarica 57:187–201

Verma ML, Kanwar SS (2012) Harnessing the potential of thermophiles: the variants of extremophiles. Dyn Biochem Process Biotechnol Mol Biol 6(1):28–39

Verma ML, Azmi W, Kanwar SS (2008) Microbial lipases: at the interface of aqueous and non-aqueous media. Acta Microbiol Immunol Hungarica 55:265–293

Verma ML, Azmi W, Kanwar SS (2009) Synthesis of ethyl acetate employing celite-immobilized lipase of *Bacillus cereus* MTCC 8372. Acta Microbiol et Immunol Hungarica 56:229–242

Vervoort Y, Linares AG, Roncoroni M, Liu C, Steensels J, Verstrepen KJ (2017) High throughput system wide engineering and screening for microbial biotechnology. Curr Opin Biotechnol 46:120–125. https://doi.org/10.1016/j.copbio.2017.02.011

Wang HH, Isaacs FJ, Carr PA, Sun ZZ, Xu G, Forest CR, Church GM (2009) Programming cells by multiplex genome engineering and accelerated evolution. Nature 460:894–898. https://doi.org/10.1038/nature08187

Wei R, Zimmermann W (2018) Microbial enzymes for the recycling of recalcitrant petroleum-based plastics: how far are we? Microb Biotechnol 10:1308–1322. https://doi.org/10.1111/1751-7915.12710

Wei T, Cheng BY, Liu JZ (2016) Genome engineering *Escherichia coli* for L-DOPA overproduction from glucose. Sci Rep 6:30080. https://doi.org/10.1038/srep30080

Williams PP (1977) Metabolism of synthetic organic pesticides by anaerobic microorganisms. Residue Rev 66:63–135. https://doi.org/10.1007/978-1-4612-6352-4_3

Wright AV, Nunez JK, Doudna JA (2016) Biology and applications of CRISPR systems: harnessing nature's toolbox for genome engineering. Cell 164:29–44. https://doi.org/10.1016/j.cell.2015.12.035

Xie X, Luo P, Han J, Chen T, Wang Y, Cai Y, Liu Q (2019) Horseradish peroxidise immobilized on the magnetic composite microspheres for high catalyticability and operational stability. Enzym Microb Technol 122:26e35

Xu F (1996) Catalysis of novel enzymatic iodide oxidation by fungal laccase. Appl Biochem Biotechnol 59:221–230. https://doi.org/10.1007/BF02783566

Ye X, Peng T, Feng J, Yang Q, Pratush A, Xiong G, Huang T, Hu Z (2019) A novel dehydrogenase 17b-HSDx from Rhodococcus sp. P14 with potentialapplication in bioremediation of steroids contaminated environment. J Hazard Mater 362:170e177. https://doi.org/10.1016/j.jhazmat.2018.09.023

Yoshida S (1998) Reaction of manganese peroxidase of *Bjerkanderaadusta*with synthetic lignin in acetone solution. J Wood Sci 44:486–490. https://doi.org/10.1007/BF00833415

Zdarta J, Meyer AS, Jesionowski T, Pinelo M (2018) Developments in supportmaterials for immobilization of oxidoreductases: a comprehensive review. Adv Colloid Interf Sci 258:1e20. https://doi.org/10.1016/j.cis.2018.07.004

The manufacturer's authorised representative in the EU is Springer
Nature Customer Service Centre GmbH, Europaplatz 3, 69115 Heidelberg,
Germany. If you have any concerns regarding our products, please
contact ProductSafety@springernature.com

Printed and bound by CPI Group (UK) Ltd, Croydon, CR0 4YY
30/09/2026
02225247-0011